Consumer-Driven Technologies in Healthcare:

Breakthroughs in Research and Practice

Information Resources Management Association
USA

IGI Global
DISSEMINATOR OF KNOWLEDGE

Published in the United States of America by
IGI Global
Medical Information Science Reference (an imprint of IGI Global)
701 E. Chocolate Avenue
Hershey PA, USA 17033
Tel: 717-533-8845
Fax: 717-533-8661
E-mail: cust@igi-global.com
Web site: http://www.igi-global.com

Library of Congress Cataloging-in-Publication Data

Names: Information Resources Management Association, editor.
Title: Consumer-driven technologies in healthcare : breakthroughs in research and practice / Information Resources Management Association, editor.
Description: Hershey, PA : Medical Information Science Reference, [2019] | Includes bibliographical references.
Identifiers: LCCN 2018004084| ISBN 9781522561989 (hardcover) | ISBN 9781522561996 (ebook)
Subjects: | MESH: Biomedical Technology | Delivery of Health Care | Consumer Health Information | Medical Informatics Applications | Patient Participation
Classification: LCC R858 | NLM W 82 | DDC 610.285--dc23 LC record available at https://lccn.loc.gov/2018004084

British Cataloguing in Publication Data
A Cataloguing in Publication record for this book is available from the British Library.

The views expressed in this book are those of the authors, but not necessarily of the publisher.

For electronic access to this publication, please contact: eresources@igi-global.com.

List of Contributors

Table of Contents

Section 3
Privacy and Security

Section 4
Surveys and Case Studies

Preface

The everchanging landscape surrounding the diverse applications of technology within the healthcare field can make it very challenging to stay on the forefront of innovative research trends. That is why IGI Global is pleased to offer this one-volume comprehensive reference that will empower students, researchers, practitioners, and academicians with a stronger understanding of consumer-driven technologies in healthcare.

This compilation is designed to act as a single reference source on conceptual, methodological, and technical aspects, and will provide insight into emerging topics including but not limited to patient-centered healthcare, telemedicine, chronic disease management, cryptography, and big data analytics. The chapters within this publication are sure to provide readers the tools necessary for further research and discovery in their respective industries and/or fields.

Consumer-Driven Technologies in Healthcare: Breakthroughs in Research and Practice is organized into four sections that provide comprehensive coverage of important topics. The sections are:

1. Accessibility and Mobility;
2. Patient Engagement;
3. Privacy and Security; and
4. Surveys and Case Studies.

The following paragraphs provide a summary of what to expect from this invaluable reference source:

Section 1, "Accessibility and Mobility," opens this extensive reference source by highlighting the latest trends in smart health and mobile healthcare. Through perspectives on electronic health records, smart cards, and data security, this section explores the application of smartphones, cloud computing, and other technologies to advance patient monitoring and preventative care. The presented research facilitates a better understanding of how mobile health systems and devices are shaping modern healthcare.

Section 2, "Patient Engagement," includes chapters on emerging innovations in healthcare service management. Including discussions on virtual care, patient perception, and personal health systems, this section presents research on the use of technology in medicine to provide quality healthcare to patients. This inclusive information assists in advancing current practices in patient-doctor communications and patient satisfaction.

Section 3, "Privacy and Security," presents coverage on how to protect and secure patient data in healthcare settings. Through innovative discussions on electronic health records, radio frequency identification, and big data, this section highlights the changing landscape of securing and storing personal health information using healthcare management systems. These inclusive perspectives contribute to the available knowledge on dual encryption methods and DNA computing.

Section 4, "Surveys and Case Studies," discusses coverage and research perspectives on healthcare information management systems and e-health systems. Through analyses on user satisfaction, web healthcare applications, and physical therapy, this section contains pivotal information on the latest healthcare technological advances and uses in the Kingdom of Bahrain, Poland, Malaysia, China, Canada, and the United States. The presented research facilitates a comprehensive understanding of healthcare quality and medical information systems from a regional perspective.

Although the primary organization of the contents in this work is based on its four sections, offering a progression of coverage of the important concepts, methodologies, technologies, applications, social issues, and emerging trends, the reader can also identify specific contents by utilizing the extensive indexing system listed at the end.

Section 1
Accessibility and Mobility

Chapter 1
Mobile Health Care:
A Technology View

Veda Prakash Mangu
Graphene Semiconductors Ltd., India

ABSTRACT

This chapter gives a high-level view of the technology involved in the solution of Mobile Health Care with Cloud Computing as back-bone. It emphasizes on Hardware elements, Computation requirements when the solution covers huge scope of medical problems at the mega scales across wide areas. This chapter discusses sub-systems of the solution, that include Smart Phones, Computation Engines, High End Transportation Systems, Multi-Specialty Hospitals, Smart Phones/Digital Personal Assistants used by Medical Practitioners. Discusses on the accuracies, bandwidth requirements and latencies present in the systems, also emphasizes on the required accuracies as the problem area is Human Life. To address the challenges that arises when the solution gets high degree of maturity, this chapter proposes review of the current day protocols in the systems. Also proposes to integrate intelligent applications and different eco-systems like Big Data, Data Analytics and Internet of Things, and best adaptability of these areas with Nano-technologies to result in increased average life time of humans.

1. INTRODUCTION

Technology became a must for human beings in many ways over last 50 years. Starting from micron nodes to today's very advanced nodes of nanometer (Nano Technology) scales, technology has been impacting human life to make it more and more safer and speed. Along with its software models, technology has been delivering so much value proposition in Computational Mathematics for all engineering fields, Medical Innovation, Educational Research, Defense Organizations, Automobiles, Mass Communications, Consumer Electronics. And especially in last decade electronics became a social need in the form of mobile devices. These mobile devices are embedding much computation power into it adding intelligence day by day. This added intelligence will help people to improve the productivity in multi fold.

DOI: 10.4018/978-1-5225-6198-9.ch001

This chapter, gives a hypothesis of Mobile Health Care, that means how a mobile device can be best used for the taking care of our health. And all the sub-systems involved into it from a preliminary view to high-end application usage.

In a simple view of this system, each user of mobile phone is connected with a health care center for immediate medical treatment and first aid. But there are multiple complex sub-systems take part in making it real.

This has been in practice in a very minimal level as pharmacy people keep enquiring us if we need any general medicine for seasonal diseases like cold and cough and also the regular medicine which we keep using as per prescription. Same model can be used for little more complex medical problems with the request initiated from an end user in the usage model. As per the criticality of the situations, medical conditions either first aid can be given or respective specialist can visit the patient with automated means of message processing systems.

2. HIGH LEVEL VIEW OF THE SOLUTION

Figure 1 is the pictorial view of the Mobile Health Care System, which shows all sub-systems, groups participating in the model. Each system on its own is again a complex system which needs accuracy and guaranteed service delivery model.

The system can be viewed as below:

1. End Users
2. High End Servers sitting cloud
3. Physicians and Hospitals
4. Mobile Services connected to hospitals or groups of hospitals.

Figure 1. System level view of the solution

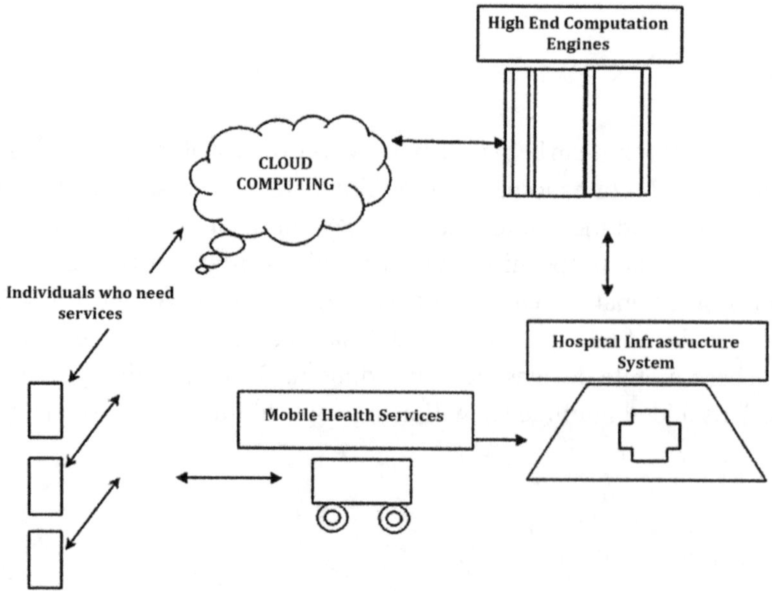

End Users

All individuals who generates the requests for consulting services on different health problems. The requests either can be generated manually by the user or automatically by the sensors embedded into the mobile devices they carry. The sensors and respective intelligence embedded into the devices will continuously monitor different health parameters of the individual and warn the user to consider for visiting a doctor. The software applications running on the mobile devices will process the collected data and send consulting signals to the master controller.

High End Servers

This is the place where high computation engine process the data received from the end user and find out the hospital or network of hospitals where the requests can be services. These engines will keep track of huge amount of data of all the people falling under its visibility. There can be multiple Clouds in the system to cater the requirements of all areas in a state according to the population. All the requests coming from different mobile devices are collected, stored and processed in this building block. This system needs high accuracy in delivery of the services. For enabling this guaranteed delivery, it also receives, stores the data from different hospitals in the area it covers. This is for ensuring the availability of respective specialists corresponding to a problem noticed from certain areas. High End Computation Engines sitting in the clouds, continuously monitors all the incoming lines from end user and also from hospitals. Along with hospitals, it also has to manage the Health Vehicles data and co-ordinate with specialists available with a particular vehicle.

Physicians and Hospitals

This block includes all the hospitals participating in the mobile health care. For this all the hospitals should be upgraded with the required facilities, Hardware and Software systems to enable the mobile health care services. All doctors should carry the mobile device which is enabled with required software. The doctors can communicate with applications running on the high-end servers for taking required actions-either he can accept the call from the cloud and immediately take one mobile-vehicle and attend the patient with in the safe time. Along with the existing health care models, all hospitals need to support the mobile care by dedicating the required resources.

Mobile Health Services

This block includes the ambulance services, mobile ICUs with specialists. This ensures immediate treatment to be started for a patient. This blocks integrates the multi-specialty services, first-aid, specialist consulting services for all kinds of emergency and critical care.

On top of all these building blocks of the said Mobile Health Care System, all the devices in those are having well defined intelligence for certain services like parameter measurement of 'N' number of health conditions, data extraction from the numbers measures, analysis of the extracted data, giving a brief based on the analysis done, and talk to corresponding specialists and try to bring up an environment where the patient can be started taking defined treatment. This involved all the existing subjects like- SW Engineering, Electronics Involved in each of the device, Networking of all the medical devices along with the mobile an individual carry, Data Analytics, Internet of Things.

3. DETAILED DESCRIPTION OF THE SUB-SYSTEMS

The following sections will give more details on the hardware present in each building block of the solution.

Mobile Devices

The devices which an individual carry can better be used to text, call or video conferencing with the nearby health care center in the simplest form of the solution. In addition to that, the devices can integrate different biometric sensors into it for collecting the continuous samples of health parameters like, Blood Pressure, Rate of the Heart Beat etc. We also have smart wearable devices which can replace the mobiles we use today. We also can integrate or connect different PDA (Personal Digital Assistants) which embed sensors to do more measurements like Sugar Levels in blood, Complete Blood Picture and any other new bio-metric measurements with the mobiles where health care applications are running. All these electronics devices integrated with the sensors, will be Wi-Fi enabled or also can communicate with our mobile via Blue Tooth interfaces for transferring the measured raw data. These data are collected by the mobile devices registered in the Mobile Health Care Solution and there its converted into a manageable data structure for communicating with other sub-systems of the solution.

Figure 2 shows sample images to describe on integration of sensors into mobile devices, PDAs which in turn can connected with mobiles without wires.

Likewise, sensors to monitor continuous the health conditions of individuals will be integrated into the mobile devices and the health care applications running in mobile processors, will collect, process and construct the LSD (Life Saving Data) and share with the solution. On receiving this data, Medical Assistants will observe the parameters, coordinate with the specialists and signal the ambulance systems in emergency cases, or directly interact with the individual for knowing further information and give medical advice appropriately. If not the Ambulance Systems can automatically reach the point of action and give first aid and brings to the hospitals.

Figure 2. Bio sensors integrated with mobile devices

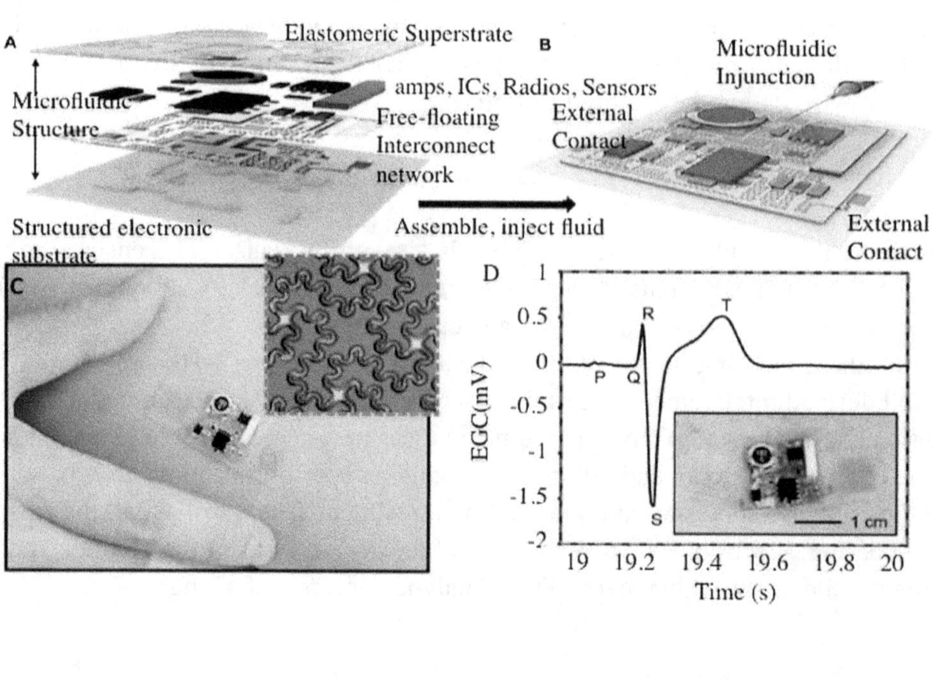

High End Servers

As this solution need to process and route the service requests, Life Saving Data Structures with high degree of accuracies and speed, this Sub-System in the solution (High End Servers) need to provide high throughput and computing density along with built-in virtualization and extreme scalability. These includes highly efficient platforms for deploying large-scale, mission-critical applications. This versatility, along with powerful, bundled virtualization capabilities, makes them an ideal platform on which to consolidate large numbers of applications and databases.

As this sub-system is in the supply chain of a health care solution, these must satisfy requirements of Reliability, Availability, and Serviceability characteristics that make high service levels possible. Delivering RAS capabilities means much more than just having reliable components. It includes a combination of hardware and software features combined with advanced, integrated management and monitoring.

The servers provide an optimal solution for all database workloads, ranging from scan-intensive, complex data types, to highly concurrent online transaction processing applications. This on line processing of the transactions incurred in the solution also is a feature for addressing all financial part of the service model.

Mobile Devices Usage by Physicians and in Hospitals

The use of mobile devices by health care professionals (HCPs) has transformed many aspects of clinical practice. Mobile devices have become commonplace in health care settings, leading to rapid growth in the development of medical software applications (apps) for these platforms. Numerous apps are now available to assist HCPs with many important tasks, such as: information and time management; health record maintenance and access; communications and consulting; reference and information gathering; patient management and monitoring; clinical decision-making; and medical education and training.

Mobile devices and apps provide many benefits for HCPs, perhaps most significantly increased access to point-of-care tools, which has been shown to support better clinical decision-making and improved patient outcomes. However, some HCPs remain reluctant to adopt their use. Despite the benefits they offer, better standards and validation practices regarding mobile medical apps need to be established to ensure the proper use and integration of these increasingly sophisticated tools into medical practice. These measures will raise the barrier for entry into the medical app market, increasing the quality and safety of the apps currently available for use by HCPs.

Health Care Professions will use the advanced smart phones for managing the information they have for conducting their daily duties, time, for maintaining the Electronics Health Records, Electronic Medical Records for ease in accessing and storing them. And also they can use for scanning, sending, texting, encoding, audio and video conferencing with all required stake holders in the solution. Even Health Care Professionals will use these PDA (Personal Digital Assistants) for their reference study purposes for their personal development which in turn will help in better services to the point of actions.

Different usages of the PDA devices in hospitals are briefed below.

- **Information Management:** Physicians will make use of the PDA devices for writing and dictating notes, recording audio, taking photographs, organize information and images, for reading the reference books and also for accessing the services provided in clouds.

- **Time Management:** The devices are useful for physicians to schedule appointments, meetings and record the schedules. There can be apps developed for automating these tasks also, based on the categorization of the requests coming from individuals and also based on the type of the lifesaving data.
- **Records Maintenance and Access:** PDA are used for accessing Electronics Health Records (EHR) and Electronic Medical Records (EMR). The records cane be scanned images, scan reports, ECH reports, CT and MRI Scan Film records etc. Physicians can use the PDAs to coding of their medical notes on these electronic records for improvement purposes. And also they can use it for prescribing electronically on these records so that the newly generated records can be used in this solution in multifold.
- **Consulting and Communicating the Advices:** Physicians can use the PDA devices for voice, video calling with other specialists for discussing on certain issues on patients or, medical conditions. Even they can conduct international audio and video conferencing on chief advices of critical care conditions. Different applications running on these devices enable multimedia messaging, social networking in doctor's communities etc.
- **Reference Gathering and Educational Purposes:** PDAs are also useful on educational purposes for continuous improvement of their skills and knowledge by managing the medical text books, journals, medical literature, keep referring to the research portals on their interesting domains, Drug reference guides and also news on their related topics can be tracked very efficiently by the physicians.
- **Clinical Decision-Making:** PDAs assists physicians in clinical decision support systems, treatment guidelines, diseases diagnosis aids, medical calculations, laboratory tests ordering, interpretation of the test results etc.
- **Patient Monitoring:** Monitoring patient's health, location, rehabilitation, collecting the clinical data, continuous monitoring of different parameters of patient health conditions.
- **Medical Education and Training:** Medical practisoners can continue their research programs or higher-education by efficiently using these Digital Devices. Conduct certain case studies. He can take assessment tests. Work on different surgical simulations when the problems are in very sensitive organs like heart or brains.

Mobile Health Care Systems

The current ambulance systems are a best example in the simplest view of the solution. Not only the immediate lifesaving medicines and first-aid services, this mobile system services are based on the service request type, and emergency conditions of the point of actions etc. Starting from first aid this system provides multi-specialty medical services.

Figure 3 can give brief on how ambulances can be integrated with all advanced lifesaving facilities of Intensive Care Units (ICUs) and be helpful for critical care conditions of the patients. Based on the services requests generated by the point of actions can exactly route to the very much required medical advisors automatically and in no time or allowable safer time periods, these advanced ambulances can come down to the point of actions and start giving treatment so that the patient survives.

Figure 3. Transportation systems where reaching point of action is difficult

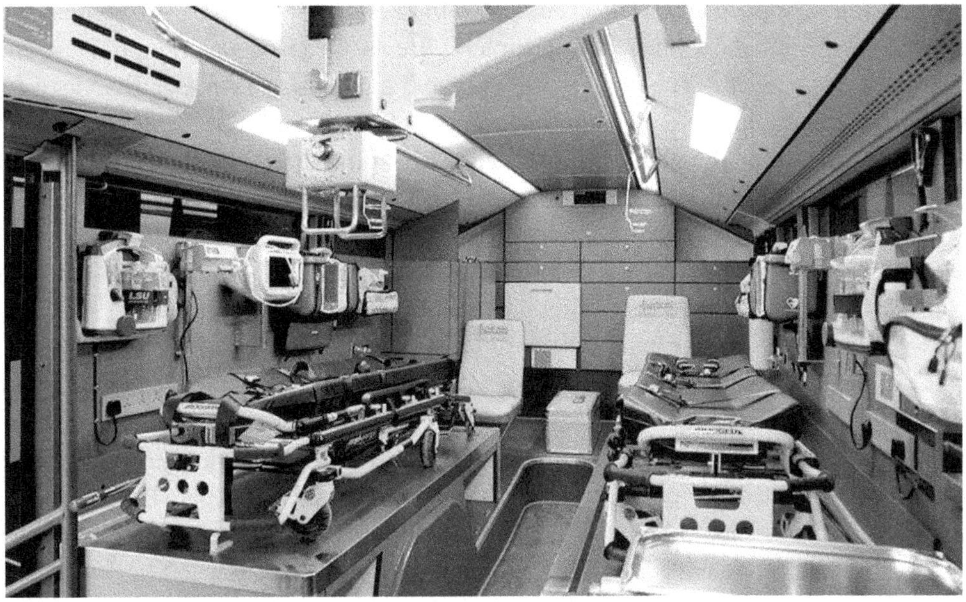

4. PROBLEMS ASSOCIATED WITH THE SYSTEM

First of all, the points, the said system is going to be a player where "Human Life" is being to be main aspect of business. Given that fact, it is very much important to understand different limitations and bounded parameters of different electronics components participating in the solution. Major and known limitations of the current day technologies are Limited Bandwidths, Security Features, Quality of Services, Device Power.

Along with the limitations of technologies, limitations imposed by current infrastructure which is going to fit in the solution in medical services is also important.

Below is the list of limitations of technology:

1. **Design Difference Between Specification to Implementation:** +/- 5% of the tolerance in the design implementation to design specifications is going to play a major consideration while designing the sub-systems of the solution. All services must be designed with this deviation in consideration with mobile health care solution. There can be errors incurred into Life Saving Data while in air between the sub-systems.
2. **Probability of Errors Injected in to Data:** As the basic back-bone of the solution is wire-less technologies, there is always a probability of errors injected into the Life Saving Data Structure in air or within the electronic components of the sub-systems of solution. Each and every protocol will have this provision in definition itself, by making use of which the electronic components can be designed to limit the errors to very minimal. There must be error detecting and correcting algorithms to be implemented in each component of the sub-systems by making use of the protocol provisions.

3. **Limited Bandwidth in Data Communication:** All systems will be functioning with fixed frequencies that in turn pose a limitation in delays in delivery of services. But, given this system is for saving human life where very high accuracies are required, this can impose the boundaries or limitations of the services. This means, for example a patient who got cardiac arrest may not be given lifesaving guaranteed services in the moment of occurrence. Because the inaccuracies in all the involving systems in the above pictorial view are additive in nature, and hence it may take its own time for rendering the defined services and result in known delays. Because of this limitation, the solution needs to define the timeliness of the delivery of the services accordingly.
4. **Data Security:** Security of the integrity of the data is another concern in this solution. Because, it imposes a lot of issues when the solution gets maturity and start providing the services to a wider area networks. In this case, the basic data structures and SW Elements need to be defined with high degree of robustness to avoids security issues of the Life Saving Data. There are QOS (Quality of Services) aspects added in definition of all the protocols which involved in electronic components in the solution. These aspects are to be taken into consideration in defining the services, service limitations, services boundaries that results in high degree of security. Data Encryption, Decryption, 128-bit Security Keys should be implemented in handling the data integrity part of the solution.

All the above issues have to be mitigated in defining the solution.

5. **Infrastructure Facilities:** In addition to above technological limitations, there are few more problems associated with solution implementation. As the solution integrates multiple stake holders or sub-systems, current day infrastructure also imposes a problem here. Say, at highest degree of solution maturity, we need highly sophisticated transportation networks purely dedicated for these services. For example, an ambulance with high-en technology need to reach to point of service delivery within a stipulated time as per definition of the service.
6. **People Awareness:** For best utilization of the services of the solution, people also need to be knowledgeable to some extent. They have to be in a position to understand the steps and people involved in the system so that they provide necessary support. When there are multiple sensors integrated (we are going to discuss on sensors in next sections) into the mobile device, the individual need to understand how to make use of them. Though there are easy-to-understand human interface devices are defined, individuals should have minimum levels of technology awareness for best services. For this, governments have to work towards 100% literacy among societies. At that juncture, this kind of solutions gain huge utilization factor.

5. SOLUTIONS FOR ABOVE PROBLEMS

This author had discussed the problems associated with the solution in above section and also tried to give solution while discussing the problems. When the system gets improvised over the period, these inaccuracies of tolerance, delays, data securities, bandwidth requirements can be avoided by mitigating them in design of every level of the sub-systems in the solution.

When we think of mitigating them, we should have a strong foundation in definition of the service primitives. Different stake holders from core medical researchers, marketing people, technology developers, Software Developers have to work on a common platform. This common platform has to be defined in an open source model and layered approach.

If we understand more on mitigating above problems and solutions into the mobile health care application of cloud computing, the author coins below terms that need to be addresses with more attention.

New Life Saving Data

The data which is generated, communicated, processed, serviced with this mobile health care solution has to be very robust enough that caters very wider life-span of the solution and services. Just like other Audio, Video and Text Data Type, this data has to be defines as a New Life Saving Data (LSD) and this should be well defined with all possible variations in the data in accordance with all possible health care services. Meaning, there should be a high-level data structure that addresses all kinds of general outpatient, emergency, critical care services required for all kinds of diseases, medical emergencies, accidents etc., This single new data structure keeps improvised over the period when many of medical services are brought under this solution. This way, all the Software, Hardware and System developers should be in sync to define the new data structure and deliver models, communication channels and services definition in exiting satellite communication.

All variations of the data structures should be defined by considering all the characteristics and parameters of all the medical emergencies, like different diseases, accidents, critical care requirements, allergies and general health problems. This set of defined packets can be categorized accordingly the allowable latencies in the treatments.

Irrespective of the data-type definition, end of the day this data is going to be transmitted and received over the wire-less and wired interfaces with in the solution. For this reason, the current protocols like Internet Protocol (IP), Wireless Lan Protocol (802.11), PCI Express Protocol (PCIe) need to be thoroughly analyzed and re-framed to fit in the parameters of this New Life Saving Data structures to cater needs of all medical problems, all wider area networks.

Quality of Services (QoS)

This solution will not compromise on the Quality of Services. Every request from any corner of the defined Safe Area of the solution must be communicated and given common resources for accurate and speed transfer of the Life Saving Data. In one abstractions (Open System Interface (OSI) Reference Model) all electronics systems generally designed into a layered architecture and each layer will have its own quality of services. All the existing designed will have to be re-visited and the respective functionality of each layer at each level in the solution should consider the LSD Type and its attached QOS Definition also at the very root level of the data types and communication model of the same.

Not only at the protocol level of the communication, the QOS aspects at each sub-system High-End Servers, Mobiles, Hospitals, Transport Means should be defined accordingly and ensure the delivery of the services with respect to the LSD Data Packets.

Internet of Things

Just like we connected to rest of our colleagues, known business connections, friend's circles, relatives and all known people every electronic gadget also gets connected to many other devices used by us. Every chord of this connectivity results in a specialized solution for solving different problems that human faces or provides a luxury of computation power to different data that humans need to know. Internet of things is becoming a new standard for smart cities, smart buildings, smart transportation, smart energy, smart industries, smart health and all together smart living. Application I.O.T can energize the Mobile Health Care Solution in a level beyond current imagination (Figure 4).

Smart Health: I.O.T

With I.O.T Eco System in place, all the electronic gadgets not necessarily the smart phones will have a communication channel and can accommodate future enhancements of the solution in current focus of this chapter. The Electronic Gadgets embeds various Bio Metric Sensors with defined intelligence with their SW models will generate more quality medical data of the person that it is carried by and can pass of those data structures to local master command center where it can further be processed by and do take necessarily actions as required by the received Life Saving Data structure. It will co-ordinate with other sub-systems that are Advanced Ambulance Systems, Multi-Specialty Hospital Systems, Advanced Health Care Professionals in a regular programmable interval. The programmable intervals are also derived from the statistical data analysis of the services rendered and health patterns of the particular areas.

Eco System of Internet of Things in mind, we can best visualize the maturity levels of the Mobile Health Care solution with the advanced developments happening in the field of Biosensors and their

Figure 4. Applications of internet of things

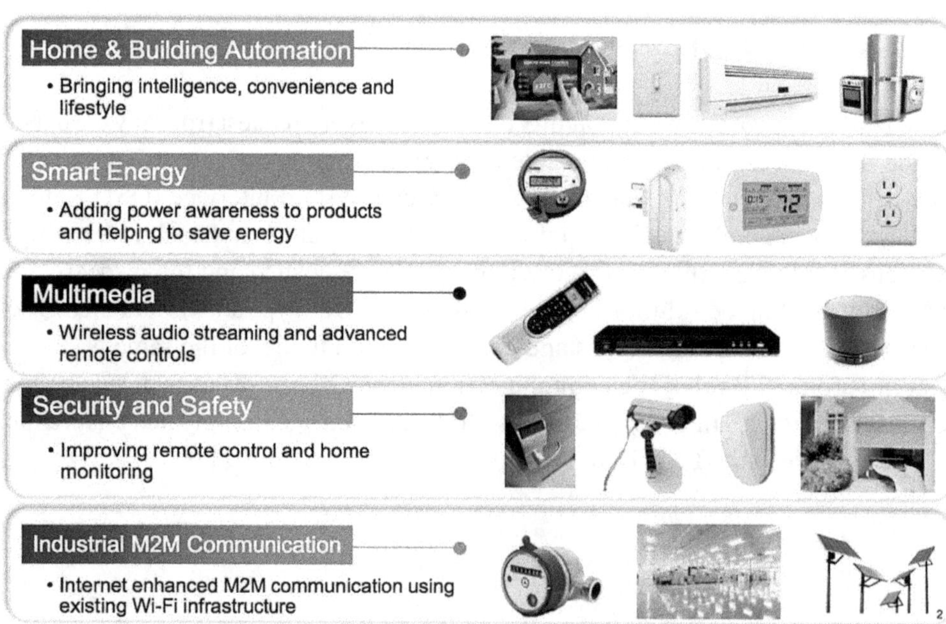

integration into medical devices. We can propose, the integration of these Biosensors and devices which integrates these sensors with today's smart phones with the I.O.T as back bone of the networking of all the devices.

Biosensor Device

The diagram in Figure 5 explains the basic principle in the functionality of the Biosensors.

As shown in above diagram, Biosensor Device is an analytical device that converts biological reactions into measurable signals like an electrical signal which is proportional to Target Analyte concentration in the input samples. A typical biosensor consists of two elements: biological sensing element in recognition layer and a transducer for the detection of Target Analyte concentration. Biological reaction takes place in close contact with the Transducer to ensure that most of the biological reaction is detected. A third element reference can be added along with the two elements which produce a small reference signal which can be an electrical or light signal. This signal serves as a control parameter for the representation of the observations. These observations help to decide on certain interpretations of health conditions and helps to take all necessary medical supporting actions.

Different Target Analytes which are studied with the biosensors in the Health Domain include for example an enzyme, an antibody, or a microorganism present in human body cells or blood samples.

The signal processing elements shown in the diagram are embedded in the Hardware portions of the smart phones which runs certain computational algorithms with a well-defined Software (or programming model). Both of these Software and Hardware elements are tightly coupled with a High Performance, High Speed Embedded Microprocessors sitting in the final electronic product.

As silicon-based chips are functional on the principles of materials science, biosensor transducers are additionally subject to biological principles. The target analyte in above diagram connected to an electronic element.

With the above explanation, we can define a biosensor as a biosensor is a self-contained integrated device, which is capable of providing specific quantitative or semi-quantitative analytical information using a biological recognition element (biochemical receptor) which is retained in direct spatial contact

Figure 5. Basic principle of biosensor

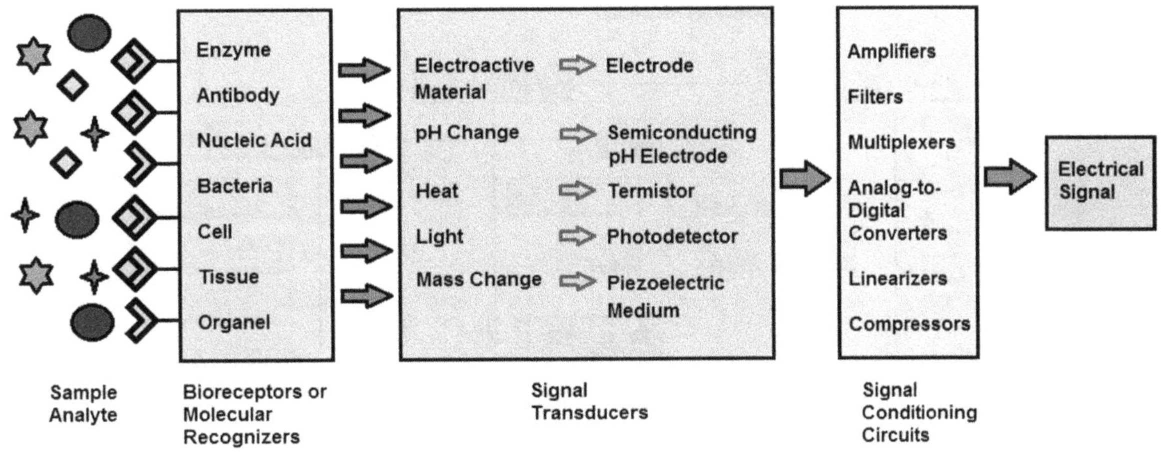

with an electrochemical transduction element. The first of its kind biosensor was developed in 1962 by Clark and Lyon as an „enzyme electrode, for the measurement of glucose levels in blood samples.

Taking above emphasis the Biosensor principle can be represented by Figure 6 with more details.

As per the advancements in the Transducer and in turn Biosensor technologies, the sensors can be implanted into human body in different places with a robust communication model to integrate with the Smart Phones. These biosensors are functionally useful for measuring different health parameters of the individuals who carry or implants those devices, and share the generated data with the Smart Phone in regular intervals. The Decision Making algorithms keep running in the smart phones will keep generating the necessary commands or data structures. These are the data structures which we have discussed as Life Saving Data Types.

Biosensor Signal Detection Methods

Below is a short list of some of the detection methods used in biosensors. The advantages and disadvantages can be point of importance for using these methods for suitability of adapting the method in our Health Care Solution. In our solution, main point we look at if the possibility of the miniaturization of the sensor technology.

- **Amperometry:** Is operated at a given applied potential between the working electrode and the reference electrode, and the generated signal is correlated with the concentration of target compounds. In the amperometric detection, the current signal is generated as a function of the reduction or oxidation of an electro-active product on the surface of a working electrode.
- **Conductometry:** Is a technique depending on the conductivity change in the solution due to the production or consumption of ions, for example, by the metabolic activity of microorganisms. The measurement of conductance can be fast and sensitive, making conductometric microbial biosensors very attractive. Such biosensors are suitable for miniaturization since they require no reference electrode in the system.

Figure 6. Biosensor components

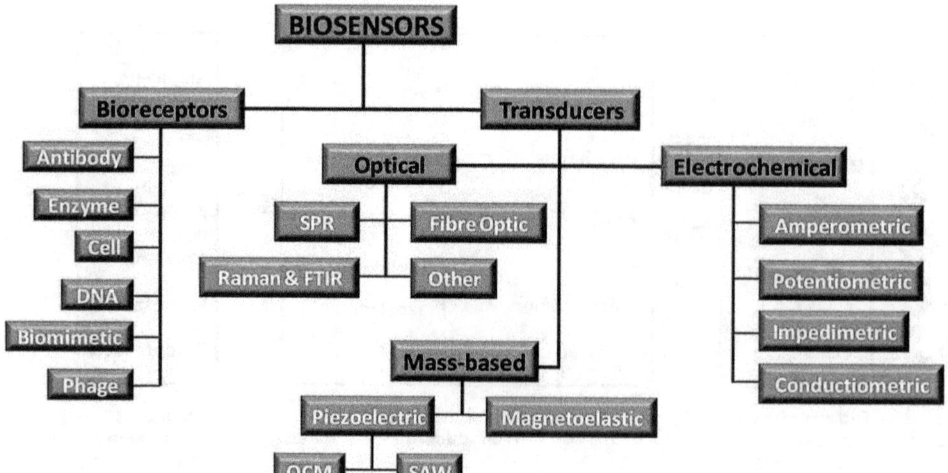

- **Potentiometry:** Involves the measurement of the potential difference between the working electrode and the reference electrode which is dependent on concentration-related behavior. The transducer employed in the potentiometric technique is usually a gas-sensing electrode or an ion-selective electrode. The sensitivity and selectivity of potentiometric biosensor are outstanding due to the species-selective working electrode used in the system. However, a stable and accurate reference electrode is required.

When the scope of the medical problems assisted with this solution becomes wider and area of service delivery is wider, there will be huge amount of data generated in the solution. The processing of this huge amount of data becomes a tedious task when we think of the required accuracies, security and guaranteed delivery of the service. This problem can be better answered with the application of upcoming area of Big Data and Data Analytics in data processing. And we also have high computation engines sitting in clouds for helping us in solving this problem.

Smart Phones (PDAs) Where the Service Requests Are Generated From Individuals

These electronics gadgets must be designed for functioning with very low power requirements, so that battery life can be extended, so that no single case will miss the services because of drained battery in the gadget. Not only this, the data communication model needs re-definition for this to happen. This means, now all the data which is being transmitted from each electronics gadget will either be an isochronous data or bulk data. Meaning, isochronous being time sensitive, it will either be audio or video data and bulk being some kind of text files. And all these data types are having its own delivery models and guaranteed bandwidth allocations in from the SW and HW point of views.

But when we integrate the Mobile Health Care, systems will be re-visited for defining all the data transfer models from considering a new data type which is a Life Saving Data. Though it is Isochronous/Bulk Type in nature, all together it is going to be a new data type from HW and SW stand point. The delivery models, arbitration of the common resources in this solution must be re-defined and tuned for addressing the new requirements of this Life Saving Data. Not only the new Data Type definition, re-definition of the existing communications systems, but also the Transport Means need to be reviewed and there must be a dedicated Transport Road Ways with good quality for safer and speeder mobility of the patients where next level of critical care is delivered to him.

In the hardware level implementations every protocol has definitions of Device Classes, QoS Models, Transaction Types etc., attributes into it. This special Life Saving Data structures should be treated and consider those provisions and can solve inaccuracies, delays etc in existing frame-works.

Data Analytics and Big Data

In comparison to products, services and physical means of business, present age can be treated as age of Digital Data. All the business starting with the very preliminary form of requirements are taking shape of the in Digital Data. Every second humans generating huge amount of data in the digital form and using that for correlation, populating the required information out from millions of bytes of digital data. This area is coming up as Data Analytics where Software Engineers are trying to abstract all the truths, myths into a set of data structures so that they can easily store, retrieve, process and run their analysis algorithms to bring out conclusions based on the numbers residing in the digital data.

The same can equally applies to our Mobile Health Care Solution in its novel version where huge amount of Life Saving Data needs high speed processing to bring out immediate medical solutions of any patient in any area in a predefined boundary. The same solution can recursively be applied to all defined bounded areas and cover entire area of interest, let us say City, Town, Village, District, State in an ideal level even entire nation can be covered with the solution.

6. FUTURE SCOPE

The advanced technologies like Internet of Things, Machine Intelligence, Artificial Intelligence, Robotics, Bi Data and Data Analytics play a major role when the solution gets matured to a level where the individuals who face any health issues will automatically route to the specialty treatment centers using this solution in well before the allowable safe time. At its high degree of maturity, the solution ultimately is going to reduce sudden deaths of human beings and in effect increases the average life time of humans. The solution at its maturity will drastically improves the healthy and quality life of humans. At its high degree of maturity, many biosensors with accessible, reliable, well defined machine intelligence can be integrated with today's mart phone in miniaturized way. Even different medical hand held devices can best be integrated with hand held smart phones in a well-defined communication channels.

7. CONCLUSION

Given today's advancements in miniaturization of the devices, high degree of integration of computing power, social penetration of mobile devices and different applications available, all together a sophisticated and efficient health care systems can be built around these technologies. This integration of all the aspects of Health Related Matters of human beings results in the best and in time diagnosis, accurate medical treatments, world class specialty medical and surgical services. With the focus of further improvements in medical procedures, Theory of Genes Decoding and Encoding, and their integration of next generation systems like High Density Computing, Robotics, Internet of Things, Data Analytics etc., will result in sophisticated health management and that in turn results in increasing the average life span of humans.

ADDITIONAL READING

Introduction to Biosensors. (n.d.). Retrieved from www.sirebi.org

Yoo, E.-H., & Lee, S.-Y. (2010). Glucose biosensors: An overview of use in clinical practice. *Sensors (Basel, Switzerland)*, *10*, 4558–4576. doi:10.3390100504558

This research was previously published in Cloud Computing Systems and Applications in Healthcare edited by Chintan M. Bhatt and S. K. Peddoju, pages 1-18, copyright year 2017 by Medical Information Science Reference (an imprint of IGI Global).

Chapter 2
Mobile Health Systems and Electronic Health Record:
Applications and Implications

Kijpokin Kasemsap
Suan Sunandha Rajabhat University, Thailand

ABSTRACT

This chapter reveals the overview of mobile health systems; the adoption of mobile health systems; mobile health systems and patient monitoring; the overview of mobile health technology; the advanced issues of Electronic Health Record (EHR); and the challenges of EHR in global health care. Mobile health helps deliver the health care services with quality care, improved workflow, and increased patient interaction while minimizing complexity and cost to achieve the desired goals in health care settings. EHR systems are the real-time and patient-centered records that make information available instantly and securely to authorized users. The chapter argues that applying mobile health systems and EHR has the potential to improve health care efficiency and gain sustainable competitive advantage in global health care.

INTRODUCTION

Nowadays, information and communication technology (ICT) affects the health system development across many developing countries, particularly through the application of mobile communications (Nisha, Iqbal, Rifat, & Idrish, 2016). Mobile health is an extension of electronic health in which health care services can be accessed through smart mobile devices (Anshari & Almunawar, 2015), while guaranteeing the mobility of patients for their free activity (Lee, 2016). Mobility is the indispensable part of life today (Gürsel, 2016). Due to the attractive features (e.g., cost-effective sensors and wireless communication capabilities), mobile devices have received great attention in the health care context (Sood, Gururajan, Hafeez-Baig, & Wickramasinghe, 2017).

The potential for mobile health to transform formal health care provision, especially in the geographically remote areas, is huge (Hampshire et al., 2015) regarding many aspects of human's life (Dias, Ribeiro, & Furtado, 2016). The mobile health field focuses on the utilization of mobile health technol-

DOI: 10.4018/978-1-5225-6198-9.ch002

ogy to support hospital care, healthy behavior, patient monitoring, and educational awareness (Househ, Borycki, Kushniruk, & Alofaysan, 2012). Mobile health technology can alert the patients and health care providers about health-related information and emergencies through text messaging on mobile devices (Ikhu-Omoregbe & Azeta, 2012). Telemedicine is the provision of diagnosis and health care from a distance using media, such as interactive computer programs and information technology (Kasemsap, 2017a). Mobile phone-based telemedicine has proven to be an effective approach for information exchange and for providing feedback between patients and their health care providers (Goyal et al., 2016).

EHR is defined as an electronic record of health-related information on an individual that conforms to the nationally recognized interoperability standards and can be created, managed, and consulted by authorized clinicians and staff across health care organizations (Kamath & Donahoe-Anshus, 2012). Health information technology (HIT) is the area of IT involving the design, development, creation, utilization, and maintenance of information systems for the health care industry (Kasemsap, 2017b). EHR is the part of the larger national initiative focusing on HIT, which is the exchange of health information in an electronic environment (D'Abundo, 2013). EHR systems are very important in health care settings and have the potential to transform the health care system from a mostly paper-based industry to the one that utilizes the clinical data and other pieces of information to assist health care providers in delivering the higher quality of care to their patients (Kasemsap, 2017c).

EHR is considered as the significant method of improving the efficiency of health care system (Gibson & Abrams, 2010). EHR includes health information, such as observations, laboratory tests, diagnostic imaging reports, treatments, therapies, drugs administered, and allergies (de la Torre Díez, Sánchez, Coronado, & López Gálvez, 2010). EHR attracts the particular concern about the unauthorized access and disclosure of personal information contained in the records (Ries, 2011). A fundamental requirement for achieving continuity of care is recognized as the integration and interoperability of clinically oriented systems toward the realization of EHR (Kitsiou, 2009). Through EHR, patients are provided with tools to help them manage their health care, clinicians are able to access the up-to-date patient information, and governments are showing transparency to the public by reporting health information on their websites (Protti, 2008). The suitable utilization of EHR requires the realistic concept of electronic health by all the involved professions (Ceruti, Geninatti, & Siliquini, 2016).

This chapter is based on a literature review of mobile health systems and EHR. The extensive literature of mobile health systems and EHR provides a contribution to practitioners and researchers in order to maximize their impact in global health care.

BACKGROUND

Health care systems experience a wide variety of challenges with the integration of mobile and ubiquitous technology (Olla & Tan, 2009). The use of mobile devices (e.g., smartphones, pagers, tablets, and Wi-Fi phones) can accelerate the admit-and-discharge process and promote the direct communication between clinicians and patients (Moghimi & Wickramasinghe, 2017). While designing mobile health systems, the focal point of research is concentrated on the design of innovative developments for improving the practice of health care and the increase in well-being with a strong focus on the functional requirements (Mayora et al., 2016).

Mobile health is the new edge on health care innovation (Silva, Rodrigues, de la Torre Diez, Lopez-Coronado, & Saleem, 2015) and is about the ability to deliver and manage the health care services and information through wireless networks (Bamigboye, 2012). Many patients seek the health-related answers for their specific individual needs from any place, at any time (Koumpouros & Georgoulas, 2017). The patient-centered health care paradigm allows for the increased quality of care and quality of life for patients while increasing personal freedom to be connected to health care providers and health care services (Falchuk, Famolari, Fischer, Loeb, & Panagos, 2010).

EHR has been adopted in many countries due to its ability to enhance and raise the health care quality (Nassar, Othman, Hayajneh, & Ali, 2015). EHR is a method of organizing the patient data-making use of the advances in the field of IT (Galani & Nikiforou, 2006) and provides each individual with a secure and private lifetime record of their key health history and care within the health care system (Pullen & Al-Hakim, 2016). Health care organizations are encouraged to adopt EHR, which are expected to improve the quality of health care by providing the accurate and up-to-date patient information, to improve patients' and providers' adherence to treatment guidelines, and to allow health information to be shared among authorized parties within and outside an organization (Zhang, Yu, & Shen, 2012).

EHR systems lead to the improved legibility of physician's notes (Shachak, Hadas-Dayagi, Ziv, & Reis, 2009) and the better access to patient charts (Walji, Taylor, Langabeer, & Valenza, 2009). Service innovations have been linked to the changes in the way that health care organizations organize their work (Bhuyan, Zhu, Chandak, Kim, & Stimpson, 2014). The inappropriate implementation or adoption of EHR can contribute to medical errors that cause serious injuries or even deaths (Institute of Medicine of the National Academies, 2012).

SIGNIFICANT PERSPECTIVES ON MOBILE HEALTH SYSTEMS AND ELECTRONIC HEALTH RECORD

This section presents the overview of mobile health systems; the adoption of mobile health systems; mobile health systems and patient monitoring; the overview of mobile health technology; the advanced issues of EHR; and the challenges of EHR in global health care.

Overview of Mobile Health Systems

In the developed world, mobile health systems can increase the dissemination of health information and actionable public health information for improving the ability to diagnose, treat, and track many chronic diseases (Siddiqui et al., 2015). Coinciding with the rise of the aging population, the prevalence of both chronic diseases and health care costs has skyrocketed, making it challenging to improve the quality and the reach of health care to those in need (Lan, Zhang, & Lu, 2016). Through mobile health systems, patients will not only utilize the online health care services (e.g., making appointments, viewing health records, and having consultations through their mobile devices), but they can also participate and contribute to the discussion in the knowledge-sharing forums (Almunawar, Anshari, & Younis, 2015).

Mobile health systems and its corresponding mobility functionalities have a strong impact on the typical health care monitoring and alerting systems, clinical and administrative data collection, record maintenance, health care delivery programs, medical information awareness, and prevention systems (Zuehlke, Li, Talaei-Khoei, & Ray, 2009). Mobile health systems have the potential to create the health care-related future by translating the existing applications and by creating the new ideas (Tamposis, Pouliakis, Fezoulidis, & Karakitsos, 2017). The use of mobile phones in health care settings can potentially deliver many benefits, because of their ability to improve the access to information resources (Chib, 2010). The advancement in mobile health technology and the increasing number of aging people have given rise to a need for the better understanding on how mobile phones can be used to the advantage of the aging population (Nikou, 2015).

Mobile health describes the utilization of a broad range of telecommunication and multimedia technologies within wireless care delivery design (Moumtzoglou, 2016) and provides a medium to efficiently transfer the health-related information (Redha, Hartwick, & Sikka, 2015). Mobile health appears to offer a way for health care delivery to revolutionize itself and to emphasize the significant areas of access, quality, and value (Wickramasinghe & Goldberg, 2007). Mobile health systems can play an important role in supporting health care by providing the health care applications that access the health care records and reduce the paperwork for clinical physicians, nurses, and health care providers (Archer, 2010).

Mobile health strategies include the use of mobile phones for data collection, access to training, communication among health care providers, and the promotion of health behaviors in the community (Agarwal, Perry, Long, & Labrique, 2015). Through mobile health systems, patient engagement in self-health is recognized as an important goal in improving health outcomes (Mitra & Padman, 2014). Health care providers can maximize the benefits of electronic tools by educating themselves to better understand the potential uses, challenges, and benefits of mobile health systems (Folami, 2014).

Adoption of Mobile Health Systems

Mobile health can be used as a powerful health behavior change tool for health prevention and self-management as they are ubiquitous, carried on the person and are capable of computational capacity (Ben-Zeev et al., 2013). The adoption of mobile health necessitates the complex buying behavior, and consumers must integrate various ideas to justify their decision to receive the mobile health services (Shareef, Ahmed, Kumar, & Kumar, 2015). Technological, behavioral, and social beliefs of mobile health systems' functional, organizational, and professional's benefits can enhance the attitude toward using them (Shareef, Kumar, Kumar, & Hasin, 2013). The intention to use the mobile health systems significantly leads to the actual acceptance behavior (Dwivedi, Shareef, Simintiras, Lal, & Weerakkody, 2016).

The developing world faces many challenges in realizing the technical expertise required to adopt the mobile health applications (Yadav, Aliasgari, & Poellabauer, 2016). The technical difficulties encountered in using mobile health technology are an inadequate physical infrastructure, quality of service (QoS) issues, and insufficient access by the user to the hardware/software communication infrastructure (Oddershede & Carrasco, 2008). Lack of data accuracy and information integrity in mobile health systems can cause the serious harm to patients and limit the benefits of such promising technology (Sako, Karpathiou, Adibi, & Wickramasinghe, 2017).

Mobile Health Systems and Patient Monitoring

Mobile health is considered as a way to emphasize prevention through mobile monitoring devices and to reduce the overall cost of health care (Ciaramitaro & Skrocki, 2012). Recording of physiological vital signs in patients' real-life environment (Testa, Coronato, Cinque, & de Pietro, 2016) is useful in the management of chronic disorders, such as for heart failure, hypertension, diabetes, chronic pain, and severe obesity (Jasemian, 2009). Monitoring patients in diverse environments, through mobile health systems, is one of the major benefits of this approach (Jasemian, 2009).

The design of mobile health monitoring systems has attracted the interest of large communities from industry and academia (Cinque, Coronato, & Testa, 2012). The remote monitoring and mobile applications for health care are designed to enable the consumers to enjoy the improved health care delivery (Edirisinghe, Stranieri, & Wickramasinghe, 2017). Remote sensors are linked to mobile phones, which are used to facilitate data transmission to health care providers (Skorin-Kapov, Dobrijevic, & Piplica, 2014).

Measurement-based care is the important component of collaborative care for depression, a depression care model that has been shown to improve the depression outcomes for primary care patients in dozens of clinical trials (Woltmann et al., 2012). Measurement practices in routine clinical practice are likely to occur less frequently or consistently than in research settings or than guidelines recommend (Pfeiffer et al., 2015). The utilization of measurement-based care allows patients to receive the same quality of care and outcomes in primary care settings as can be achieved in specialty mental health care settings.

Overview of Mobile Health Technology

With the increasing popularity of mobile devices and development of wireless communication network technologies, more and more studies integrate both mobile devices and context-aware technology for developing the health care service platforms to enhance the personalized health management toward gaining the effectiveness of preventive medicine (Wiesner & Pfeifer, 2014). The developments in the wireless and mobile markets are capitalized by the medical device industry (Angelidis, 2009). The effective intervention tool in self-management and care is the patients' use of mobile-based devices (Misra, 2016). Despite its great potential, more research is still needed as to how to effectively integrate the mobile health technology and apply it to health care services (Wang, Chen, Kuo, Chen, & Shiu, 2016).

Mobile health technology significantly changes the way enterprises, institutions, and people utilize the current software systems (Stavros, Vavoulidis, & Nasioutziki, 2016). Mobile health technology has yielded many activities that include emergency response systems through toll-free access numbers; disease surveillance and control; human resource coordination, management, and supervision; decision support to remote clinicians; and decision support information available at the point of care (Mechael, 2010). In low-income regions, mobile phone–based tools can improve the scope and efficiency of field health workers (DeRenzi et al., 2011).

Smartphones and other mobile devices are utilized to perform mobile diagnostic tests, to access EHR, to support clinical decisions, and to provide new method of medical education among health care-related activities (Queirós et al., 2016). Physicians and health professionals are more likely to use mobile software applications that provide them with access to references to health care information, such as guidelines and information found in journal articles. These mobile health applications provide information to health professionals that can be used in their clinical decision making (Tin, Cummings, & Borycki, 2014).

Mobile health applications have become increasingly widespread, such as developing the health system to providing diabetes health education information, as well as encouraging users to talk to physicians through the mobile information system (Wangberg, Arsand, & Andersson, 2006). Mobile health applications can support dissemination and uptake of evidence to reduce and ultimately eliminate the health care-associated infections (Schnall & Iribarren, 2015). Many mobile health applications are being used by or recommended to patients and health care providers with little understanding of their functionality or ability to integrate data into health care systems (IMS Institute for Healthcare Informatics, 2013).

Mobile health information services allow them to obtain the useful health information and guidance to achieve the better health self-management (Deng, Mo, & Liu, 2014). Through the utilization of mobile health information services, health care administrators can enhance their management system, thus reducing the operating costs (Wang et al., 2016). Through mobile health information services, it is possible to provide the remote health monitoring system with reduced cost and improved efficiency (Khalid, Muhammad, Patrick, & Hossam, 2012).

Advanced Issues of Electronic Health Record

EHR performs a multitude of functions utilizing different types of data (Hayrinen, Saranto, & Nykanen, 2008). EHR data can facilitate the modern clinical decision support (Prokosch & Ganslandt, 2009), conduct the biomedical association studies (Lang, Burkle, Laumann, & Prokosch, 2008), improve the EHR security (Chen, Lorenzi, Nyemba, Schildcrout, & Malin, 2014), and evaluate the effectiveness of cost treatments (Muranaga, Kumamoto, & Uto, 2007). EHR data can be utilized to efficiently support the learning health care system, where information about health care operations is translated into knowledge for the evidence-based clinical practice and positive change (Etheredge, 2014).

The amount of information in the digital format is found to be excessive (Whittaker, Aufdenkamp, & Tinley, 2009), therefore how the information is organized in EHR systems becomes the critical issue in information retrieval and information utilization (Saleem et al., 2009). Advantages of EHR data collection through automated data capture include the access to vast amounts of existing data, which may be acquired faster and more reliably (Effler et al., 1999) and with a greater capacity for growth over time. Establishing an interface for data harvesting requires institution-specific implementation and potentially considerable IT resources given the complexities and heterogeneities of current EHR (Broberg et al., 2015).

Challenges of Electronic Health Record in Global Health Care

Creating EHR that support various needs of health care presents the important challenges (Senathirajah, Bakken, & Kaufman, 2014). Although decades of research have focused on defining the requirements for health care systems that support physicians and their tasks, studies suggest that health care information systems often fail to support the efficient clinical decision making and completion of relevant tasks (Greenhalgh, Potts, Wong, Bark, & Swinglehurst, 2009). EHR implementations are widespread and have been recognized as the costly investments (Nguyen, Bellucci, & Nguyen, 2014). Health care systems may fail to take into consideration the significant variability of medical information needs that differ concerning health care context, specialty, role, individual patient, and health care organizations (Senathirajah et al., 2014).

EHR can obstruct the patient-centeredness and communication with patients (Shachak et al., 2009). Communication and collaboration needs are frequently not met because health care systems are designed only for an individual's linear workflow without the ability to share the features that support collaboration (Han, Carcillo, & Venkataraman, 2005). Accessing large amounts of information through the finite screen space necessitates negotiating the multiple screens. The management of information on the poorly organized display may create a burden on the limited human cognitive resources (Horsky, Kaufman, Oppenheim, & Patel, 2003).

To train health care providers how to utilize EHR, simulation training should be considered as the effective method of teaching prior to the implementation of EHR in health care institutions (Vuk et al., 2015). Simulation training provides learners an opportunity to manage the health care systems, with focused learning objectives and immediate feedback about their performance, before interacting with the real patients (Okuda et al., 2009). Simulation training in which health care providers organize their workflow by interacting with standardized patients, lay people trained to report the patient history and convey the symptoms, can improve the skills of health care providers to utilize EHR and to increase patient safety (Patterson et al., 2013).

FUTURE RESEARCH DIRECTIONS

The classification of the extensive literature in the domains of mobile health systems and EHR will provide the potential opportunities for future research. Reducing medical errors, increasing patient safety, and improving the quality of health care are the major goals in the health care industry (Kasemsap, 2017d). Telemedicine brings the health care value through its ability for the remote visits with patients, immediate access to health care professionals, real-time access to health data, and health monitoring capabilities (Kasemsap, 2017e). An examination of linkages among mobile health systems, EHR, and telemedicine toward reducing medical error in global health care should be further studied.

CONCLUSION

This chapter indicated the overview of mobile health systems; the adoption of mobile health systems; mobile health systems and patient monitoring; the overview of mobile health technology; the advanced issues of EHR; and the challenges of EHR in global health care. Mobile health helps deliver the health care services with quality care, improved workflow, and increased patient interaction while minimizing complexity and cost to achieve the desired goals in health care settings. Mobile health provides users the best treatment with optimum care, anywhere and anytime. Mobile health brings users the advanced mobile patient care tools to enhance quality treatment and streamline workloads. Through mobile health, health care providers can monitor patient's condition and respond proactively before it leads to any chronic disorder. Mobile health simplifies the health care provider's complexity by providing patient's history promptly when they need to make patient care decisions toward improving quality of care and reducing medical errors.

EHR systems are the real-time and patient-centered records that make information available instantly and securely to authorized users. While an EHR does contain the medical and treatment histories of patients, an EHR system is built to go beyond standard clinical data collected in a provider's office and can be inclusive of a broader view of a patient's care. EHRs can contain a patient's medical history, diagnoses, medications, treatment plans, immunization dates, allergies, radiology images, and test results; allow access to evidence-based tools that providers can use to make decisions about a patient's care; and automate the health care provider workflow. Applying mobile health systems and EHR has the potential to improve health care efficiency and gain sustainable competitive advantage in global health care.

REFERENCES

Agarwal, S., Perry, H. B., Long, L. A., & Labrique, A. B. (2015). Evidence on feasibility and effective use of mHealth strategies by frontline health workers in developing countries: Systematic review. *Tropical Medicine & International Health*, *20*(8), 1003–1014. doi:10.1111/tmi.12525 PMID:25881735

Almunawar, M. N., Anshari, M., & Younis, M. Z. (2015). Incorporating customer empowerment in mobile health. *Health Policy and Technology*, *4*(4), 312–319. doi:10.1016/j.hlpt.2015.08.008

Angelidis, P. (2009). Mobile telemonitoring insights. In J. Tan (Ed.), *Medical informatics: Concepts, methodologies, tools, and applications* (pp. 107–112). Hershey, PA: IGI Global. doi:10.4018/978-1-60566-050-9.ch010

Anshari, M., & Almunawar, M. N. (2015). Mobile health (mHealth). In M. Khosrow-Pour (Ed.), *Encyclopedia of information science and technology* (3rd ed.; pp. 5607–5614). Hershey, PA: IGI Global. doi:10.4018/978-1-4666-5888-2.ch553

Archer, N. (2010). Mobile e-health: Making the case. In K. Khoumbati, Y. Dwivedi, A. Srivastava, & B. Lal (Eds.), *Handbook of research on advances in health informatics and electronic healthcare applications: Global adoption and impact of information communication technologies* (pp. 446–457). Hershey, PA: IGI Global. doi:10.4018/978-1-60566-030-1.ch027

Bamigboye, A. (2012). Mobile healthcare: Challenges and opportunities. In B. Ciaramitaro (Ed.), *Mobile technology consumption: Opportunities and challenges* (pp. 77–98). Hershey, PA: IGI Global. doi:10.4018/978-1-61350-150-4.ch006

Ben-Zeev, D., Kaiser, S. M., Brenner, C. J., Begale, M., Duffecy, J., & Mohr, D. C. (2013). Development and usability testing of FOCUS: A smartphone system for self-management of schizophrenia. *Psychiatric Rehabilitation Journal*, *36*(4), 289–296. doi:10.1037/prj0000019 PMID:24015913

Bhuyan, S. S., Zhu, H., Chandak, A., Kim, J., & Stimpson, J. P. (2014). Do service innovations influence the adoption of electronic health records in long-term care organizations? Results from the U.S. national survey of residential care facilities. *International Journal of Medical Informatics*, *83*(12), 975–982. doi:10.1016/j.ijmedinf.2014.09.007 PMID:25453201

Broberg, C. S., Mitchell, J., Rehel, S., Grant, A., Gianola, A., Beninato, P., ... Sahn, D. J. (2015). Electronic medical record integration with a database for adult congenital heart disease: Early experience and progress in automating multicenter data collection. *International Journal of Cardiology*, *196*, 178–182. doi:10.1016/j.ijcard.2015.05.140 PMID:26142077

Ceruti, M., Geninatti, S., & Siliquini, R. (2016). Use and reuse of electronic health records: Building information systems for improvement of health services. In *E-health and telemedicine: Concepts, methodologies, tools, and applications* (pp. 961–975). Hershey, PA: IGI Global. doi:10.4018/978-1-4666-8756-1.ch049

Chen, Y., Lorenzi, N., Nyemba, S., Schildcrout, J. S., & Malin, B. (2014). We work with them? Health workers interpretation of organizational relations mined from electronic health records. *International Journal of Medical Informatics*, *83*(7), 495–506. doi:10.1016/j.ijmedinf.2014.04.006 PMID:24845147

Chib, A. (2010). The Aceh Besar midwives with mobile phones project: Design and evaluation perspectives using the information and communication technologies for healthcare development model. *Journal of Computer-Mediated Communication*, *15*(3), 500–525. doi:10.1111/j.1083-6101.2010.01515.x

Ciaramitaro, B. L., & Skrocki, M. (2012). mHealth: Mobile healthcare. In B. Ciaramitaro (Ed.), Mobile technology consumption: Opportunities and challenges (pp. 99–109). Hershey, PA: IGI Global. doi:10.4018/978-1-61350-150-4.ch007

Cinque, M., Coronato, A., & Testa, A. (2012). Dependable services for mobile health monitoring systems. *International Journal of Ambient Computing and Intelligence*, *4*(1), 1–15. doi:10.4018/jaci.2012010101

D'Abundo, M. L. (2013). Electronic health record implementation in the United States healthcare industry: Making the process of change manageable. In V. Wang (Ed.), *Handbook of research on technologies for improving the 21st century workforce: Tools for lifelong learning* (pp. 272–286). Hershey, PA: IGI Global. doi:10.4018/978-1-4666-2181-7.ch018

de la Torre Díez, I., Sánchez, R. H., Coronado, M. L., & López Gálvez, M. I. (2010). Electronic health records in a tele-ophthalmologic application with Oracle 10g. In A. Lazakidou (Ed.), *Biocomputation and biomedical informatics: Case studies and applications* (pp. 89–105). Hershey, PA: IGI Global. doi:10.4018/978-1-60566-768-3.ch005

Deng, Z., Mo, X., & Liu, S. (2014). Comparison of the middle-aged and older users' adoption of mobile health services in China. *International Journal of Medical Informatics*, *83*(3), 210–224. doi:10.1016/j.ijmedinf.2013.12.002 PMID:24388129

DeRenzi, B., Borriello, G., Jackson, J., Kumar, V. S., Parikh, T. S., Virk, P., & Lesh, N. (2011). Mobile phone tools for field-based health care workers in low-income countries. *Mount Sinai Journal of Medicine: A Journal of Translational and Personalized Medicine, 78*(3), 406–418. doi:10.1002/msj.20256

Dias, C. M., Ribeiro, A. G., & Furtado, S. F. (2016). An overview about the use of healthcare applications on mobile devices. In M. Cruz-Cunha, I. Miranda, R. Martinho, & R. Rijo (Eds.), *Encyclopedia of e-health and telemedicine* (pp. 285–298). Hershey, PA: IGI Global. doi:10.4018/978-1-4666-9978-6.ch024

Dwivedi, Y. K., Shareef, M. A., Simintiras, A. C., Lal, B., & Weerakkody, V. (2016). A generalised adoption model for services: A cross-country comparison of mobile health (m-health). *Government Information Quarterly*, *33*(1), 174–187. doi:10.1016/j.giq.2015.06.003

Edirisinghe, R., Stranieri, A., & Wickramasinghe, N. (2017). A taxonomy for mHealth. In N. Wickramasinghe (Ed.), *Handbook of research on healthcare administration and management* (pp. 596–615). Hershey, PA: IGI Global. doi:10.4018/978-1-5225-0920-2.ch036

Effler, P., Ching-Lee, M., Bogard, A., Ieong, M. C., Nekomoto, T., & Jernigan, D. (1999). Statewide system of electronic notifiable disease reporting from clinical laboratories: Comparing automated reporting with conventional methods. *Journal of the American Medical Association*, *282*(19), 1845–1850. doi:10.1001/jama.282.19.1845 PMID:10573276

Etheredge, L. M. (2014). Rapid learning: A breakthrough agenda. *Health Affairs*, *33*(7), 1155–1162. doi:10.1377/hlthaff.2014.0043 PMID:25006141

Falchuk, B., Famolari, D., Fischer, R., Loeb, S., & Panagos, E. (2010). The mHealth stack: Technology enablers for patient-centric mobile healthcare. *International Journal of E-Health and Medical Communications, 1*(1), 1–17. doi:10.4018/jehmc.2010010101

Folami, F. F. (2014). M-health technology as a transforming force for population health. In B. Adeoye & L. Tomei (Eds.), *Effects of information capitalism and globalization on teaching and learning* (pp. 256–262). Hershey, PA: IGI Global. doi:10.4018/978-1-4666-6162-2.ch021

Galani, O., & Nikiforou, A. (2006). Electronic health records. In A. Lazakidou (Ed.), *Handbook of research on informatics in healthcare and biomedicine* (pp. 1–8). Hershey, PA: IGI Global. doi:10.4018/978-1-59140-982-3.ch001

Gibson, C. J., & Abrams, K. J. (2010). Will privacy concerns derail the electronic health record? Balancing the risks and benefits. In S. Kabene (Ed.), *Healthcare and the effect of technology: Developments, challenges and advancements* (pp. 178–196). Hershey, PA: IGI Global. doi:10.4018/978-1-61520-733-6.ch011

Goyal, S., Morita, P., Lewis, G. F., Yu, C., Seto, E., & Cafazzo, J. A. (2016). The systematic design of a behavioural mobile health application for the self-management of type 2 diabetes. *Canadian Journal of Diabetes, 40*(1), 95–104. doi:10.1016/j.jcjd.2015.06.007 PMID:26455762

Greenhalgh, T., Potts, H., Wong, G., Bark, P., & Swinglehurst, D. (2009). Tensions and paradoxes in electronic patient record research: A systematic literature review using the meta-narrative method. *The Milbank Quarterly, 87*(4), 729–788. doi:10.1111/j.1468-0009.2009.00578.x PMID:20021585

Gürsel, G. (2016). Mobility in healthcare: M-health. In A. Panagopoulos (Ed.), *Handbook of research on next generation mobile communication systems* (pp. 485–511). Hershey, PA: IGI Global. doi:10.4018/978-1-4666-8732-5.ch019

Hampshire, K., Porter, G., Owusu, S. A., Mariwah, S., Abane, A., Robson, E., ... Milner, J. (2015). Informal m-health: How are young people using mobile phones to bridge healthcare gaps in Sub-Saharan Africa? *Social Science & Medicine, 142*, 90–99. doi:10.1016/j.socscimed.2015.07.033 PMID:26298645

Hayrinen, K., Saranto, K., & Nykanen, P. (2008). Definition, structure, content, use and impacts of electronic health records: A review of the research literature. *International Journal of Medical Informatics, 77*(5), 291–304. doi:10.1016/j.ijmedinf.2007.09.001 PMID:17951106

Horsky, J., Kaufman, D., Oppenheim, M., & Patel, V. (2003). A framework for analyzing the cognitive complexity of computer-assisted clinical ordering. *Journal of Biomedical Informatics, 36*(1/2), 4–22. doi:10.1016/S1532-0464(03)00062-5 PMID:14552843

Househ, M., Borycki, E. M., Kushniruk, A. W., & Alofaysan, S. (2012). mHealth: A passing fad or here to stay? In J. Rodrigues, I. de la Torre Díez, & B. Sainz de Abajo (Eds.), Telemedicine and e-health services, policies, and applications: Advancements and developments (pp. 151–178). Hershey, PA: IGI Global. doi:10.4018/978-1-4666-0888-7.ch007

Ikhu-Omoregbe, N., & Azeta, A. (2012). Design and deployment of a mobile-based medical alert system. In M. Watfa (Ed.), *E-healthcare systems and wireless communications: Current and future challenges* (pp. 210–219). Hershey, PA: IGI Global. doi:10.4018/978-1-61350-123-8.ch010

IMS Institute for Healthcare Informatics. (2013). *Patient apps for improved healthcare: From novelty to mainstream*. Parsippany, NJ: IMS Institute for Healthcare Informatics.

Institute of Medicine of the National Academies. (2012). *Health IT and patient safety*. Washington, DC: The National Academies Press.

Jasemian, Y. (2009). Patient monitoring in diverse environments. In P. Olla & J. Tan (Eds.), *Mobile health solutions for biomedical applications* (pp. 129–142). Hershey, PA: IGI Global. doi:10.4018/978-1-60566-332-6.ch007

Kamath, J. R., & Donahoe-Anshus, A. L. (2012). Electronic health record: Adoption, considerations and future direction. In A. Kolker & P. Story (Eds.), *Management engineering for effective healthcare delivery: Principles and applications* (pp. 309–332). Hershey, PA: IGI Global. doi:10.4018/978-1-60960-872-9.ch015

Kasemsap, K. (2017a). Telemedicine and electronic health: Issues and implications in developing countries. In K. Moahi, K. Bwalya, & P. Sebina (Eds.), *Health information systems and the advancement of medical practice in developing countries* (pp. 149–167). Hershey, PA: IGI Global. doi:10.4018/978-1-5225-2262-1.ch009

Kasemsap, K. (2017b). Analyzing the role of health information technology in global health care. In N. Wickramasinghe (Ed.), *Handbook of research on healthcare administration and management* (pp. 287–307). Hershey, PA: IGI Global. doi:10.4018/978-1-5225-0920-2.ch017

Kasemsap, K. (2017c). Mastering electronic health record in global health care. In N. Wickramasinghe (Ed.), *Handbook of research on healthcare administration and management* (pp. 222–242). Hershey, PA: IGI Global. doi:10.4018/978-1-5225-0920-2.ch014

Kasemsap, K. (2017d). The perspectives of medical errors in the health care industry. In M. Riga (Ed.), *Impact of medical errors and malpractice on health economics, quality, and patient safety* (pp. 113–143). Hershey, PA: IGI Global. doi:10.4018/978-1-5225-2337-6.ch005

Kasemsap, K. (2017e). The importance of telemedicine in global health care. In N. Wickramasinghe (Ed.), *Handbook of research on healthcare administration and management* (pp. 157–177). Hershey, PA: IGI Global. doi:10.4018/978-1-5225-0920-2.ch010

Khalid, E., Muhammad, A., Patrick, M., & Hossam, S. H. (2012). Ubiquitous health monitoring using mobile web services. *Procedia Computer Science*, *10*, 332–339. doi:10.1016/j.procs.2012.06.044

Kitsiou, S. (2009). Overview and analysis of electronic health record standards. In A. Lazakidou & K. Siassiakos (Eds.), *Handbook of research on distributed medical informatics and e-health* (pp. 84–103). Hershey, PA: IGI Global. doi:10.4018/978-1-60566-002-8.ch005

Koumpouros, Y., & Georgoulas, A. (2017). mHealth R&D activities in Europe. In I. Management Association (Ed.), Gaming and technology addiction: Breakthroughs in research and practice (pp. 758–789). Hershey, PA: IGI Global. doi:10.4018/978-1-5225-0778-9.ch034

Lan, T., Zhang, J., & Lu, Y. (2016). Transforming the blood glucose meter into a general healthcare meter for in vitro diagnostics in mobile health. *Biotechnology Advances*, *34*(3), 331–341. doi:10.1016/j.biotechadv.2016.03.002 PMID:26946282

Lang, M., Burkle, T., Laumann, S., & Prokosch, H. U. (2008). Process mining for clinical workflows: Challenges and current limitations. *Studies in Health Technology and Informatics*, *136*, 229–334. PMID:18487736

Lee, T. (2016). Mobile healthcare computing in the cloud. In *Mobile computing and wireless networks: Concepts, methodologies, tools, and applications* (pp. 1412–1432). Hershey, PA: IGI Global. doi:10.4018/978-1-4666-8751-6.ch061

Mayora, O., Frost, M., Arnrich, B., Gravenhorst, F., Grunerbl, A., Muaremi, A., & ... . (2016). Mobile health systems for bipolar disorder: The relevance of non-functional requirements in MONARCA project. In *E-health and telemedicine: Concepts, methodologies, tools, and applications* (pp. 1395–1405). Hershey, PA: IGI Global. doi:10.4018/978-1-4666-8756-1.ch070

Mechael, P. (2010). Opportunities and challenges of integrating mHealth applications into rural health initiatives in Africa. In M. Cruz-Cunha, A. Tavares, & R. Simoes (Eds.), *Handbook of research on developments in e-health and telemedicine: Technological and social perspectives* (pp. 704–727). Hershey, PA: IGI Global. doi:10.4018/978-1-61520-670-4.ch034

Misra, R. (2016). The benefits and challenges of using mobile-based tools in self-management and care. In A. Dwivedi (Ed.), *Reshaping medical practice and care with health information systems* (pp. 1–13). Hershey, PA: IGI Global. doi:10.4018/978-1-4666-9870-3.ch001

Mitra, S., & Padman, R. (2014). Engagement with social media platforms via mobile apps for improving quality of personal health management: A healthcare analytics case study. *Journal of Cases on Information Technology*, *16*(1), 73–89. doi:10.4018/jcit.2014010107

Moghimi, H., & Wickramasinghe, N. (2017). The development of a secure hospital messaging and communication platform. In N. Wickramasinghe (Ed.), *Handbook of research on healthcare administration and management* (pp. 243–267). Hershey, PA: IGI Global. doi:10.4018/978-1-5225-0920-2.ch015

Moumtzoglou, A. S. (2016). The nexus of m-health and self-efficacy. In A. Moumtzoglou (Ed.), *M-health innovations for patient-centered care* (pp. 341–365). Hershey, PA: IGI Global. doi:10.4018/978-1-4666-9861-1.ch017

Muranaga, F., Kumamoto, I., & Uto, Y. (2007). Development of site data warehouse for cost analysis of DPC based on medical costs. *Methods of Information in Medicine*, *46*(6), 679–685. PMID:18066419

Nassar, D. A., Othman, M., Hayajneh, J. A., & Ali, N. (2015). An integrated success model for an electronic health record: A case study of Hakeem Jordan. *Procedia Economics and Finance*, *23*, 95–103. doi:10.1016/S2212-5671(15)00526-2

Nguyen, L., Bellucci, E., & Nguyen, L. T. (2014). Electronic health records implementation: An evaluation of information system impact and contingency factors. *International Journal of Medical Informatics*, *83*(11), 779–796. doi:10.1016/j.ijmedinf.2014.06.011 PMID:25085286

Nikou, S. (2015). Mobile technology and forgotten consumers: The young-elderly. *International Journal of Consumer Studies*, *39*(4), 294–304. doi:10.1111/ijcs.12187

Nisha, N., Iqbal, M., Rifat, A., & Idrish, S. (2016). Mobile health services: A new paradigm for health care systems. In *E-health and telemedicine: Concepts, methodologies, tools, and applications* (pp. 1551–1567). Hershey, PA: IGI Global. doi:10.4018/978-1-4666-8756-1.ch078

Oddershede, A. M., & Carrasco, R. A. (2008). Perception of mobile technology provision in health service. In W. Huang, Y. Wang, & J. Day (Eds.), *Global mobile commerce: Strategies, implementation and case studies* (pp. 345–364). Hershey, PA: IGI Global. doi:10.4018/978-1-59904-558-0.ch019

Okuda, Y., Bryson, E. O., DeMaria, S., Jr., Jacobson, L., Quinones, J., Shen, B., & Levine, A. I. (2009). The utility of simulation in medical education: What is the evidence?. *Mount Sinai Journal of Medicine: A Journal of Translational and Personalized Medicine, 76*(4), 330–343. doi:10.1002/msj.20127

Olla, P., & Tan, J. (2009). *Mobile health solutions for biomedical applications*. Hershey, PA: IGI Global. doi:10.4018/978-1-60566-332-6

Patterson, E. S., Zhang, J., Abbott, P., Gibbons, M. C., Lowry, Z. S., Quinn, T. M., ... Brick, D. (2013). Enhancing electronic health record usability in pediatric patient care: A scenario-based approach. *Joint Commission Journal on Quality and Patient Safety*, *39*(3), 129–135. doi:10.1016/S1553-7250(13)39019-9 PMID:23516763

Pfeiffer, P. N., Bohnert, K. M., Zivin, K., Yosef, M., Valenstein, M., Aikens, J. E., & Piette, J. D. (2015). Mobile health monitoring to characterize depression symptom trajectories in primary care. *Journal of Affective Disorders*, *174*, 281–286. doi:10.1016/j.jad.2014.11.040 PMID:25527999

Prokosch, H. U., & Ganslandt, T. (2009). Perspectives for medical informatics: Reusing the electronic medical record for clinical research. *Methods of Information in Medicine*, *48*(1), 38–44. PMID:19151882

Protti, D. (2008). A comparison of how Canada, England, and Denmark are managing their electronic health record journeys. In A. Kushniruk & E. Borycki (Eds.), *Human, social, and organizational aspects of health information systems* (pp. 203–218). Hershey, PA: IGI Global. doi:10.4018/978-1-59904-792-8.ch012

Pullen, T., & Al-Hakim, L. (2016). Shared electronic health records as innovation: An Australian case. In L. Al-Hakim, X. Wu, A. Koronios, & Y. Shou (Eds.), *Handbook of research on driving competitive advantage through sustainable, lean, and disruptive innovation* (pp. 500–533). Hershey, PA: IGI Global. doi:10.4018/978-1-5225-0135-0.ch021

Queirós, A., Silva, A. G., Ferreira, A., Caravau, H., Cerqueira, M., & Rocha, N. P. (2016). Assessing mobile applications considered medical devices. In M. Cruz-Cunha, I. Miranda, R. Martinho, & R. Rijo (Eds.), *Encyclopedia of e-health and telemedicine* (pp. 111–127). Hershey, PA: IGI Global. doi:10.4018/978-1-4666-9978-6.ch010

Redha, W., Hartwick, K., & Sikka, N. (2015). Mobile health in emergency care. In Z. Yan (Ed.), *Encyclopedia of mobile phone behavior* (pp. 825–838). Hershey, PA: IGI Global. doi:10.4018/978-1-4666-8239-9.ch068

Ries, N. M. (2011). Legal issues in health information and electronic health records. In *Clinical technologies: Concepts, methodologies, tools and applications* (pp. 1948–1961). Hershey, PA: IGI Global. doi:10.4018/978-1-60960-561-2.ch708

Sako, Z. Z., Karpathiou, V., Adibi, S., & Wickramasinghe, N. (2017). Data accuracy considerations with mHealth. In N. Wickramasinghe (Ed.), *Handbook of research on healthcare administration and management* (pp. 1–15). Hershey, PA: IGI Global. doi:10.4018/978-1-5225-0920-2.ch001

Saleem, J. J., Russ, A. L., Justice, C. F., Hagg, H., Ebright, P. R., Woodbridge, P. A., & Doebbeling, B. N. (2009). Exploring the persistence of paper with the electronic health record. *International Journal of Medical Informatics*, *78*(9), 618–628. doi:10.1016/j.ijmedinf.2009.04.001 PMID:19464231

Schnall, R., & Iribarren, S. J. (2015). Review and analysis of existing mobile phone applications for health care–associated infection prevention. *American Journal of Infection Control*, *43*(6), 572–576. doi:10.1016/j.ajic.2015.01.021 PMID:25748924

Senathirajah, Y., Bakken, S., & Kaufman, D. (2014). The clinician in the driver's seat: Part 1 – A drag/drop user-composable electronic health record platform. *Journal of Biomedical Informatics*, *52*, 165–176. doi:10.1016/j.jbi.2014.09.002 PMID:25240253

Shachak, A., Hadas-Dayagi, M., Ziv, A., & Reis, S. (2009). Primary care physicians' use of an electronic medical record system: A cognitive task analysis. *Journal of General Internal Medicine*, *24*(3), 341–348. doi:10.100711606-008-0892-6 PMID:19130148

Shareef, M. A., Ahmed, J. U., Kumar, V., & Kumar, U. (2015). Effect of mobile phone SMS on m-health: An analysis of consumer perceptions. In P. Thomas, M. Srihari, & S. Kaur (Eds.), *Handbook of research on cultural and economic impacts of the information society* (pp. 284–296). Hershey, PA: IGI Global. doi:10.4018/978-1-4666-8598-7.ch012

Shareef, M. A., Kumar, V., Kumar, U., & Hasin, A. A. (2013). Application of behavioral theory in predicting consumers adoption behavior. *Journal of Information Technology Research*, *6*(4), 36–54. doi:10.4018/jitr.2013100103

Siddiqui, M., Islam, M. Y. U., Mufti, B. A. I., Khan, N., Farooq, M. S., Muhammad, M. G., ... Kazi, A. M. (2015). Assessing acceptability of hypertensive/diabetic patients towards mobile health based behavioral interventions in Pakistan: A pilot study. *International Journal of Medical Informatics*, *84*(11), 950–955. doi:10.1016/j.ijmedinf.2015.08.009 PMID:26321485

Silva, B. M. C., Rodrigues, J. J. P. C., de la Torre Diez, I., Lopez-Coronado, M., & Saleem, K. (2015). Mobile-health: A review of current state in 2015. *Journal of Biomedical Informatics*, *56*, 265–272. doi:10.1016/j.jbi.2015.06.003 PMID:26071682

Skorin-Kapov, L., Dobrijevic, O., & Piplica, D. (2014). Towards evaluating the quality of experience of remote patient monitoring services: A study considering usability aspects. *International Journal of Mobile Human Computer Interaction*, *6*(4), 59–89. doi:10.4018/ijmhci.2014100104

Sood, V. R., Gururajan, R., Hafeez-Baig, A., & Wickramasinghe, N. (2017). Adoption of mobile devices in the Australian healthcare: A conceptual framework approach. In N. Wickramasinghe (Ed.), *Handbook of research on healthcare administration and management* (pp. 662–685). Hershey, PA: IGI Global. doi:10.4018/978-1-5225-0920-2.ch040

Stavros, A., Vavoulidis, E., & Nasioutziki, M. (2016). The use of mobile health applications for quality control and accreditational purposes in a cytopathology laboratory. In A. Moumtzoglou (Ed.), *M-health innovations for patient-centered care* (pp. 262–283). Hershey, PA: IGI Global. doi:10.4018/978-1-4666-9861-1.ch013

Tamposis, I., Pouliakis, A., Fezoulidis, I., & Karakitsos, P. (2017). Mobile platforms supporting health professionals: Need, technical requirements, and applications. In *Medical imaging: Concepts, methodologies, tools, and applications* (pp. 1020–1043). Hershey, PA: IGI Global. doi:10.4018/978-1-5225-0571-6.ch041

Testa, A., Coronato, A., Cinque, M., & de Pietro, G. (2016). Services and monitors for dependability assessment of mobile health monitoring systems. In *E-health and telemedicine: Concepts, methodologies, tools, and applications* (pp. 602–618). Hershey, PA: IGI Global. doi:10.4018/978-1-4666-8756-1.ch030

Tin, E. E., Cummings, E., & Borycki, E. (2014). Review of the consumer perspective framework for healthcare applications. In M. Househ, E. Borycki, & A. Kushniruk (Eds.), *Social media and mobile technologies for healthcare* (pp. 1–15). Hershey, PA: IGI Global. doi:10.4018/978-1-4666-6150-9.ch001

Vuk, J., Anders, M. E., Mercado, C. C., Kennedy, R. L., Casella, J., & Steelman, S. C. (2015). Impact of simulation training on self-efficacy of outpatient health care providers to use electronic health records. *International Journal of Medical Informatics*, *84*(6), 423–429. doi:10.1016/j.ijmedinf.2015.02.003 PMID:25746460

Walji, M. F., Taylor, D., Langabeer, J. R., & Valenza, J. A. (2009). Factors influencing implementation and outcomes of a dental electronic patient record system. *Journal of Dental Education*, *73*(5), 589–600. PMID:19433534

Wang, S. L., Chen, Y. L., Kuo, A. M. H., Chen, H. M., & Shiu, Y. S. (2016). Design and evaluation of a cloud-based mobile health information recommendation system on wireless sensor networks. *Computers & Electrical Engineering*, *49*, 221–235. doi:10.1016/j.compeleceng.2015.07.017

Wangberg, S. C., Arsand, E., & Andersson, N. (2006). Diabetes education via mobile text messaging. *Journal of Telemedicine and Telecare*, *12*(S1), 55–56. doi:10.1258/135763306777978515 PMID:16884582

Whittaker, A. A., Aufdenkamp, M., & Tinley, S. (2009). Barriers and facilitators to electronic documentation in a rural hospital. *Journal of Nursing Scholarship*, *41*(3), 293–300. doi:10.1111/j.1547-5069.2009.01278.x PMID:19723278

Wickramasinghe, N., & Goldberg, S. (2007). The Wi-INET model for achieving m-health success. In D. Taniar (Ed.), *Encyclopedia of mobile computing and commerce* (pp. 1004–1010). Hershey, PA: IGI Global. doi:10.4018/978-1-59904-002-8.ch168

Wiesner, M., & Pfeifer, D. (2014). Health recommender systems: Concepts, requirements, technical basics and challenges. *International Journal of Environmental Research and Public Health, 11*(3), 2580–2607. doi:10.3390/ijerph110302580 PMID:24595212

Woltmann, E., Grogan-Kaylor, A., Perron, B., Georges, H., Kilbourne, A. M., & Bauer, M. S. (2012). Comparative effectiveness of collaborative chronic care models for mental health conditions across primary, specialty, and behavioral healthcare settings: Systematic review and meta-analysis. *The American Journal of Psychiatry, 169*(8), 790–804. doi:10.1176/appi.ajp.2012.11111616 PMID:22772364

Yadav, N., Aliasgari, M., & Poellabauer, C. (2016). Mobile healthcare in an increasingly connected developing world. *International Journal of Privacy and Health Information Management, 4*(2), 76–97. doi:10.4018/IJPHIM.2016070106

Zhang, Y., Yu, P., & Shen, J. (2012). The benefits of introducing electronic health records in residential aged care facilities: A multiple case study. *International Journal of Medical Informatics, 81*(10), 690–704. doi:10.1016/j.ijmedinf.2012.05.013 PMID:22749424

Zuehlke, P., Li, J., Talaei-Khoei, A., & Ray, P. (2009). *A functional specification for mobile eHealth (mHealth) systems*. Paper presented at the 11th International Conference on e-Health Networking, Applications and Services (Healthcom 2009), Sydney, Australia.

ADDITIONAL READING

Andrews, L., Gajanayake, R., & Sahama, T. (2014). The Australian general public's perceptions of having a personally controlled electronic health record (PCEHR). *International Journal of Medical Informatics, 83*(12), 889–900. doi:10.1016/j.ijmedinf.2014.08.002 PMID:25200198

Bandyopadhyay, T., & Zadeh, B. (2015). Mobile health technology in the US: Current status and unrealized scope. In *Economics: Concepts, methodologies, tools, and applications* (pp. 1664–1681). Hershey, PA: IGI Global. doi:10.4018/978-1-4666-8468-3.ch088

Black, C., Tagiyeva-Milne, N., Helms, P., & Moir, D. (2015). Pharmacovigilance in children: Detecting adverse drug reactions in routine electronic healthcare records. A systematic review. *British Journal of Clinical Pharmacology, 80*(4), 844–854. doi:10.1111/bcp.12645 PMID:25819310

Bote, J. (2013). Electronic health record proposal for long-term preservation. In M. Cruz-Cunha, I. Miranda, & P. Gonçalves (Eds.), *Handbook of research on ICTs for human-centered healthcare and social care services* (pp. 529–548). Hershey, PA: IGI Global. doi:10.4018/978-1-4666-3986-7.ch028

Casey, M. M., Moscovice, I., & McCullough, J. (2014). Rural primary care practices and meaningful use of electronic health records: The role of regional extension centers. *The Journal of Rural Health, 30*(3), 244–251. doi:10.1111/jrh.12050 PMID:24118180

Dorflinger, L. M., Gilliam, W. P., Lee, A. W., & Kerns, R. D. (2014). Development and application of an electronic health record information extraction tool to assess quality of pain management in primary care. *Translational Behavioral Medicine, 4*(2), 184–189. doi:10.100713142-014-0260-5 PMID:24904702

Ervin, K. (2016). Legal and ethical considerations in the implementation of electronic health records. In *E-health and telemedicine: Concepts, methodologies, tools, and applications* (pp. 1432–1444). Hershey, PA: IGI Global. doi:10.4018/978-1-4666-8756-1.ch072

Gajanayake, R., Sahama, T., & Iannella, R. (2016). The role of perceived usefulness and attitude on electronic health record acceptance. In *E-health and telemedicine: Concepts, methodologies, tools, and applications* (pp. 49–59). Hershey, PA: IGI Global. doi:10.4018/978-1-4666-8756-1.ch003

Hagar, Y., Albers, D., Pivovarov, R., Chase, H., Dukic, V., & Elhadad, N. (2014). Survival analysis with electronic health record data: Experiments with chronic kidney disease. *Statistical Analysis and Data Mining: The ASA Data Science Journal*, *7*(5), 385–403. doi:10.1002am.11236

Harvey, M. J., & Harvey, M. G. (2014). Privacy and security issues for mobile health platforms. *Journal of the Association for Information Science and Technology*, *65*(7), 1305–1318. doi:10.1002/asi.23066

Iltchev, P., Śliwczyński, A., Szynkiewicz, P., & Marczak, M. (2016). Mobile health applications assisting patients with chronic diseases: Examples from asthma care. In A. Moumtzoglou (Ed.), *M-health innovations for patient-centered care* (pp. 170–196). Hershey, PA: IGI Global. doi:10.4018/978-1-4666-9861-1.ch009

Jacquinet, M., & Curado, H. (2016). Opportunities and challenges for electronic health record: Concepts, costs, benefits, and regulation. In M. Cruz-Cunha, I. Miranda, R. Martinho, & R. Rijo (Eds.), *Encyclopedia of e-health and telemedicine* (pp. 969–975). Hershey, PA: IGI Global. doi:10.4018/978-1-4666-9978-6.ch075

King, J., Patel, V., Jamoom, E. W., & Furukawa, M. F. (2014). Clinical benefits of electronic health record use: National findings. *Health Services Research*, *49*(1pt2), 392–404. doi:10.1111/1475-6773.12135 PMID:24359580

Krishnamoorthy, P., Gupta, D., Chatterjee, S., Huston, J., & Ryan, J. J. (2014). A review of the role of electronic health record in genomic research. *Journal of Cardiovascular Translational Research*, *7*(8), 692–700. doi:10.100712265-014-9586-0 PMID:25119857

Kuo, M. H., Wang, S. L., & Chen, W. T. (2016). Using information and mobile technology improved elderly home care services. *Health Policy and Technology*, *5*(2), 131–142. doi:10.1016/j.hlpt.2016.02.004

Lehnbom, E. C., Brien, J. E., & McLachlan, A. J. (2014). Knowledge and attitudes regarding the personally controlled electronic health record: An Australian national survey. *Internal Medicine Journal*, *44*(4), 406–409. doi:10.1111/imj.12384 PMID:24754689

Marco-Ruiz, L., Moner, D., Maldonado, J. A., Kolstrup, N., & Bellika, J. G. (2015). Archetype-based data warehouse environment to enable the reuse of electronic health record data. *International Journal of Medical Informatics*, *84*(9), 702–714. doi:10.1016/j.ijmedinf.2015.05.016 PMID:26094821

McAlearney, A. S., Hefner, J. L., Sieck, C. J., & Huerta, T. R. (2015). The journey through grief: Insights from a qualitative study of electronic health record implementation. *Health Services Research*, *50*(2), 462–488. doi:10.1111/1475-6773.12227 PMID:25219627

Medhanyie, A. A., Moser, A., Spigt, M., Yebyo, H., Little, A., Dinant, G., & Blanco, R. (2015). Mobile health data collection at primary health care in Ethiopia: A feasible challenge. *Journal of Clinical Epidemiology, 68*(1), 80–86. doi:10.1016/j.jclinepi.2014.09.006 PMID:25441699

Menon, S., Singh, H., Meyer, A. N. D., Belmont, E., & Sittig, D. F. (2014). Electronic health record-related safety concerns: A cross-sectional survey. *Journal of Healthcare Risk Management, 34*(1), 14–26. doi:10.1002/jhrm.21146 PMID:25070253

Menshawy, M. E. L., Benharref, A., & Serhani, M. (2015). An automatic mobile-health based approach for EEG epileptic seizures detection. *Expert Systems with Applications, 42*(20), 7157–7174. doi:10.1016/j.eswa.2015.04.068

Purkayastha, S. (2013). Design and implementation of mobile-based technology in strengthening health information system: Aligning mHealth solutions to infrastructures. In *User-driven healthcare: Concepts, methodologies, tools, and applications* (pp. 689–713). Hershey, PA: IGI Global. doi:10.4018/978-1-4666-2770-3.ch034

Rush, J., Postelnick, M., & Schulz, L. (2015). Use of electronic health record clinical decision support tools in antimicrobial stewardship activities. *Current Treatment Options in Infectious Diseases, 7*(2), 90–100. doi:10.100740506-015-0042-8

Schumaker, R. P., & Reganti, K. P. (2016). Implementation of electronic health record (EHR) system in the healthcare industry. In *E-health and telemedicine: Concepts, methodologies, tools, and applications* (pp. 1001–1016). Hershey, PA: IGI Global. doi:10.4018/978-1-4666-8756-1.ch051

Schwartz, P. H., Caine, K., Alpert, S. A., Meslin, E. M., Carroll, A. E., & Tierney, W. M. (2015). Patient preferences in controlling access to their electronic health records: A prospective cohort study in primary care. *Journal of General Internal Medicine, 30*(1), 25–30. doi:10.100711606-014-3054-z PMID:25480721

Seçkin, G., & Kahana, E. (2015). Smart phone health applications. In Z. Yan (Ed.), *Encyclopedia of mobile phone behavior* (pp. 898–905). Hershey, PA: IGI Global. doi:10.4018/978-1-4666-8239-9.ch073

Sittig, D. F., Gonzalez, D., & Singh, H. (2014). Contingency planning for electronic health record-based care continuity: A survey of recommended practices. *International Journal of Medical Informatics, 83*(11), 797–804. doi:10.1016/j.ijmedinf.2014.07.007 PMID:25200197

Speciale, A. M., & Freytsis, M. (2013). mHealth for midwives: A call to action. *Journal of Midwifery & Womens Health, 58*(1), 76–82. doi:10.1111/j.1542-2011.2012.00243.x PMID:23317302

Straub, H., Adams, M., & Silver, R. K. (2014). Can an electronic health record system be used for preconception health optimization? *Maternal and Child Health Journal, 18*(9), 2134–2140. doi:10.100710995-014-1461-8 PMID:24627232

Tavares, J., & Oliveira, T. (2016). Electronic health record portals definition and usage. In M. Cruz-Cunha, I. Miranda, R. Martinho, & R. Rijo (Eds.), *Encyclopedia of e-health and telemedicine* (pp. 555–562). Hershey, PA: IGI Global. doi:10.4018/978-1-4666-9978-6.ch043

Victores, A. J., Coggins, K., & Takashima, M. (2015). Electronic health records and resident workflow: A time-motion study of otolaryngology residents. *The Laryngoscope*, *125*(3), 594–598. doi:10.1002/lary.24848 PMID:25059224

Wills, M. J. (2013). Individual, organizational, and technological barriers to EHR implementation. In S. Sarnikar, D. Bennett, & M. Gaynor (Eds.), *Cases on healthcare information technology for patient care management* (pp. 35–54). Hershey, PA: IGI Global. doi:10.4018/978-1-4666-2671-3.ch002

KEY TERMS AND DEFINITIONS

Data: The information in a pattern suitable for storing and processing by a computer.

Electronic Health Record: The digital version of a patient's paper chart.

Health Care: The act of taking preventative or necessary medical procedures to improve a person's well-being.

Information: The data that can be stored in and retrieved from a computer.

Information Technology: The development, installation, and implementation of computer systems and applications.

Interoperability: The ability of a system or component to effectively execute with other systems.

Mobile Health: The different practices of delivering the medical services through the use of mobile technologies.

Physician: A person who is licensed to practice medicine, or a person who is skilled in healing.

Technology: The application of science, especially to the industrial or commercial objectives.

This research was previously published in Next-Generation Mobile and Pervasive Healthcare Solutions edited by Jose Machado, António Abelha, Manuel Filipe Santos, and Filipe Portela, pages 67-85, copyright year 2018 by Medical Information Science Reference (an imprint of IGI Global).

Chapter 3
Big Data and mHealth:
Increasing the Usability of Healthcare Through the Customization of Pinterest – Literary Perspective

Nancy Shipley
Towson University, USA

Joyram Chakraborty
Towson University, USA

ABSTRACT

Technology has greatly improved healthcare. Research has shown that improved user experience can vastly increase the volume and quality of knowledge gained from the accessible resources. Electronic medical records have been mandated by the U.S. government transitioning the medical field to computer based applications with the goal of efficiency and meaningful use. From this vast data, the end user could potentially increase their healthcare understanding through a centralized mobile-platform that brings the expertise from medical professionals, smart health and well-being best practices, alternative medicine and social media such as crowd-sourcing materials. In this chapter, we present a mHealth solution incorporating the features of Pinterest in order to provide a single portal for the dissemination of healthcare information.

INTRODUCTION

Although technology in healthcare has increased greatly over the past decade, the user experience of the combined available resources can unquestionably be improved (Blumenthal, 2009). Many Healthcare functions have been transferred to computer-based applications such as the institution of electronic health records (Jha et al., 2009). However, the end user could conceivably increase their healthcare knowledge through a platform that brings the knowledge from medical experts, smart health and wellbeing proven practices, alternative medicine and social media and crowdsourcing material (Alyass et al., 2015). In the past decade technology has also improved rapidly as cloud computing and big data have broadened the

DOI: 10.4018/978-1-5225-6198-9.ch003

scope of technical offerings increasing the ultimate experience for the end users. However, this technology has not been maximized, especially for health care uses (Bizer et al., 2012). The use of enhanced visualization could increase the user experiences for healthcare best practices and implementation. Employing a new frame work that could be quickly accessed via mobile devices could offer the patient an improved user experience with their medical information queries as they fight against diseases and sustain healthy living. Building on the discovery principal that Pinterest has established through their superior search capability and enriched user interface, customization can occur to offer a single source for patients to research, medical diagnosis, smart health and wellbeing, alternative medicine and social media solutions. A carefully designed empirical study could determine if customizing Pinterest would improve the use of Healthcare Big Data by expanding the patients' knowledge concerning their health issues via a mHealth solution. The specific contributions of this chapter are an identified solutions gap in the eHealth literature that can be addressed by using our proposed solution of customizing the Pinterest application.

The healthcare community is confronting an almost insurmountable amount of health and healthcare-related content generated from numerous patient care points of contact, highly technical medical instruments, and web-based health communities (Arunasalam, 2013). The contemporary healthcare big data areas that are taking shape are genomics-driven big data covering genotyping, gene expression, sequencing data and payer–provider big data including electronic health records, insurance records, pharmacy prescription, patient feedback and responses (Raghupathi & Raghupathi, 2014). Over the past decade, electronic health records (EHR) have been extensively implemented in hospitals and medical facilities giving way to significant clinical knowledge and a deeper understanding of patient disease patterns from such computer based collections. Using this vast healthcare data, doctors and patients alike may be more readily able to diagnosis the issues and determine treatments to remedy their health problems. Utilizing lessons learned from the immense healthcare data, the end user could increase their healthcare knowledge. Expanding on a unified mobile accessible platform, the healthcare big data could offer patients a wealth of awareness via a mHealth application. Medical professionals have been publishing their expertise in journals that can be tapped to expand the patient health understanding. In addition, there are many websites that offer smart health and wellbeing proven practices. Physical ailments can also be addressed through alternative medicine. Social media offers many remedies to health issues as those seeking wellness write blogs and effectively crowdsource to seek healthy solutions. These offerings along with many others could be bundled together to offer a superior distinct healthcare tool.

After decades of implementing computer-based applications in almost every field of study, a by-product taking shape is immense data which is now known as big data (Peters et al., 2014). This vast amount of data now collected and accessible via mobile devices via cloud computing technology allows for humans to move more deeply into the exploration of the information age. Not harnessed to its fullest, the big data technology now being researched will give way to superior growth of understanding for decision making and everyday life especially in healthcare (Jee & Kim, 2013) . Challenges are apparent in this new technology of big data on cloud computing. Although they cover all ranges, the main few areas are data capture, data storage, data analysis, and data visualization (Chen, 2014). Traditional data was mapped and placed in organized relational databases to be searched and presented in software applications for the various end users to perform their daily tasks. Unlike traditional data, big data is readily available on the internet, from sensors, and even disjointed databases with none of it organized for customary relational database viewing methods (Ma et al., 2015). The user friendly platform for searching and analyzing this great amount of big data sources is still in the beginning stages.

BACKGROUND

Presently, the process of visualization in big data projects is most often the image of dots on a map or graphical views. By enhancing the view of big data, the user experience will most likely increase and additional knowledge will be obtained. To present this information, a search for an optimal framework has occurred exposing that Pinterest can offer big data information in the most succinct manner to the end user. Not only will the user be able to access the big data platform from traditional PCs and laptops, but mobile access using smartphones and tablets will be the ultimate end goal. A potential solution is streamlining big data interface that can be accessed from anywhere allowing the end user to easily obtain healthcare information to better their lifestyle.

Through the power of cloud computing and big data, the end users can now demand more from software applications. This superior technology can be exploited for healthcare uses to more easily sift through medical data discovering greater medical solutions. Several studies have articulated the benefits of improving visualization of healthcare big data to increase the users' experience as they move toward a better understanding of healing remedies and healthy living guidelines (Zhang et al., 2008). Employing a new framework accessed by mobile devices, patients could harness this knowledge as they combat diseases and achieve healthy living. Fully utilizing the discovery principal available through Pinterest by utilizing their comprehensive search and powerful user interface, customization can occur to offer a unified software mHealth application for patients to research medical diagnosis, healthy living, alternative medicine and social media options.

A continued pursuit took place where many in the millennial generation were sought for their opinions to find popular sites. Time and time again, the young adults said they preferred Pinterest. Reviewing the About page on the Pinterest site, they classify themselves as the world's catalog of ideas. Their mission is to help people discover the things they love, and inspire their users to go do those things in their daily lives. Back in March 2010, when they co-founded Pinterest, Ben Silbermann, Evan Sharp and Paul Sciarra sought out to create a site about discovery of all the relevant and interesting information that can be found on the internet. Using the website, users discover and organize things they like. From a technology view, the company is based on big data, where the software captures rich metadata around each image and stores them for latter access. Considering users love the site and the big data search technology is powerful, the idea began of customizing Pinterest to meet patient's healthcare needs.

IMPROVING HEALTHCARE SOLUTIONS

Healthcare Information Sources

Through numerous patient care points of contact, highly technical medical instruments, health related payment systems and web-based health communities, our society now has access to a vast amounts of health and healthcare-related internet based content (Avinash et al., 2005). Many healthcare big data projects have been put in place to utilize the information including genomics-driven big data covering genotyping, gene expression (O'Driscoll, 2013); sequencing data and payer–provider big data including electronic health records (Rind et al., 2011); insurance records, pharmacy prescription, patient feedback and responses, and health and well-being best practices (Stroetmann, 2013). Still studies have shown that the end users' healthcare needs are not fully met (Harrison & Mort, 1998).

HEALTHCARE TECHNOLOGY

Electronic health records (EHR) have been widely implemented in hospitals and medical facilities allowing patients' access to their records online and opening clinical research areas increasing understanding of patients' diseases patterns from healthcare big data collections (Ludwick & Doucette, 2009). Access to this healthcare data is greatly expanding the knowledge of patients and medical professionals alike (Murdoch & Detsky, 2013). Clinical improvements are an example where analyzing patient characteristics in conjunction with the cost and outcomes of care work to identify the most clinically cost effective treatments are occurring to help doctors' aid patients (Arunasalam, 2013). Reduction of costs to identify, predict and minimize fraud is another area of interest which utilizes big data technology (Bates et al., 2014). Efficiency of processes to help patients locate providers, manage their care, and improve their health is yet another sample of healthcare technology possible improvements (Neff, 2013). Other areas that are being explored include genomics, bioinformatics (Swan, 2013), pharmacogenomics, dentistry, HIV epidemiology, cardiovascular research (Pah et al., 2015) and much more. Given the technology available and the data being collected, doctors and patients may be more readily able to diagnosis the issues and determine treatments to remedy their health problems (Vaitsis et al., 2014). The medical expertise of various professionals can now be shared through the technology offerings (Philips et al., 2004). The collection of all the medical knowledge could help patients with their health needs especially patients with chronic health issues (Wyber et al., 2015).

In addition to the typical medical advice, patients are seeking alternative medicine to address all of their health needs (Eisenberg et al., 1998). Smart health and well-being best practices can also be instituted by patients to address their health concerns outside of offerings from basic medicine offerings. While it holds true that most patients are not alone with their health problems, very often they are not able to connect with a community of people who have had the same issues. Through Social Media, patients could help each other morally, technically through suggestions for medical care, and specifically through healthy practices such as diet and exercise as they share their challenges and needs (Wen-ying et al., 2009). Another avenue of this health knowledge can be obtained from the mobile access that the user is utilizing to extract the data (Kumar et al., 2013). As the application obtains the patients geographical location, data presented can be altered based on health care needs in the area (Free et al., 2013). For example, virus breakouts can be warned to travelers, detailing of best practices to ward off diseases.

The application of technology in healthcare is increasing overtime (Yao et al., 2015). Although the rate may not be as fast as society's expectance of the digital practical applications, the health-based profession is finding usage for technology to replace their paper based systems. Driving down costs, like most other businesses, is the main factor for the adoption of technology in healthcare (Hayhurst, 2015). In health insurance claims, detecting fraud, abuse, waste, and errors, thereby reducing recurrent losses and facilitating enhanced patient care has been a main focus for healthcare technology. In addition, for medical facilities directly streamlining patient related applications such as medical records, patient registration systems, and clinical based applications has greatly increased the efficiency of hospitals and medical practices. Big data applications have been identified to assist in reducing costs in the areas of high-cost patients, readmissions, triage, decompensation, adverse events, and treatment optimization for diseases (Kuiken, 2013). For the ease of use that technology offers, medical professionals also find that the use of software applications assist in their everyday use through clinical applications, research studies of diseases, and overall trending analysis.

MEDICAL EXPERTISE

Medical professionals utilize software applications to assist their day to day operations in areas such as research of genomics, bioinformatics, pharmacogenomics, HIV epidemiology, chronic disease research and much more (Schmidt et al., 1990). For example, in the field of genomics, researchers are producing genome-wide data sets on ever-expanding study populations. Because of technology, broad access to these data, stored samples, and EMRs are accelerating society's understanding of the role of genes, environment, and behavior in health and disease. With this information, new knowledge is gained by improving diagnostics, targeting drug development, and exploration into new insights about how to prevent and treat disease (Curtis et al., 2014). As an interdisciplinary field, bioinformatics is another example of technology in healthcare where methods are developed using software tools for understanding biological data. Bioinformatics combines computer science, statistics, mathematics, and engineering to analyze and interpret biological data. Digital uses are found very important in the study of pharmacogenomics which identifies how genes affect a person's response to drugs. In addition, chronic diseases are studied and identified using technology to sort through the vast amount of data points. Patients, as the final end users of the healthcare technology implementations, have various applications that can be used to assist in their healthcare knowledge (Lin, 2015). However, as technology improves, combining all of these areas will also assist to offer more sophisticated software usages. Within a single source search area, having access to the medical expertise information would greatly help patients and doctors diagnosis issues and determine best practices to stay healthy. The big data platform is the best technology available to offer the medical expertise documentation in a single search application.

DISPERSED HEALTHCARE INFORMATION SOURCES

Patients often have many healthcare solutions available to them. However, they are located at many dispersed sources. There is healthcare technology that can offer them insight into their actual vital signs. Their medical records are now electronically stored and can be accessible to review. Medical experts have a plethora of information which is reachable at many different locations on the internet. Alternative medicine may also offer healthy solutions to address the patient's needs. Feeding into the health information sources, smart health and wellbeing regiments are available online. In addition, social media offerings can also offer peer support as others also are affected with similar diseases. Figure 1 shows how health knowledge is scattered and dispersed throughout various points of access.

Big Data and the Technical Revolution

The large swathes of data being generated through the many computer-based applications continuously can be queried and reachable through cloud computing technology (Rayport & Heyward, 2011). This big data technology has potential to offer everyone greater knowledge to improve every aspect of their lives particularly in healthcare. Building on cloud computing, the big data technology is uniquely different than previous database technology in that with only simple configuration systems can be expanded to calculate extensible algorithms on terabytes of data.

Figure 1. This diagram shows health knowledge is scattered and dispersed throughout various points of access

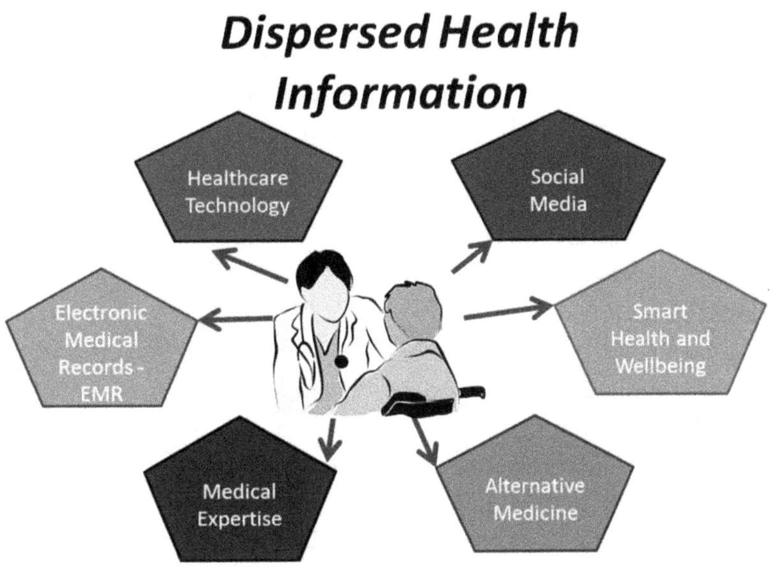

Health knowledge is scattered and dispersed throughout various points of access.

BIG DATA OVERVIEW

At any time, data is being gathered and generated across the internet (Xia et al., 2012). There is digital data that is emitted from sensors such as mobile phone usage. There is data generated from paper-based forms as processes become almost completely computer based such as Tax Forms (Emani et al., 2015). There is personal data such as address and contact information found with a quick review across the internet. There is social media data as people go online to state their opinions. The amount of data available via the internet is almost endless. The issue lies in how to present meaningful results from the data. As defined by Gartner (Gartner, 2014), big data is high-volume, high-velocity and high-variety information assets that demand cost-effective, innovative forms of information processing for enhanced insight and decision making. Taking this data and being able to create progressive decisions based off of the analytical results is the main goal of big data (Demirkan, 2013). The term "Big" is applied because traditional data processing cannot be utilized as the data is too large and it is often unstructured which makes it a challenge to organize in customary formatting (Gandomi & Haider, 2015).

BIG DATA TECHNOLOGY

Utilizing the Cloud Computing infrastructure offers a platform for the big data to finally begin to be managed and analyzed (Andreolini et al., 2015). As best defined by Hashem and his team, "Cloud computing and big data are conjoined. Big data provides users the ability to use commodity computing to process distributed queries across multiple datasets and return resultant sets in a timely manner" p. 102 (Hashem, 2015). Cloud computing provides the underlying engine through the use of Hadoop Distributed

File System (HDFS) (Reyes-Ortiz et al., 2015), a class of distributed data-processing platforms (Feller et al., 2015). Large data sources from the cloud and Web are stored in a distributed fault-tolerant database and processed through a programming model for large datasets with a parallel distributed algorithm in a cluster (Hu et al., 2014). Retrieving this data, NoSQL is a type of database that allows for management of large distributed data through a scheme free, easy replication, and simple API. At the programming layer, MapReduce sanctions the processing of large amounts of datasets stored in parallel in the cluster (Darji & Waghela, 2014). From there, views of the data can be compiled and visualized for end users via APIs (Maitrey & Jha, 2015).

BIG DATA USABILITY

Big data usability was not the focus when big data technology was first established. Not implemented like most technologies with the attention on end user, big data has become a technology more because of an outcome and not just because end users made requirements (Kambatla et al., 2014). Therefore, the existing big data usability is a view of datasets in graphs and maps. There are not requirements and use cases that can be identified for the creation of big data usability (Kum et al., 2015). The data is simply there, and usability experts need to now figure out how best to use the overwhelming amount of information (Walji et al., 2013).

A good example of big data usability comes from the database designer field. To address the need for more comprehensive views of big data, Macneil & Elmqvist (MacNeil & Elmqvist, 2013) set out to create visual interfaces that could easily interrupt large data sets from a web based view. The overall concept is to make large data more readily understandable to the end user. The key feature the team determines is not the ability to represent all data in a single view, but rather the capability for the end user to be able to easily create new views and coordinate these views to support comparison and correlation. To address the viewing need, the team created a visualization mosaic which they describe as a hierarchical container structure for visual representations in tiles that present a predefined data set. The main features of mosaics are space-filling layout, visual transformations, mosaic management and formal notation. Although the team explained various applications that are similar, only their solution was built on Java for Web based solutions. Examples of their efforts are described as they built a crime scene map and rock climbing view. Each process was analyzed as the overall steps were noted on how the implementation could be executed to view a map in both scenarios of data elements of the items. Crime blots were shown in Seattle and a rock climbing map was shown for Tibetan Himalaya region. This usability allowed certain sets of data to be displayed in visual usage, related to mapping features.

BIG DATA IN HEALTHCARE

In the world of medicine, the first great revolution was the gathering of research to improve medical practices. The next great medical revolution is here with the incorporation of biomedical sciences and the data sciences (O'Driscoll, 2013). That is to say with "big data" now widely used, research can be compiled and analyzed giving way to much more accurate medical decisions with the amalgamation of the computer sciences (DJ, 2014). At Johns Hopkins Hospital, new research is moving forward incorporating big data processes into specific areas of medicine. These studies hope to improve medical practices and create better outcome for the patients (Al-Jumeily et al., 2015). Areas of focus include HIV (Young, 2015), genomics (Fan & Liu, 2013), rheumatology, radiology, and cancer research. Questions are asked that research hopes to address such as should all cancer patients be treated, can the medical staff predict how often patients will visit the doctors, and does the correlation between office visits and treatments relate to overall health (Calyam et al., 2016). Utilizing big data analytics together with the medical experts and the computer sciences professionals the goal is to enhance the usage for patients given the wealth of medical data (Marx, 2013).

The Proposed Portal Experience

The user experience for big data is in only the beginning stages. Although there are some types of views of big data and limited search availability, the overall user experience can be greatly increased. This limited user experience for big data in health care is greatly lacking. In addition, there are many areas that can offer more options for the end user (Suchanek & Weikum, 2013). Data visualization is the present most common means for the industry to access big data technology (Wright et al., 2013). Analytics transforms the data in to a visual depiction of the data offering map views or other high-level charts and graphs from the data summation. The end-user for these processes is generally upper management or decision makers trying to determine a path forward for their business or government agency (GE et al., 2014). However, big data has more to offer than simple hints towards future plans. The general public already enjoys big data searches through Google and other search engines to find knowledge to enrich their lives. Elevating these search offerings could certainly improve the user experience for big data technology. Building on an existing search framework, the user experience could be increased with only incremental changes. This framework must be available on all devices for the best user experience including desktops, laptops, smart phones and all mobile devices. Ensuring that the interface is simple and yet easy to use is extremely important for the best user experience.

VISUALIZATION

Data visualization is defined as the presentation of data in a pictorial or graphical format (Tam & Song, 2016). The objective is to enable decision makers to view data analytics presented visually with the expectation to take the data summation and understand difficult concepts or identify new patterns (Lee et al., 2013). Allowing the user to control the views, many companies employ interactive visualization tools to maneuver through the data rapidly customizing charts and graphs to gain more knowledge from the data (Liu et al., 2013). Using pictures to understand data has been utilized for centuries, using maps and graphs as early as the 1600s and then pie charts in the early 1800s (Miksch & Aigner, 2014). Superior technology now allows for even more improvement and usage of data visualization (Tatarchuk et al., 2008). Data Servers now make it possible to process enormous amounts of data at extremely fast speed (Shvachko et al., 2010). Data visualization has become a rapidly evolving blend of science and art that feasible because of data analytics (Thorvaldsdóttir et al., 2013). These computers are processing extensive algorithms to meet the views that the users are requesting (Yang et al., 2015). With this big data technology there is potential for great opportunity; the next phases of data visualization are researching how to expand the present visual interfaces that are offered of primarily charts, maps and contextual searches (Krämer & Senner, 2015).

Visualization methods have offered increased benefits to many organizations over the past decade beginning with business intelligence and moving to full visual analytics for many subjects (Sagiroglu & Sinanc, 2013). The developing field of visual analytics concentrates on making sense of the big data through integrating human judgement using visual representations and interaction techniques in the analysis process (Keim et al., 2008). The Healthcare field is using visual analytics to extract insights from Electronic Health Records (Wang et al., 2011), gain knowledge from medical professionals (Zhang et al., 2008), and view how patients are searching for diagnosis on the web (Brownstein et al., 2009). Building on these applications of visualization and visual analytics, big data technology can move forward in its functionality and uses. Still, big data visualization has much improvement ahead to become a superior user friendly experience (Thomas & Kielman, 2009).

USER EXPERIENCE

Research into big data technology user experiences are only at the very beginning stages (Chan, 2015). In fact, research has found that the user interface of the most big data projects is an afterthought (Fisher et al., 2012). The bulk of the big data projects are centered on setting up the platform, gathering the data and then at some stage a user requests to see the data. Only at that point, the technical team discovers that the data is not as useful for true end user knowledge and have a very little view of how the users can actually benefit from the terabytes of data that is available for searching. However, for a truly good user experience of any software application, the end user requirements must be established first, and not only as an addendum.

To improve big data applications, the users need to be contacted and surveyed to determine how they could use an application and what tools would best suit their needs. The issue with the big data platforms is no one, not even the user, truly knows what they want or could even have from this new technology. Modern end users enjoy the benefits of having rich data at their hand through their mobile devices. However, improving the software interface could be improved to offer them a better view of the data (Marchionini, 1989). Search engines and some few applications such as Facebook, Instagram and Pinterest offer users some satisfaction for probing the vast array of data to meet their needs. Unfortunately, there have been few studies to truly understand how users could utilize big data and how to improve their user experience (Senkowski & Branscum, 2015). A simple incremental improvement to a modern application may offer enhancements that may meet the users' needs (Hassenzahl & Tractinsky, 2006). With true user feedback on specific application information, user experience could be improved (Kim & Chang, 2007). Design changes could be implemented with the end user as the central goal (Kopetz, 2011). Thereby, step by step the user experience of big data technology will ultimately improve.

FRAMEWORK

To make the most of the user experience, a framework to view the big data technology works best. Already users are comfortable with maneuvering through software systems that function with set features available in all screens (Cohen & Oviatt, 2000). For example, many well established applications display buttons along the top and left side for changing the views and acting on the data such as Outlook, Internet Explorer, and Microsoft Word. Having an existing framework will allow the users to more easily transition their existing interface ability to the new big data interface usages (Marchionini & Komlodi, 1998). Not only should the user interface be based on an existing framework, the architecture should be building upon best practices of solid infrastructure conventions (Ghezzi et al., 2015). Maintaining both a framework for the user interface and the technical platform will allow incremental changes that improve the big data offerings.

MOBILE TECHNOLOGY

Mobile application offerings are the most user friendly technology of the present day. It is hard to walk down the street without seeing somebody looking at their handheld devices for many of their daily tasks or for simply just socializing (Cifuentes, 2013). Thanks in part to Steve Jobs, the mobile device technology has changed the world for ever. Using a standard framework, software engineers can create software applications called "apps" that can be easily downloaded to a mobile device for almost any purpose. The application is installed on the mobile device, but primarily works off of the wireless network hooking the application up to servers in data centers (Han et al., 2015). Now with cloud computing being accessible to anyone for inexpensive monthly fees, applications can fully utilize cloud technology and big data architecture with relative low overhead (Laurila et al., 2012).

Not to be left behind, the healthcare industry has many uses for mobile applications (Strack et al., 2015). In fact, the need is growing so rapidly that there is a new word for mobile health which is entitled mHealth, meaning the practice of medicine and public health supported by mobile devices. "For the healthcare industry, mobile applications provide a new frontier in offering better care and services to patients, and a more flexible and mobile way of communicating with suppliers and patients. Mobile applications will provide important real time data for patients, physicians, insurers, and suppliers. In addition, it will revolutionize the way information is managed in the healthcare industry and redefine the doctor – patient communication." (Siau & Shen, 2006) To begin, the mobile device can be used to encourage a patient to exercise and increase their baseline of healthiness (Monroe et al., 2015). The goal, of course, would be to create an app that could assist users to manage their exercise and weight (Gowin et al., 2015). From there, mobile applications give users the knowledge they need when they are moving through their everyday life activities. Mobile users want to be able get the information they need at their fingertips, and that includes more healthcare knowledge (Miller et al., 2015). Patients are looking for answers for their medical issues such as common infectious diseases, fending off heart disease (Logan, 2013) and even researching mental health issues (Bardram et al., 2013).

There are many existing mobile device applications monitor physical activity, manage in-home care, and assist medical professionals in collecting patient data assessment to address their medical needs. For example, the software application Ginger.io (Kuiken, 2013) offers a mobile application in which diabetic patients agree to be tracked through their mobile phones and assisted with behavioral health treatments. Utilizing the sensors in smartphones, the application records location, texting, calling, and even movement in conjunction with patient surveys delivered to their smartphones. The Ginger.io application incorporates the patient's health details with public research from NIH and other public sources to create a wealth of behavioral health data for each patient. Together with their doctors the patients gain helpful insights pointing out changes in activity, food intake and general health care to improve diabetic issues (Kuiken, 2013). A patient could be traveling to another country and not able to speak the language, however, with the help of a mHealth application, finding native health care for emergencies and having warnings documented concerning local health issues could be available at the finger tips (Paul & Dredze, 2011). At some point in the future, the end user may find this technology invaluable.

SOLUTIONS AND RECOMMENDATIONS

Solutions are few to meet this mHealth single search solution. The literature review suggests there is a gap where end users are lost as they try to meet their healthcare needs. The use of Google is one of the best most utilized search engines to answer health related questions. Patients presently utilize the superior search of big data offerings to sort through the vast internet to try to answer their health concerns. The issue that occurs is easily sorting and categorizing the new information for easy retrieval at another time. Accessing the big data easily allows the user to transition from simple data information to knowledge. Therefore, the examination continued to increase the user experience of the resulting set offered.

With the goal of a superior mobile platform for big data searching that already has a satisfied user-base; Pinterest was selected as the best option to improve a big data offering for the healthcare industry. Pinterest can be customized to meet the needs of the ever growing user base offering a superior mHealth solution. Easily searchable and retrievable big data offerings can now be accessed routinely. As shown in Figure 2, the mHealth Pinterest App can be the Portal for the various healthcare software offerings. A customized Pinterest application can offer end users the ability to research the various healthcare technology solutions available, view their own medical diagnosis via Electronic Medical Records, search the assessments of medical professionals, review alternative medicine, research healthy living possibilities, and explore social media options.

Figure 2. This diagram shows how dispersed health knowledge solutions in various software applications can be brought together through a big data mHealth Portal app

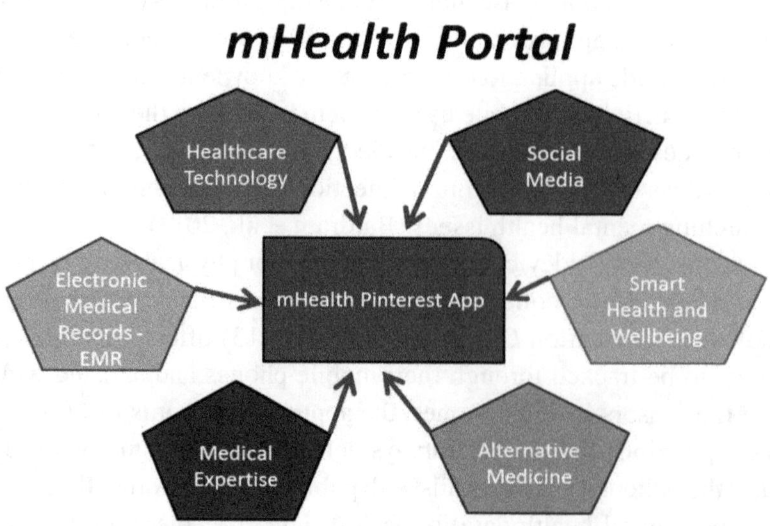

Dispersed health knowledge brought together through a big data mHealth App.

Many users simply enjoy Pinterest to share ideas such as home improvements, food and recipes or crafts. However, the user profile can be customized to present topics that have specific interests such as exercise, healthcare, and healthy living. With specific health related topics established and reputable health content "Pinned", the Pinterest interface can offer a superior search engine for reviewing healthy topics that could help injured or sick patients find solutions to their illnesses and even the everyday person find methods to staying healthy longer by extending fit habits. Patients could use Pinterest to research any health issues that they are facing such as the good recipes for the common flu, how to ward off malaria when traveling aboard, or treatments and social support for Chronic Obstructive Pulmonary Disease (Paige et al., 2015). For the reasons of big data usage, superior search capability, an easy to use mobile platform, and high end user acceptance, Pinterest is a good big data solution from which to build upon to offer a mHealth solution.

Although there are many ways to address healthcare needs, one possible solution worthy of study is to implement the popular big data search engine Pinterest to determine if it can be customized to improve the patients' understanding of their healthcare issues. There is very little research performed on the Pinterest application especially in the field of healthcare. Therefore, it may be that only minor changes to Pinterest could address the needs of the patients' population. Figure 3 shows a view of the proposed Portal Screenshot. Using a healthcare Pinterest application as their source of greater knowledge, patients may find their ideas moving them forward in healthcare remedies and smart wellbeing choices.

Figure 3. This image is a view of the proposed Portal Screenshot to address the big data mHealth Portal app solution

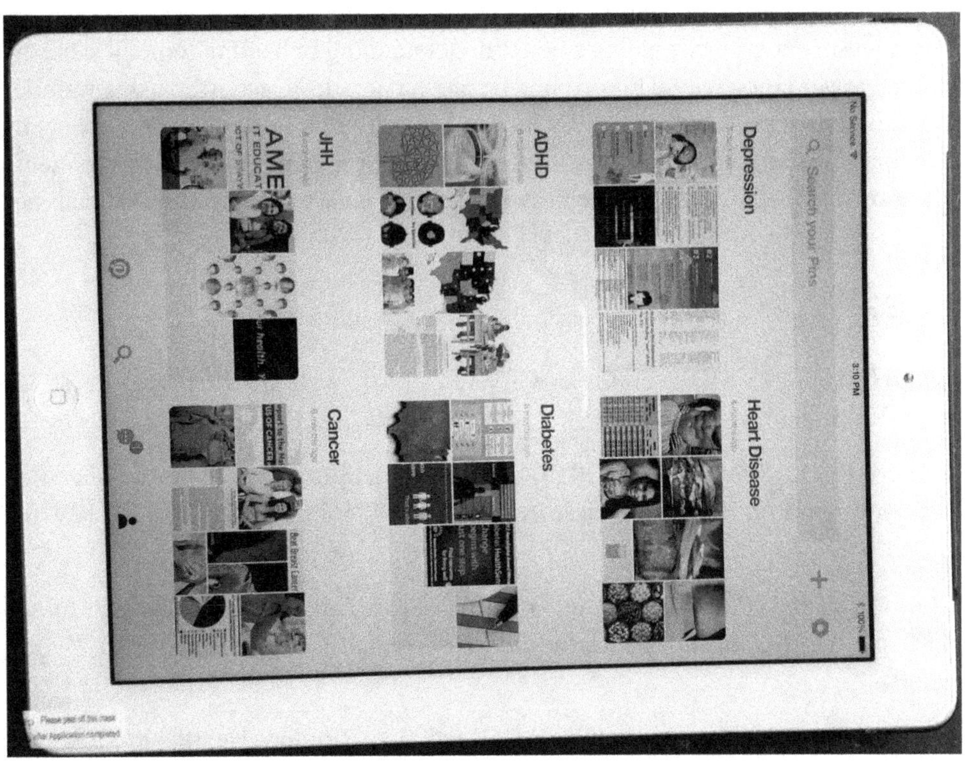

FUTURE RESEARCH DIRECTIONS

There is an increasing need for universally accessible initiative in healthcare technology, especially in healthcare. Big data in healthcare is rapidly evolving and has much to offer even the older population and accessibility should be considered in the design stages of every visualization application project to enhance the usability for all. Further research is needed to understand the outcomes of improvements in big data software usage. This research offers a first step in understanding how incremental changes to an interface can offer much greater satisfaction to the patients and health seekers. Beginning with the Pinterest application, small changes could be implemented to make the vast amount of data useful to all. Decisions could be made more rapidly once the patient has full access to the insight the data can foretell.

CONCLUSION

The user experience of viewing healthcare big data can be greatly improved gaining superior knowledge from the vast array of information readily available (Raghupathi & Raghupathi, 2014). With the expansion of electronic medical records and much of the medical industry digitized, boundless information is at the patient's finger tips. From this immense data, a platform that brings the expertise from medical professionals, smart health and wellbeing proven practices, alternative medicine and social media could increase the end user's healthcare understanding. With a technological breakthrough in cloud computing and big data, mHealth software applications can offer substantially more information to the end users. This innovative technology can be utilized for healthcare uses to more easily search medical related information discovering greater medical solutions. Cultivating healthcare big data visualization could intensify the user experience gaining a better understanding of healing remedies and healthy living guidelines leading toward longevity. This original framework could be accessed using mobile devise such as tablets in assisting patients to attain healthy living. This research study may prove that employing the discovery principals available through a customized Pinterest application can offer patients the ability to research their own medical diagnosis, healthy living, alternative medicine and social media options.

REFERENCES

Al-Jumeily, D., Hussain, A., Mallucci, C., & Oliver, C. (2015). *Applied Computing in Medicine and Health*. Morgan Kaufmann.

Alyass, A., Turcotte, M., & Meyre, D. (2015). From big data analysis to personalized medicine for all: Challenges and opportunities. *BMC Medical Genomics*, *8*(1), 33. doi:10.118612920-015-0108-y PMID:26112054

Andreolini, M., Colajanni, M., Pietri, M., & Tosi, S. (2015). Adaptive, scalable and reliable monitoring of big data on clouds. *Journal of Parallel and Distributed Computing*, *79*, 67–79. doi:10.1016/j.jpdc.2014.08.007

Arunasalam, U. S. a. B. (2013). Leveraging Big Data Analytics to Reduce Healthcare Costs. *IT Pro*, 21-28.

Avinash, G., Liu, R., & Roehm, S. (2005). *System and method for integrated learning and understanding of healthcare informatics*. Google Patents.

Bardram, J. E., Frost, M., Szántó, K., Faurholt-Jepsen, M., Vinberg, M., & Kessing, L. V. (2013). Designing mobile health technology for bipolar disorder: a field trial of the monarca system. *Proceedings of the SIGCHI Conference on Human Factors in Computing Systems*. 10.1145/2470654.2481364

Bates, D. W., Saria, S., Ohno-Machado, L., Shah, A., & Escobar, G. (2014). Big data in health care: Using analytics to identify and manage high-risk and high-cost patients. *Health Affairs*, *33*(7), 1123–1131. doi:10.1377/hlthaff.2014.0041 PMID:25006137

Bizer, C., Boncz, P., Brodie, M. L., & Erling, O. (2012). The meaningful use of big data: Four perspectivesfour challenges. *SIGMOD Record*, *40*(4), 56–60. doi:10.1145/2094114.2094129

Blumenthal, D. (2009). Stimulating the adoption of health information technology. *The New England Journal of Medicine*, *360*(15), 1477–1479. doi:10.1056/NEJMp0901592 PMID:19321856

Brownstein, J. S., Freifeld, C. C., & Madoff, L. C. (2009). Digital disease detection—harnessing the Web for public health surveillance. *The New England Journal of Medicine*, *360*(21), 2153–2157. doi:10.1056/NEJMp0900702 PMID:19423867

Calyam, P., Mishra, A., Antequera, R. B., Chemodanov, D., Berryman, A., Zhu, K., ... Skubic, M. (2016). Synchronous Big Data analytics for personalized and remote physical therapy. *Pervasive and Mobile Computing*, *28*, 3–20. doi:10.1016/j.pmcj.2015.09.004

Chan, A. (2015). Big data interfaces and the problem of inclusion. *Media Culture & Society*, *37*(7), 1078–1083. doi:10.1177/0163443715594106

Chen, C., & Zhang, C.-Y. (2014). Data-intensive applications, challenges, techniques and technologies: A survey on Big Data. *Information Sciences*, *275*, 314–347. doi:10.1016/j.ins.2014.01.015

Cifuentes, J. D. M. (2013). Context modelling for serendipitous discoveries in exploratory mobile search over social media. *Fifth BCS-IRSG Symposium on Future Directions in Information Access (FDIA 2013)*.

Cohen, P., & Oviatt, S. (2000). Multimodal interfaces that process what comes naturally. *Communications of the ACM*, *43*(3), 45–33. doi:10.1145/330534.330538

Curtis, L. H., Brown, J., & Platt, R. (2014). Four health data networks illustrate the potential for a shared national multipurpose big-data network. *Health Affairs*, *33*(7), 1178–1186. doi:10.1377/hlthaff.2014.0121 PMID:25006144

Darji, A., & Waghela, D. (2014). Parallel Power Iteration Clustering for Big Data using MapReduce in Hadoop. *International Journal of Advanced Research in Computer Science and Software Engineering*, *4*(6).

Demirkan, H., & Delen, D. (2013). Leveraging the capabilities of service-oriented decision support systems: Putting analytics and big data in cloud. *Decision Support Systems*, *99*(1), 412–421. doi:10.1016/j.dss.2012.05.048

Eisenberg, D. M., Davis, R. B., Ettner, S. L., Appel, S., Wilkey, S., Van Rompay, M., & Kessler, R. C. (1998). Trends in alternative medicine use in the United States, 19901997: Results of a follow-up national survey. *Journal of the American Medical Association*, *280*(18), 1569–1575. doi:10.1001/jama.280.18.1569 PMID:9820257

Emani, C. K., Cullot, N., & Nicolle, C. (2015). Understandable big data: A survey. *Computer Science Review*, *17*, 70–81. doi:10.1016/j.cosrev.2015.05.002

Fan, J., & Liu, H. (2013). Statistical analysis of big data on pharmacogenomics. *Advanced Drug Delivery Reviews*, *65*(7), 987–1000. doi:10.1016/j.addr.2013.04.008 PMID:23602905

Feller, E., Ramakrishnan, L., & Morin, C. (2015). Performance and energy efficiency of big data applications in cloud environments: A Hadoop case study. *Journal of Parallel and Distributed Computing*, *79*, 80–89. doi:10.1016/j.jpdc.2015.01.001

Fisher, D., DeLine, R., Czerwinski, M., & Drucker, S. (2012). Interactions with big data analytics. *Interactions, 19*(3), 50-59.

Free, C., Phillips, G., Galli, L., Watson, L., Felix, L., Edwards, P., ... Haines, A. (2013). The effectiveness of mobile-health technology-based health behaviour change or disease management interventions for health care consumers: A systematic review. *PLoS Medicine*, *10*(1), e1001362. doi:10.1371/journal.pmed.1001362 PMID:23349621

Gandomi, A., & Haider, M. (2015). Beyond the hype: Big data concepts, methods, and analytics. *International Journal of Information Management*, *35*(2), 137–144. doi:10.1016/j.ijinfomgt.2014.10.007

Gartner, I. (2014). *Glossary Big data*. Academic Press.

Ge, B., Ge, S., & Minard, T. (2014). Visualizations make big data meaningful. *Communications of the ACM*, *57*(6).

Ghezzi, A., Balocco, R., & Rangone, A. (2015). A fuzzy framework assessing corporate resource management for the mobile content industry. *Technological Forecasting and Social Change*, *96*, 153–172. doi:10.1016/j.techfore.2015.01.004

Gowin, M., Cheney, M., Gwin, S., & Franklin Wann, T. (2015). Health and fitness app use in college students: A qualitative study. *American Journal of Health Education*, *46*(4), 223–230. doi:10.1080/19325037.2015.1044140

Han, Q., Liang, S., & Zhang, H. (2015). Mobile cloud sensing, big data, and 5G networks make an intelligent and smart world. *IEEE Network*, *29*(2), 40–45. doi:10.1109/MNET.2015.7064901

Harrison, S., & Mort, M. (1998). Which champions, which people? Public and user involvement in health care as a technology of legitimation. *Social Policy and Administration*, *32*(1), 60–70. doi:10.1111/1467-9515.00086

Hashem, I., Yaqoob, I., Anuar, N. B., Mokhtar, S., Gani, A., & Khan, S. U. (2015). The rise of big data on cloud computing: Review and open research issues. *Information Systems*, *47*, 98–115. doi:10.1016/j.is.2014.07.006

Hassenzahl, M., & Tractinsky, N. (2006). User experience-a research agenda. *Behaviour & Information Technology*, *25*(2), 91–97. doi:10.1080/01449290500330331

Hayhurst, C. (2015). Mining for Answers from Big Data. *Biomedical Instrumentation & Technology*, *49*(2), 84–92. doi:10.2345/0899-8205-49.2.84 PMID:25793337

Hu, H., Wen, Y., Chua, T.-S., & Li, X. (2014). Toward scalable systems for big data analytics: A technology tutorial. *IEEE Access*, *2*, 652–687. doi:10.1109/ACCESS.2014.2332453

Jee, K., & Kim, G.-H. (2013). Potentiality of big data in the medical sector: focus on how to reshape the healthcare system. *Healthcare Informatics Research, 19*(2), 79-85.

Jha, A. K., DesRoches, C. M., Campbell, E. G., Donelan, K., Rao, S. R., Ferris, T. G., ... Blumenthal, D. (2009). Use of electronic health records in US hospitals. *The New England Journal of Medicine*, *360*(16), 1628–1638. doi:10.1056/NEJMsa0900592 PMID:19321858

Kambatla, K., Kollias, G., Kumar, V., & Grama, A. (2014). Trends in big data analytics. *Journal of Parallel and Distributed Computing*, *74*(7), 2561–2573. doi:10.1016/j.jpdc.2014.01.003

Keim, D. A., Mansmann, F., Schneidewind, J., Thomas, J., & Ziegler, H. (2008). *Visual analytics: Scope and challenges. In Visual data mining* (pp. 76–90). Springer.

Kim, D., & Chang, H. (2007). Key functional characteristics in designing and operating health information websites for user satisfaction: An application of the extended technology acceptance model. *International Journal of Medical Informatics*, *76*(11), 790–800. doi:10.1016/j.ijmedinf.2006.09.001 PMID:17049917

Kopetz, H. (2011). *Real-time systems: design principles for distributed embedded applications*. Springer Science & Business Media. doi:10.1007/978-1-4419-8237-7

Krämer, M., & Senner, I. (2015). A modular software architecture for processing of big geospatial data in the cloud. *Computers & Graphics*, *49*, 69–81. doi:10.1016/j.cag.2015.02.005

Kuiken, P. G. B. K. D. K. S. V. (2013). *The 'big data' revolution in healthcare.* Center for US Health System Reform Business Technology Office.

Kum, H.-C., Stewart, C. J., Rose, R. A., & Duncan, D. F. (2015). Using big data for evidence based governance in child welfare. *Children and Youth Services Review*, *58*, 127–136. doi:10.1016/j.childyouth.2015.09.014

Kumar, S., Nilsen, W. J., Abernethy, A., Atienza, A., Patrick, K., Pavel, M., ... Spruijt-Metz, D. (2013). Mobile health technology evaluation: The mHealth evidence workshop. *American Journal of Preventive Medicine*, *45*(2), 228–236. doi:10.1016/j.amepre.2013.03.017 PMID:23867031

Laurila, J. K., Gatica-Perez, D., Aad, I., Bornet, O., Do, T.-M.-T., Dousse, O., . . . Miettinen, M. (2012). *The mobile data challenge: Big data for mobile computing research.* Paper presented at the Pervasive Computing.

Lee, T.-Y., Tong, X., Shen, H.-W., Wong, P. C., Hagos, S., & Leung, L. R. (2013). Feature tracking and visualization of the Madden-Julian Oscillation in climate simulation. *IEEE Computer Graphics and Applications*, *33*(4), 29–37. doi:10.1109/MCG.2013.36 PMID:24808057

Lin, W., Dou, W., Zhou, Z., & Liu, C. (2015). A cloud-based framework for Home-diagnosis service over big medical data. *Journal of Systems and Software*, *102*, 192–206. doi:10.1016/j.jss.2014.05.068

Liu, Z., Jiang, B., & Heer, J. (2013). *imMens: Real-time Visual Querying of Big Data.* Paper presented at the Computer Graphics Forum. 10.1111/cgf.12129

Logan, A. G. (2013). Transforming hypertension management using mobile health technology for telemonitoring and self-care support. *The Canadian Journal of Cardiology*, *29*(5), 579–585. doi:10.1016/j.cjca.2013.02.024 PMID:23618506

Ludwick, D. A., & Doucette, J. (2009). Adopting electronic medical records in primary care: Lessons learned from health information systems implementation experience in seven countries. *International Journal of Medical Informatics*, *78*(1), 22–31. doi:10.1016/j.ijmedinf.2008.06.005 PMID:18644745

Ma, Y., Wang, L., Liu, P., & Ranjan, R. (2015). Towards building a data-intensive index for big data computing–A case study of Remote Sensing data processing. *Information Sciences*, *319*, 171–188. doi:10.1016/j.ins.2014.10.006

MacNeil, S., & Elmqvist, N. (2013). *Visualization mosaics for multivariate visual exploration.* Paper presented at the Computer Graphics Forum.

Maitrey, S., & Jha, C. (2015). MapReduce: Simplified Data Analysis of Big Data. *Procedia Computer Science*, *57*, 563–571. doi:10.1016/j.procs.2015.07.392

Marchionini, G. (1989). Information-seeking strategies of novices using a full-text electronic encyclopedia. *Journal of the American Society for Information Science*, *40*(1), 54–66. doi:10.1002/(SICI)1097-4571(198901)40:1<54::AID-ASI63.0.CO;2-R

Marchionini, G., & Komlodi, A. (1998). Design of interfaces for information seeking. *Annual Review of Information Science & Technology*, *33*, 89–130.

Marx, V. (2013). Biology: The big challenges of big data. *Nature*, *498*(7453), 255–260. doi:10.1038/498255a PMID:23765498

Miksch, S., & Aigner, W. (2014). A matter of time: Applying a data–users–tasks design triangle to visual analytics of time-oriented data. *Computers & Graphics*, *38*, 286–290. doi:10.1016/j.cag.2013.11.002

Miller, T., Chandler, L., & Mouttapa, M. (2015). A needs assessment, development, and formative evaluation of a health promotion smartphone application for college students. *American Journal of Health Education*, *46*(4), 207–215. doi:10.1080/19325037.2015.1044138

Monroe, C. M., Thompson, D. L., Bassett, D. R. Jr, Fitzhugh, E. C., & Raynor, H. A. (2015). Usability of Mobile Phones in Physical Activity–Related Research: A Systematic Review. *American Journal of Health Education*, *46*(4), 196–206. doi:10.1080/19325037.2015.1044141

Murdoch, T. B., & Detsky, A. S. (2013). The inevitable application of big data to health care. *Journal of the American Medical Association*, *309*(13), 1351–1352. doi:10.1001/jama.2013.393 PMID:23549579

Neff, G. (2013). Why big data won't cure us. *Big Data, 1*(3), 117-123.

ODriscoll, A., Daugelaite, J., & Sleator, R. D. (2013). Big data, Hadoop and cloud computing in genomics. *Journal of Biomedical Informatics*, *46*(5), 774–781. doi:10.1016/j.jbi.2013.07.001 PMID:23872175

Pah, A., Rasmussen-Torvik, L., Goel, S., Greenland, P., & Kho, A. (2015). Big data: What is it and what does it mean for cardiovascular research and prevention policy. *Current Cardiovascular Risk Reports*, *9*(1), 1–9. doi:10.100712170-014-0424-3

Paige, S. R., Stellefson, M., Chaney, B. H., & Alber, J. M. (2015). Pinterest as a resource for health information on Chronic Obstructive Pulmonary Disease (COPD): A social media content analysis. *American Journal of Health Education*, *46*(4), 241–251. doi:10.1080/19325037.2015.1044586

Paul, M. J., & Dredze, M. (2011). You are what you Tweet: Analyzing Twitter for public health. *ICWSM*, *20*, 265–272.

Peters, D. P., Havstad, K. M., Cushing, J., Tweedie, C., Fuentes, O., & Villanueva-Rosales, N. (2014). Harnessing the power of big data: Infusing the scientific method with machine learning to transform ecology. *Ecosphere*, *5*(6), 1–15. doi:10.1890/ES13-00359.1

Philips, Z., Ginnelly, L., Sculpher, M., Claxton, K., Golder, S., Riemsma, R., . . . Glanville, J. (2004). *Review of guidelines for good practice in decision-analytic modelling in health technology assessment*. Academic Press.

Raghupathi, W., & Raghupathi, V. (2014). Big data analytics in healthcare: Promise and potential. *Health Information Science and Systems*, *2*(1), 1. doi:10.1186/2047-2501-2-3 PMID:25825667

Rayport, J., & Heyward, A. (2011). *Envisioning the cloud: The next computing paradigm and its implication for technology policy*. Accessed May.

Reyes-Ortiz, J. L., Oneto, L., & Anguita, D. (2015). Big data analytics in the cloud: Spark on hadoop vs mpi/openmp on beowulf. *Procedia Computer Science*, *53*, 121–130. doi:10.1016/j.procs.2015.07.286

Rind, A., Wang, T. D., Aigner, W., Miksch, S., Wongsuphasawat, K., Plaisant, C., & Shneiderman, B. (2011). Interactive information visualization to explore and query electronic health records. *Foundations and Trends in Human-Computer Interaction*, *5*(3), 207–298. doi:10.1561/1100000039

Sagiroglu, S., & Sinanc, D. (2013). *Big data: A review*. Paper presented at the Collaboration Technologies and Systems (CTS), 2013 International Conference on. 10.1109/CTS.2013.6567202

Schmidt, H., Norman, G., & Boshuizen, H. (1990). A cognitive perspective on medical expertise: Theory and implication. *Academic Medicine*, *65*(10), 611–621. doi:10.1097/00001888-199010000-00001 PMID:2261032

Senkowski, V., & Branscum, P. (2015). How college students search the internet for weight control and weight management information: An observational study. *American Journal of Health Education*, *46*(4), 231–240. doi:10.1080/19325037.2015.1044139

Shvachko, K., Kuang, H., Radia, S., & Chansler, R. (2010). *The hadoop distributed file system*. Paper presented at the 2010 IEEE 26th symposium on mass storage systems and technologies (MSST). 10.1109/MSST.2010.5496972

Siau, K., & Shen, Z. (2006). Mobile healthcare informatics. *Medical Informatics and the Internet in Medicine, 31*(2), 89–99. doi:10.1080/14639230500095651 PMID:16777784

Skiba, D. J. (2014). The Connected Age: Big Data & Data Visualization. *Nursing Education Perspectives, 35*(4), 267–269. doi:10.5480/1536-5026-35.4.267 PMID:25158424

Strack, R. W., Orsini, M. M., Fearnow-Kenney, M., Herget, J., Milroy, J. J., & Wyrick, D. L. (2015). Developing a web-based tool using information and communication technologies to expand the reach and impact of Photovoice. *American Journal of Health Education, 46*(4), 192–195. doi:10.1080/19325037.2015.1044585 PMID:27642378

Stroetmann, K. A. (2013). Achieving the integrated and smart health and wellbeing paradigm: A call for policy research and action on governance and business models. *International Journal of Medical Informatics, 82*(4), e29–e37. doi:10.1016/j.ijmedinf.2012.05.008 PMID:22727880

Suchanek, F., & Weikum, G. (2013). Knowledge harvesting in the big-data era. *Proceedings of the 2013 ACM SIGMOD International Conference on Management of Data*. 10.1145/2463676.2463724

Swan, M. (2013). The quantified self: Fundamental disruption in big data science and biological discovery. *Big Data, 1*(2), 85-99.

Tam, N. T., & Song, I. (2016). *Big Data Visualization. In Information Science and Applications (ICISA) 2016* (pp. 399–408). Springer. doi:10.1007/978-981-10-0557-2_40

Tatarchuk, N., Shopf, J., & DeCoro, C. (2008). Advanced interactive medical visualization on the GPU. *Journal of Parallel and Distributed Computing, 68*(10), 1319–1328. doi:10.1016/j.jpdc.2008.06.011

Thomas, J., & Kielman, J. (2009). Challenges for visual analytics. *Information Visualization, 8*(4), 309–314. doi:10.1057/ivs.2009.26

Thorvaldsdóttir, H., Robinson, J. T., & Mesirov, J. P. (2013). Integrative Genomics Viewer (IGV): High-performance genomics data visualization and exploration. *Briefings in Bioinformatics, 14*(2), 178–192. doi:10.1093/bib/bbs017 PMID:22517427

Vaitsis, C., Nilsson, G., & Zary, N. (2014). Visual analytics in healthcare education: Exploring novel ways to analyze and represent big data in undergraduate medical education. *PeerJ, 2*, e683. doi:10.7717/peerj.683 PMID:25469323

Walji, M. F., Kalenderian, E., Tran, D., Kookal, K. K., Nguyen, V., Tokede, O., ... Stark, P. C. (2013). Detection and characterization of usability problems in structured data entry interfaces in dentistry. *International Journal of Medical Informatics, 82*(2), 128–138. doi:10.1016/j.ijmedinf.2012.05.018 PMID:22749840

Wang, T. D., Wongsuphasawat, K., Plaisant, C., & Shneiderman, B. (2011). Extracting insights from electronic health records: Case studies, a visual analytics process model, and design recommendations. *Journal of Medical Systems, 35*(5), 1135–1152. doi:10.100710916-011-9718-x PMID:21541691

Wen-ying, S. C., Hunt, Y. M., Beckjord, E. B., Moser, R. P., & Hesse, B. W. (2009). Social media use in the United States: Implications for health communication. *Journal of Medical Internet Research, 11*(4), e48. doi:10.2196/jmir.1249 PMID:19945947

Wright, H., Mathers, C., & Walton, J. (2013). Using visualization for visualization: An ecological interface design approach to inputting data. *Computers & Graphics*, *37*(3), 202–213. doi:10.1016/j.cag.2013.01.013

Wyber, R., Vaillancourt, S., Perry, W., Mannava, P., Folaranmi, T., & Celi, L. A. (2015). Big data in global health: Improving health in low-and middle-income countries. *Bulletin of the World Health Organization*, *93*(3), 203–208. doi:10.2471/BLT.14.139022 PMID:25767300

Xia, F., Yang, L. T., Wang, L., & Vinel, A. (2012). Internet of things. *International Journal of Communication Systems*, *25*(9), 1101–1102. doi:10.1002/dac.2417

Yang, Y., Zhang, L., Zhen, Y., & Ji, R. (2015). Learning for visual semantic understanding in big data. *Neurocomputing*, (169): 1–4.

Yao, Q., Tian, Y., Li, P.-F., Tian, L.-L., Qian, Y.-M., & Li, J.-S. (2015). Design and development of a medical big data processing system based on hadoop. *Journal of Medical Systems*, *39*(3), 1–11. doi:10.100710916-015-0220-8 PMID:25666927

Young, S. D. (2015). A big data approach to HIV epidemiology and prevention. *Preventive Medicine*, *70*, 17–18. doi:10.1016/j.ypmed.2014.11.002 PMID:25449693

Zhang, J., Wolfram, D., Wang, P., Hong, Y., & Gillis, R. (2008). Visualization of health-subject analysis based on query term co-occurrences. *Journal of the American Society for Information Science and Technology*, *59*(12), 1933–1947. doi:10.1002/asi.20911

KEY TERMS AND DEFINITIONS

App: An app or software application is typically a small, specialized program downloaded onto mobile devices.

Big Data: Big data is high-volume, high-velocity and high-variety information assets that demand cost-effective, innovative forms of information processing for enhanced insight and decision making.

Cloud Computing: Cloud Computing is the practice of using a network of remote servers hosted on the Internet to store, manage, and process data, rather than a local server or a personal computer.

Crowdsourcing: Crowdsourcing is the process of obtaining (information or input into a particular task or project) by enlisting the services of a number of people, either paid or unpaid, typically via the Internet.

Electronic Medical Records (EMR): Electronic Medical Records is a digital version of the traditional paper-based medical record for an individual. The EMR represents a medical record within a single facility, such as a doctor's office or a clinic.

Hadoop Distributed File System (HDFS): The Hadoop Distributed File System is a Java-based file system that provides scalable and reliable data storage, and it was designed to span large clusters of commodity servers. HDFS has demonstrated production scalability of up to 200 PB of storage and a single cluster of 4500 servers, supporting close to a billion files and blocks.

MapReduce: A MapReduce job usually splits the input data-set into independent chunks which are processed by the map tasks in a completely parallel manner. The framework sorts the outputs of the maps, which are then input to the reduce tasks. Typically both the input and the output of the job are stored in a file-system.

NoSql: A NoSQL (originally referring to "non SQL", "non relational" or "not only SQL") database provides a mechanism for storage and retrieval of data which is modeled in means other than the tabular relations used in relational databases.

Visual Analytics: Visual analytics is an outgrowth of the fields of information visualization and scientific visualization that focuses on analytical reasoning facilitated by interactive visual interfaces.

Visualization: Visualization is any technique for creating images, diagrams, or animations to communicate a message. Visualization through visual imagery has been an effective way to communicate both abstract and concrete ideas since the dawn of humanity.

This research was previously published in Next-Generation Mobile and Pervasive Healthcare Solutions edited by Jose Machado, António Abelha, Manuel Filipe Santos, and Filipe Portela, pages 46-66, copyright year 2018 by Medical Information Science Reference (an imprint of IGI Global).

Chapter 4
mHealth in Maternal, Newborn, and Child Health Programs around the World

Briana Britton
Healthcare Innovation Technology Lab, USA

Laura Pugliese
Healthcare Innovation Technology Lab, USA

Stan Kachnowski
Healthcare Innovation Technology Lab, USA

ABSTRACT

The incorporation of mobile devices into the delivery of healthcare, known as mHealth, is changing the way care is delivered in the 21st century. The impact of mHealth is particularly salient in low and middle income countries (LMICs), where mHealth poses the opportunity to increase access and quality of healthcare in systems where supportive infrastructure is otherwise lacking. This approach is well-suited to target issues of maternal and child health, permitting an increase in health education, communication, monitoring, and care to what are often vulnerable and hard-to-access populations. Employing mHealth tactics that target such populations can improve the overall access and quality of maternal and child health in the developing world - a priority for the United Nations as reflected in the Millennium Development Goals (MDGs). While the field of mHealth is new and still developing, many programs and thought-leaders have already successfully applied mHealth strategies in interventions to improve maternal and child health through health education, preventive care, emergency response, biometric data collection, and training healthcare workers.

INTRODUCTION

mHealth refers to "mobile health:" the incorporation of mobile devices like cell phones into the delivery of healthcare. Mobile communication employed in the delivery of healthcare are used to improve health outcomes by increasing access to healthcare and health information – offering the utility to disperse up-to-date public health information directly to patients, issue timely reminders or referrals for treatments, or allow health workers to monitor the spread of disease (*mHealth for Development*, 2009).

DOI: 10.4018/978-1-5225-6198-9.ch004

OVERVIEW

The United Nation's Millennium Development Goals to reduce child mortality and improve maternal health offer a metric mHealth programs can work to meet. Due to the rapid penetration of mobile phones, particularly in low- and middle-income countries (LMICs), there is a great opportunity to harness mobile technologies in the integration of healthcare delivery for maternal, infant and child health.

Current Scientific Knowledge in Maternal, Newborn, and Child mHealth

The World Health Organizations' mHealth Technical and Evidence Review Group (mTERG) for reproductive, newborn, and child health, is a pioneer in the field of mHealth. The coalition is a compilation of experts from the fields of health, academia, and research institutions. The WHO mTERG recognizes that decision-makers in low and middle-income countries are faced with a multitude of information about mHealth tools and systems with little instructions on how to select the best approach of implementing mHealth tactics. The coalition provides guidance on selecting appropriate mHealth initiatives. Included in mTERG is Dr. Marleen Temmerman, a medical doctor and Belgian Senator. As a gynecologist, Dr. Temmerman is an international leader in maternal healthcare. Through her work on assessing the impact of syphilis in pregnancy outcomes in Kenya, Dr. Temmerman champions for improvements in healthcare of disadvantaged populations and for the reproductive rights of women (Temmerman, 2000).

Aaltje Camielle Noordam, a scholar from Maastricht University in the Netherlands and affiliated with UNICEF, creates an evidence-base for mHealth projects, focusing mainly on the delay in care and how to provide more efficient healthcare access (Noordam, 2011).

A professor of Global Health Policy at New York University, Karen Grepin researches opportune avenues for improvements to maternal and child healthcare, particularly in low and middle-income countries (Grepin, 2013).

Dr. Seth Noar, a professor at the School of Journalism and Mass Communication at the University of North Carolina at Chapel Hill, leads the charge in understanding how to best communicate with patients. Dr. Noar is testing message design theories and frameworks to best understand what types of messages will be most memorable and call to action the receiver (Noar, 2012).

Mobile Communication in the Delivery of Healthcare

The use of mobile and wireless communication technologies to support health objectives has the potential to change the face of healthcare delivery worldwide (Aahman, 2014). Driving this change are rapid advances in mobile technology, a rise in new opportunities of integrating mobile devices with user-friendly backend systems and a wider increase in mobile cellular networks.

The integration of mHealth into the delivery of care offers significant opportunities to increase access to public health information in low and middle income countries (LMICs), which often struggle with the largest burdens of disease, poverty, and large population growth rates (*Compendium of ICT Applications*, 2007). Increasing penetration and reach of mobile technologies in these regions enables interventions delivered by phone to reach populations in ways that circumvent challenges and constraints of under-resourced healthcare systems (*Compendium of ICT Applications*, 2007). mHealth strategies have been effectively employed in LMICs to increase access to antenatal services, increase access to education in breastfeeding, improve follow-up visits for mothers and their children and facilitate the delivery of malarial treatment regimens (*Compendium of ICT Applications*, 2007).

According to The United Nations Foundation and Vodafone Foundation Technology Partnership and *Special Delivery: An analysis of mHealth in maternal and newborn health programs and their outcomes around the world*, there are six key applications for mHealth (*mHealth for Development*, 2009):

1. Increasing education and awareness of public health issues.
2. Preventative care.
3. Increasing communication in emergency situations.
4. Data collection.
5. Biometric readings.
6. Training for health workers.

Applying mHealth Tactics to Maternal and Infant Health

Traditional maternal and infant healthcare delivers services to support womens health through the continuum of adolescent sexual health, family planning, pregnancy, postpartum health, and infant health of their child. Health awareness, education and monitoring are tactics employed by mHealth programs to support such traditional services. For example, mHealth tactics have been delivered in programs to increase communication between sexual health clinics and teens or provide contraception reminders via short messaging service (SMS) text (Lim, 2008). Programs like *Text4Baby* in the United States and Russia sends text messages to women who are pregnant or have children younger than 1 year old to provide them with information and reminders to improve their health and the health of their children (Parker, 2012).

While mHealth has positive implications for individuals around the globe, the rise of mHealth for maternal and infant care has a focus on LMICs, where rates of maternal and infant mortality and morbidity are highest (Aahman, 2014). Poverty and in-access to healthcare is a main factor affecting maternal health, where barriers including long distance to clinics, high cost of transportation, low access to health information and low perceived quality of care restrict women from seeking timely and sufficient care (Aahman, 2014). Almost 40% of the estimated 9.7 million child-under-five deaths occur in the child's neonatal stage of development (*Maternal and Newborn Health,* 2012). The largest absolute number of newborn deaths occurs in South Asia, and the highest national rates of neonatal mortality occur in sub-Saharan Africa (*Maternal and Newborn Health*, 2012). One of the most significant predictors of a child's health and survival is the health of the mother, and ninety nine percent of maternal and newborn mortality happens in low- and middle-income countries where more than 50% of women deliver without skilled personnel (*Maternal and Newborn Health*, 2012).

The Millennium Development Goals: A Potential Task for mHealth

The Millennium Development Goals were established by the United Nations in 2000 to launch the world into a new era of health, equality, and sustainability. Millennium Development Goal 5, to improve maternal health, has seen positive progress: as of 2010, maternal mortality has declined by nearly half since 1990, though the goal is to decrease maternal mortality by three quarters by 2015. The U.N. cites that poverty and lack of access to education perpetuate high adolescent birth rates, and inadequate funding for family planning is a major failure for fulfilling the commitment to improving women's reproductive health (*Compendium of ICT Applications*, 2007).

Due to these underlying structural constraints that limit achieving the Millennium Development Goals (MDG), many individuals in the public health field are looking toward innovative, cost effective initiatives like those of mHealth (Tamrat, 2012). Studies show that taking advantage of mHealth tactics can effectively support the Millennium Development Goals 4 and 5, which aim to reduce child mortality and improve maternal health (Tamrat, 2012).

How mHealth Can Promote Maternal and Child Health in Low-Resource Settings

While mHealth has applications in high-income countries, many individuals in the field have a focus towards developing interventions targeting regions with low income levels and resource restrictions. There are 78 mobile-cellular subscriptions per 100 inhabitants in the world – signaling a ubiquity of mobile phones and a novel opportunity to communicate with patient populations like never before (Parker, 2012). LMICs are exhibiting a rapid emergence of mobile phone use. With 64% of all mobile phone users presently located in the developing world, mobile technology offers a unique opportunity to efficiently deliver care in remote regions and low-resource countries (*Compendium of ICT Applications*, 2007). Wireless signals cover 85% of the world's population, allowing communication to extend far beyond the limits of the electrical grid (*mHealth*, 2011). Estimates show that by 2012, half of all individuals in remote areas will have access to mobile phones (mHealth for Development, 2009). There is promise in utilizing mHealth programs in LMICs to alleviate the limitations in access to health services that currently act as key drivers of poor health outcomes (Cormick, 2012).

Applications of mHealth

1. Education and Awareness

SMS texting offers a low-cost, scalable outreach platform for healthcare providers to communicate with their patient population (*mHealth for Development*, 2009). Sending SMS can be used effectively in education and awareness applications. Messages are directly sent to users' phones from their providers or public health personnel. The United States based intervention *Text4Baby* uses SMS texting to deliver health messages and engage pregnant women and new mothers in healthy behaviors. *Text4Baby* sends three free messages a week to pregnant women offering evidence-base health information relevant to their stage of pregnancy. After the child is born, the woman still receives three text messages per week, though they are customized to provide information regarding healthcare tips for their newborn baby (Parker, 2012).

Studies and anecdotal evidence show that SMS text messages have a greater impact on influencing behavior compared to radio or television campaigns (*mHealth for Development*, 2009). SMS alerts have particular promise in targeting hard-to-reach populations in rural areas where the absence of clinics, dearth of healthcare providers, and lack of available healthcare information prevent women from making informed choices regarding their pregnancy and/or the health of their newborn child.

2. Preventive Care

Though traditional communication mediums such as radio and television have merit to disseminating public health education, SMS texting stands out as having several unique advantages: cost effectiveness, convenience, broad reach, and widespread popularity in low-resource settings (*mHealth for Development*, 2009). As in the *Text4Baby* campaign, the Mobile Midwife program (currently implemented in Ghana, India and Cambodia) sends text messages containing education and health visit reminders to pregnant women and recent mothers. The Mobile Midwife program provides new mothers and their families with SMS and/or voice messages that are time-specific to the stage of their pregnancy or the age of their child. Community nurses use the Nurse Application to collect patient data and upload patient health records to centralized databases, allowing them to track the care of their patients and issue reminders when patients miss scheduled antenatal or postnatal visits (*Maternal and Infant Health*, 2012). The text messages include relevant health information about their stage of pregnancy or the age of their infant, as well as reminders to visit their healthcare providers for antenatal and postnatal care.

3. Emergency Situations

mHealth can be leveraged for maternal, infant and child health by breaking down communication barriers to more effectively expedite emergency situations (Tamrat, 2012). One of the first studies to show the efficacy of mobile communication in the field of maternal health in the developing world is the Rural Extended Services and Care for Ultimate Emergency Relief (RESCUER) program launched in 1996 rural Uganda. Community-based volunteer traditional birth attendants (TBAs) were trained to understand complications from childbirth and were advised on program protocols for how to respond by utilizing walkie-talkies linked to health units. Following the intervention, the maternal mortality rate in areas of where the program was implemented decreased by approximately 50% due to the increase in referrals to health facilities (Tamrat, 2012).

4. Data Collection

mHealth reaches beyond participant-facing health programs to the use of mHealth to actively collect health data. After collection of data, researchers can evaluate potential health interventions to best serve that particular community. Crucial to understanding the effectiveness of existing policies and programs, data collection can sometimes be challenging to perform in the field prior to the advent of mobile technologies. With the use of mobile devices, remote data collection allows researchers to communicate with populations who may be geographically isolated from hospitals or clinics – traditional locations of study data collection (*mHealth for Development*, 2009). Today the data collection process is more efficient and reliable when conducted via mobile device as compared to submitting paper-based surveys, which requires manual entering into health databases. Utilizing mobile devices not only by-passes extra human resources required to manually enter paper-based information, but mobile devices also offer real-time basis and accessibility for individuals who otherwise would not be able to physically travel to the study site. Streamlining data collection by the use of mobile technologies allows for cheaper, more effective studies and interventions.

5. Biometric Readings

Though the majority of mobile technology is used for communication in the delivery of care, with the proliferation of smartphones (cell phones with operating systems), harnessing mobile technology can go beyond communication to the use of biometric readings. Three students in Kampala Uganda invented a smartphone application that can measure a fetal heartbeat. The app is called WinSenga – "Senga" is a local term for an aunt who helps during pregnancy, implying that the app is as helpful as a family member in promoting maternal health. The smartphone app couples with a tiny microphone in a plastic horn (Evans, 2014). The horn is based on the Pinard horn, a type of stethoscope often made of wood or metal used by midwives since the 19th century (Davis-Floyd, 1997). Though midwives in Uganda have used Pinard horns before, the smartphone app records the fetal heartbeat and analyzes the sound. Depending on the rhythm of the heartbeat, the phone prompts the midwife of possible suggestions on how to best follow up with patient care.

The device can be used particularly in understaffed hospitals in remote parts of Uganda. Since the app records the fetal heartbeat, obstetricians operating in a resourced constrained condition could maximize her time with each patient by replaying the recording while she was out attending to another patient. Though the smartphone app should not be used without an individual who has training in maternal health, the diagnostic feature could have positive implications in assisting midwives.

6. Training for Health Workers

Studies conducted in Egypt, Cameroon, and Malaysia demonstrate that mobile phones can assist health workers in decision making, allowing health workers in the community to consult with higher-skilled members of the workforce (*Why mobile phones make a difference*, 2013).

FUTURE RESEARCH DIRECTIONS

Expectations for the future impact of mHealth are high, but certain limitations need to be addressed to ensure the tactic can be employed to reach its fullest potential. A recent survey indicates that roughly one-half of patients believe that mHealth will improve the convenience, cost and quality of their healthcare in the coming years (Figgis, 2014), but some experts warn that clinicians and patients should be aware of the limitations of the devices employed, of type information provided and how well information can be conveyed. Future directions in mHealth research are focused in three key areas:

1. Identify concrete value to stakeholders. Benefits of innovation in the field of maternal and child mHealth must be clearly defined for individuals to underwrite its development.
2. Think globally. Individuals who will benefit the most from mHealth interventions are in the developing world. As such, most innovation is targeted at LMICs, but once innovations are established in the developing world, such mHealth tools and strategies can be applied to developed countries as well.
3. Avoid the trap of seeing mHealth as separate from healthcare delivery. mHealth's greatest value will be how it integrates with established health systems in the future. One-on-one patient and physician interaction cannot be completely replaced, but mHealth will revolutionize the manner in which healthcare is delivered.

REFERENCES

Aahman, E., Begg, S., Boschi-Pinto, C., & Chatterji, S. (n.d.). *The global burdens of disease*. World Health Organization.

Compendium of ICT Applications on Electronic Government – Volume 1. Mobile Applications on Health and Learning. (2007). *United Nations*. Retrieved June 10, 2014, from http://unpan1.un.org/intradoc/groups/public/documents/un/unpan030003.pdf

Cormick, G., Kim, N., Rodgers, A., Gibbons, L., & Buekens, P. (2012, August 6). Interest of pregnant women in the use of SMS (short message service) text messages for the improvement of perinatal and postnatal care. *Reproductive Health, 9*(9).

Davis-Floyd, R., & Fishel Sargent, C. (1997). *Child birth and authoritative knowledge: Cross-cultural perspectives*. University of California Press. Retrieved June 5, 2014, from http://content.ub.hu-berlin.de/monographs/toc/ethnologie/BV011542334.pdf

Evans, J. (2014, May). Pre-natal care goes mobile in Uganda. *VOA Learning English*. Retrieved June 5, 2014, from http://m.learningenglish.voanews.com/a/pre-natal-care-goes-mobile-uganda/1903737.html

Figgis, P. (2014). *Emerging mHealth: Paths for growth*. Retrieved August 6, 2014, from http://www.pwc.com/en_GX/gx/healthcare/mhealth/assets/pwc-emerging-mhealth-full.pdf

Grepin, K. A., & Klugman, J. (2013, May 18). Maternal health: A missed opportunity for development. *Lancet, 381*(9879), 1691–1693. doi:10.1016/S0140-6736(13)60981-2 PMID:23683617

Lim, M., Hocking, J., Hellard, M., & Campbell, A. (2008, May). SMS STI: A review of the uses of mobile phone text messaging in sexual health. *International Journal of STD & AIDS, 19*(5).

Maternal and Infant Health. (2012, September). *Grameen Foundation*. Retrieved June 15, 2014, from http://www.grameenfoundation.org/what-we-do/health/maternal-and-infant-health

Maternal and Newborn Health. (2012, June 12). *UNICEF*. Retrieved June 3, 2014, from http://www.unicef.org/health/index_maternalhealth.html

mHealth. (2011). *World Health Organization*. Retrieved June 11, 2014, from http://www.who.int/goe/publications/goe_mhealth_web.pdf

mHealth and the Future of Medicine. (n.d.). *Rockefeller Foundation*. Retrieved August 6, 2014, from http://www.rockefellerfoundation.org/uploads/files/9fd8b605-f89f-4ddd-9851-3541c5d196fd-silos-to.pdf

mHealth for Development: The Opportunity of Mobile Technology for Healthcare in the Developing World. (2009). *Vodafone*.

Noar, S. M., & Harrington, N. G. (2012). *Ehealth application*. New York, NY: Routledge.

Noordam, A., Kuepper, B. M., Stekelenburg, J., & Milen, A. (2011, February). Improvement of maternal health services through the use of mobile phones. *Tropical Medicine & International Health, 16*(5), 622–626. doi:10.1111/j.1365-3156.2011.02747.x PMID:21342374

Parker, R., Dmitrieva, E., Froloy, S., & Gazmararian, J. (2012). Text4Baby in the United States and Russia: An opportunity for understanding how mHealth affects maternal and child health. *Journal of Health and Communication*.

Singh, P. (2013, December 13). Why mobile phones make a difference to healthcare in Sub-Saharan Africa. *One Million Community Health Workers*. Retrieved June 15, 2014, from http://www.globalcitizen.org/Content/Content.aspx?id=65bf3de6-b552-460c-8ca7-f7e2d235f557

Tamrat, T., & Kachnowski, S. (2012, July). Special delivery: an analysis of mHealth in maternal and newborn health programs and their outcomes around the world. *Maternal and Child Health Journal, 16*(5).

Temmerman, M., Gichangi, P., Fonck, K., Apers, L., Claeys, P., Van Renterghem, L., & Kiragu, D. (2000). Effect of syphilis control programme on pregnancy outcomes in Nairobi, Kenya. *Sexually Transmitted Infections*. Retrieved August 12, 2014, from http://sti.bmj.com/content/76/2/117.full

ADDITIONAL READING

Konschak, C. (2013). MHealth. Global Opportunities and Challenges (2ndnd ed.). Virginia Beach, VA: Convurgent Publishing.

Malvey, D., & Slovensky, D. J. (2014). *MHealth: Transforming Healthcare*. New York, NY: Springer. doi:10.1007/978-1-4899-7457-0

Ranck, J. (2012). *Connected Health: How Mobile Phones, Cloud, and Big Data Will Reinvent Health Care*. San Francisco, CA: Gigaom Books.

KEY TERMS AND DEFINITIONS

Biometric Device: A device that measures physiological characteristics such as finger print or blood pressure.

Low and Middle Income Countries: Referred to as developing countries with gross national income per capita of $4,085 or less as of 2013. The term is broad reaching and is not intended to imply that the economies have reached the final stages of development.

Maternal and Child Health: the health of women during pregnancy, childbirth, and the postpartum period; and the health of children during the newborn stage through childhood.

mHealth: An abbreviation for "mobile health", or the practice of medicine and mobile health supported by mobile devices.

Public Health: The health of the population as a whole, especially as monitored, regulated, and promoted by the state.

SMS Text Message: Short Message Service (SMS) is a text messaging service component of phone, Web, or mobile communication systems. It uses standardized communications protocols to allow fixed line or mobile phone devices to exchange short text messages.

United Nations: An intergovernmental organization created to promote international cooperation following the Second World War to prevent another such conflict.

World Health Organization: A specialized agency of the United Nations that is concerned with international public health.

This research was previously published in the Encyclopedia of Mobile Phone Behavior edited by Zheng Yan, pages 818-824, copyright year 2015 by Information Science Reference (an imprint of IGI Global).

Chapter 5
The Use of Smart Card Technology in Health Care

Jonathan Gates
Kent County Hospital, USA

Kathryn J. Jervis
University of Rhode Island, USA

Valerie Maier-Speredelozzi
University of Rhode Island, USA

Koray Ozpolat
University of Rhode Island, USA

Jyh-Hone Wang
University of Rhode Island, USA

Su-Hsiu Wu
University of Rhode Island, USA

INTRODUCTION

This article describes a health information technology (IT) system for patient-centered access to health information through smart cards as an enhancement to Electronic Health Records (EHR) systems currently becoming implemented in the United States (U.S.) and around the world. Smart card technology provides patients with a virtual assistant accessed through data readers. This system allows patients and health care providers to enter patient information such as demographics, health history, health insurance, or symptoms onto the smart card. Patients and providers can access the patient's EHR through a cloud-based Internet platform, or with a smart card through data readers embedded in kiosks at provider locations. Patients could easily access their own health records with easy-to-understand explanations about their health condition and treatment. While smart card technology is widely used outside the U.S.,

DOI: 10.4018/978-1-5225-6198-9.ch005

the U.S. health care system may significantly benefit from this type of health IT to better coordinate patient care and control costs. This article explains the benefits of smart card technology, the use of self-service kiosks, comparisons to cloud-based Internet access for patient health care data, and privacy and security issues for this technology.

BACKGROUND

At present, the lack of a wholly integrated health information system leads to redundancy for the patient and a struggle for providers to coordinate care and control costs. Patients become frustrated to repeat complex sets of information and health history to multiple providers, such as emergency personnel, hospital admission staff, and primary care doctors, often within one episode of care. Patients and providers may lose faith in the medical system. Information that patients convey may not be properly captured. Providers should document the data correctly once and share among providers to assist in patient compliance with prescribed care, while better managing costs of care. Several agencies advocate for health IT to help health care workers communicate more effectively in care coordination for patient safety, better efficiency, and cost reduction in the U.S. and the world (National Quality Forum, 2011; IOM, 2011).

Health IT also extends geographic access to low- and middle-income countries around the world where there are critical shortages of health care providers (Lewis et al., 2012). To control costs while maintaining quality, healthcare has adapted systems improvement initiatives, such as PDCA (Plan, Do, Check, Act) cycles, TQM (Total Quality Management) methods, and Six Sigma and Lean Systems. Breakthrough technological advances such as Electronic Health Records (EHR) and Electronic Medical Records (EMR) are increasingly used to boost the capacity and assure sustainability of healthcare systems particularly with regards to the vast amount of patient data needed to manage care and control costs.

In the U.S., the increase in EHR adoptions arose from pressure exerted by the Obama Administration who offered $2.5 billion in incentives to increase the use of these systems (U.S. Department of Health & Human Services, 2010). As a result, the number of office-based physicians applying for EHR certification through the Office of the National Coordinator (ONC) for Health Information Technology grew to over 100,000 nationwide (ONC, 2012). A National Ambulatory Medical Care Survey (NAMCS) of physicians found that 57% nationally use some type of EHR/EMR system (Hsiao et al., 2011). Hence, the expanding use of EHR and EMR systems may still not completely satisfy patient and provider needs.

While EHR is supposedly capable of interoperability between organizations, EMR does not communicate with other EHRs or EMRs (Garrett & Seidman, 2011). In 2012, EHR and EMR adoptions markedly increase redundant manual data entry needs. The inability to consistently share accurate data remains a substantial barrier to realization of return on investment for EMR and EHR adopters. Some systems may deliver unmodified content to the patient, who may not understand the terminology. Instead, smart card technology, cloud-based Internet access, self-service kiosks and data readers would provide the patient a human-factors approach to explaining technical information in layman's terms. Patient-identified 'inaccuracies' can be systematically shared with providers to correct real or perceived gaps in data and improve patient compliance with effective therapies, providing a natural bridge to shared decision making (Heisler et al., 2009; Stacey et al., 2011).

Outside of health care, the public routinely uses computers and smart phones to securely access information and perform tasks for banking, home management, travel reservations, and more. Patients seek similar ease in accessing health information, although health care has been slow to develop patient information tools of comparable functionality (Krist & Woolf, 2011). The addition of smart card technology to EHR systems may overcome limitations of those systems while retaining patient control with convenient, familiar technology similar to that already used for other functions.

SMART CARD TECHNOLOGY

Issues, Controversies, Problems

Smart Cards

As a mature technology, smart cards offer independence and reliability at low cost, and they are widely used by healthcare, finance, government, mass transit, and other sectors in developing and developed countries (Frost & Sullivan, 2012). The smart card is called an "integrated circuit card" by the International Standards Organization (ISO) (Smart Card Tutorial, 1992). A smart card is defined as a "card incorporating one or more integrated [electronic] circuits within its thickness" (Xiao & Yu, 2009). The integrated circuit is mainly a memory unit but could also include a microprocessor (Smart Card Tutorial, 1992). Xiao and Yu (2009) explain that the smart card has tamper-proof microcontrollers that encrypt all information. The smart card is read by a card reader, connects to a computer, or is accessed wirelessly.

Cards that have physical contact with a reader are called contact cards. A typical contact smart card operates under the ISO/IEC 7816 standard, a plastic card body, a chip embedded in the body, and an often gold-plated contact point. When a contact card is inserted into a smart card reader, commands, data, and card status are transmitted over the physical contact points (Smart Card Alliance, 2012). A contactless smart card is powered by its own antenna through magnetic field induction initiated by the antenna in a card reader no more than 4 inches away for a passive card or less than 2 feet for an active card that requires a battery to enhance the card's power. Data on contactless cards are updated under the international standard of ISO/IEC 14443, which operates at 13.56 MHz and specifies "the physical characteristics, radio frequency power and signal interface, initialization and anti-collision protocols and transmission protocol" (Smart Card Alliance, 2012). A dual-interface card implements contact and contactless interfaces on one card that interoperates with both types (Smart Card Alliance, 2003).

Contact or contactless smart cards can run on Java card platforms that are device and operating system independent, e.g., Windows, Mac OS, Linux, Unix and others (Ortiz, 2003). Smart cards hold substantial amounts of data and are expected to last about ten years (Xiao & Yu, 2009). Smart cards used in health care are typically processor-enabled through semiconductor technology where the data flow is bi-directional with read and write capability (Shelfer & Procaccino, 2002).

Smart cards are used in several applications in addition to health care. For example, since 1987, Turkey implemented smart cards for driver's licenses to track unsafe driving behaviors of commercial heavy vehicle drivers (Linxens, 2012). Contact smart cards are used in Europe (i.e. France) as prepaid phone cards in public phones (Smart Card Tutorial, 1992). In Malaysia, every citizen above 12 years of age has a smart card called MyKad, which includes a photo and finger-print identification for citizenship, driver's license, ATM and digital certificate purposes (Knight, 2001).

In health care, France, Germany, and Taiwan implemented the smart card as an efficient and cost-effective tool for patient identification, insurance payments, medical records, and prescription fulfillment. In 2004, ten European countries implemented healthcare smart cards (Sembritzki, 2004). In 2002, over

80 million smart cards were in use in Germany's healthcare system (Shelfer & Procaccino, 2002). In Germany, private nonprofit companies provide insurance and are required to unconditionally include all citizens (Reid, 2010). The card contained health insurance data and was read by card readers or computers where data was serially transmitted to insurance forms. In 2011, Germany deployed 2.5 million second-generation eHealthcare cards and another 15 million cards in 2012 (Gemalto, 2012).

France launched the "Sesam Vitale" smart card program between 1997 and 2001 for citizens' use within the country (Smart Card Alliance, 2006). Administered by the Ministry of Health, this system was the first fully automatic system used in health care (Sembritzki, 2004). The system requires two cards, one for the patient and one for the provider to reduce administrative costs with faster payments from insurers to providers reducing reimbursement time from two months to a few days. (Smart Card Alliance, 2006). The French government is able to track healthcare spending, and improve patient service quality and access because the card holds the patient's medical history and prescription information (Sembritzki, 2004). Over 50 million French citizens use the card to save over 1 billion Euros each year (Smart Card Alliance, 2006).

In 1999, the Taiwan government focused on privacy and security, EHRs, and a health insurance smart card that enabled efficient access for health providers that cut down fraud and duplicate care (Li, 2012; Jian et al., 2011). Taiwan's population is 22.5 million and 96% of its population is part of the national health insurance (NHI) in a single, government-run system with mandatory participation (Reid, 2008). Taiwan has the lowest administrative costs in the world because the government insurance office pays medical bills automatically (Reid, 2008).

Many countries, like France, Germany, Taiwan, provide much more universal health coverage than the U.S. (Reid, 2010; WHR, 2000), which may make use of smart cards easier to implement. Canada has a single-payer system, similar to the British Beveridge model where the government acts as the single payer. Countries with better resources such as Switzerland, Denmark, and the Scandinavian countries (WHR, 2000) may also adopt smart card use more easily than lower GDP per capita countries like India, where most medical costs are paid out-of-pocket. Countries like Egypt that relied partially on general taxation for health care, with high out-of-pocket payments (WHR, 2000) seem unlikely to adopt smart cards, particularly in light of government unrest. Yet, in 2013, one company, Identive Group, Inc., is supplying 300,000 smart card readers to support a national electronic ID card program in the Middle East to decrease identify fraud, and manage access to public services online (GlobeNewswire, 2013). Others, like Saudi Arabia, Russia and South Africa, are reluctant to implement health IT solutions (Frost & Sullivan, 2012).

Smart card use in the U.S. may not be as widespread as in other countries for a variety of reasons. Some smart card critics in the U.S. prefer a cloud-based health records system to a card-based system, because "carrying around a card, which is common in Europe, is not our culture," said John Halamka, chairman of the U.S. Healthcare Information Technology Standards Panel (Vezina, 2011). Some providers are concerned about inaccurate card information, confidentiality issues, and forgotten or lost cards (Rosli et al., 2009). U.S. legislative inquiries questioned the high cost of smart card implementation and security and privacy issues (Martin & Rice, 2010).

To alleviate these concerns, smart cards may be secured in multiple ways. One method is to authenticate the cardholder by a password or PIN. Patient data could be stored in an encrypted card to protect against professional hacker attacks if the card is lost. Using biometric authentication technologies (i.e. fingerprint, iris recognition) could further enhance strong verification and identity authentication (Smart Card Alliance, 2011). Emergency room (ER) facilities or ambulance crews could be granted special, logged access rights to these smart cards to bypass the cardholder's password. Digital certificates could be instrumental in keeping a log of any provider modification or update to the card data. To maintain integrity of the patient data, any modification/addition to the card by a healthcare provider cannot be repudiated later except with the approval of another healthcare provider. Additionally, patients should be allowed to mark data for discussion with providers at a later date.

Cloud-Based EHR

Some prefer cloud-based EHR over smart card technology to manage patients' health care data (Vezina, 2011). Cloud computing is a new concept for accessing and managing data, typically through the Internet. Cloud-based EHRs require less up-front costs than a client-server-based system, better support for small/ solo practices and no in-house hardware/ software maintenance (Polack, 2012). Cloud-based EHRs also offer advantages for meeting "meaningful use" requirements, which means using the technology to achieve health and efficiency goals to improve health care quality, safety and care coordination (Haughton, 2012; Smart Card Alliance, 2011).

Despite its benefits, we find evidence of problems with cloud-based EHR. A serious drawback is the loss of patient data if a hosting vendor goes out of business (Polack, 2012). Additionally, there is always a trade-off between the system security and system efficiency. Any cloud-based system will have to manage a large number of friendly and unfriendly logins at any time, a major risk to its security. In April, 2011, an Amazon Cloud failure took down hundreds of websites, some for days. An important question is whether or not patients in need can wait while the cloud server, Internet, or EHR is down for extended periods. Instead of using paper and pens as a backup plan, smart cards can be used as a temporary EHR and storage of new patient data during an EHR outage.

Other problems with cloud-based EHR are variations in broadband connectivity, connection speed and cost. Much variation in connection speed occurs between the 50 states in the U.S. (Akamai, 2012). By country, South Korea and Hong Kong report much higher connection speeds than the U.S. The New America Foundation, a non-profit nonpartisan think tank, published "The Cost of Connectivity" report in July, 2012 to compare high-speed Internet prices in 22 cities worldwide, which is also a concern for cloud-based EHR usage (Hussain et al., 2012).

The results indicate that U.S. consumers in major cities tend to pay higher prices for slower speeds compared to consumers abroad. ... The results add weight to a growing body of evidence that suggests that the U.S. is lagging behind many of its international counterparts, most of whom have much higher levels of competition and, in turn, offer lower prices and faster Internet service (Hussain et al., 2012).

With such a large range in broadband adoption among the 50 U.S. states, possibly due to cost concerns and lack of sufficient infrastructure, it is perhaps unrealistic to adopt a cloud-based EHR system for the nation.

An additional concern of cloud-based EHR is special needs patients, particularly as the population ages in western countries with associated increases in Alzheimer's disease. In such situations, a smart card would be much easier for special needs population patients to carry and use than cloud-based EHR. As an example, in acute healthcare situations, patients are often physically or mentally unable to interact with a cloud-based interface - hence the success of medic-alert type bracelets.

Loss of access to the Internet renders all cloud-based applications useless. A portable, distributed, expansive, efficient, and economic smart card health system would be a much better solution to natural and man-made disasters that require evacuations of patients and their medical records (Hartocollis & Bernstein, 2012). In such emergencies, health providers could use the patient's smart card with essential health records and a portable card reader connected with or embedded in a laptop, which can be standard equipment in every ambulance. A solar adaptor charger would provide power for the laptop. One other technology that may address these problems is self-service kiosks located in physicians' offices or hospitals that may be utilized with or without smart cards.

Touch-Screen Self-Service Kiosks

Because it takes time for patients to answer questions at clinics or hospitals, and repeat information to different healthcare workers, providers may add self-service kiosks. Self-service health kiosks increase capacity of healthcare workers and avoid data entry errors. Kiosks can also be used for health education while patients wait for service. Kiosks can interface with smart cards, bar codes, and magnetic strip cards. Kiosks are more expensive than smart cards alone, but allow patients to view and possibly print their own health records, or even to potentially update certain data. Required primary care questions about a patient's pain or mental state can be answered through a kiosk at the hospital or clinic before the patient sees the nurse or physician. In fact, one study finds evidence that socially difficult questions may be answered more truthfully when patients are responding to a computer (Webb et al., 1999).

Kiosks are already in use in many locations in the U.S. Blue Cross/Blue Shield of Massachusetts began using on-site multimedia kiosks in 1995 to allow users to access information about insurance services and health care and medical issues (Raghupathi & Tan, 2002). The U.S. Department of Veterans Affairs (VA) developed a kiosk-based system for patients at mental health facilities who lack cognitive or computer skills to complete online self-assessments. It combines audio cues with a web-based interface. The collected data converts to Health Level 7 (HL7) international format for use overseas (U.S. Department of Veterans Affairs, 2010). For patients of the VA, EHR can be accessed through the open source system called V*ist*A (VA, 2010; Mosquera, 2010), which could later add an interface with smart cards. The Heritage Valley Health System in Beaver, PA developed and implemented a patient registration system with bar-code access Care Cards and 40 kiosks in 2006, which reduces the average check-in time from 10 to 2 minutes and waiting time from 37 to 24 minutes (AHRQ Health Care, 2006).

Some providers of kiosks are entering the Health IT systems market. DynaTouch markets kiosks for $15,000-$24,000 depending on the degree of customization required and number of kiosk stations needed. DynaTouch also works with a software integration company, SCI Solutions, to program the communications protocol between the kiosk and an EHR system. An alternative provider of kiosk and printer equipment is Zebra Technologies, who partners with Optical Phusion. For software needs, the system interfaces among smart cards, card readers, kiosks, and EHRs need to be integrated and managed in a functional and secure manner. The Health Level 7 (HL-7) International is the global authority on standards for interoperability of health IT in over 55 countries (HL7, 2013).

Software designers can address interoperability issues among EHR systems. In the U.S., all Office of the National Coordinator (ONC) Health Information Technology certified EHR/EMR products can be found online through its website (ONC, 2012). The U.S. Veteran Administration's WorldVista EHR 2.0 is open source and complete for both ambulatory and inpatient practice settings. Another software source is OpenVista CareVue 1.5 by Medsphere Systems, which is a modular EHR. Clinicians designed this free, open-source EHR software that became very popular for use in some American community hospitals and in other countries around the world, e.g., in the United Kingdom and Taiwan (Open Health News, 2013a, 2013b; Cruickshank, 2013; Liu, 2011; Smart Card Alliance, 2011). In March 2013, the HL7 International decided to license their primary standards and other selected intellectual properties at no charge (HL7, 2013). Hopefully, this decision can promote HL7 standards worldwide and reduce interoperability problems in healthcare in the near future (Ouellette, 2012).

Privacy HIPAA Issues

Kiosks, smart cards, and the systems that interact with them and their environments must maintain privacy standards according to the Health Insurance Portability and Accountability Act (HIPAA) and other applicable organizational, state and federal regulations. A smart card and kiosk system should generate information analogous to Personal Health Records (PHR) that is more accurately maintained. We must ensure that the HIPAA "privacy Rule still regulates how an individual's health information held by a HIPAA covered entity enters the PHR" (HHS, 2012). This is also true for smart card and self-service kiosk technology.

Solutions and Recommendations

The combination of smart cards and kiosks is an important innovation for medical care. The technology can reduce complexities in intake and insurance reimbursement processes, and can enhance patient compliance with recommended therapies. Electronic registration can improve patient access to care, clarify patient financial responsibility, and avoid provider reimbursement errors. The use of smart cards and kiosks can increase access to health care for emergency patients, and those unable to speak for themselves because of language problems or mental incapacity. Smart cards may mitigate the effects of demographic differences possibly inherent in traditional registration procedures. For therapies, smart card access could help patients track blood sugar or blood pressure, symptom mapping, diet and exercise, or compliance with testing and specialist visits. Overall, this technology should improve health services utilization.

FUTURE RESEARCH DIRECTIONS

Smart card utilization studies may discover whether health care efficiency, effectiveness and costs improve with this technology. Researchers may conduct empirical research about improvements in access to care, coordination of care, or assurance of privacy and security. Already, some international studies empirically examined privacy and security concerns with smart cards and found that citizens are concerned about government surveillance of personal identifying information (Martin & Rice, 2010; Rosli et al., 2009). Simultaneously, the researchers also cite many studies that found a substantial reduction in identity theft and fraud with smart card technology. The U.S. Department of Health & Human Services regularly calls for grant proposals to study health care information technology and its impact.

CONCLUSION

This article focuses on the use of smart card technology to enhance the flow of patient information and critical data to health care providers. We describe uses of the technology and compare smart cards to cloud-based EHR systems, and explain the interaction between smart cards and self-service kiosks. We consider issues and solutions that may arise with use of this technology. Ultimately, with careful attention to engendering public confidence and trust, smart card technology may proliferate in positive ways in the U.S. health care system, similar to its widespread use overseas.

REFERENCES

Akamai (2012). The state of the Internet 2nd Quarter, *2012 Report, 5*(2). Retrieved November 13, 2012 from http://www.akamai.com/stateoftheInternet/

Cruickshank, J. (2013). *A paperless NHS? The NHS is swapping best practice and ideas with the US Veteran Health Administration to make it happen.* Retrieved April 9, 2013 from http://www.2020health.org/2020health/Press/Press-Releases-2013/Making-Connections.html

Frost & Sullivan. (2012). Electronically enhanced healthcare fosters improved HC delivery in EMEA countries. *PR Newswire*. Retrieved February 1, 2014 from http://www.prnewswire.com/news-releases/frost--sullivan-electronically-enhanced-healthcare-fosters-improved-hc-delivery-in-emea-countries-151098525.html

Garrett, P., & Seidman, J. (2011). EMR vs. EHR – what is the difference? [Blog]. Retrieved February 1, 2014 from http://www.healthit.gov/buzz-blog/electronic-health-and-medical-records/emr-vs-ehr-difference/

Gemalto (2012). *Gematlto to issue an additional 15 million eHealth cards in 2012 for AOK insures.* Retrieved April 10, 2013 from http://www.smartcardalliance.org/articles/2012/03/20/gemalto-to-issue-an-additional-15-million-ehealth-cards-in-2012-for-aok-insurees

GlobeNewswire. (2013). Identive smart card readers selected to support Middle Eastern national ID program. Retrieved February 1, 2014 from http://www.identive-group.com/en/our-company/press-releases/2013/identive-smart-card-readers-selected-to-support-middle-eastern-national-id-program

HL7 International. (2013). HL7's standards licensed at no cost. Retrieved April 9, 2013 from http://www.hl7.org/implement/standards/nocost.cfm?ref=nav

Hartocollis, A., & Bernstein, N. (2012). At Bellevue, a Desperate Fight to Ensure the Patients' Safety. *New York Times*. Retrieved November 13, 2012 from http://www.nytimes.com/2012/11/02/nyregion/at-bellevue-a-desperate-fight-to-ensure-the-patients-safety.html?pagewanted=all

Haughton, J. A. (2012). Perspective on: Cloud-based modular EHRs offer advantages for meaningful use. *HIMMS News*. Retrieved February 1, 2014 from http://www.himss.org/News/NewsDetail.aspx?ItemNumber=4052

Health Care Innovations Exchange, A. H. R. Q. (2006). *Technology-enabled streamlining of hospital registration process reduces check-in times, waiting times, and costs.* Retrieved August, 15, 2012 from http://www.innovations.ahrq.gov/content.aspx?id=1862

Heisler, M., Tierney, E., Ackermann, R. T., Tseng, C., Narayan, K. M., Crosson, J., & ... (2009). Physicians' participatory decision-making and quality of diabetes care processes and outcomes: results from the triad study. *Chronic Illness, 5*(3), 165–176. doi:10.1177/1742395309339258 PMID:19675116

Hsiao, C., Hing, E., Socey, T.C., & Cai, B. (2011). Electronic health record systems and intent to apply for meaningful use incentives among office-based physician practices: United States, 2001 – 2011. *NCHS Data Brief, 79.*

Hussain, H., Kehl, D., Lennett, B., Li, C., & Lucey, P. (2012). The cost of connectivity. Retrieved November 13, 2012 from http://newamerica.net/publications/policy/the_cost_of_connectivity

Institute of Medicine (IOM). (2011). *Health IT and patient safety: Building safer systems for better care. Committee on Patient Safety and Health Information Technology.* Washington, DC: National Academies Press.

Jian, W. S., Wen, H. C., Scholl, J., Shabbir, S. A., Lee, P., Hsu, C. Y., & Li, Y. C. (2011). The Taiwanese method for providing patients data from multiple hospital EHR systems. *Journal of Biomedical Informatics, 44*(2), 326–332. doi:10.1016/j.jbi.2010.11.004 PMID:21118726

Knight, W. (2001). Malaysia pioneers smart cards with fingerprint data. Retrieved March 15, 2013 from http://www.newscientist.com/article/dn1331

Krist, A., & Woolf, S. (2011). A Vision for Patient-Centered Health Information Systems. *Journal of the American Medical Association, 305*(3), 300–301. doi:10.1001/jama.2010.2011 PMID:21245186

Lewis, T., Synowiec, C., Lagomarsino, G., & Schweitzer, J. (2012). E-health in low- and middle-income countries: Findings from the Center for Health Market Innovations. *Bulletin of the World Health Organization.* Retrieved April 12, 2012 from http://www.who.int/bulletin/volumes/90/5/11-099820/en/index.html#

Li, Y.C. (2012). Taiwan HIT case study. *The National Bureau of Asian Research.* Retrieved August 15, 2012 from http://www.pacifichealthsummit.org/downloads/hitcasestudies/economy/taiwanhit.pdf

Linxens. (2012). Linxens e-government. Retrieved March 15, 2013 from http://www.linxens.com/products/smartcards/applications/e-government

Liu, C. T. (2011). Development of an interoperability infrastructure for exchange of electronic health records among hospitals in Taiwan. In Proceedings of the 12th International HL7 Interoperability Conference. Disney World, Florida. Retrieved April 9, 2013 from http://www.hl7.org/events/ihic2011/papers/saturday/S_Q1_3_20110514_IHIC_LIU_EHR_Infrastructure.pdf

Martin, N. J., & Rice, J. L. (2010). Building better government IT: Understanding community beliefs and attitudes towards smart card technologies. *Behaviour & Information Technology, 29*(4), 433–444. doi:10.1080/01449291003767938

Mosquera, M. (2010). VA working on 'AViVA' – next generation of VistA. Retrieved March 11, 2010 from http://www.healthcareitnews.com/news/va-working-aviva-next-generation-vista

National Quality Forum (NQF). (2011). U.S. Department of Health & Human Services. Retrieved February 12, 2012 from http://www.qualityforum.org/Events/Webinars/Patient_Safety_Webinars/Patient_Safety_Webinars.aspx

Office of the National Coordinator for Health Information Technology (ONC). (2012). Retrieved February 12, 2012 from http://onc-chpl.force.com/ehrcert

Open Health News. (2013a). Petition to have DoD adopt the EHR used by the VA posted on the White House web site. Retrieved April 9, 2013 from http://www.openhealthnews.com/hotnews/petition-have-dod-adopt-ehr-used-va-posted-white-house-web-site

Open Health News. (2013b). *Open source VistA system offers major benefits for small community & critical access hospitals (CAH).* Retrieved April 9, 2013 from http://www.openhealthnews.com/hotnews/open-source-vista-system-offers-major-benefits-small-community-critical-access-hospitals-cah

Ortiz, C. E. (2003). *An introduction to Java card technology – Part 1.* Retrieved April 10, 2013 from http://www.oracle.com/technetwork/java/javacard/javacard1-139251.html

Ouellette, P. (2012). *Will free HL7 standards stifle interoperability issues?* Retrieved April 9, 2013 from http://ehrintelligence.com/2012/09/04/will-free-hl7-standards-stifle-interoperability-issues/

Polack. P. J. (2012). *Pros and cons of cloud based or web based EMR systems.* Retrieved August 15, 2012 from http://www.kevinmd.com/blog/2011/05/pros-cons-cloud-based-web-based-emr-systems.html

Raghupathi, W., & Tan, J. (2002). Strategic IT applications in health care. *Communications of the ACM, 45*(12), 56–61. doi:10.1145/585597.585602 PMID:12238525

Reid, T. R. (2008). *Taiwan takes fast track to universal health care.* Retrieved March 28, 2013 from http://www.npr.org/templates/story/story.php?storyId=89651916

Reid, T. R. (2010). *The healing of America: A global quest for better, cheaper, and fairer health care.* New York: Penguin Books.

Rosli, R. M., Taylor, D. M., Knott, J. C., Das, A., & Dent, A. W. (2009). Health smart cards: Differing perceptions of emergency department patients and staff. *Australian Health Review, 33*(1), 136–143. doi:10.1071/AH090136 PMID:19203343

Sembritzki, J. (2004). Use and development of health cards in Europe. In *Proceedings of the IFHRO congress & AHIMA Convention.* Retrieved April 2, 2013 from http://library.ahima.org/xpedio/groups/public/documents/ahima/bok3_005516.hcsp?dDocName=bok3_005516

Shelfer, K. M., & Procaccino, J. D. (2002). Smart card evolution. *Communications of the ACM, 45*(7), 83–88. doi:10.1145/514236.514239

Smart Card Alliance. (2003). Contactless payment and the retail point of sale: Applications, technologies and transaction models.*Smart card alliance report*. Retrieved April 11, 2013 from http://www.smartcardalliance.org/resources/lib/Contactless_Pmt_Report.pdf

Smart Card Alliance. (2006). *Sesam Vitale*. Retrieved March 28, 2013 from http://www.smartcardalliance.org/resources/pdf/Sesam_Vitale.pdf

Smart Card Alliance. (2011). *Smart card and biometrics*. Retrieved February 12, 2012 from http://www.smartcardalliance.org/pages/publications-smart-cards-and-biometrics

Smart Card Alliance. (2012). \Publication Number: HCC-12002. Retrieved March 28, 2013 from http://www.smartcardalliance.org/resources/pdf/Smart_Card_Technology_in_Healthcare_FAQ_FINAL_096012.pdf

Smart Card Tutorial. (1992). Retrieved March 30, 2013 from http://www.smartcard.co.uk/tutorials/sct-itsc.pdf

Stacey, D., Bennett, C. L., Barry, M. J., Col, N. F., Eden, K. B., Holmes-Rovner, … Thomson, R. (2011). Decision aids for people facing health treatment or screening decisions. *Cochrane Consumers and Communication Group*.doi: 10.1002/14651858.CD001431.pub4

U.S. Department of Health & Human Services (HHS). (2010). *Secretary Sebelius announces final rules to support 'meaningful use' of electronic health records*. Retrieved February 1, 2014 from http://www.cms.gov/Newsroom/MediaReleaseDatabase/Press-Releases/2010-Press-Releases-Items/2010-07-13.html

U.S. Department of Veterans Affairs (VA). (2010). *Veterans health information systems and technology architecture (VistA)*. Retrieved February 12, 2012 from V*ist*A related links: http://www.va.gov/vista_monograph/

Vezina, K. (2011). Medical Smart Cards find their niche. *MIT Technology Review*. Retrieved from http://www.technologyreview.com/printer_friendly_article.aspx?id=37773

Webb, P. M., Zimet, G. D., Fortenberry, J. D., & Blythe, M. J. (1999). Comparability of a computer-assisted versus written method for collecting health behavior information from adolescents. *The Journal of Adolescent Health*, *24*, 383–388. doi:10.1016/S1054-139X(99)00005-1 PMID:10401965

World Health Report (WHR). (2000). *Health systems: improving performance*. Switzerland: World Health Organization. Retrieved March 29, 2013 from http://www.who.int/whr/2000/en/

Xiao, C., & Yu, A. (2009). Medical Smart Card System for Patient Record Management. *Bears Breaking Boundaries 2009 Science, Technology, and Energy Policy White Paper Competition*. Retrieved March 26, 2013 from http://step.berkeley.edu/White_Paper/Xiao_Yu.pdf

ADDITIONAL READING

deSilva, P. (2008) Time for an NHS Smart Card? *British Medical Journal, 336*(7634), (January 5).

Fong, S., Zhuang, Y. (2012). Using medical history embedded in biometrics medical card for user identity authentication: data representation by AVT hierarchical data tree. *Journal of Biomedicine Biotechnology,* published online April 4.

Horowitz, B. T. (2012). Smart card use surging in health care, government. Retrieved April 8, 2013 from http://www.eweek.com/c/a/Health-Care-IT/Smart-Card-Use-Surging-in-Health-Care-Government-698623/

Jimison, H., Gorman, P., Woods, S., et al. (2008). Barriers and drivers of health information technology use for the elderly, chronically ill and underserved. Rockville (MD): Agency for Healthcare Research and Quality; 2008. Evidence Report/Technology Assessment No. 175. AHRQ Publication No. 09-E004.

Karsh, B.-T., Weinger, M., Abbott, P., & Wears, R. L. (2010). Health information technology: fallacies and sober realities. *Journal of the American Medical Informatics Association, 17*(6), 617–623. doi:10.1136/jamia.2010.005637 PMID:20962121

Mantas, G., Lymberopoulos, D., & Komninos, N. (2012). PKI security in large-scale health care networks. *Journal of Medical Systems, 36*(3), 1107–1116. doi:10.100710916-010-9573-1 PMID:20827570

Rhoads, J., & Drazen, E. (2009). Touchscreen check-in: kiosks speed hospital registration. Retrieved November 1, 2012 from http://www.chcf.org/~/media/MEDIA%20LIBRARY%20Files/PDF/T/PDF%20TouchscreenCheckInKiosks.pdf

Spongenberg, H. (2010). Hurdles confront e-health across Europe. Retrieved March 31, 2013 from http://euobserver.com/news/29445

Stead, W., & Lin, H. (Eds.). (2009). *Computational technology for effective health care: immediate steps and strategic directions*. Washington, D.C.: National Academies Press.

Svigals, J. (2012). The long life and imminent death of the mag-stripe card. *Spectrum Magazine, IEEE* (June).

Terhune, C. (2012). Patient data outage exposes risks of electronic medical records. Retrieved August 15, 2012 from http://www.latimes.com/business/la-fi-hospital-data-outage-20120803,0,5302779.story

Terry, K. (2011). Smart card plan proposed to combat Medicare fraud. Retrieved October 5, 2012 from http://www.fiercehealthit.com/story/smart-card-plan-proposed-combat-medicare-fraud/2011-09-14

Versel, N. (2009). What would you do if an EMR goes down? Retrieved November 13, 2012 from http://www.fierceemr.com/story/what-would-you-do-if-emr-goes-down/2009-11-19

Wilson, C., & Peterson, A. (2010). Managing personal health information: an action agenda. Rockville (MD): Agency for Healthcare Research and Quality; 2010. AHRQ Publication No. 10-0048-EF.

KEY TERMS AND DEFINITIONS

Cloud-Based Electronic Health Records (EHR): The electronic health records stored and transferred over the Internet.

Electronic Health Records (EHR): The digital data through network connections containing patient medical history and demographic information.

Electronic Medical Records (EMR): The digital data located at a provider's location that contain patient medical history and demographic information.

Health Information Exchange (HIE): The movement of electronic patient data across health care network systems.

Health Information Technology (HIT): Electronic systems designed to manage patient health data and its exchange across health care providers.

Health Insurance Portability and Accountability Act (HIPAA): 1996 U.S. regulation to protect security and privacy of patient health data.

HL7 International Format: Health Level Seven® International (HL7): the international standards for interoperability of health IT, which includes the integration, exchange, sharing, and retrieval of electronic health information. Messaging standards define data exchange protocols for packaging and communicating information.

Personal Health Record (PHR): A patient's collection of individual medical data on paper, electronically, or web-based records.

Smart Cards: A credit card-like device that holds data on electronic strip (unidirectional) or processes data through a microchip (bidirectional).

This research was previously published in the Encyclopedia of Information Science and Technology, Third Edition edited by Mehdi Khosrow-Pour, pages 3458-3467, copyright year 2015 by Information Science Reference (an imprint of IGI Global).

Chapter 6
Patient Accessible EHR is Controversial:
Lack of Knowledge and Diverse Perceptions Among Professions

Isabella Scandurra
Orebro University, Sweden

Anette Jansson
Region Orebro County, Sweden

Marie-Louise Forsberg-Fransson
Region Orebro County, Sweden

Ture Ålander
Uppsala University, Sweden

ABSTRACT

In Sweden, a national eHealth service providing Patient Accessible Electronic Health Records is now being widely deployed, with 400 000 users in January 2016. Although the Patient Data Act states that patients have a right to take part of their health records, the introduction has been controversial. Results from a pre-deployment questionnaire to record-keeping care professions in a healthcare region indicate that perceptions and knowledge differ not only between the professions but, more importantly, that knowledge about current eHealth development and action plans needs to increase as implementation will affect their work processes. Staff perceptions and knowledge are considered being some of the most important issues to handle during the implementation of eHealth services aiming to provide healthcare information and communication tools for patients and relatives. To cover the gaps, specific training is needed, and all record-keeping professionals need to be more involved in the implementation of such eHealth services.

DOI: 10.4018/978-1-5225-6198-9.ch006

INTRODUCTION

The number of eHealth services for patients and relatives is rapidly increasing as many countries are launching such services as a means to manage an ageing population, to increase efficiency in healthcare, and to empower patients. One of these services is the patients' online access to their own EHR (Delbanco, Walker, Darer, Elmore & Feldman 2010). Patient accessible electronic health records (PAEHRs) enable patients to access and sometimes also manage personal health information that is made available to them by their healthcare providers. It is thought that the shared management nature of medical record access improves patient outcomes and patient satisfaction (Jilka, Callahan, Sevdalis, Mayer, & Darzi, 2015). However, direct access to a health record of a patient is not seen as entirely positive. In the current debate, a discrepancy is noted between legislation, policies and decision ambitions versus the care professionals' preferences and knowledge regarding the issue.

In the case of the Swedish eHealth service "Journalen", a national study points out that healthcare professionals did not favor the reform (Scandurra, Jansson, Forsberg-Fransson & Ålander, 2015). They expressed concerns in relation to their work environment (e.g. anticipation of increased workload), that patients would mostly be worried by reading their health records and that such service would be of little value for the patient (Grünloh, Cajander & Myreteg, 2016, Scandurra et al., 2015). Consequently, the adoption of these services by healthcare professionals have often not been as successful as anticipated by politicians and vendors (de Luisignan et al., 2013) and to date, services aiming to provide healthcare information and communication tools for patients and relatives do not always provide the expected effect (Erlingsdottir & Lindholm, 2014).

Patients, on the other side, want to communicate and interact more with their healthcare provider, which can be facilitated and streamlined by information technology (Pagliari, Shand & Fischer, 2012). Patients also often strive to manage their own health and illnesses and want to find ways to get copies of their electronic health records (EHRs); to share with their physicians, load into apps, submit to researchers, link to their genomic data, or have it on hand just in case (Mandl & Kohane, 2016).

Aware of the discrepancy of ideas and the controversy during the development in the Swedish pilot county (Lyttkens, 2015) this paper aims to investigate knowledge and perceptions among different healthcare professions in a region that not yet has deployed the service. The outcome of this study may also support practice in targeting activities to the staff in order to facilitate the implementation and acceptance of the service.

Research Rationale

Physicians have traditionally regarded the health record as a tool for themselves and other healthcare professionals (Grühnloh et al., 2016), while patient representatives have argued that patients should own their own information (Docteur & Coulter 2012). Docteur and Coulter (2012) reported that patients had a difficulty in getting access to their own health records, due to records not being available electronically, and that healthcare providers could be slow and reluctant in responding to patient requests for disclosure of record copies (Docteur & Coulter, 2012, p. 74).

This handling is opposite towards the Swedish legislation: Patients have the right to take part of their health records, dating back more than 60 years, and information shall be submitted as soon as possible Moreover, health records should be written in the Swedish language and clearly designed so they are understandable to patients (Patient Data Act, 2008). Another study examined the approach of physicians to documentation (Allvin & Kvist, 2011). One-third of 139 physicians responded to the survey. All but one respondent replied that they wrote the documentation "to a very large extent" to their colleagues. Regarding writing to the patient, the same amount of physicians replied "to a very low extent" as "to a very high extent." The article concluded e.g. that physicians rarely or never adapt record documentation to the patient (Allvin & Kvist, 2011).

The controversy of PAEHRs has been discussed from different research perspectives, sometimes based on organizational theory. Being a multidisciplinary arena and a complex "professional organization", other players have difficulties to control the clinical professionals (Hallin & Siverbo, 2003, p. 187). This is partly due to the fact that appreciation of the colleagues is more fundamental to the professional than to get an acceptance by management (Thylefors, 2007, p. 53-54). When the eHealth services are to be implemented in the healthcare context, organizational theory by e.g. Mintzberg (1992) and Söderström (2005) regarding clarification on how multi-cultural professional organizations work, could provide guidance in how the implementation should be performed. To gain knowledge of further implementation, preferences and knowledge of healthcare professionals in this pre-deployment study are discussed in the light of these concepts.

DEVELOPMENT AND DEPLOYMENT OF PAEHRS

Internationally, there is a drive towards providing patient eHealth services as PAEHRs, but it has been limited in part by professional resistance and concerns about security and privacy (Mandl, Szolovits & Kohane, 2001; Wiljer et al., 2008), legal constraints (Tiik, 2010) and low uptake of other online resources for patients. In a systematic review from 2014, a lack of evidence from high-quality studies about the impact of online access was noticed, yet it was clear that the tensions between the growing consumer demand to access data and a healthcare system not yet ready to meet these demands have increased in recent years (de Lusignan et al., 2014). Many of the studies identified in the review originated from the USA, from large health plan-based programmes, whereas a minority of studies originated from Europe. The review indicated that online patient access to their EHR and other services offer increased convenience and satisfaction (de Lusignan et al., 2014).

However, professionals were often concerned about the impact on workload and risk to privacy, which was also the case in Sweden (Grünloh et al., 2016; Scandurra et al., 2015). The authors of the review article concluded that there might need to be a redesign of the business process to engage health professionals in online access and of the PAEHR in making it easier to use and provide equity of access to a wider group of patients. Another review published in 2015 called for more empirical testing regarding the effect of PAEHRs on health outcomes for patients and healthcare providers (Jilka et al., 2015).

National PAEHRs in the Northern European Countries

The conditions for investments in public eHealth services are good in the Northern European countries, and the Scandinavian countries are leading innovators in the eHealth services. Regarding the introduction of region-wide EHR systems, Sweden has reached far; county councils have introduced EHRs to 100% of hospitals, primary care, and psychiatry. In Europe, the corresponding figures are 65% in total and for hospitals alone, 81% (CeHIS, 2012).

A communication from the European Commission in 2010 made clear that the objective was that all Europeans should have access to their health records online by 2020. In "A Digital Agenda for Europe" the European Commission stated in Key Action 13 to: "Undertake pilot actions to equip Europeans with secure online access to their medical health data by 2015 and to achieve by 2020 widespread deployment of telemedicine services" (European Commission 2010, p. 29) which also includes PAEHRs (European Commission, 2012).

Adhering to the eHealth Action Plan of the European Commission (2012), all Nordic countries and a number of other European countries (e.g. Estonia, Spain, the Netherlands) already have national services for patient access to the EHR. Important to note is that accomplishment of the Health Action Plan takes different expressions: the deployment differs from the pilot or regional stage to broad national implementations. The services also show different data sets, and decision-makers have made different decisions regarding e.g. from which date they show health information to the patient, which could make use for the mobile patient a bit complicated (Scandurra, Lyttkens & Eklund, 2016).

Denmark is prominent in promoting the use of ICT in healthcare. Already in 2008 parts of the National eHealth strategy in Denmark was realized by creating direct access to healthcare services for citizens. The aim was also to improve communication between healthcare professionals. Currently, the Danish national healthcare portal (sundhed.dk) provides citizens with access to healthcare information from public hospitals, doctor's notes, lab results, prescription list, logs, consents, registration of donations and wills. (Sundhed, 2008). Based on the strategy that regards 2013-2019 new tools are developed, e.g. mobile apps and telehealth services for patients (Sundhed, 2015).

In Sweden, the National eHealth Strategy states that access to personal health data should be given via Internet. One of the objectives of the National Strategy for eHealth is that "ICT should be used as a tool to create a benefit for citizens, healthcare staff, and decision makers". (Ministry of Health and Social Affairs, 2010) For the individual, this means for example access to quality-assured information about health. In addition, the individual should have access to documentation from previous visits and treatments, and should be offered customized service and interactive eHealth services to exercise participation and self-determination (CeHis, 2012).

Gradual Deployment and Increase in Use of the Swedish PAEHR

Already in 1997, Uppsala County Council in Sweden started a project with the aim to give patients access to their medical data (Eklund & Joustra-Enquist, 2004). The fast digital development and experiences from the Sustains project pushed forward a change of the Swedish legislation in 2008 (Patient Data Act, 2008), which permitted healthcare organizations to give patients direct access to their EHRs including laboratory values and the doctor's notes. As part of the EU-project Sustains (Lyttkens, 2015) during 2012-2014, Uppsala County Council extended the deployment of public eHealth services to a national pilot. In November 2012, the 350 000 inhabitants were given access to their health records through the National Internet patient portal "1177.se, the Healthcare Guide" (www.1177.se) and two years later, in November 2014, almost 60 000 patients had used the Uppsala PAEHR (Inera, 2015).

The PAEHR is an important realization of the National eHealth Strategy and from 2015 the national eHealth organization (Inera) has taken the responsibility for the public eHealth service. The Action Plan for 2013-2018 (CeHis, 2012) building upon the eHealth Strategy states that "all citizens should have access to the entire health record online by 2017." To follow this line, all regions are committed to providing their citizens online access to their electronic health record (EHR) as part of the public eHealth service (CeHis, 2012), although they all have different implementation plans (Scandurra et al., 2016). In June 2016 16 of 21 counties upload parts of their EHRs to the national PAEHR service (Inera, 2016). Other public eHealth services in which the citizens themselves can ask questions and receive answers, order and renew prescriptions, manage their own appointments online, etc. are also being developed and deployed (Lyttkens, 2015). Currently, almost 3 100 000 citizens (30,4% of the Swedish population) have chosen to create and administer their own account on the patient portal 1177.se, and the PAEHR has approximately 400 000 active users (Inera, 2016).

Regional PAEHRs in the United Kingdom and the United States of America

In comparison with the Nordic countries, deployment of PAEHRs in the United Kingdom (UK) and the USA regards healthcare regions rather than the entire country.

The United Kingdom (UK) has the policy to enable citizens to access their EHRs (Fischer, 2010). An article from the UK presented several studies where having access to patients' health records resulted in improved self-care, better adherence to treatment in chronic diseases and improved relationship between physician and patient (Bhavnani, Fisher, Winfield & Seed, 2011). Their own study also revealed that patients experienced increased security, increased confidence and trust in the care when they had access to their health records. Approximately 33% (n = 56/169) of the respondents reported difficulties in understanding their records, and 63 patients (n = 63/178, 38%) stated that they found errors. The errors consisted of improper documentation of medical events, the absence of documentation of procedures and incorrect information about allergies and health conditions (Bhavnani et al., 2011). The results should be interpreted according to a response rate of 35% (213 responses of a total of 610.)

Another study from the UK indicated increased efficiency when patients interacted with health professionals via direct access to detailed electronic medical records. Most professionals (physicians, administrators) and patients considered that the service had increased the mutual trust and communication between patients and professionals. The patients' knowledge about health and health behavior were also improved. Negative factors reported were concerns about safety, responsibility and demands of increased resources (Pagliari et al., 2012).

In an American qualitative study regarding Personal Health Records (PHRs) where the patient is coproducer of his record, 21 people were interviewed, from various professions in healthcare, regarding their perception about PHRs (containing information about medications, test results, exercise and diet, created and managed by the patient, sometimes in combination with the patient's healthcare provider) (Huba & Zhang, 2012). Results showed that PHRs were supported by several parties such as physicians, private healthcare providers and policy makers. However, there was also a lack of full understanding of the different professions in healthcare regarding PHRs and how PHRs could be used in healthcare today.

The majority of the professionals considered it would give them valuable information from patients and would recommend the patients to enter information in their own PHR. The study also stated that information usually provided by the patient is needed, in order to effectively treat a patient. Patients in the study wanted to take part of their EHRs and wanted to have an active role in the shaping of personal health information (Huba & Zhang, 2012).

METHODS

This study emanates from 2012 in Örebro County Council, where a survey was carried out regarding staff knowledge of privacy, logging, the patient's right to receive part of the logs and to publish those as an eHealth service on the healthcare portal (Forsberg-Fransson, Jansson & Jensen, 2012). 364 people responded the survey, whose result demonstrated that there was a large need for information and training regarding privacy, logging, and control of log files. The respondents had inadequate knowledge about new legislation and regulations, particularly regarding logging. Further, there was a connection between knowledge and perceptions: staff who had knowledge of legislation and regulations were more willing to make the log excerpts readily available to patients via the eHealth service (Forsberg-Fransson et al., 2012). A conclusion based on these findings was that it is important to continue studying preferences, perception and knowledge of the care professionals, in order to among other things facilitate implementation of new work routines and procedures. The study presented in this article is part of a larger context, the national research consortium DOME, Deployment of Online Medical Records and eHealth services (DOME, 2016) and the study domain was chosen was the Region Örebro County again. Opposite to many other studies that contain perceptions of physicians only, this study wanted to include all registered professions in healthcare that shall document in a health record.

Research Question

This pre-deployment study intended to investigate whether there is a gap between legislation as well as what decision-makers want regarding PAEHR versus knowledge and preferences of the healthcare staff. Sub-questions to build the questionnaire regarded:

- Staff acknowledgement about ownership of the health record, national eHealth strategy and action plans for implementation
- Perceptions of the healthcare staff regarding the PAEHR functionality of e.g. giving direct access versus using a respite time
- Potential differences regarding whether the PAEHR is a good eHealth service
    - within and between different professions
    - depending on work experience
    - within and between different workplaces
- Potential impact on the work practice

Data Collection

Primary data collection of the pre-deployment study was based on responses from a questionnaire proportionately distributed to 100 respondents via email. The 100 respondents, permanent and temporary employees of the Region Örebro County, were randomly selected among 6919 employees operating in five selected departments with a legal demand to keep records. Regarding age and gender, the demography of the study was similar to the population of employees. The survey was sent to employees from various professions such as physicians, nurses/midwives, physiotherapists, occupational therapists, medical

Table 1. Disposition of the questionnaire

Content	No of Questions	Question ID of Closed Response	Likert-Style Scale	Free Text Field
Attribute questions	6	1, 2, 3, 4, 5, 6		
Knowledge questions	4	8, 9, 10, 11		
Perception statements	8	16, 17	7, 12,13, 14, 15, 21	
Behavioural questions	2	19, 20		
Own perceptions	2			18, 22

social workers, speech therapists and medical secretaries, which all work daily with documentation in the EHR. Secondary data consisted of legislation, articles, theses, books and other reading materials.

Although this study should be carried out as a pilot with only a small number of employees, the proportional stratified sample was chosen as it provided a greater opportunity to get a breadth of views from the various professional categories. This may lead to a more valid full-scale survey as every stratum (subgroup) will be represented by its exact share of the population. To validate the questionnaire, five people were asked to complete it and the questions were approved by the respondents. The pre-deployment survey was conducted in October 2013 through an internal web survey tool, esMaker, version NX2, v2.0 (Entergate, 2013), in the Region Örebro County. This was prior to any deployment of PAEHR in the region, but a year after the first implementation in another county (Uppsala). Reminders were automatically sent three times after the first mailing on a three days' interval.

Data Analysis

Using esMaker, the questionnaire was first created and customized concerning e.g. the anonymity of respondents and later data was analyzed using the same tool. Ethical approval was conducted in accordance with the principles of the Declaration of Helsinki (World Medical Association, 2008). The study also had the approval of the Director of Region Örebro County. Descriptive statistics was used processing the results as well as free text field reporting for two of the statements used.

In the questionnaire, closed response questions created a majority of the response types (19 of 22 questions) of which five questions (7, 12,13, 14, 15, 21) used a Likert-style ranging from "very low extent" to "very high extent", plus "do not know" and "not applicable". The statements used in this study were inspired by other statements from surveys directed towards registered nurses and physicians performed by the DOME consortium e.g. (Scandurra et al., 2015). The Cronbach alpha for the questionnaire design in those surveys was calculated, and many statements were identical in order to more easily compare respondents or populations. The questionnaire was constructed with different types of reply options to maintain the motivation of the respondents, and a possibility to provide your own perception in free text (Table 1).

In the missive, the respondents were informed that participation in the study was voluntary and that all information was treated confidentially, as well as that results of this survey can lead to improvements in their current work situation.

RESULTS

The intention was to test the questionnaire as a cross-sectional survey yielding a result of the current situation in a particular population, but results of this study cannot be generalized. The proportional stratified sample of 100 employees implied an opportunity to obtain input from various professional categories. The response rate was 45% (45 people of which one respondent did not complete due to "too many inquiries"). The external loss was 55 people (55%).

The results are based on the research sub-questions and are presented by descriptive statistics in four sections: staff acknowledgment; perceptions of PAEHR functionality; PAEHR as a good eHealth service; and potential impact of the work as well as demography and quotations.

Demography

The majority of respondents (n=44) in the study were women (82%), and over 45 years (66%). 22 people worked in speciality hospital clinics (50%), in administration; eight people (18%), in primary care; 9 people (20%), and in a psychiatric ward; five people (11%) and one person worked at another workplace (2%). 34/43 of the respondents had a work experience of more than ten years (79%). Half of the respondents were nurses (22/44). All prospective professionals were not represented in the outcomes; midwives and speech-language pathologist were not reached. For a more detailed description of the various occupational categories represented, see Table 2.

Staff Knowledge

Three questions (number 8, 9, 11) regarded the knowledge of eHealth among the respondents:

1. Ownership of the health records. In Sweden, the healthcare provider is the owner of the health record.
2. The national eHealth strategy.

Table 2. Respondents' professions

Profession	Number
Nurse	22
Secretary	7
Physician	5
Physiotherapist	3
Occupational therapist	2
Counselor (Medical social worker)	2
Manager	2
Administrator	1
Total	**44**

3. The national action plans for eHealth implementation. The goals are that patients in Sweden can get access online to parts of their EHRs in 2014 and the entire EHR in 2017. The results are presented in Table 3 and 4.

It was a remarkable number of respondents who replied that the patient is the owner of the health record. No variation between different professions can be seen, i.e. this analysis cannot conclude that any professional group had more knowledge than the others regarding the ownership of the health record.

Regarding the question, whether the respondents had knowledge of the action plans for eHealth implementation e.g. PAEHR, the 9/44 respondents with a positive answer were divided into the following professions: 5 nurses, two secretaries, one physician and one manager.

Question number 10 was an open follow-up question to number 9 where the two respondents (manager and secretary) who knew about the national eHealth strategy could extend their answers.

Perceptions of PAEHR Functionality

It clearly emerged that the respondents found it important that "the notes are signed" before publication in the PAEHR, and that it is important to have a "respite," i.e. time to review and signing/correcting errors, before publication online.

Signed Notes

Regarding the question about providing the patient access _or_ direct access to the notes in the PAEHR, a majority 34/43 found it very important that the notes be signed before publication online. Two respondents answered that it was of some importance, one that it was not important at all. The remaining six did not know.

Respite Time

19/44 (43%) of the respondents found the functionality of "having a respite, or adjustment time, before online publication of a note" to be important 'to a high extent' and 11/44 (25%) to 'a very high extent.' One found it to be of no importance. The other 6 answered 'do not know' or 'not applicable'.

Table 3. Knowledge of ownership of the health record

	Patient	Care Provider (Correct)	Patient and Care Provider	Patient and Responsible Personnel	Patient and Care Provider	Patient, Care Provider and Responsible Personnel
No	18	14	8	2	1	1

Table 4. Respondents' knowledge of national eHealth strategy and action plans for PAEHR

	Yes	No	Total
Knowledge of the national eHealth strategy	2	42	44
Knowledge of action plans for PAEHR	9	35	44

PAEHR is Important for the Patient and Increases the Quality of Care

There was a spectrum of the answers regarding whether "PAEHR is important for the patient" as well as whether the service "increases the quality of care for the patient".

The statement "important for the patient" yielded that almost half of the respondents were positive to the statement. Eleven (11/44) of the respondents agreed to a very high extent (5) with the statement, and seven (7/44) of the respondents agreed to a high extent (4). This can be compared to fourteen (14/44) of the respondents did not agree at all, or agreed to a low extent, i.e. indicated estimation options one and two. The 'Agree to a very high extent' was chosen by six nurses, two secretaries, one physiotherapist, one occupational therapist and one manager, but none of the physicians.

The other statement "increases the quality of care for the patient" yielded that 14/44 chose the mid alternative (3), and 8/44 chose 'do not know'. The other 22 respondents were distributed equally on the side of agree (4, 5) and disagree (1, 2).

A sub-analysis regarding the statements "important for the patient" and "PAEHR as a good eHealth service" revealed that the 11 respondents that agreed to a very high extent regarding the "importance for patients" were less positive to the statement "a good eHealth service": Five respondents agreed to 'a very high extent=5' and five agreed to 'a high extent=4', while one did not agree at all (1).

PAEHR as a Good eHealth Service

Concerning the question of "PAEHR as a good eHealth service", the responses were diverse: 14/44 found the eHealth service providing 'very little'=1 or 'little'=2 value and 17/44 found it being of 'high'=4 or 'very high'=5 value. The rest were of no perception or in between (Table 5).

Differences Due to Profession

Due to the low number of respondents, the results are not generalizable. However, within the nursing group, which was the largest group of respondents (22/44), twelve (12/22) of the respondents said they agree that PAEHR is "a good e-Health Service." No nurse selected the option 'to a very low extent'. Three (3/22) nurses answered that they agreed with the statement 'to a low extent.' In these results there was a variation between different professions:

Table 5. Question 12, 13, 14 and 15: Perceptions of the respondents regarding functionality and effects of the eHealth service; "To which extent do you think the PAEHR"

	Very Low Extent	2	3	4	Very High Extent	Do Not Know	N/A	Missing
"Signed notes" is important	1	-	-	2	34	3	3	1
"Respite time" is important	1	4	2	11	19	4	3	-
Increases the quality of care for patients	5	6	14	7	4	8	-	-
Is important for the patient	7	7	6	7	11	4	2	-
Is a good eHealth service	7	7	2	10	7	8	3	-

Of the seven (7/44) of the respondents who agreed that "PAEHR is a good e-health service" to a high extent, six (6/7) were nurses and one (1/7) secretary.

Twelve of the nurses found PAHER being "a good eHealth service" to 'a high'=4 or 'very high extent'=5 while five nurses answered 'do not know'. One physician responded positively, agreeing that "PAEHR is a good eHealth service" to a 'high extent'=4. Of the seven (7/44) of the respondents who answered that the service is not at all ('to a very low extent'=1) a good eHealth Service, 3 were physicians (of a total of 5). The others were: two (2/7) secretaries, one (1/7) counselor and one (1/7) an occupational therapist (see Table 6).

The nursing group tends to be more favorable to the introduction of PAEHR than other professional groups.

Differences Due to Workplace

No significant difference. Personnel working in psychiatry has reported lower values and expressed some concerns in the free text; see the quote in the section below. Nurses in somatic hospital ward also reported more positive values.

Differences Due to Work Experience

No difference could be seen between personnel with work duration more or less than ten years in their profession.

Potential Impact on the Work Practice

Healthcare staff expects the working conditions to change with the introduction of eHealth services in general and specifically when introducing a service like PAEHR (Jilka et al., 2015; Scandurra et al., 2015; Huba & Zhang, 2012).

Nevertheless, the statement "To which extent do you feel that PAEHR can contribute to increased workloads" resulted in seventeen (17/44) of the respondents answering the middle option (3). Ten (10/44) of the respondents answered that the workload will increase to a high extent, and two (2/44) to a very high extent. A total of ten (10/44) of the respondents did not agree with the statement (4 to a very low extent=1, and six to a low extent=2).

Table 6. Responses by profession to the question "is PAEHR a good eHealth Service?"

Profession	Very Low Extent	2	3	4	Very High Extent	Do Not Know	Not Involved N/A
Physician=5	3	1		1			
Nurse n=22		3	1	6	6	5	1
Secretary n=8	2	2		1	1	1	1
Paramed* n=7	2	1	1	1		2	
Manager n=2				2			

*Paramed = physiotherapist, occupational therapist, counselor. N = 44

The statement "I will have to explain expressions/test results in the record when the record becomes available online," had twenty-five (25/44) of the respondents agreeing to a high and very high extent. Four (4/44) of respondents did not agree at all, or at low levels. Seven (7/44) estimated a third which is a middle option between the extremes and not to a great extent. Four (4/44) of respondents answered 'do not know', and four (4/44) did not answer the question.

Although this block of statements regarding changes yielded a higher number of the fields of 'do not know' and 'not involved,' the results demonstrate that the expected changes are difficult to estimate. Consequently, the statements here get a 'mid option'=3 by eight persons regarding all tasks that inquire a change.

The statement "explain the expression/lab results" yielded that twenty-five respondents (25/44) agreed with the statement 'to a high extent' and 'to a very high extent. Meanwhile only four persons agreed to the statement with a low or a very low extent (see Table 7).

Quotations

The questionnaire provided two possibilities of free text writing, to share the perceptions and the thoughts of the respondents by their own words. Question 18 related to this eHealth service in particular: "If PAEHR is implemented in this region, do you have expectations/concerns?" Question 22 related to the deployment of eHealth services in general: "Any other perception regarding eHealth services?"

Physicians, counselors, and therapists took the opportunity to write comments. In general, respondents who are physicians left more comments in free text, than other professions. 13 respondents expressed their thoughts, of which five can be referred to the administration of information. Here is a quote mirroring their concerns:

If I will write differently: it depends whether it is expected of us to document as usual, or in a way that the patient understands everything. Then it will take infinitely longer time to write notes. This will not benefit the patient because I will spend most of the time to documenting instead of having plenty of time to meet with the patient and have a conversation about his physical limitation and treatment. In the meeting with the patient, there is the opportunity for participation and understanding of the patient regarding their situation and treatment! I can see both pros and cons of having easier access to the health records; we can more easily create notions and problems that would not otherwise occur, but also speed up the information exchange and then reduce the anxiety that is often built up in anticipation.

Table 7. Question 12, 13, 14, 15: Perceptions of the respondents regarding the potential impact on working conditions when introducing PAEHR (n=44); "To which extent do you think the PAEHR will lead to..."

	Very Low Extent	2	3	4	Very High Extent	Do Not Know	Not Involved N/A
Increased workload	4	6	17	10	2	5	
I will have to explain expressions and lab results	1	3	7	11	14	4	4
Changes in my working environment	8	6	8	5	7	6	4
Changes in frequency of phone contact	5	4	8	11	7	4	5
Changes in consultation content	4	6	8	9	7	4	5

Nine (9/13) of the quotes mention the risk of patients misunderstanding the record:

There is a risk of much misunderstanding, as there may be a lot of medical text.

May be good for some patients but can also create anxiety for some.

Easy to misunderstand. I think the doctor will weigh the words in a different way.

A quote from the psychiatry ward expresses a concern...

(I am) Working with young people with psychiatric diagnoses. It can be difficult for the young people to read what parents say and vice versa. Could be that you do not dare to say the truth in order not to hurt your family.

while other respondents also expect advantages:

I think it is an advantage if the patent is informed before (the consultation). This provides a safer patient.

More time will be spent on record writing, I guess. Perhaps each note will get longer and more detailed to allow a patient to comprehend it more easily. For a patient, this is beneficial, and it makes healthcare more accessible. Fewer phone calls and it is easier for the patient and relatives to follow the care flow/ planning.

ANALYSIS AND DISCUSSION

Healthcare is an arena where many interests meet. This small-scale study served to test the questionnaire and the technique in order to get a reliable and valid questionnaire to use at a later stage for an increased sample size. Nevertheless, the results of the study reveal a significant lack of knowledge among care professional regarding ownership of the health records as well as the national eHealth strategy. New work practices and procedures containing information and training activities can be needed to improve the introduction and management of the PAEHR for the entire staff. The results indicated which type of information to focus on. The results can also help creating guidelines for how to motivate staff regarding the introduction of new eHealth services.

Results

In this pre-deployment study, a gap is noted between legislation and ongoing development regarding PAEHR and the knowledge and preferences of the healthcare staff regarding this matter. This poses questions to be answered: Who owns the health record? Moreover, whom is it made for? Is it mainly a tool for the staff? According to the Patient Data Act (2008) the healthcare provider has the responsibility for all established records; therefore, neither patient nor the staff can claim that the records belong to them. The purpose of this legislation is that “Information management in healthcare should be organized in order to cater patient safety and quality and to promote cost” (Patient Data Act, 2008:355, Section 2). Although the legislation states that the health record belongs to the healthcare providers, most of the healthcare personnel perceived that patients or patients together with healthcare providers, or staff responsible for documentation owned the health record. Only 14 of 44 respondents knew that the healthcare provider in Sweden is the owner of the health record. Staff knowledge about the national eHealth strategy and its action plans was also poor, although all county councils jointly work towards those goals. Only 2/44 and 8/44 were acquainted with the national strategy or the action plan, respectively. The different domains of the healthcare organization, as explained by Hallin & Siverbo (2003, p. 55), in which three different groups operate in their respective domain: politicians, administrators and health professionals, could be an explanation for this discrepancy.

Another remarkable finding is that almost all personnel agreed that a respite time is important, before the notes are published online. It was also important that only signed notes are published. It is obvious that healthcare personnel want the medical record to be correct before it is presented to the patients. However, the fact is that not all medical notes are signed (8% are never signed, according to Inera, (2016)) but nevertheless used in healthcare. After notification from the local physician union, the district court has established that healthcare notes in the records are "public documents" and as such, the patients have the right to read their own notes before they are signed. As a result of the use of the PAEHR in Uppsala, the possibility for the patient to choose was introduced; whether (1) the patient wants to read the notes directly or (2) wait the 14 days of respite time after the notes were written or (3) if the patient only wants to read signed notes (Scandurra et al., 2016). The user interface of the application shows whether the note is signed or not. 98% chose to see the notes directly, without respite time, sometimes meaning before the notes are signed (Lyttkens, 2015).

The general question whether PAEHR is a good eHealth service indicated different opinions between nurses and physicians. The nurses rated PAEHR higher than physicians and secretaries. The answers from other professions were too few to analyze. When it comes to duration of time spent in the profession, no obvious difference could be seen. Regarding different work places, staff in psychiatric ward seemed to be more concerned and nurses in the hospital generally more positive. This is in line with previous studies (e.g. Scandurra et al., 2015) which provides three comparable results. Firstly, as personnel gets more acquainted with the eHealth service, their perceptions changes. This is valid also for physicians when they get personal experience by using the service for themselves or relatives. Secondly, professionals with experience of PAEHR were significantly more positive about the eHealth service as a reform compared to their peers without experience: nurses outside the pilot county were less positive compared to nurses that work in the Uppsala County where the patients have been able to read their record online since 2012. Thirdly, there were also differences between the professions: nurses with experience of the service were generally more positive compared to physicians in the same region. (Scandurra et al., 2015)

Such experiences are interesting to reflect upon, especially as this pre-deployment study revealed perceptions that the health record still is, or should be viewed as, a tool for the staff and that the record belongs to and should be used by healthcare professionals only:

PAEHR is a rare half-baked idea that will bring increased costs of administration, increased worries of the patient and family, increased workload on all parts of the healthcare organization. Risk of increased confusion. A record is a medical work tool to support the medical treatment, not a channel for information to the patient.

This quote was in line with results from a report stressing that physicians traditionally regard the health record as their working tool and that caregivers can be slow in providing the copies of one's record when asked for (Docteur & Coulter, 2012, p 74). The quotation leads the thoughts to healthcare being a complex organization, difficult to manage as many actors come together in the same arena. One group of actors who is getting a stronger position in healthcare is the patient group (Hallin & Siverbo 2003, p. 25). Patients are enforced by legislation, policies and regulations by e.g. The European Commission and different authorities and organizations on national and international levels, as the example of policy makers' clear objectives about making the PAEHR available as described in the introduction. In the professional groups, on the other hand, there is scepticism that seems hard to influence. This could be due to the fact that physicians traditionally are central actors in operations management and development in healthcare (Hallin & Siverbo 2003, p. 55). This is further explained by healthcare being referred to as a "professional bureaucracy", with characteristics as the dominance of one or more strong professional groups where management is not expected to interfere with the ongoing work (Mintzberg, 1992). The structure requires a vertical obedience, while the adaptation of the professional organization can be described as horizontal (Mintzberg, 1992). Moreover, healthcare is a multi-professional organization characterized by being dependent on a number of different professional groups, which are associated with strong and value-based traditions (Söderström, 2005, p. 3). The professionals have special knowledge basis and values often leading to their own perspective of the world and the business they work in (Söderström, 2005, p. 9). In the case of PAEHR, this is demonstrated by the Swedish Medical Association that has given its views on this issue in a policy document. Their proposals argue that health records should be accessible online only after 14 days of respite (Swedish Medical Association, 2011). Further, in October 2013 chairman of the Medical Association's Council for pharmaceuticals, IT, and medical technology states that unsigned entries should not be made available to patients online. This is due to the need of, among other things, the freedom of exchange of ideas and peripheral hypotheses expressed in writing for internal professional communication without creating unwarranted concerns of patients (Båtelson, 2013). This demand was taken care of in the pilot county by building a new field in the EHR (feeder system) called "early hypothesis" which is not uploaded to the patient accessible view. The Swedish Medical Association has not changed its viewpoint and still accounts for these statements. Since the pilot county started its deployment project representatives of the physicians' local union, have expressed a distrust of the PAEHR. On the other hand, there is the union of nurses, midwives and other health professionals, the Swedish Association of Health Professionals, embracing the development and deployment of such eHealth services. Now in 2016 they even sharpened their statements and went out with the directive: "Stop patronizing the patient. Provide all information to the patients and let them make all decisions regarding their own information by themselves" (Olsson, 2016).

It is possible that the differences in the perceptions held by the unions on a governing level shine through in the responses by their respective professionals. Here the authors also note the risk of a power conflict unless the implementation of PAEHR is accompanied with clear information about which legislation, guidelines and decided actions that apply, and that the deployment work is characterized by openness and participation of all stakeholders.

Methods

A weakness of this pre-deployment study was the moderate response rate, 45% and the small sample size, a limitation set by the region. Strengths of the study were the proportional stratified sample of employees in the Region Örebro County and that age and gender were similar to the population of the employees. Methods used, for collecting data as well as the analysis method, are repeatable when performing a survey with an increased sample size.

Here no general conclusions can be drawn as the sample was small. A great variety of responses was noted, which as well made it difficult to see indications that met the research questions. There are other articles on the pros and cons of making records available to patients via Internet-based services. However, motivation to publish papers like this is that there is still a lack of evidence-based studies with a focus on use and effects of patient and professional views as well as health outcomes of the PAEHR solution. These results, although based on a low number of employees, can indicate ways for improved working conditions, in order to provide a more secure patient health and facilitate the implementation of new eHealth systems.

FUTURE WORK

Future work would be to perform national pre- and post-deployment surveys, where various care professionals from different regions respond in parallel. Extending this pre-deployment study to a region-wide survey, or at least one where statistically significant answers could be found, would also be interesting. Physicians generally left more comments in free text compared to other professions. The content of the quotes could form a basis for a qualitative follow-up study, to understand the underlying patterns of the perceptions of the health professionals and thereby get a comprehensive and in-depth description of the issue. In the DOME consortium, there is a possibility to perform cross-survey analyses of different survey data-sets (e.g. Scandurra et al., 2015) in order to delve better into differences of the professions or regions. More results will also be published based on other research questions, e.g. use of PAEHR in relation to patients' participation in the care, comparisons between the perceptions of the care professionals and the patient's own expectations, demands and experiences, e.g. by building on results from a PAEHR pre-study.

CONCLUSION

As patients have the right by legislation to take part of their health information, it is important that healthcare professionals be knowledgeable about this law and how it is applied. Results from this pre-deployment questionnaire to various care professions in a healthcare region demonstrate that perceptions differ. Main findings of this study were that the health professionals lacked insight in the national eHealth strategy, future activity planning and legislation regarding health records. The differences in perceptions between professional groups regarded e.g. the degree of PAEHR as a good eHealth service. Answers of the respondents indicated that the workload might increase with the introduction of the

PAEHR. Results from this pre-deployment study can to some extent validate results from larger surveys, e.g. demonstrating that nurses tend to be more positive to PAEHR as an eHealth service, compared to physicians. To cover the gaps, specific training is needed, and all record-keeping professionals need to be more involved in the implementation of such eHealth services as it will affect their work processes.

ACKNOWLEDGMENT

The action research project "DOME - Deployment of Online Medical Records and eHealth Services" was financed by VINNOVA (2012-02233) – Swedish Governmental Agency for Innovation Systems. The SUSTAINS implementation project was supported by the European Commission (No 297206). We would also like to give our gratitude to the Region Örebro County for supporting the distribution of the web questionnaire to the employees. Special thanks to the respondents who participated.

REFERENCES

Allvin, H., & Kvist, M. (2011). Lättillgängliga journaler– rättighet med problem? *Läkartidningen 6*(108), 256-257. Retrieved from http://www.lakartidningen.se/store/articlepdf/1/15954/LKT1102s256_257.pdf

Båtelson, K. (2013). Journal på nätet kräver etiska hänsyn. *Läkartidningen 22*:110 CCY4. 2013-05-28. Retrieved from http://www.lakartidningen.se/Opinion/Signerat/2013/05/Journal-pa-natet-kraver-etiska-hansyn/

Bhavnani, V., Fisher, B., Winfield, M., & Seed, P. (2011). How patients use access to their electronic GP record – a quantitative study. *Family Practice,* 28(2), 188-194. Retrieved from http://fampra.oxfordjournals.org.lib.costello.pub.hb.se/content/28/2/188

CeHis. (2012). Action plan 2013-2018, eHealth cooperation of county counties, regions and municipalities. In Swedish. Retrieved from http://www.inera.se/Documents/OM_OSS/handlingsplan_2013_2018.pdf

de Lusignan, S., Mold, F., Sheikh, A., Majeed, A., Wyatt, J.C., Quinn, T., Cavill, M., Gronlund, T.A., Franco, C., Chauhan, U., Blakey, H., Kataria, N., Barker, F., Ellis, B., Koczan, P., Arvanitis, T.N., McCarthy, M., Jones, S. & Ra, I. (2013). Patients' online access to their electronic health records and linked online services: a systematic interpretative review. *BMJ Open, 4*(9).

Delbanco, T., Walker, J., Darer, J. D., Elmore, J. G., & Feldman, H. J. (2010). Open Notes : Doctors and Patients Signing On. *Annals of Internal Medicine, 153*(2), 121–125. doi:10.7326/0003-4819-153-2-201007200-00008 PMID:20643992

Docteur, E., & Coulter, A. (2012). Patientcentrering i svensk hälso- och sjukvård. En externutvärdering och sex rekommendationer för förbättring. *Care Services Analysis Report, 2012*, 5.

DOME Consortium. (2016). Retrieved from https://projectdome.wordpress.com

Eklund, B., & Joustra-Enquist, I. (2004). Sustains--direct access for the patient to the medical record over the Internet. *Studies in Health Technology and Informatics*, 100, 182–189. PMID:15718578

Entergate. (2013). esMaker. Retrieved from www.entergate.se

Erlingsdottir, G., & Lindholm, C. (2014). When patient empowerment encounters professional autonomy: The conflict and negotiation process of inscribing an eHealth service. Scandinavian Journal of Public Administration, 19(2), 7-28.

European Commission. (2010). A Digital Agenda for Europe. Communication from the Commission to the European parliament, the Council, the European Economic and Social Committee and the Committee of the regions. Retrieved from https://ec.europa.eu/digital-agenda/

European Commission. (2012). eHealth Action Plan 2012-2020 - Innovative healthcare for the 21st century. Communication from the Commission to the European parliament, the Council, the European Economic and Social Committee and the Committee of the regions. Retrieved from http://ec.europa.eu/health/ehealth/docs/com_2012_736_en.pdf

Fischer, B. (2010). *Online medical records*. Retrieved from http://www.centreforwelfarereform.org/library/by-date/online-medical-records.html

Forsberg-Fransson, M.-L., Jansson, A., & Jensen, C. (2012). *Loggar i e-journalen - varje patients rättighet att lätt få tillgång till dem* [B Thesis]. Business School, Örebro: Örebro Universitet.

Grünloh, C., Cajander, Å., & Myreteg, G. (in press). "The Record is our Work Tool!" - Physicians' Framing of a Patient Portal in Sweden. *J. Med. Internet Res.*

Hallin, B., & Siverbo, S. (2003). *Styrning och organisering inom hälso- och sjukvård. 8. uppl.* Lund: Studentlitteratur AB.

Huba, N., & Zhang, Y. (2012). Designing Patient-Centered Personal Health Records Health Care Professionals' Perspective on Patient-Generated Data. *Journal Medicine System, 36*(6), 3893-3905. Retrieved from http://link.springer.com.lib.costello.pub.hb.se/content/pdf/10.1007%2Fs10916-012-9861-z

Inera. (2015). Statistics about My health record. Retrieved from http://www.lul.se/sv/Extranat/For_vardgivare/MOT-PATIENTEN/Sjukvard1/Journal-via-natet-min-journal-i-U-a-lan/Statistik/

Inera. (2016). Journalen. Retrieved from http://www.inera.se/TJANSTER--PROJEKT/Journalen/

Jilka, S. R., Callahan, R., Sevdalis, N., Mayer, E. K., & Darzi, A. (2015). Nothing About Me Without Me: An Interpretative Review of Patient Accessible Electronic Health Records. *Journal of Medical Internet Research, 17*(6), e161. doi:10.2196/jmir.4446 PMID:26123476

Lyttkens, L. (2015). Support USers To Access INformation and Services Document D6.3 Final Pilot Evaluation Uppsala County Council, v 1.0. EU project report: ICT PSP – Empowering patients. Retrieved from http://www.sustainsproject.eu/

Mandl, K. D., & Kohane, I. S. (2016). Time for a Patient-Driven Health Information Economy? *The New England Journal of Medicine, 374*(3), 205–208. doi:10.1056/NEJMp1512142 PMID:26789868

Mandl, K. D., Szolovits, P., & Kohane, I. S. (2001). Public standards and patients' control: How to keep electronic medical records accessible but private. *BMJ (Clinical Research Ed.)*, *322*(7281), 283–287. doi:10.1136/bmj.322.7281.283 PMID:11157533

Ministry of Health and Social Affairs. (2010). National eHealth – the Strategy for Accessible and Secure Information in Health and Social Care S2010.020. Retrieved from http://www.regeringen.se/sb/d/108/a/148429

Mintzberg, H. (1992). *Structure in Fives: Designing Effective Organizations*. Prentice Hall.

Olsson, A. (2016). Samtliga landsting inför samma typ av nätjournal. *Vårdfokus, journal of the Swedish Association of Health Professionals*. Retrieved from http://www.vardfokus.se/webbnyheter/2016/februari/samtliga-landsting-infor-samma-typ-av-natjournal/

Pagliari, C., Shand, T., & Fischer, B. (2012). Embedding online patient access in JK primary care: a survey os stakeholder experiences. *Record, Journal of the Royal Society of Medicine Short Reports*, 3(34). DOI: Retrieved from http://shortreports.rsmjournals.com/content/3/5/34.short10.1258/shorts.2012.012009

Patient Data Act. (2008). SFS 2008:355. Retrieved from http://www.riksdagen.se/sv/Dokument-Lagar/Lagar/Svenskforfattningssamling/Patientdatalag-2008355_sfs-2008-355/

Scandurra, I., & Jansson, A. Forsberg-Fransson. M-L. & Ålander, T. (2015). Is 'Patient's Online Access to Health Records' a Good Reform? – Opinions from Swedish Healthcare Professionals Differ. *Procedia Computer Science 64*, 964–968 doi: Retrieved from http://www.sciencedirect.com/science/article/pii/S187705091502749010.1016/j.procs.2015.08.614

Scandurra, I., Lyttkens, L., & Eklund, B. (in press). Implications of Swedish National Regulatory Framework of the Patient Accessible Electronic Health Record. *Proceedings of the Stud. Health Technol. Inform 2016 Medical Informatics Europe/HEC2016*, München.

Söderström, M. (2005). Ledning och ledarskap i multiprofessionella organisationer (MPO). *SYNOPSIS Aktuell forskning i statsförvaltningen nr 10.*

Sundhed. (2008). Sammenhængende Digital Sundhed i Danmark. National strategi for digitalisering af sundhedsvæsenet 2008 – 2012. Retrieved from www.sundhed.dk/content/cms/80/4380_national-strategi-2008-2012.pdf

Sundhed (2015). Sundhed.dk- direktoer vinder internationellt eHealth Award. Retrieved from https://www.sundhed.dk/service/presserum/pressemeddelelser/sundheddk-direktoer-vinder/

Swedish Medical Association. (2011). *Patientens delaktighet i sin vård.* Retrieved from https://www.slf.se/upload/Lakarforbundet/Patientens%20delaktighet%20i%20sin%20v%C3%A5rd2011.pdf

Thylefors, I. (2007). *Ledarskap i human service organisationer* (2 uppl.). Stockholm: Natur och Kultur.

Tiik, M. (2010). Rules and access rights of the Estonian integrated e-Health system. *Studies in Health Technology and Informatics*, *156*, 245–256. PMID:20543358

Wiljer, D., Urowitz, S., Apatu, E., DeLenardo, C., Eysenbach, G., Harth, T., ... Leonard, K. J. (2008). Canadian committee for patient accessible health records. Patient accessible EHR: Exploring recommendations for successful implementation strategies. *Journal of Medical Internet Research, 10*, e34. doi:10.2196/jmir.1061 PMID:18974036

World Medical Association. (2008). WMA Declaration of Helsinki – Ethical Principles for Medical Research Involving Human Subjects. Retrieved from http://www.wma.net/en/30publications/10policies/b3/

This research was previously published in the International Journal of Reliable and Quality E-Healthcare (IJRQEH), 6(1); edited by Anastasius Moumtzoglou, pages 29-45, copyright year 2017 by IGI Publishing (an imprint of IGI Global).

Chapter 7
Promoting Better Healthcare for Patients in Critical Condition:
An IoT-Based Solution to Integrate Patients, Physicians, and Ambulance Services

Itamir Barroca
UFRN, Brazil

Gibeon Aquino
UFRN, Brazil

Maria Alzete Lima
UFRN, Brazil

ABSTRACT

The high cost of healthcare services, the aging population and the increase of chronic disease is becoming a global concern. Several studies have indicated the need to minimize the process of hospitalization and the high cost of patient care. A promising trend in healthcare is to move the routines of medical checks from a hospital to the patient's home. Moreover, recent advances in microelectronics have boosted the advent of a revolutionary model involving systems and communication technology. This new paradigm, the Internet of Things (IoT), has a broad applicability in several areas, including healthcare. Based on this context, this chapter aims to describe a computer platform based on IoT for the remote monitoring of patients in critical condition. Furthermore, it is planned to approach the current advances and challenges of conceiving and developing a set of technology-centric, targeting issues relevant to underdeveloped countries, particularly in regards to Brazil's health infrastructure.

DOI: 10.4018/978-1-5225-6198-9.ch007

INTRODUCTION

The high cost of healthcare services, the aging population and the increase of chronic disease are becoming a global concern. Several studies have indicated the need for strategies to minimize the institutionalization process and the effects of the high cost of patient care (Hochron, & Goldberg, 2015). A promising trend in health treatments is to move the routines of hospital medical checks to the patient's home. However, the public health system in Brazil still lacks accessible technologies that facilitate the patient monitoring process in this situation (Mendes, 2010).

On the other hand, recent advances in microelectronics, wireless, sensing and information have fueled the advent of a revolutionary model involving systems and communication technology, enabling smarter ways to "make things happen". This new paradigm, known as the Internet of Things (IoT), has a broad applicability in several areas, including health. In this trend, it is estimated that by 2020 there will be around 20 billion "things" connected (Gartner, 2015) and uniquely identifiable (Gubbi et al.., 2013). These "things" promote the basic idea of IoT that is pervasive computing around this range of devices, such as RFID tags, sensors, actuators, mobile phones, etc. (Atzori et al.., 2010). It can act as the backbone of the health system for information sharing (Abinaya & Swathika, 2015). The field of health in particular is expected to see the development and application of this trend as part of its future because it has the ability to allow hospitals to operate more efficiently and patients to receive better treatment. A type of healthcare application which will be focused on in conjunction with this new paradigm is the application of mobile health. The main objective of mobile health is to allow for the remote monitoring of the health status and the treatment of patients from anywhere (López, Fernández, Jara & Skarmeta, 2013).

In this context, the potential for change in the quality of life that can be promoted by IoT is unquestionable. Creating integrated utilities will lead to a qualitative change in the services to integrate information systems, computing and communication with extensive control (Yinong, 2016). Therefore, there is an urgent need for the development of technologies and applications related to IoT infrastructure for health care. Currently, the process of early hospital discharge of critically ill patients has enjoyed the support of this new technology to ensure the case management through remote monitoring of patients who remain under home confinement and are still considered critical. Therefore, the rapid recognition of changes in conditions that are associated with complications is crucial to improving the survival of patients by the possibility of hemodynamic instability (Jiang et al., 2016)

Although they remain under health team care or are receiving regular follow-up visits, patients in home hospitalization are encouraged to perform a daily evaluation with self-monitoring of their vital signs and patterns related to their health condition (Kotsimbos, Williams, & Anderson, 2012). Moreover, although everyone receives instructions for detecting and reporting changes in the critical state for home self-monitoring, many patients find it challenging to identify the critical threshold values, i.e. the lower or upper limits of blood pressure, and recognize changes in their own personal baselines (Dabbs et al., 2009). Considering the amount of monitored data, it is hard to engage the patient in self-management; it is essential, therefore, to send this information to the care providers (Hendriks & Rademakers, 2014; Coulter, 2012).

Thus, Brazilian researcher groups seek to develop a computational solution in order to improve the remote care of the patient in a critical situation, which is expected to extend the health care of patients in hospital critical situation for their homes. The hypothesis put forward is that the technology will help maintain the supported self-care and remote management case. It therefore foresees overcoming the constraints identified in the process of institutionalization of people in a critical health situation. Finally, this chapter describes a computer platform based on IoT for the remote monitoring of patients in critical condition. It involves embedding sensors in patients, physicians, clinical staff, medical equipment and physical spaces in order to monitor, track and alert. Moreover, it is planned to approach the current advances and challenges of conceiving and developing a set of technology-centric strategies with the aim of making possible the effective application of affordable technologies in this kind of monitoring, targeting issues relevant to underdeveloped countries, particularly in regards to Brazil's current health infrastructure. The following sections describe the background of this research, the challenges and opportunities related to this topic, the proposed technology-based healthcare platform, future research directions and the conclusion.

BACKGROUND

Global projections indicate the need for strategies to minimize the effects of chronic health problems that can lead the individual to lose the ability to perform daily routine activities. There is a need for health promotion behaviors that offer new remote management options for patient cases with chronic diseases in critical condition. In Brazil, more than 70% of spending on health care is for the treatment of chronic diseases, which is seen as most alarming since they have the greatest potential for an increase in government spending (Brazil, 2013).

The services which serve patients in critical health situation are often overcrowded. Another problem is the growth of patients who require specialized care, such as intensive care, resulting in an insufficient number of vacancies (Brazil, 2013). These people need continuous monitoring to assess multiparameters from pulse pressure to oxygen saturation and electrocardiographic trace. However, insufficient material resources are considered the main difficulty in these sectors (Zandomenighi, Moorish, Oliveira & Martins, 2014). The strategies identified by developing countries is to build more hospitals, train an adequate number of doctors and equip health facilities with high-quality diagnostic instruments. However, this approach is a process that takes time and requires many resources (Hindia, Rahman, Ojukwu, Hanafi, & Fattouh, 2016). The best approach would be to propose a remote monitoring system with the ability to automatically communicate with the service providers, produce alerts to assist whenever necessary (Benlamri, & Docksteader, 2010; Constantinescu, Kim, & Feng, 2012; Pitsillides et al.., 2006;. Gorp, & Comuzzi, 2014), and integrate with emergency medical services.

The search for solutions has indicated that new devices designed for use in health service as mobile apps showed a reduced response time of nurses and doctors from thirty to five minutes, and a 20% reduction in waiting time in the emergency and medical observation units that saved $720,000, among other benefits (Hochron, & Goldberg, 2015). Thus, if a device with a sensor detecting conditions which reflect the individual's health status is configured to communicate with a portable computing device, an all care schedule can be operated efficiently. Currently, most people have access to these mobile devices, and these devices have become quite affordable (Hochron, & Goldberg, 2015). It is possible to think of a "surveillance system in mobile health," so you can easily reach large numbers of people with better planning of medical care. The problem goes beyond questions related to healthcare spending and is becoming a common concern, since with the global aging population and the increase of chronic diseases, people will demand more robust equipment to survive (Pang, 2013). The solution found by developed countries is to reduce the number of rooms in hospitals and increase the proportion of home health treatments (Prescher et al., 2014).

In Brazil, the Home Care is inserted in the proposed National Health System as a consolidated strategy in primary care and currently takes dimensions complementary to hospital care. It is characterized by a set of health promotion, prevention, and treatment of home provided diseases, ensuring continuity of care and integrated into health care networks (Brazil, 2013). In accordance with the logic of a health service organization, it seeks to insert, in the home environment, the remote monitoring system consisting of the patient's health parameters, daily measurements and its transmission to mobile phones belonging to the responsible healthcare professionals (Prescher et al., 2014; Brennan et al., 2010). Adherence to this complementary monitoring modality by the patient can be considered as a strategy for the adoption of continued care. Still, with regards to the Brazilian context, groups of researchers have made efforts to integrate a data mobile platform related to critical patients with minimally invasive monitoring. They seek strategies to help to promote health in order to suit the modern quotidian permeation of technology, trying to offer easy access to remote monitoring technology, promoting better integration between healthcare networks, care in the home environment and operationalizing the remote monitoring of patients in critical condition. Thus, it is intended to reduce public spending by reducing admissions numbers.

Despite concerns about security and privacy risks, remote monitoring is considered a tool to promote patient-centered care (Filkins, et al., 2016;. Sieverink, Siemons, Braakman-Jansen & Van Gemert-Pijnen, 2016). The use of mobile technology in healthcare can lead to a great revolution in digital health. Currently, the concept of mobile health (or mHealth) broadly describes the use of telecommunication technologies that deliver health actions to promote the welfare of patients (Dwivedi, 2016).

CHALLENGES AND OPPORTUNITIES

The health care environment is facing a challenge of raising the quality of services alongside new requirements for minimal resource utilization. At the same time, society is requiring more services due to demographic changes and the prevalence of chronic disease which puts the individual into critical health conditions. In the United States, it is expected that the growth in direct costs related to cardiac insufficiency will grow from $21 billion in 2012 to $53 billion in 2030 (Heidenreich et al., 2015). Most of these costs are related to hospitalization, which accounts for 4,800,000 hospital admissions annually (Piccini et al., 2016).

The technology information and communication systems were considered to be important tools to help solve these challenges (Chen, Kennedy, Sales, & Hofer, 2013; Jogh et al., 2012). However, traditional remote management by telephone failed to improve results and reduce readmission (Chaudhry et al., 2014). In contrast, the results of automatic remote monitoring (MR) have been more favorable, showing better outcomes (Hindricks et al., 2014; Saxon et al., 2015; Varma et al., 2015; Portugal et al., 2016). In previous analyses regarding the ambulatory treatment efficiency, the use of remote monitoring has reduced medical and nurse actions and the costs associated with displacement of patients, as compared to the traditional method (Heidbuchel et al., 2015; Varma et al., 2015).

However, the benefits of remote monitoring can go beyond efficiency in patient care, since this modern technology enables early intervention to prevent the morbidity of the patient and prevent or potentially reduce hospitalization, with profound implications for health costs, time in the trial, clinical decisions and a shorter length of stay. In a recent cohort study of 490,000 patients, where 37% (34,259) used remote monitoring compared with the traditional method, there was a 31% reduction in the cost of hospitalization and 23% fewer readmissions within 30 days (Piccini et al., 2016).

Despite evidence that remote monitoring offers several potential advantages to patients, including early detection of arrhythmia, a malfunction or organ system failure (Crossley et al., 2011; Varma et al., 2010), initiatives using remote monitoring are still scarce. In particular, patients with atrial fibrillation history, stroke and early diagnosis of myocardial infarction are the biggest beneficiaries of this method. Oral anticoagulation, for example, is often underutilized in patients with atrial fibrillation history; the possibility of early detection of more episodes of this fibrillation increases with the appropriate use of oral anticoagulation and the patients' chance of survival increases as a consequence (Piccini et al., 2016).

In order to face the challenges in the management of seriously ill patients' cases and deal with the issue of cost reduction in the hospitalization process with the support of continuous monitoring services, the paradigm of the architecture used to build smart applications that perform data processing through mobile devices is used. These devices are incorporated into the components used to support the decision-making process based on the professionals' knowledge. Mobile devices may allow an accurate and continuous analysis of the patient's health status to be performed locally, minimizing the transmission network, avoiding communication delays or interruptions and maintaining appropriate levels of security and privacy (Minutolo, Esposito, & Pietro, 2016). Therefore, it can act as a decision support system, capable of automatically processing and correlating large volumes of monitored physiological parameters, detecting suspicious changes and providing alarms in response to the worsening of the patient's condition, as well as suggestions on possible actions. It can facilitate diagnosis and personalized treatments, predict the patient's condition based on the observation of several monitored levels and empower patients to actively participate in their health (Minutolo, Esposito, & Peter, 2016).

It is important to note that there are many barriers to the implementation of patient management for the remote system, including the refusal of the patient, the lack of access to technology for remote monitoring and access to telecommunications, lack of infrastructure in health services, lack of resources and institutional support. Eliminating these hurdles with an awareness of the professional on the importance of continuous remote monitoring is essential to improving care in health and meeting the new challenges imposed by the advances of technology and health.

A TECHNOLOGY-BASED HEALTHCARE PLATFORM

This section presents the technology-based healthcare platform describing the aspects of design and requirements, the actors and use cases that guided its development. It also explains the platform's architecture considering its modules, their relationship, components and protocols.

Design Issues and Requirements

The main goal of the proposed technology-based healthcare platform is to provide remote monitoring for patients in a critical situation, and it was developed considering the necessity to transfer the healthcare from the hospital (hospital-centric) to home (home-centric). This platform is IoT-based and integrates patients, physicians and ambulance services in order to promote better care and rapid preventive and reactive urgent actions. Regarding requirements, this platform has *Remote Patient and Ambient Monitoring, Patient Healthcare Data Management, Patient Health Condition Management and Emergency and Crisis Management.* Figure 1 presents the relationship between these requirements:

Figure 1. The relationship between requirements

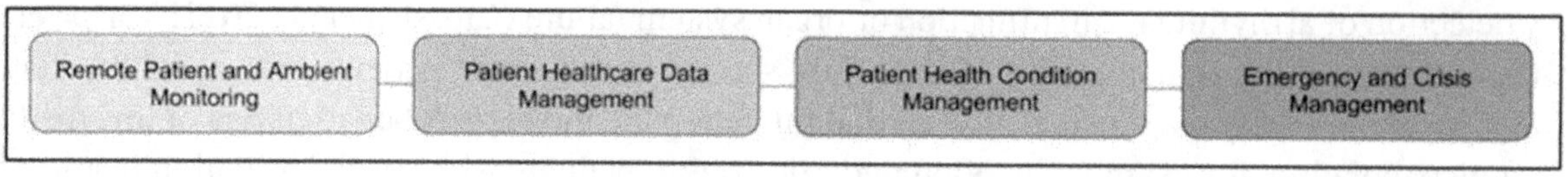

The *Remote Patient and Ambient Monitoring* involves the acquirement of data from sensors attached to the patient's body and in the ambient (patient's home). The acquired data from the sensors are used by clinical staff (physician and nurses) for healthcare treatment and emergency alert purposes. Thus, the sensors attached to the patient body provide information about:

1. **Electrocardiogram (ECG)** (Maglogiann, 2012; Jara et al., 2013; Yang et al., 2014; Maksimović et al., 2015): Recording of the electrical activity of the heart in the form of specific waves. The ECG monitoring can be used to monitor the heart rate of a patient, assess the effects of an illness or injury on the function of the pacemaker, and evaluate the response after a physician's procedure. The ECG can give information about the orientation of the heart, conduction disturbances, electrical effects of medications and electrolytes, the mass of the heart muscle and the presence of ischemic damage. However, to evaluate the effectiveness of the mechanical activity of the heart the pulse and blood pressure of the patient is evaluated (Aehlert, 2011).
2. **Blood Pressure** (Jara et al., 2013; Raad et al., 2015; Maksimović et al., 2015): Recorded as a ratio between two numbers, systolic - the top number, which is also the higher of the two numbers, measures the pressure in the arteries when the heart beats (when the heart muscle contracts); diastolic - the bottom number, which is also the lower of the two numbers, measures the pressure in the arteries between heartbeats (when the heart muscle is resting between beats and refilling with blood) (American Heart Association, 2016). For example: Read as 140X90 mmHg (millimeters of mercury). Can be decisive in the early identification of cardiac and vascular problems (Vidal-Petiot et al., 2016). In cases of high pressure, their control reduces the risk of cardiovascular events and death (Zanchetti, Thomopoulos, & Parati, 2015; Ettehad et al., 2016). Similarly, a reduction in blood pressure may not be compatible with survival and vasoactive drugs need to be administered urgently (Vidal-Petiot et al., 2016).
3. **Blood Glucose** (Poenaru, et al., 2013): Monitoring is the main tool you have to check patient diabetes control (American Diabetes Association, 2016). Population data indicate that 30-40% of people with type 1 diabetes experience an average of 1 to 3 episodes of severe hypoglycemia each year. With self-monitoring and patient education and care, the patient may benefit from a controlled glycemic rate with individual goals set by the team of health professionals. During the last decade, the introduction of continuous glucose monitoring to facilitate the self-administration has shown an improvement in glucose control and reduced exposure to hypoglycemia (Bolinder et al. 2016). Experience shows the beneficial effect of continuous monitoring of blood glucose (Thabit, Bally, & Hovorka, 2016).
4. **Heart Rate** (van der Valk et al., 2015; Raad et al., 2015; Gia et al., 2014; Khattak et al., 2014): The number of heartbeats per unit of time, usually per minute. The heart rate is based on the number of contractions of the ventricles (the lower chambers of the heart) (Medicinet, 2016). It refers to the number of heartbeats per unit time, usually expressed as beats per minute (bpm) (American Heart

Association 2016). The heart rate variability has been used as a noninvasive means of assessing the neural control of the heart and is used to identify hemodynamic problems. In other words, the pulse rate translates cardiac function; its monitoring is important because even in sinus rhythm it can face a paroxysmal fibrillation which can lead to widespread uncontrolled cardiac electrical activity and impair heart function (Port et al., 2015). Importantly, there are variations as to its value, for what may be appropriate for one patient may not be suitable for others, for example, heart failure patients with left ventricular ejection usually preserved need a low heart rate (Gelder et al., 2016). This allows the care team to plan individualized intervention, including programming the alarm system.

5. **Oxygen Saturation** (Maglogiann, 2012; Sebestyen et al., 2014; Chiuchisan et al., 2014; Raad et al., 2015): It is especially useful to detect hypoxemia associated with critical problems such as cardiovascular ones (Ewer, 2014). The oxygen uptake occurs primarily in the lungs, constituting the first step in the process of oxygen to the tissues. The oxygen taken up in the lungs is transported in the blood in two ways: by dissolving in plasma and also combined with hemoglobin. Hemoglobin is capable of carrying 98-99% of all oxygen in the blood and can be viewed through the oxygen saturation measured by pulse oximetry. The arterial oxygen saturation is determined as a percentage, on average it is in the range of 95% to 100%. There may be some changes and false readings of oxygen saturation, which are usually caused by chills, hypotension, low perfusion and edema (Bezerbashi et al., 2016).
6. **Temperature** (Ray et al., 2015; Tabish et al., 2014): The human being is homeothermic, i.e. has the ability to maintain body temperature within a certain predetermined range despite variations in the thermal environment - thermal homeostasis (Gasparrini et al., 2015). Increased body temperature may indicate increased cell metabolism, consumption of O2 and CO2 production, demands on the heart and lungs and additional stress to the cardiopulmonary system and infectious processes, and therefore may justify a continued investigation in critically ill patients (Cahill, & Prendergast, 2016). Measured in degrees Celcius (° C), it is taken by means of a catheter close to the skin in the axillary region (Hall, 2006) continuously.
7. **Breathing Rate** (Castillejo et al., 2013; Chiuchisan, et al., 2014): It is measured by the respiratory motion for one minute, measured in rpm. It demonstrates not only lung function but can denote problems in other systems, such as neurological and cardiac (Cahill, & Prendergast, 2016; Goligher, Ferguson, & Brochard, 2016; Ferguson et al., 2013).
8. **Capnography**: Capnography is a graphical display of the carbon dioxide concentration in exhaled and inhaled function of time, and is used to monitor ventilatory support. There is a growing recognition that capnography is rich in information regarding lung and circulatory physiology and provides insights into many diseases and treatments (Sweeney & McAuley 2016). These include inadequate matching conditions of ventilation and perfusion, such as pulmonary embolism and obstructive pulmonary diseases and adequacy of chest compressions during cardiac arrest or responsiveness in patients in shock (Nassar & Schmidt 2016; Langhan et al., 2014).

The sensors from the ambient provide information about environment temperature (Maglogiann, 2012; Yang et al., 2013; van der Valk et al., 2015), location (Maglogiann, 2012) with latitude and longitude, and humidity (Yang et al., 2013). This is important because the control of the ambient temperature and humidity can directly affect the patient's treatment. Regarding the location, it assists in the rapid response of the ambulance service. Therefore, as the patient in critical condition is at home and not in a hospital, which is a more controlled environment, this ambient information is of greater importance for effective healthcare and enriches the remote monitoring provided by this platform.

The *Patient Healthcare Data Management* records the data about the patient: name, gender, birth, contacts, address, family information, physician information (name and contacts), health insurance information, health situation and the history of monitoring sensors and emergency alerts. These data are important to physicians and nurses to understand the current situation and history of patients, and also to facilitate the accurate monitoring of health treatment.

The *Patient Health Condition Management* considers the patient healthcare data, especially the health situation and history of sensors monitoring data, to allow the definition of critical levels to the values of the sensors that are important to the rapid response in case of emergency. It also defines rules to actions considering the settled critical levels for a patient and the related alerts.

Finally, the *Emergency and Crisis Management* address the information about the patient's health condition and the services that should be alerted in case of emergency with a monitored patient in a critical situation. As this patient is at home and not in a hospital, the efficiency of a rapid response in an emergency case can be the deciding factor between life and death.

To achieve the presented requirements, this platform is composed of ten use cases, presented by the use case diagram in Figure 2 and it indicates four actors: *the hospital operator, physician and nurses, ambulance service operator and the patient and family.*

Considering the hospital operator actor, it interacts with the use cases related to patient, health insurance and clinical staff data, that are:

1. **Patients Data Management**: It allows the register of data related to *Patient Healthcare Data Management* cited before;
2. **Clinical Staff Data Management**: It allows the register of data related to the clinical staff (physician and nurse). These data include: name, contacts and specialty from physician and nurses;
3. **Health Insurance Data Management**: It allows the register of data from the health insurance;

Figure 2. Healthcare platform use case diagram

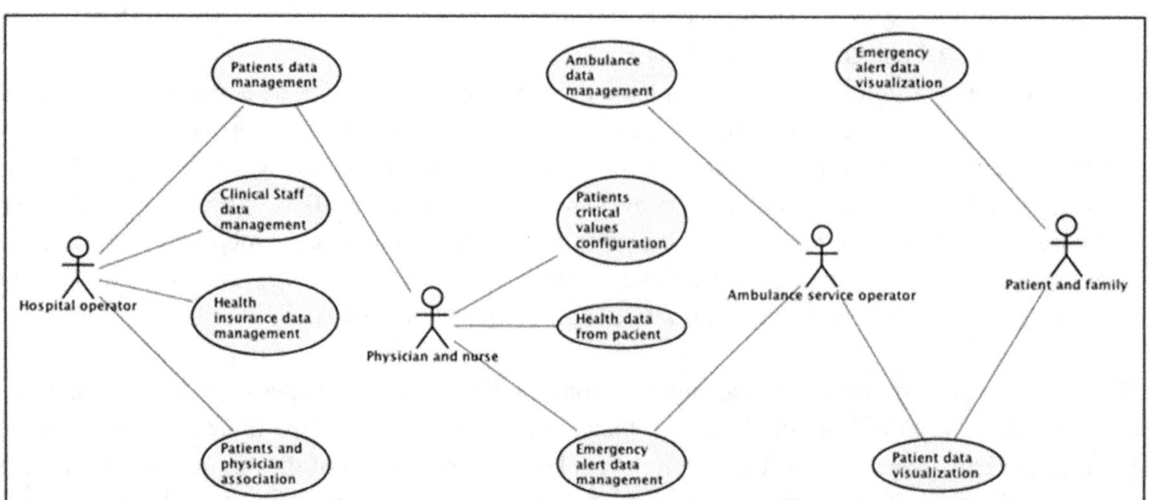

4. **Patient and Physician Association**: It allows the register of the responsibility of a physician with a patient.

Regarding the physician and nurse actors, they can use the patients data management, and interact with others use cases related to *Remote Patient and Ambient Monitoring*, *Patient Healthcare Data*, *Patient Health Condition Management* and *Emergency and Crisis Management*, that are:

1. **Patients Critical Values Configuration**: it allows the definition of critical levels for the values of the sensors attached to the patient body that are considered in the alerts and notifications;
2. **Health Data From Patient**: It allows the visualization of real-time health data from the sensors deployed in any patient body and ambient;
3. **Emergency Alert Data Management**: It allows the notification and alerts to be presented and managed by physicians and nurses.

The ambulance service operator actor uses the emergency alert data management use case and interacts with others use cases related to *Emergency and Crisis Management*, that are:

1. **Ambulance Data Management**: It allows the management of data from the ambulances, such as real-time location and activation of an ambulance to an emergency.
2. **Patient Data Visualization**: It allows the visualization of real-time health data from the sensors deployed in a single patient body and ambient, besides his home location.

Finally, the patient and family actor uses the patient data visualization use case and interacts with the emergency alert data visualization, related to *Emergency and Crisis Management*. The emergency alerts and patient data regards only to itself, in a case of a patient, or his familiar, in a case of a family.

Thus, the proposed platform was developed considering the presented requirements, actors, and use cases. It provides integration between patients, physician and ambulance services, for efficient patients' healthcare in critical condition. In the following subsection, the architecture overview is presented with details of these platform modules.

Architecture Overview

The healthcare platform architecture, presented in Figure 3, is composed of five modules: *Patient's Home, Cloud Health Infrastructure, Hospital, Family's Home and Ambulance Service*. These modules address the solution's functional requirements and work together to achieve the goal of remote monitoring and efficient healthcare for patients in critical condition.

Considering the *Patient's Home module*, it is mainly composed of sensors that provide body and ambient remote monitoring. The sensors attached to the patient body, described in the previous section, are part of a multi-parameter portable patient monitoring, which continuously measures his vital signs. This monitoring is configured by the clinical staff at the patient's home and does not require his intervention. Regarding security and essential performance issues, the monitor is in agreement with the standard IEC 60601-1-11:2015 (IEC 60601, 2015) that defines the basic safety and essential performance of medical electrical equipment and medical electrical systems for use in the home healthcare environment. The sensors of the ambient monitoring, described in the previous section, are also deployed at the patient home.

Figure 3. IoT-based platform to integrate patients, physicians, and ambulance services

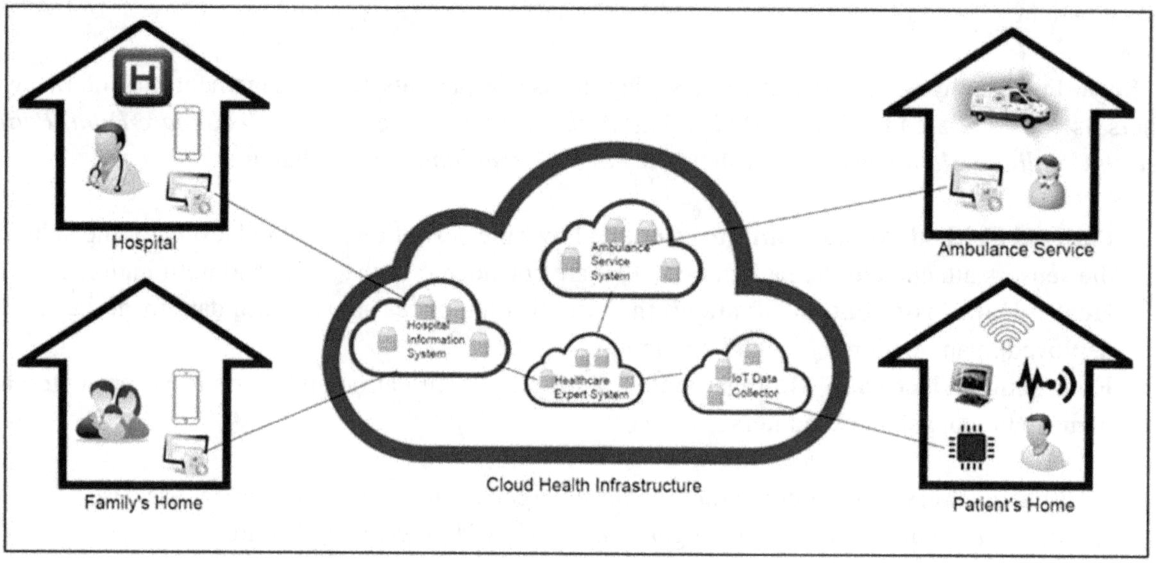

Thus, this multi-parameter monitor is connected through the Internet to the *Cloud Health Infrastructure* module, as presented in Figure 4. The ambient sensors are connected to a gateway through 6LoWPAN protocol. Regarding 6LoWPAN, it is a protocol for Wireless Sensor Networks (WSNs) defined to enable IPv6 packets to be carried on top of low power wireless networks, specifically exploiting IEEE 802.15.4 protocol (Khattak et al., 2014). This gateway is also connected by the Internet to the IoT Platform of the *Cloud Health Infrastructure* module. The reason for a gateway to the sensors is because they do not have interfaces for direct connection to the Internet.

Figure 4. The connections between the multi-parameter monitor and ambient sensors

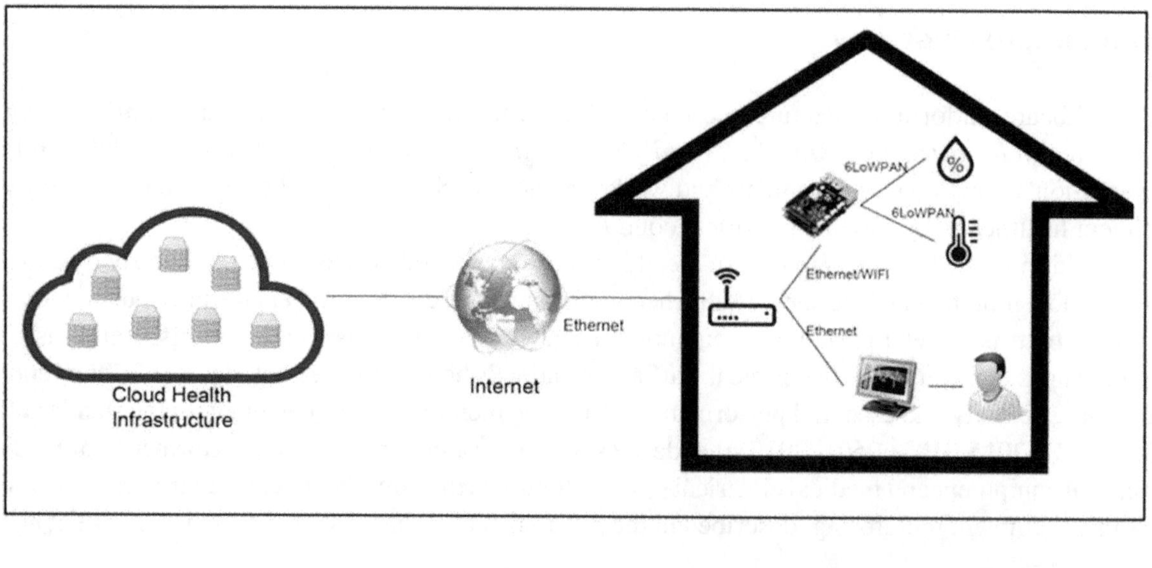

Regarding the *Cloud Health Infrastructure module*, it is composed by the IoT Data Collector, the Healthcare Expert System, the Hospital Information System and the Ambulance Service System. These systems implement the use cases described previously. The features of each system are:

1. **IoT Data Collector**: Responsible for receiving sensors' data. This is a challenge for realizing IoT due to the huge number of devices integrated into this component and their diversity in terms of data formats, protocols, nature of components, etc. (Delicato et al., 2013).
2. **Healthcare Expert System**: Configuration of the patients' critical values for alerts that are displayed in the Hospital Information and Ambulance System. These critical values are used in the rules defined by the physicians and nurses. This system provides standard mapped rules, presented in Table 1, and realizes machine learning and analytics to assist the physician in defining the appropriate critical values for each patient. Regarding the rule, it is composed of a type of sensor, a value, a critical level and its action. This action is associated with a color, according to the Manchester Triage System, that are displayed on the related information systems.
3. **Hospital Information System**: Data management of patients, clinical staff and health insurance, and the association between patients and physicians. It also provides mobile views for patient data and alerts visualization.
4. **Ambulance Service System**: Data management of emergency alert and ambulances. It also provides visualization of the patient's data.

Table 1. Rules of the expert system

Sensor	Value	Critical Level	Action
ECG	Alteration of the electrocardiographic trace.	Irregular electrical activity, ventricular fibrillation supra or infra of ST asystole.	Red alarm issued to the medical staff on the Hospital Information System and Physician mobile app. If asystole, it alerts the ambulance service system and the ambulance mobile app.
Blood Pressure	Normal is 120 X 80 mmHG	Over 140X90mmHg or under 90X60mmHg.	Green alert issued to the family mobile app. Considering the patient history, a red alert can be issued to the medical staff in the Hospital Information System and Physician mobile app.
Blood Glucose	Normal between 100 and 126mgdL	Over 200mg/dL or under 60 mg/dL.	Green alert issued to the family mobile app. Considering the patient history, a red alert can be issued to the medical staff in the Hospital Information System and Physician mobile app.
Heart rate	Normal between 60 and 100bpm	Under to 60bpm or over a 100 bpm	Red alarm issued to the medical staff on the Hospital Information System and Physician mobile app. Considering the patient history, it can alert the ambulance service system and the ambulance mobile app.
Breathing rate	Normal between 16rpm and 20rpm	Under to 12rpm or over to 35rpm	Green alert issued to the family mobile app. Considering the patient history and clinical staff programming, it can alert the ambulance service system and the ambulance mobile app.
Temperature	Normal between 36ºC and 37.5ºC	Over 37.5°C	Green alert issued to the family mobile app. Considering the patient history and in cases of desirable hypothermia, clinical staff can program the threshold.
Capnography	Normal between 35mmHg and 45mmHg	Under to 35mmHg or over to 55mmHg	Red alarm issued to the medical staff on the Hospital Information System, Physician mobile app, the ambulance service system and the ambulance mobile app.

The data received by the IoT Data Collector are used by the Healthcare Expert System, which contains rules and creates derived information to be used by the Hospital Information System and the Ambulance Service System. The Healthcare Expert System uses techniques of Machine Learning and Analytics based on the huge amount of received data to produce knowledge about the patient's health behavior. This knowledge is then stored in the computer and users call upon the computer for specific advice as needed. The computer can make inferences and arrive at a specific conclusion (Liao et al., 2005). For example, the critically ill patients, particularly those with hemodynamic instability signals, need a diagnosis and immediate treatment. This condition presents itself with signs of tissue perfusion and impaired tissue oxygenation, which is usually detected by macrocirculatory parameters or global hemodynamic measurements such as blood pressure and oxygen saturation in arterial blood (Bazerbashi et al., 2014). When a critical value is captured by this healthcare expert system, it automatically generates an alert message to the Hospital Information System, the physician, the patient's family and ambulance service system providing support for a specific decision on when and how to intervene. Thus, the patient's state of the classification system is issued together with the monitoring values. Another example is that if the data from the sensors show that the patient's heart rate is zero, it can translate this as a heart attack. Therefore, this Healthcare Expert System also notifies the Hospital Information System, the physician, the patient's family and Ambulance Service System.

Moreover, the Healthcare Expert System also provides an API to make the patient's information available to authorized third party systems, taking into account privacy and ethics. This API is composed of RESTFul Web Services (Oracle, 2016) and the uses JSON (JSON, 2016). The purpose of this API is to facilitate the development of new solutions with the use of this data to promote innovation in the healthcare area. As a result, companies and researchers can benefit from this use.

The *Hospital module* is used by physicians, nurses and clinical staff, and it uses the hospital information system. This system contains the records of patients including information about age, gender, name, contacts, family contact. It also provides the real-time remote monitoring of the patients in critical condition. Integrated with this Hospital Information System, there is a mobile app, where in a case of any problem with a patient, it notifies the physician responsible for him. With this notification, this mobile app also presents the real-time situation and data from the sensors, such as ECG, blood pressure, blood glucose, heart rate, oxygen saturation, temperature, and breathing rate.

The *Family's Home module* is used by the patient's family and is connected with the Hospital Information System, however the information presented is unique to their related patient. It provides a mobile app that displays real-time monitoring from the sensors connected to the patient. Finally, this mobile app displays less information than the version used by the physicians and clinical staff because some of the data from sensors require medical expertise to be understood.

Regarding the ambient monitoring provided by this platform, all data from the sensors - temperature, location, and humidity - are presented at the hospital information system and its mobile apps, and at the ambulance service system. This monitoring is important because the control of the ambient temperature and humidity can directly affect the patient health treatment.

The *Ambulance Service module* is connected to the Ambulance Service System and is used by operators. The ambulances have a mobile app that is connected to the Ambulance Service System, using 3G/4G, to receive real-time information about the body and ambient monitoring sensors, emergency alarms and patient's situation. It is important to emphasize that the monitored location data from the patient is a key point for an effective response to the ambulance service.

Considering the presented modules of this platform and it purpose, there are important requirements that need to be addressed: privacy (Maglogiann, 2012; Sebestyen et al., 2014; Ray, 2015), security (Świątek & Rucinski, 2013; Fan et al., 2014; Hassan, 2014; Gao et al., 2015), interoperability (Jara et al., 2013; Castillejo et al., 2013; Gia et al., 2015), scalability (Maglogiann, 2012; Jara et al., 2013; Mohammed et al., 2014), reliability (Świątek & Rucinski, 2013; Hassan et al., 2014; Gao et al., 2015), robustness (Jara et al., 2013), ubiquity (López et al., 2013), portability (Mohammed et al., 2014), performance (Mohammed et al., 2014), availability (Maglogiann et al., 2012; Jara et al., 2013; Świątek & Rucinski, 2013) and integrity (Maksimović, et al., 2015). Once all the information transmitted from the sensors to the systems and mobile apps are sensible, there is a need for privacy in this communication. To assure privacy, this platform uses encryption. There is also the special need for security, which is guaranteed mainly by authentication (Maksimović et al., 2015). The IoT platform from the Cloud Health Infrastructure module assures scalability, integrity, portability and interoperability between the different types of the connected monitoring devices. The expert system of this module addresses the need for ubiquity, considering the defined rules and the customization feature. Finally, the proposed platform and its modules organization aim to achieve good performance, robustness, reliability, and availability regarding the information about patient monitoring.

FUTURE RESEARCH DIRECTIONS

With the use of a technology-based healthcare approach, there is an unprecedented opportunity for improving the quality and efficiency of healthcare. Consequently, this will improve the wellness of all citizens and lead to a better application of governments' financial resources. However, to effectively implement this strategy there are still big challenges that need to be addressed by research and innovation.

To effectively implement and adopt a technological approach based on IoT for remote healthcare, several non-functional issues need to be addressed. One of the most important aspects of healthcare is the safety of the solution. The safety in this context involves several factors, particularly: reliability, availability, security, integrity within a long-time-period. In software engineering, the ability to deal with these requirements is known as dependability. The exploration of dependability issues in the context of IoT is a very young and fertile research topic, creating an excellent opportunity to develop science and innovation involving both industry and academics. The engineering challenges - of conceiving and developing a set of technology-centric strategies to address the dependability requirement in healthcare applications based on the IoT infrastructure - is an open field for academics. Moreover, how to address dependability in healthcare applications based on the IoT infrastructure and targeted at underdeveloped countries, for example Brazil, is a big challenge due to the absence of basic health infrastructure that is generally found in developed countries.

Another key issue related to the proposed approach is the challenge of handling a huge volume of data produced by the different elements of the platform. In particular, the patient's home component has the capacity of collecting data from several sensors repeatedly during the day in a short period of time. It might produce a huge volume of data if we consider several patients being monitored at the same time. That produced data must be evaluated in real-time in order to detect a patient's health anomalies. Moreover, a historical data analysis of patient's health measures must be performed in order to detect patterns and anomalies, build a health profile of critical patients and make predictive analyses to prevent critical situations and more. Due to the abovementioned characteristics involving the proposed platform, it represents a rich field to apply, evaluate and evolve the techniques of the recent area known as Big Data.

Interoperability is also an open research problem, mainly when faced information technology systems in medical context. This ability allows different information systems, healthcare devices, and software applications to communicate, exchange data, and use the shared information across physicians, hospital, pharmacy, ambulance, and patient regardless of the application or application vendor. Despite the clear benefits of this capacity, i.e. improving the quality, safety, efficiency, and efficacy of healthcare delivery, the reality shows that currently, it is far from being achieved. Several issues must be addressed involving communication protocols, connectivity interfaces, open standards for healthcare data to reach an effectivity level of interoperability to allow the electronic exchange of patient's measures among all IT parties involved.

Finally, the use of remote monitoring is emerging as an opportunity to address resource shortages, health workers' overload, and an increased prevalence of chronic disease. However, assuring the long term effectiveness of remote monitoring and its impact on health indicators is a challenge to be overcome by experimental research on usability and clinical effects, to support medical decisions that can be subsidized based on remote monitoring.

CONCLUSION

To effectively implement this proposed platform in Brazil there are still big challenges not generally observed in developed countries that need to be addressed by research and innovation. Although the Brazilian constitution declares that healthcare is an official government responsibility and that it is the right of every citizen to have free public access to it, nevertheless government hospitals are often crowded, waiting times for appointment and treatment are lengthy, and the facilities are often inadequate. Indeed, healthcare is one of the worst problems in Brazil, ranking above violence and education.

On the other hand, the recent advances in IoT applied to healthcare is promoting a powerful environment for a disruptive innovation in the health research field as well as in the computer science area. This current scenario creates several opportunities to improve the effectiveness and efficiency of the use of national health infrastructure. Improving the quality and efficiency of the care for patients in a critical situation through the exploitation of technological "smart" solutions would promote substantial results in terms of welfare for the patients. The deployment of remote monitoring solutions for healthcare is not science fiction. The current advances in technology demonstrate that its development and deployment are viable and, in specific scenarios, it is already a reality. Some companies are already focused on the development of solutions for remote patient monitoring (RPM) and, at the same time, several types of research are being conducted to develop this field.

The proposed platform was described focusing on the patient's home care, but it can be easily evolved to also consider the patients inside the hospital in addition to other use cases involving health professionals, medical devices, information systems and other actors. With the use of a technology-based healthcare approach, there is an unprecedented opportunity for improving the quality and efficiency of the public health system in Brazil. Consequently, this will improve the wellness of all citizens and a better application of government financial resources. Moreover, the study, development, and deployment of appropriate technology-focused solutions addressing the specific issues and peculiarities of public

Brazilian healthcare is an excellent opportunity to develop science and innovation as well as inducing economic and social development.

Although patient monitoring technology is introduced as one solution to increase the quality and efficiency of the public health system, there is a real barrier in Brazil due to the absence (or inadequate availability) of basic technological infrastructure, which is the premise for this proposal. For this reason, the biggest challenge in an underdeveloped country is related to the influencing of public policy to make available the basic elements needed to make feasible remote healthcare for patients in a critical condition.

REFERENCES

Abinaya, V. K., Kumar, V., & Swathika. (2015). Ontology Based Public Healthcare System in Internet of Things (IoT). *Procedia Computer Science*, *50*, 99–102. doi:10.1016/j.procs.2015.04.067

Aehlert, B. (2011). *ACLS Advanced Cardiac Life Support, Study Guide* (4th ed.). Mosby.

American Diabetes Association. (2016). *Checking Your Blood Glucose*. Retrieved September 14, 2016, from http://www.diabetes.org/living-with-diabetes/treatment-and-care/blood-glucose-control/checking-your-blood-glucose.html

American Heart Association. (2016). *Understanding Blood Pressure Readings*. Retrieved September 14, 2016, from http://www.heart.org/HEARTORG/Conditions/HighBloodPressure/AboutHighBloodPressure/Understanding-Blood-Pressure-Readings_UCM_301764_Article.jsp

Atzori, L., Iera, A., & Morabito, G. (2010). The internet of things: A survey. *Computer Networks*, *54*(15), 2787–2805. doi:10.1016/j.comnet.2010.05.010

Bazerbashi, H., Merriman, K. W., Toale, K. M., Chaftari, P., Carreras, M. T. C., Henderson, J. D., ... Rice, T. W. (2014). Low tissue oxygen saturation at emergency center triage is predictive of intensive care unit admission. *Journal of Critical Care*, *29*(5), 775–779. doi:10.1016/j.jcrc.2014.05.006 PMID:24973103

Benlamri, R., & Docksteader, L. (2010). MORF: A mobile health-monitoring platform. *IT Professional*, *3*(3), 18–25. doi:10.1109/MITP.2010.3

Bolinder, J., et al. (in press). Novel glucose-sensing technology and hypoglycaemia in type 1 diabetes: a multicentre, non-masked, randomised controlled trial. *Lancet*.

Brasil. (2013). *Ministério da Saúde. Secretaria de Atenção à Saúde. Departamento de Atenção Especializada. Manual instrutivo da Rede de Atenção às Urgências e Emergências no Sistema Único de Saúde (SUS)*. Brasília, DF: Editora do Ministério da Saúde.

Brennan, T. (2010). Gestão da doença para promover o controle da pressão arterial entre os afro-americanos. *Popul Manag Saúde*, *13*(2), 65–72.

Cahill, T. J., & Prendergast, B. D. (2016). Infective endocarditis. *Lancet*, *387*(10021), 882–893. doi:10.1016/S0140-6736(15)00067-7 PMID:26341945

Castillejo, P., Martinez, J. F., Rodriguez-Molina, J., & Cuerva, A. (2013). Integration of wearable devices in a wireless sensor network for an E-health application. *IEEE Wireless Communications*, *20*(4), 38–49. doi:10.1109/MWC.2013.6590049

Chaudhry, S. I., Mattera, J. A., Curtis, J. P., Spertus, J. A., Herrin, J., Lin, Z., ... Krumholz, H. M. (2010). Telemonitoring in patients with heart failure. *The New England Journal of Medicine*, *363*(24), 2301–2309. doi:10.1056/NEJMoa1010029 PMID:21080835

Chen, L. M., Kennedy, E. H., Sales, A., & Hofer, T. P. (2013). Use of health IT for higher-value critical care. *The New England Journal of Medicine*, *368*(7), 594–597. doi:10.1056/NEJMp1213273 PMID:23363474

Chiuchisan, I., Costin, H. N., & Geman, O. (2014, October). Adopting the internet of things technologies in health care systems. In *Electrical and Power Engineering (EPE), 2014 International Conference and Exposition on* (pp. 532-535). IEEE. 10.1109/ICEPE.2014.6969965

Constantinescu, L., Kim, J., & Feng, D. D. (2012). SparkMed: A framework for dynamic integration of multimedia medical data into distributed m-health systems. *IEEE Transactions on Information Technology in Biomedicine*, *16*(1), 40–52. doi:10.1109/TITB.2011.2174064 PMID:22049371

Coulter, A. (2012). Patient engagement—what works? *The Journal of Ambulatory Care Management*, *35*(2), 80–89. doi:10.1097/JAC.0b013e318249e0fd PMID:22415281

Crossley, G. H., Boyle, A., Vitense, H., Chang, Y., & Mead, R. H. (2011). CONNECT Investigators. The CONNECT (Clinical Evaluation of Remote Notification to Reduce Time to Clinical Decision) trial: The value of wireless remote monitoring with automatic clinician alerts. *Journal of the American College of Cardiology*, *57*(10), 1181–1189. doi:10.1016/j.jacc.2010.12.012 PMID:21255955

Dabbs, A. D., Myers, B. A., Mc Curry, K. R., Dunbar-Jacob, J., Hawkins, R. P., Begey, A., & Dew, M. A. (2009). User-centered design and interactive health technologies for patients. *Computers, Informatics, Nursing*, *27*(3), 175–183. doi:10.1097/NCN.0b013e31819f7c7c PMID:19411947

Delicato, F. C., Pires, P. F., & Batista, T. (2013). *Middleware solutions for the Internet of Things*. Springer London. doi:10.1007/978-1-4471-5481-5

Dwivedi, S. (2016). Reshaping Medical Practice and Care with Health Information Systems. In R. Misra (Ed.), *The Benefits and Challenges of Using Mobile-Based Tools in Self-Management and Care* (pp. 1–13). University of Hull. doi:10.4018/978-1-4666-9870-3

Ettehad, D., Emdin, C. A., Kiran, A., Anderson, S. G., Callender, T., Emberson, J., ... Rahimi, K. (2016). Blood pressure lowering for prevention of cardiovascular disease and death: A systematic review and meta-analysis. *Lancet*, *387*(10022), 957–967. doi:10.1016/S0140-6736(15)01225-8 PMID:26724178

Ewer, A. K. (2014, August). Pulse oximetry screening: Do we have enough evidence now? *Lancet*, *384*(9945), 725–726. doi:10.1016/S0140-6736(14)60575-4 PMID:24768154

Fan, Y. J., Yin, Y. H., Da Xu, L., Zeng, Y., & Wu, F. (2014). IoT-based smart rehabilitation system. *IEEE Transactions on Industrial Informatics, 10*(2), 1568-1577.

Ferguson, N. D., Cook, D. J., Guyatt, G. H., Mehta, S., Hand, L., Austin, P., ... Meade, M. O. (2013). High-Frequency Oscillation in Early Acute Respiratory Distress Syndrome. *The New England Journal of Medicine*, *368*(9), 795–805. doi:10.1056/NEJMoa1215554 PMID:23339639

Filkins, B. L. (2016). Privacy and security in the era of digital health: What should translational researchers know and do about it?. *Am J Transl Res.*, *8*(3), 1560–1580. PMID:27186282

Gao, R., Zhao, M., Qiu, Z., Yu, Y., & Chang, C. H. (2015, June). Web-based motion detection system for health care. In *Computer and Information Science (ICIS), 2015 IEEE/ACIS 14th International Conference on* (pp. 65-70). IEEE.

Gartner. (2015). *Gartner says 6.4 billion connected ''things''will be in use in 2016, up 30 percent from 2015*. Retrieved June 10, 2016, from Http://www.gartner.com/newsroom/id/3165317

Gasparrini, A., Guo, Y., Hashizume, M., Lavigne, E., Zanobetti, A., Schwartz, J., ... Armstrong, B. (2015). Mortality risk attributable to high and low ambient temperature: A multicountry observational study. *Lancet*, *386*(9991), 369–375. doi:10.1016/S0140-6736(14)62114-0 PMID:26003380

Gelder, I. C. V. (2016). Rate control in atrial fibrillation. *Lancet*, *388*(10046), 818–828. doi:10.1016/S0140-6736(16)31258-2 PMID:27560277

Gia, T. N., Thanigaivelan, N. K., Rahmani, A. M., Westerlund, T., Liljeberg, P., & Tenhunen, H. (2014, October). Customizing 6LoWPAN networks towards Internet-of-Things based ubiquitous healthcare systems. In NORCHIP, 2014 (pp. 1-6). IEEE.

Goligher, E. C., Ferguson, N. D., & Brochard, L. J. (2016). Clinical challenges in mechanical ventilation. *Lancet*, *387*(10030), 1856–1866. doi:10.1016/S0140-6736(16)30176-3 PMID:27203509

Gorp, V. P., & Comuzzi, M. (2014). Lifelong personal health data and application software via virtual machines in the cloud. *IEEE Journal of Biomedical and Health Informatics*, *18*(1), 36–45. doi:10.1109/JBHI.2013.2257821 PMID:24403402

Gubbi, J., Buyya, R., Marusic, S., & Palaniswami, M. (2013). Internet of things (iot): A vision, architectural elements, and future directions. *Future Generation Computer Systems*, *29*(7), 1645–1660. doi:10.1016/j.future.2013.01.010

Hall, J. E. (2006). *Guyton and Hall Textbook of Medical Physiology (12th ed.)*. Elsevier.

Hassan, M. M., Albakr, H. S., & Al-Dossari, H. (2014, November). A Cloud-Assisted Internet of Things Framework for Pervasive Healthcare in Smart City Environment. In *Proceedings of the 1st International Workshop on Emerging Multimedia Applications and Services for Smart Cities* (pp. 9-13). ACM. 10.1145/2661704.2661707

Heidbuchel, H., Hindricks, G., Broadhurst, P., Van Erven, L., Fernandez-Lozano, I., Rivero-Ayerza, M., ... Annemans, L. (2015). EuroEco (European Health Economic Trial on Home Monitoring in ICD Patients): A provider perspective in five European countries on costs and net financial impact of follow-up with or without remote monitoring. *European Heart Journal*, *36*(3), 158–169. doi:10.1093/eurheartj/ehu339 PMID:25179766

Hendriks, M., & Rademakers, J. (2014). Relationships between patient activation, disease-specific knowledge and health outcomes among people with diabetes; a survey study. *BMC Health Services Research*, *14*(1), 393. doi:10.1186/1472-6963-14-393 PMID:25227734

Hindia, M.N., Rahman, T.A., Ojukwu, H., Hanafi, E.B., & Fattouh, A. (2016). Enabling Remote Health-Caring Utilizing IoT Concept ov er LTE-Femtocell Networks. *Lista Jornal PLoS One, 11*(5).

Hindricks, G., Taborsky, M., Glikson, M., Heinrich, U., Schumacher, B., Katz, A., ... Søgaard, P. (2014). Implant-based multiparameter telemonitoring of patients with heart failure (IN-TIME): A randomised controlled trial. *Lancet*, *384*(9943), 583–590. doi:10.1016/S0140-6736(14)61176-4 PMID:25131977

Hochron, S., & Goldberg, P. (2015). Driving physician adoption of mHeath solutions. *Healthcare Financial Management*, *69*(2), 36–39. PMID:26665538

IEC 60601. (2015). *IEC 60601-1-11:2015 RLV Redline version*. Retrieved October 11, 2016, from https://webstore.iec.ch/publication/22261

Jara, A. J., Zamora-Izquierdo, M. A., & Skarmeta, A. F. (2013). Interconnection framework for mHealth and remote monitoring based on the Internet of Things. *IEEE Journal on Selected Areas in Communications*, *31*(9), 47–65. doi:10.1109/JSAC.2013.SUP.0513005

Jiang, Y., Sereika, S. M., Dabbs, A. D., Handler, S. M., & Schlenk, E. A. (2016). Using mobile health technology to deliver decision support for self-monitoring after lung transplantation. *International Journal of Medical Informatics*, *94*, 164–171. doi:10.1016/j.ijmedinf.2016.07.012 PMID:27573324

JSON. (2016). *Introducing JSON*. Retrieved September 14, 2016, from http://www.json.org/

Khattak, H. A., Ruta, M., Di Sciascio, E., & Sciascio, D. (2014, January). CoAP-based healthcare sensor networks: A survey. In *Proceedings of 2014 11th International Bhurban Conference on Applied Sciences & Technology (IBCAST)* (pp. 499-503). IEEE. 10.1109/IBCAST.2014.6778196

Kotsimbos, T., Williams, T. J., & Anderson, J. P. (2012). Update on lung transplantation: Programmes, patients and prospects. *European Respiratory Review*, *21*(126), 271–305. doi:10.1183/09059180.00006312 PMID:23204117

Langhan, M. L., Kurtz, J. C., Schaeffer, P., Asnes, A. G., & Riera, A. (2014). Experiences with capnography in acute care settings: A mixed-methods analysis of clinical staff. *Journal of Critical Care*, *29*(6), 1035–1040. doi:10.1016/j.jcrc.2014.06.021 PMID:25129575

Liao, S. H. (2005). Expert system methodologies and applications—a decade review from 1995 to 2004. *Expert Systems with Applications*, *28*(1), 93–103. doi:10.1016/j.eswa.2004.08.003

López, P., Fernández, D., Jara, A. J., & Skarmeta, A. F. (2013, March). Survey of internet of things technologies for clinical environments. In *Advanced Information Networking and Applications Workshops (WAINA), 2013 27th International Conference on* (pp. 1349-1354). IEEE. 10.1109/WAINA.2013.255

Maglogiann, D. I. (2012). Bringing IoT and Cloud Computing towards Pervasive Healthcare Charalampos. *Sixth International Conference on Innovative Mobile and InternetServices in Ubiquitous Computing*.

Maksimović, M., Vujović, V., & Perišić, B. (2015, June). A custom internet of things healthcare system. In *2015 10th Iberian Conference on Information Systems and Technologies (CISTI)* (pp. 1-6). IEEE. 10.1109/CISTI.2015.7170415

Mattern, F., & Floerkemeier, C. (2010). From the Internet of Computers to the Internet of Things. In *From active data management to event-based systems and more* (pp. 242–259). Springer Berlin Heidelberg. doi:10.1007/978-3-642-17226-7_15

Medicinet. (2016). *Definition of Heart Rate*. Retrieved September 17, 2016, from http://www.medicinenet.com/script/main/art.asp?articlekey=3674

Mendes, E.V. (2010). As redes de atenção à saúde. Ciênc. *saúde coletiva, 15*(5), 2297-2305.

Minutolo, A., Esposito, M., & Pietro, G. D. (2016). Design and validation of a light-weight reasoning system to support remote health monitoring applications. *Engineering Applications of Artificial Intelligence, 41*, 232–248. doi:10.1016/j.engappai.2015.01.019

Mohammed, J., Lung, C. H., Ocneanu, A., Thakral, A., Jones, C., & Adler, A. (2014, September). Internet of Things: Remote patient monitoring using web services and cloud computing. *Chest, 149*(2), 576–585.

Oracle. (2016). *What Are RESTful Web Services?*. Retrieved September 14, 2016, from http://docs.oracle.com/javaee/6/tutorial/doc/gijqy.html

Pang, Z. (2013). *Technologies and architectures of the internet-of-things (iot) for health and well-being.* Retrieved June 13, 2016, from http://kth.diva-portal.org/smash/get/diva2:621384/FULLTEXT01.pdf

Piccini J.P., et al. (in press). Impact of remote monitoring on clinical events and associated health care utilization: A nationwide assessment. *Heart Rhythm.*

Pitsillides, A. (2006). DITIS: A collaborative virtual medical team for home healthcare of cancer patients M-Health. Springer.

Poenaru, E., & Poenaru, C. (2013, November). A structured approach of the Internet-of-Things eHealth use cases. In E-Health and Bioengineering Conference (EHB), 2013 (pp. 1-4). IEEE. doi:10.1109/EHB.2013.6707299

Portugal, G., Cunha, P., Valente, B., Feliciano, J., Lousinha, A., Alves, S., ... Ferreira, R. C. (2016). Influence of remote monitoring on long-term cardiovascular outcomes after cardioverter-defibrillator implantation. *International Journal of Cardiology, 222*, 764–768. doi:10.1016/j.ijcard.2016.07.157 PMID:27521554

Prescher, S. (2014). Wird Telemonitoring von Patienten mit chronischer Herzinsuffizienz angenommen?. *Dtsch Patienten, 139*(16), 829–834.

Raad, M. W., Sheltami, T., & Shakshuki, E. (2015). Ubiquitous Tele-health System for Elderly Patients with Alzheimers. *Procedia Computer Science, 52*, 685–689. doi:10.1016/j.procs.2015.05.075

Ray, P. P. (2015, January). Internet of Things for Sports (IoTSport): An architectural framework for sports and recreational activity. In *Electrical, Electronics, Signals, Communication and Optimization (EESCO), 2015 International Conference on* (pp. 1-4). IEEE.

Saxon, L. A., Hayes, D. L., Gilliam, F. R., Heidenreich, P. A., Day, J., Seth, M., ... Boehmer, J. P. (2010). Long-term outcome after ICD and CRT implantation and influence of remote device follow-up: The ALTITUDE survival study. *Circulation*, *122*(23), 2359–2367. doi:10.1161/CIRCULATIONAHA.110.960633 PMID:21098452

Sebestyen, G., Hangan, A., Oniga, S., & Gál, Z. (2014, May). eHealth solutions in the context of Internet of Things. *Proc. IEEE Int. Conf. Automation, Quality and Testing, Robotics (AQTR 2014)*, 261-267. 10.1109/AQTR.2014.6857876

Sieverink, F., Siemons, L., Braakman-Jansen, A., & Van Gemert-Pijnen, L. (2016). Internet of Things & Personalized Healthcare. *Studies in Health Technology and Informatics*, *221*, 129. PMID:27071903

Sweeney, R.M., McAuley, D.F. (in press). Acute respiratory distress syndrome. *Lancet.*

Świątek, P., & Rucinski, A. (2013, October). IoT as a service system for eHealth. In *e-Health Networking, Applications & Services (Healthcom), 2013 IEEE 15th International Conference on* (pp. 81-84). IEEE. 10.1109/HealthCom.2013.6720643

Tabish, R., Ghaleb, A. M., Hussein, R., Touati, F., Mnaouer, A. B., Khriji, L., & Rasid, M. F. A. (2014, February). A 3G/WiFi-enabled 6LoWPAN-based U-healthcare system for ubiquitous real-time monitoring and data logging. In *2nd Middle East Conference on Biomedical Engineering* (pp. 277-280). IEEE 10.1109/MECBME.2014.6783258

Thabit, H., Bally, L., & Hovorka, R. (in press). Available at a flash: a new way to check glucose. *Lancet.*

van der Valk, S., Myers, T., Atkinson, I., & Mohring, K. (2015, April). Sensor networks in workplaces: Correlating comfort and productivity. In *Intelligent Sensors, Sensor Networks and Information Processing (ISSNIP), 2015 IEEE Tenth International Conference on* (pp. 1-6). IEEE.

Varma, N., Piccini, J. P., Snell, J., Fischer, A., Dalal, N., & Mittal, S. (2015). Relationship between level of adherence to automatic wireless remote monitoring and survival in pacemaker and defibrillator patients. *Journal of the American College of Cardiology*, *65*(24), 2601–2610. doi:10.1016/j.jacc.2015.04.033 PMID:25983008

Vidal-Petiot, E., et al. (in press). Cardiovascular event rates and mortality according to achieved systolic and diastolic blood pressure in patients with stable coronary artery disease: an international cohort study. *Lancet.*

Yang, C. T., Liu, J. C., Liao, C. J., Wu, C. C., & Le, F. Y. (2013, December). On Construction of an Intelligent Environmental Monitoring System for Healthcare. In *2013 International Conference on Parallel and Distributed Computing, Applications and Technologies* (pp. 246-253). IEEE. 10.1109/PDCAT.2013.45

Yang, G., Xie, L., Mäntysalo, M., Zhou, X., Pang, Z., Da Xu, L., ... Zheng, L. R. (2014). A health-IoT platform based on the integration of intelligent packaging, unobtrusive bio-sensor, and intelligent medicine box. *IEEE Transactions on Industrial Informatics, 10*(4), 2180-2191.

Yinong, C. (2016). Analyzing and visual programming internet of things and autonomous decentralized systems. *Simulation Modelling Practice and Theory*, *65*, 1–10. doi:10.1016/j.simpat.2016.05.002

Zanchetti, A., Thomopoulos, C., & Parati, G. (2015). Randomized controlled trials of blood pressure lowering in hypertension: A critical reappraisal. *Circulation Research*, *116*(6), 1058–1073. doi:10.1161/CIRCRESAHA.116.303641 PMID:25767290

Zandomenighi, R. C., Mouro, D. L., Oliveira, C. A., & Martins, E. A. P. (2014). Cuidados intensivos em um serviço hospitalar de emergência: Desafios para os enfermeiros. REME rev. min. *Enferm*, *18*(2), 404–414.

This research was previously published in Next-Generation Mobile and Pervasive Healthcare Solutions edited by Jose Machado, António Abelha, Manuel Filipe Santos, and Filipe Portela, pages 1-21, copyright year 2018 by Medical Information Science Reference (an imprint of IGI Global).

Section 2
Patient Engagement

Chapter 8
Strategic Healthcare Service Management

Bhagyashri Subhash Sangamnere
Modern Institute of Business Management (MIBM), India

Kalyani Srinivas C
PES Modern College of Engineering, India

ABSTRACT

Healthcare is a service industry and it consists of health organizations (hospitals, clinics etc.), people (patients, doctors and nurses) and health technologies. Healthcare organizations are complex in nature and need to improve quality while maintaining optimum cost. Patient is final consumer of health services and he is the customer hence healthcare service quality is nothing but perceived satisfaction by patient. Prevalent trends in healthcare industry such as emerging healthcare technology, increasing demand, changing disease patterns and growing government support are contributing to a need to achieve efficiency and set benchmarks by overcoming challenges in healthcare service sector being underserved and under-consumed. There is a scope to improve quality and efficiency using various strategies like adopting advanced technologies and positioning in order to achieve delight in delivery of healthcare services. Major healthcare players are adopting unique strategies irrespective of their diverse geographical presence and range of services from single specialty, super-specialty or multispecialty to deliver healthcare services efficiently.

INTRODUCTION

Health care is primarily a service industry, which constitutes major part in Indian economy. The structure of health care broadly includes the facilities such as clinics and hospitals, people i.e. skilled doctors and nurses, and technologies that potentially influence the quality of health care.

The health care industry comprises of various sectors based on market they serve. Hospitals constitute major part of the health care providing in-patient and out-patient services by contributing to around 70% of industry revenue, followed by pharmaceutical sector contributing 20% of the revenue producing and marketing drugs essential for medication. Remaining sectors like medical insurance, medical equipment and supplies.

DOI: 10.4018/978-1-5225-6198-9.ch008

HEALTHCARE AS A SERVICE INDUSTRY

The process of delivery of health care service is interactions between health care providers and patients over a certain period until desired result is obtained. Sequence of services is provided over time in relation to a specific patient complaint or diagnosis or various kinds of services are provided for specific health problems patient.

Hospitals: Major Part of Health Care

Hospital is complex and highly fragmented segment of health care; It is the organization that offers services (treatment) which satisfies customers (Patients). Hospital segment covers major share in health care industry followed by pharmaceuticals and others.

Hospitals can be broadly classified as public hospitals and private hospitals. The public health care system consists of health care facilities run by the central and state government, which provide services free of cost or at subsidized rates to low income group. For example Government hospitals in urban areas and Primary health centers, Ayush etc. in rural area. However, private hospitals are established and operated by doctors, trusts and corporate hospitals chains Example: Apollo Hospital, Fortis Health care operate as a chain of corporate hospitals spread across the country.

Figure 1. Healthcare market breakup

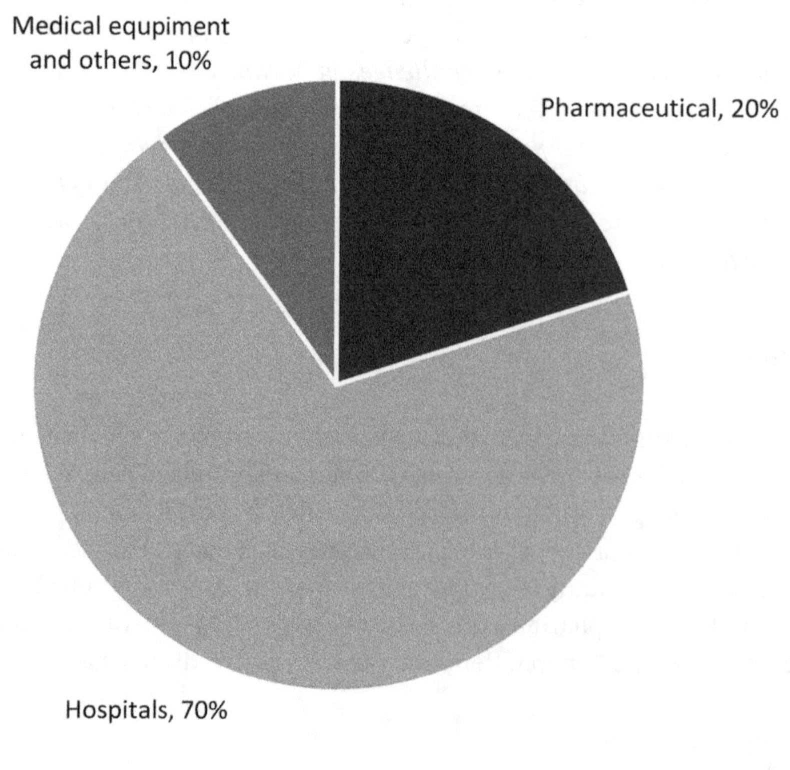

Key Highlights and Trends of Indian Healthcare Industry

1. In 2014 globally India was the sixth largest market in size.
2. Health care, pharma and biotech sector marked continuous uptrend in deal activities and witnessed approximately USD 630 million investment.
3. **Change in Disease Pattern:** With population of over one billion people which is growing at a rate of 1.6 percent per year disease pattern is shifting from communicable diseases to lifestyle diseases which demands for specialized care.
4. **Expansion of Health Care:** Geographical expansion through fulfilling demand for health in tier-II and tier-III cities. Public private partnerships (PPP) are primarily adopted with more government support.
5. **Health Care Insurance Penetration:** Improving health insurance penetration
6. **Emergence of Technology:**
    a. **Mobile Based Health Delivery:** mHealth initiatives in India are taking pace with presence of good mobile technology infrastructure and 4G internet etc.
    b. **Telemedicine:** Many major hospitals in India are adopting telemedicine services which can bridge the rural-urban divide in terms of medical facilities to remote locations. Example: Apollo and Narayan hridayalaya telemedicine centers.
    c. **Digital Health:** For standard health care service quality sector is investing heavily on various IT products and services. Example: Electronic Medical Record (EMR), Hospital Information System (HIS)

Figure 2. Healthcare structure (segments)

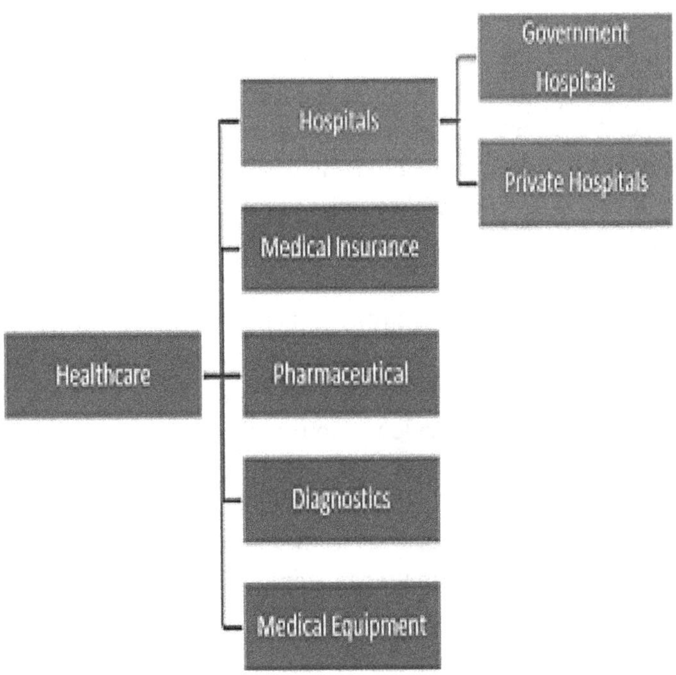

7. **Increasing Government Support:** Health care is a key focus area for government hence under the 12th five-year plan (2012–17) the Planning Commission allocated USD55 billion to the Ministry of Health and Family Welfare, which is about three times the actual expenditure under the 11th Five-Year Plan. Recently Modi's government also unveiled plans for a nationwide universal health care system known as the National Health Assurance Mission, which would provide all citizens with free drugs, diagnostic treatments, and insurance for serious ailments. In 2015, implementation of a universal health care system was delayed due to budgetary concerns (Source: Wikipedia, Planning Commission, Ministry of Health & Family Welfare)
8. **Globalization of Health Care:** The government of India has priority to develop India as a global health care hub. The health services have become tradable due to World Trade Organization and their General Agreement on Trade in Services (Smith, 2004; Smith et al,. 2009b). New element of this trade involves the flow of patients (i.e. Medical tourism), medical technology, and regulatory systems across national borders. This ease in trading health services and focus on health for all has contributed to new trends in health care service production and consumption over recent periods.

Based on these trends and structure of healthcare industry it is recognized that healthcare organizations are complex and diverse in nature. The current health care system need to fulfil the objective to improve quality and access while maintaining optimum level of cost. Main objective of the healthcare is to gain government support through public-private partnerships and grow at rapid pace through advanced technologies and health communication strategies.

BACKGROUND

Healthcare is defined as "a multitude of services rendered to individual, families or communities by health service professionals for promoting, maintaining, monitoring or restoring health" (Last, 1993)

Recently it is observed that healthcare has become one of the extremely complex industries in the world (Bertolini et al., 2011). There are an increasing number of medical specializations, complex therapies and equipments, disease burden, increasing health care quality dimensions, rapid growth in the world health care market and several service units revolve around different organizations. In Indian context as per Planning Commission Report there is a need to transform healthcare into an accountable and affordable system of quality services.

Services has been defined as either consumer services (e.g. department stores) or professional services (Eg. Healthcare) (Fitzsimmons & Sullivan, 1982; Sasser et al., 1978). Professional services, which are provided by professionals i.e. doctors or lawyers such as healthcare services, legal services etc, have been classified as close to pure services. No tangible good is exchanged between the provider and the consumer and the service is usually produced and consumed simultaneously for each customer (Woodside et al., 1989).

Health systems are undergoing rapid change and the requirements for conforming to the new challenges of changing demographics, disease patterns, emerging and re-emerging diseases. Some health systems fail to provide the essential services and some are scraping under the strain of inefficient provision of services. A number of issues including governance in health, financing of health care, human resource imbalances, access and quality of health services, along with the impacts of reforms in other areas of the economies significantly affect the ability of health systems to deliver (WHO-EMRO, 2006)

An organization's position is conveyed by exploring individuals' perceptions of that organization in comparison to other organizations with respect to specific attributes. These attributes then serve as cues for its image. It has been recognized that services when compared to products face more difficult challenges when it comes to effective positioning strategies (Kemp, Jillapalli & Becerra, 2014; Blankson & Kalafatis, 1999)

HEALTHCARE SERVICE CASE STUDY

Apollo Hospitals Enterprise Limited (AHEL)

An India-based publically listed holding company, which provides healthcare services primarily in India, Mauritius, Bangladesh, and Kuwait. Apollo is specialized in providing tertiary care. The Company owns and manages 55 hospitals in and around India. It also offers telemedicine services, education and training programs, and research services.

Positioning of Apollo Hospitals' Health Care Service

1. **Product:** Apollo Hospital provides quality health care services through 69 branches and is prominently known for knee and hip replacement surgeries besides other major ailments. The specialties at the hospital are Heart, Orthopedics, Spine, Cancer Care, Gastroenterology, Neurosciences, Nephrology and Urology Critical Care. Eg. 27,000 heart surgeries with a success rate of 99.6% on par with global standards and 70% success rate in Bone Marrow Transplant. AHEL also runs India's largest network of pharmacies and international consulting services
2. **Price:** AHEL provides high quality services at premium price owing to its brand value, assurance to patients, cost advantage and positioning; yet prices are surprisingly low by global standards. Eg. A liver transplant that cost ~$300,000 in the United States is $45,000 at an Apollo hospital.
3. **Place:** As of March 31st, 2016, the company own 9,554 beds, 1,822 pharmacies, 100 primary care and diagnostic clinics, and 100 telemedicine units in 9 countries. It is located in 15 different places across India which include Ahmedabad, Bangalore, Bhubaneshwar, Chennai, Delhi, Hyderabad, Kolkata, Madurai, Mauritius, Mysore Noida.
4. **Promotion:** Apollo partnered with SITACARE a division of Kuoni Travel (India) to promote international medical tourists through marketing campaigns (www.sitacare.com) along with international promotions Apollo focus on domestic market through the Community Initiatives like:
    a. **SACH:** Save a Childs Heart.
    b. **CURE:** Extends preventive as well as rehabilitative cancer treatment to the economically backward.
    c. **SAHI:** Society to Aid the Hearing Impaired.
    d. **DISHA:** Distance Healthcare Advancement Project.
5. **People:** Apollo has approximately 19,000 doctors, nurses, paramedics, clinical staff and management professionals to manage over 9,215 beds across 55 hospitals in India and abroad. The Apollo has approximately 50 per cent physicians hat are Indians who had returned home from careers in the United States and other developed nations. Apollo is competitive and also facilitate innovation through research facilities.

Figure 3. Apollo healthcare service marketing mix

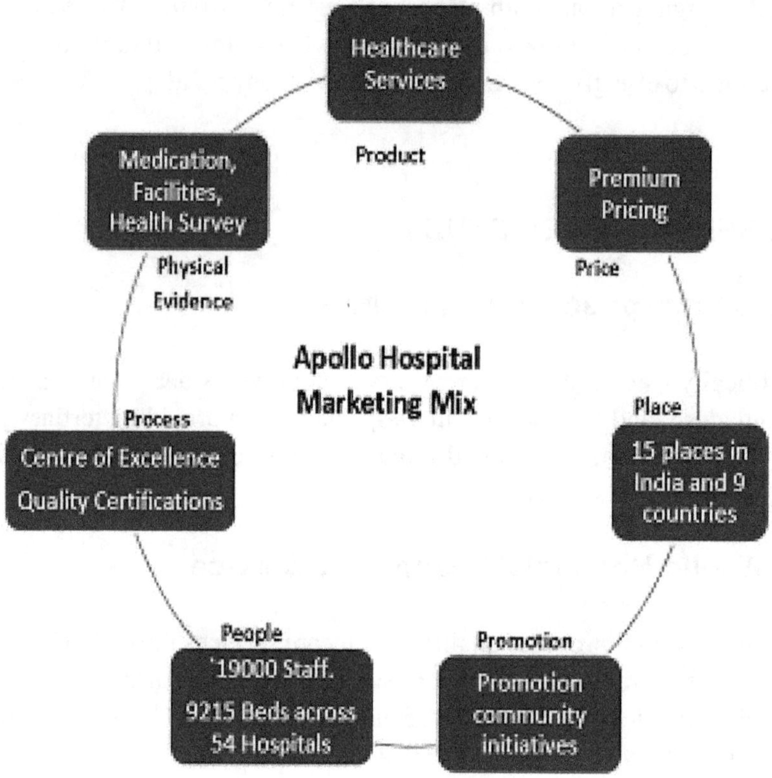

6. **Process:** The largest achievement of the Apollo Group has been to take quality health care to across the length and breadth of India. Apollo hospital is NABH, NABL accredited and has ISO 9002 and ISO 14001 certifications; Declared as a 'Centre of Excellence' by the Government of India. This operation in itself involves very established procedures and documentation.
7. **Physical Evidence:** Apollo Hospital is known for its quality health care services provision for all the ailments & diseases, assuring good medication, best facilities. Apollo Hospitals conducts a rigorous site survey process as well to take care of various parameters in all their hospitals.

Strategies Adopted by Other Players in Healthcare

The large private healthcare services providers i.e. peers of Apollo hospital like Fortis hospital, Care hospitals, Narayana Hridayalaya, Sterling hospital, Max hospital, Medanta medicity have adopted various strategies for consistent growth. They are actively seeking their goals of tapping dynamic competitive market by enhancing their reach across the country through the building new hospitals, acquisition and upgrade of existing hospitals. These major players are adopting of strategies like increasing their investments, expansion of hospital network, creating large pool of talents and diversification.

Many of Apollo's competitors like Fortis hospital and Max healthcare worked on building integrated delivery networks ranging from primary to tertiary care services. These players have diverse geographical presence and range of services from single specialty, super-specialty or multispecialty health care.

After analyzing AHEL case let there are few questions to be answered:

1. Is positioning helpful in healthcare services?
2. How Indian key healthcare players can excel to reach competitive mass market?
3. Which strategies and tactics are useful in marketing healthcare services effectively?

INDIAN HEALTHCARE INDUSTRY AND SERVICE TACTICS

Indian Health care industry is growing at rapid pace than anticipated before due to its strengthening coverage, quality services and increasing expenditure by public and private players. It is projected to grow at CAGR of 17 per cent during 2015–20 to reach USD 280 billion by 2020.

Current Healthcare Market Size is approximately USD 100 Billion (2015E)

The rate of growth of the health care industry in India is moving forward parallel with the software industry. The growth is driven by various factors that primarily includes:

- **Growing Demand:** Demand for quality health care is increasing day by day due to various reasons like awareness among patients as well as increasing disease conditions.

Figure 4. Comparison of major players
Source: Apollo Investor Presentation (31st March 2015 and 31st March 2016). 'Fortis corporate presentation' as of April 2015.

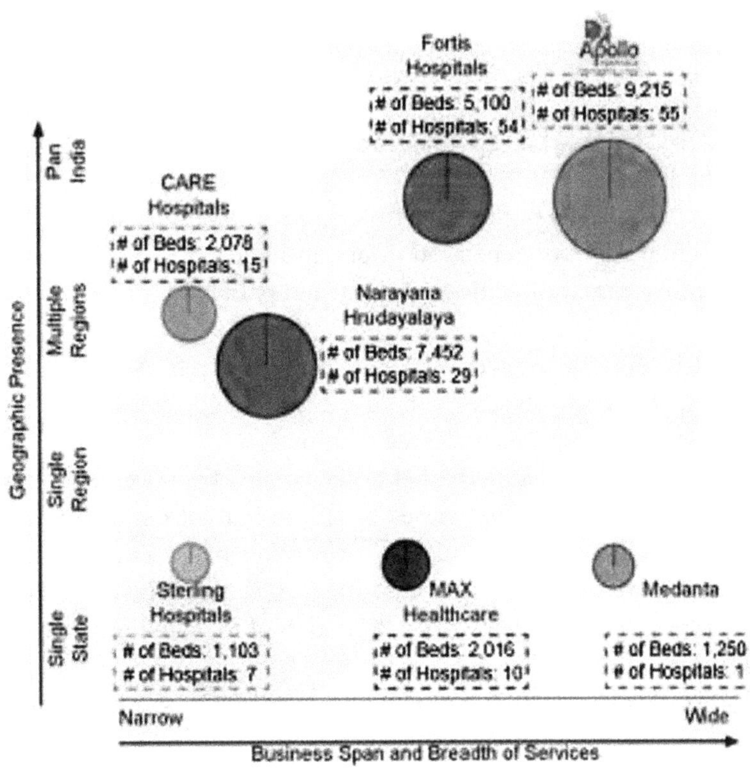

Figure 5. Indian healthcare industry growth YOY
Source: Frost & Sullivan.

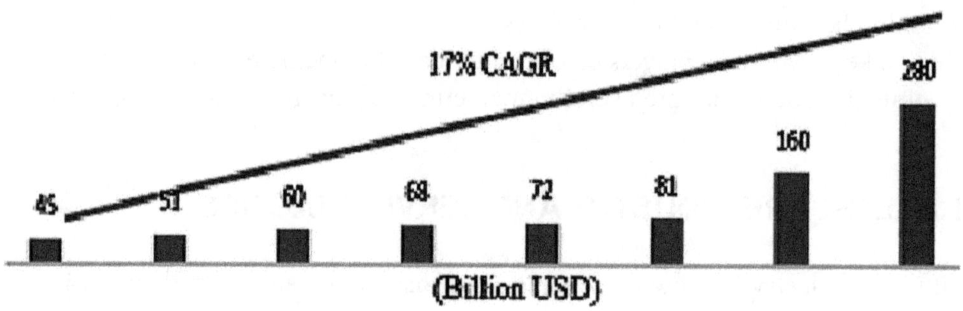

- **Government Support:** Government support to health care industry is increasing with focus on growth in health care by various government agencies through 12th five-year plan followed by a nationwide universal health care system proposed by current modi government.
- **Increasing PE and FDI Investments:** Investments in health are encouraged. Health care is witnessing both PE as well as FDI investments. The Government has permitted 100% FDI for all health services under automatic route. Developing the super specialty hospitals in India through the FDI route is very important area for future growth.
- **Innovation and Technology Introduction:** Introduction of innovative technologies in health care organizations and tremendous use of these technologies add up to the development of these services. Health research and innovation is fueled among all players. Private players are spending good percent of their profit in research to develop advanced medical treatments and techniques; which will help improve efficiency in industry and health status too.

CHARACTERISTICS OF HEALTHCARE SERVICES

Any kind of service is a combination of the intangible and the tangible aspects of these two intangible aspect dominates. A hospital as an organization offers an intangible service called healthcare that is characterized by intangibility, variability, inseparability and perishability.

- **Intangibility:** Health services are perishable and cannot be retrieved once utilized. They cannot be stored or packed. However, services can be produced and consumed instantly also it can be merged with tangible elements to increase satisfaction level.
- **Heterogeneity:** Quality of service delivered differentiates from one extreme to another. Extent and range of health services provided varies based on patient condition, disease, environment, facilities and type of treatment. Health players offer range of services under one roof and as one brand. The same service provided at different time or place may be different; This inconsistency can be cured by training the employees and automation of the work to avoid deterioration in quality over repetition. Eg. Automatic blood testing
- **Inseparability:** Healthcare services are unique and inseparable from provider. During health care service transaction, it is mandatory that provider and consumer are present at a time and at a same

place hence it is not possible that provider of one service can be available at two different places at a time.

- **Assurance, Empathy, and Responsiveness:** Health care being Patient centered. Services offered are patient centric paying attention to patient's needs and preferences. patient being final consumer and sole service receiver.
- **Timeliness:** Timed delivery of services is critical else it will negatively affect patient outcome and satisfaction. Organized healthcare minimize patient wait time and improve efficiency.
- **Adaption of Cost Leadership:** Among peers through economies and technologies; reduce overall cost. This will eventually lower the service charge and help attract patients from all over world. Low income group will easily access the health care services.
- **Selective Access to Health Care Services:** It is selective based on various factors; most of the health care expenses are paid out of pocket as only 17% of population covered under health insurance while public health care is available free for people below poverty line.

MARKETING MIX

The strategy of marketing mix is one of the core concepts of marketing theory (Ziethaml and Bitner, 2000). Kotler (2000: P15) defines the marketing mix as "the set of marketing tools that the firm uses to pursue its marketing objectives in the target market".

Services are different from products, because of their characteristics like intangibility, inseparability, heterogeneity, and perishability. A number of researchers (Ziethaml, 2000, Ahmad, 2007; Kotler, 2011) have previously argued that the traditional 4Ps of the marketing mix model are inadequate for the marketing of services marketing. Marketing mix for services is extended from 4Ps of traditional product to 7Ps adding three elements to the traditional model i.e. participants or people, physical evidence and processes (Booms and Bitner 1981). Consequently, the marketing strategy in the services should include the 7Ps of the services marketing mix and framework, which may have a crucial effect on hospital performance.

1. **Product:** Offering of commercial intent; having both tangible and intangible components including brand name, accessories and product line. Example: Arvind is brand famous for eye care. Core product is treatment of human illness including Consultations, Examinations, Surgical Operations whereas augmented products are ambience, infrastructure and equipment. Eg. Automation equipment like X-ray scanners.
2. **Price:** A particular service is acceptable to the customer at a particular price and if the price increases then the same service might not be acceptable to the customer. Service pricing follows the principles and practices of pricing of goods and therefore they are either cost based or market based. Service pricing should be such as to provide value addition and quality indication The amount extends on the category of room and the treatment or surgical procedure planned. Private sector has their own competitive, profit oriented or customer oriented price structures whereas government use price controls to ensure affordability and availability of vital drugs to masses.
    a. **Cost Based Pricing:** Price for In hospital services, this method is cumbersome because the tracking and identification of costs are difficult. Fee for services, however can be used by doctors. Notwithstanding, some hospitals in the private sector follow this method.

b. **Competition Based Pricing:** Heterogeneity of service across and within providers makes the approach complicated.
c. **Demand Based Pricing:** Cost based pricing and competition based pricing do not consider certain criteria. Demand based pricing involves price setting consistent with customer perception of value. Demand fluctuations should be successfully handle It is always a challenge to balance commercial interests with social view.

3. **Place:** Place is means by which provider get services delivered to consumers i.e. location of hospital, Clinics and Operation theatre. Public health facilities are insufficient compared to population of India. Rural or remote areas do not have enough small health service units. There are various ways in which technology and strategies are adopted to deliver services to end consumer. Telemedicine is one of the fast emerging trends of remote diagnosis and treatment of patients in rural area via video conferencing. Example: Apollo, Narayan hridyalaya and Arvind Hospitals developed public-private partnerships for telemedicine.
4. **Promotion:** India is emerging as medical tourism destination owing to state of the art private hospitals, well trained english speaking staff, advanced clinical, diagnostic facilities and relatively low cost. However generally hospitals do not go for aggressive promotion they rely on word of mouth publicity and social networks. Below are the most commonly used promotional approaches:
    a. **Health Checkups:** They plan some annual and disease specific health checkups to attract corporate clients.
    b. **Rural Camps:** Hospitals plan rural health checkup camps at very reasonable price range
    c. **Referral:** Patient referral is physician induced promotion
    d. **Advertisements:** Hospital advertise in health columns or health magazines.
5. **People:** Individuals play important role in provision of service. People are important part of the process especially in health care services. Doctors, Nurses, Administration, Management are people in health care. Skills and behavioral qualities of service personnel affects the perception of consumers and eventually the customer satisfaction. When the patient thinks of medical care he or she thinks of the physician.(Ahmad, 2007) The patient envisions medical care in terms of the people who deliver it. Thus the fifth P of marketing is the organization's people (Kotler, 2011).
6. **Process:** Process plays very important role in service sector. Process starts with activities like input, processing the input by adding value and converting it to favorable outcome. In case of health care services patient joining followed by service consumption and range of treatments, surgeries take place in sequence as: Arrival of patient → Registration → Diagnosis → Treatment → Payment → Discharge and feedback Example: Surgical Procedure, Diagnosis and treatment
7. **Physical Evidence:** Tangible component of services offered which has huge impact on customer's perception as well as evaluation. Facilities of high class hospital like infrastructure, ambience, dress code, latest technology and experienced doctors and medicines for treatment are some of the most common physical commodities in health care services.

SERVICE QUALITY AND CUSTOMER SATISFACTION

Service quality has different dimensions regarding the various service sectors However, service quality measurement enables managers to recognize quality problems and enhance the efficiency and quality of services to exceed expectations and reach customer satisfaction.

In health care setting patient is the final user hence customer. Health care service quality is perceived satisfaction by patient; it is also known to continually satisfying patient requirements. Improving quality of health care services and patient satisfaction to its population in the face of limited resources have become a major challenge for developing countries and have gained increasing attention in recent years (Badri et al., 2009; Talib et al., 2011.). Quality health care services are necessity to retain existing customers and attract new ones, increase market share and profitability. According to Institute of Medicine (2001), health care quality can be assessed from two viewpoints: patients and technical or professional. Hence important health care quality measures can be roughly categorized as:

1. **Interpersonal Aspect:** Doctor as well as staff's responsiveness, friendliness and attentiveness. Empathy and assurance are undeclared but obvious aspects that help patient evaluate the quality.
2. **Technical Competence:** Efficiency of diagnosis and treatment using advanced methods. Example: Use of Health care Information System (HIS) & EMR
3. **Amenities and Infrastructure:** Healthcare facilities, comfort and tangibles in healthcare.

Measuring Healthcare Quality

According to many studies health care quality should be measured with patient's perspective as they provide real time and valid data on quality. Patient judge health care service performance against satisfaction and expectation. Measuring quality in health care can help decision makers to improve the existing system and set control measures to maintain a level of standard.

Patient satisfaction is evaluation of distinct health care dimensions such as hospital staff behavior, doctors conduct and friendliness, availability of right service at right time, efficiency and outcomes. The ability of any organization to satisfy its customers are most easily realized when those expectations are managed so as to be consistent with the product and processes provided (Friesner et al., 2009). Many studies have demonstrated that private hospitals have higher overall health care quality than public hospitals.

Patient Satisfaction and Health Care Quality Relation

Patient satisfaction is major component and key success indicator of health care quality. It is directly proportional to quality of health services provided. Better the quality of service provided more patient is delighted. Positive word of mouth publicity by satisfied patient is very good way of health service promotion and enhanced corporate image. Satisfied patients are well responsive to the treatment hence deliver better outcomes at the end.

The effectiveness of the hospital in delivering value to its patients is reflected by Customer satisfaction (Day & Wensley, 1988). Patient satisfaction, a crucial piece in the puzzle of performance assessment, merits consideration as a performance measure appropriate for small hospitals. Patient perceptions of quality of care are increasingly central in conceptual and operational models of performance measurement (Lied & Kazandjian, 1999).

The need is increasing day by day to improve quality in health care and there is a tremendous focus on improving patient satisfaction. Ultimately, patient satisfaction and health care quality are fundamental to improving health care service performance.

Figure 6. Relation of healthcare quality to patient satisfaction

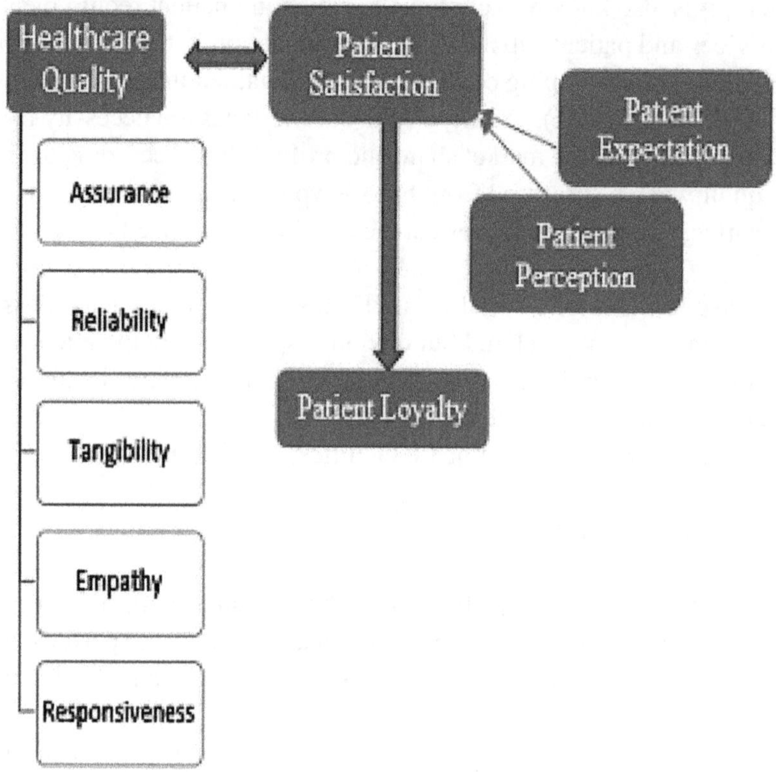

ISSUES, CONTROVERSIES, PROBLEMS

Healthcare Expenditure Statistics

India's health care expenditure as percent (%) of GDP is very low i.e. 4%. When compared to global average of 8.8% it is well below some of the emerging as well as developed countries. Whereas USA has maximum health care expenditure around 17.9% of their GDP.

Out of the total health care expenditure government's share of health spending is just 1.1% of total health spend which is at very low level hence it contributes to the poor health care infrastructure. India has proposed Rs 1.93 lakh crore ($31 bn) over the 12th Plan (2012-2017).

Considering the current state of health care infrastructure, the government needs to boost public spending on the sector to improve quality health care infrastructure and services; the government may increase public spending on health care going forward. This may open up huge growth opportunity for the private players that are focusing on rapid capacity expansion across the country.

Problems in Healthcare Industry

The demand is growing for quality healthcare however missing supply system pose a challenge. Healthcare is underserved and under-consumed. Good infrastructure and talent are both in short supply, and

Table 1. Health expenditure private and govt share comparison

Country	Health Spend % of GDP	Private Share % of Health Spend	Government Share % of Health Spend
USA	17.9	8.3	9.5
UK	9.5	1.5	8
Japan	9.4	1.6	7.8
China	5	2.3	2.7
India	4	2.9	1.1
Global	8.8	5	3.8

Figure 7. Health expenditure private and government share comparison
Source: WHO.

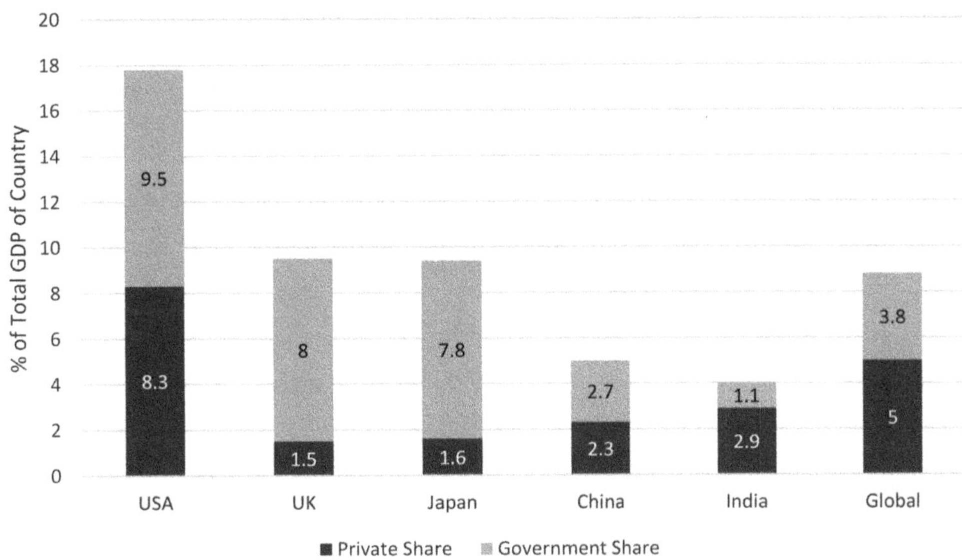

there are significant regional variations in health care delivery. Investments in primary care and public health have long been inadequate. Some of the key problems faced in healthcare industry are:

1. **Burden of New Diseases:** Fast growing population with increasing lifestyle related diseases. Non-communicable diseases such as heart disease, diabetes and cancer are spreading rapidly.
2. **Increasing Health Care Cost:** Incomes are increasing hence there is an increase in per capita expenditure, high-quality health care facilities, high use of technology and greater awareness of personal health and hygiene are contributing to increase in health costs.
3. **Low Government Funding:** Major challenge in India's health care is that government health care spending in India is among the lowest in the world. Government support needs to be increased in the form of reduced excise and customs duty, and exemption in service tax. As compared to peers' Indian health care expenditure has grown at very low pace than economy.

Table 2. Country wise comparison health workforce and infrastructure

Per 10,000 population	China	India	Indonesia	Malaysia	Singapore	Thailand	Australia	USA
Health Workforce Density								
Physicians	14.6	6.5	2.0	12.0	19.2	3.0	38.5	24.2
Nurses and midwives	15.1	10.0	13.8	32.8	63.9	15.2	95.9	98.2
Dental	0.4	0.8	0.4	1.4	3.3	0.7	6.9	16.3
Infrastructure								
Hospital beds	39	9	6	18	27	21	39	30

Source: World Health Statistics 2013.

4. **Poor Infrastructure:** There is a substantial gap in infrastructure and this problem is intensified by underutilization of existing resources. Eg. About 1.8 million beds required by 2025.
5. **Inadequate Health Care Workforce:** One of the major shortcomings for Indian health care sector is inadequate workforce. According to World Health Organization (WHO) (Source: Apollo Hospital Investor Presentation), India has a ratio of 7 doctors, 10 nurses and 9 beds per 10,000 people as compared to country like China which has ratio of 14 doctors, 15 nurses and 39 beds per 10,000 people.

Rural India, which accounts for over 70 per cent of population and is set to emerge as a potential demand source. Only 3 per cent of specialist physicians cater to rural demand. The doctor-to-patient ratio for rural India, as per the Health Ministry statistics, stands at 1:30,000 well below the WHO's recommended 1:1,000

6. **Weak Accessibility:** More and more people are demanding greater access to quality care, but cost increases make it difficult for uninsured citizens to afford care.
7. **Public Private Partnership Has Not Yet Achieved Scale:** Several small scale PPP projects were successful but they have failed to scale up so that to meet India's health care challenges.

SOLUTIONS AND RECOMMENDATION

Objective of the health care is to gain government support and grow at rapid pace. Creating public-private partnerships and inter-sectoral partnerships in healthcare industry is good solution to gain that government support. For growth, healthcare services need investment in IT and other relevant technologies to improve quality, accessibility and patient engagement. Encourage medical tourism in developing health care research areas, clinical trials and outsourcing through contract research:

1. **Accessibility and Reach:** Rural population is not accessible to most of the health care services and facilities. Quality health care services are not reachable to most part of the population. Hence, there is an opportunity to improve accessibility to these services as well as to develop the reach in health care.

Table 3. SWOT analysis of healthcare services industry

	Strength	Weakness
Strategies to be adopted in health care service industry based on SWOT matrix	• Large pool of well-trained medical professionals • Huge patient pool • Cost advantage compared to peers in other countries Eg. Cost of surgery in India is one-tenth of that in the US or Western Europe	Low level of per capita spending and poor infrastructure, hygiene.
Opportunities	**SO Strategies**	**WO Strategies**
• Telemedicine and mHealth • Medical equipment • Health care research and development: Contract research, clinical trials outsourcing • Medical tourism (Trade in services) • Vast opportunities for investment in health care infrastructure in both urban and rural India	Technology advancement and investment can boost health care research. Huge patient pool can act as target or sample in clinical trials studies. Cost advantage will develop medical tourism industry as international patients will prefer India over its peers.	Work on improving infrastructure which will eventually support technical advancement like mHealth, medical tourism and further increase investment.
Threats	**ST Strategies**	**WT Strategies**
• Burden of new diseases • Increasing health care cost • Lower Government funding • Inadequate health care workforce	Utilization of trained professionals to deal with inadequacy, burden of new diseases. Though health cost is increasing more govt fund can help maintain cost advantage.	Increase per capita spending; utilize govt funding to deal with diseases, increasing cost.

2. **Affordable and Quality Health Care:** People are becoming more conscious about health they explore new avenues and seeking quality health care. In addition, people are demanding affordable solutions that can be provided with health insurance. This perfect combination of quality and affordable health is in demand and needs to be supplied efficiently.
3. **Health Care Spending as a Percentage of GDP:** There is a scope to increase percent (%) of GDP spent on health care services. As per health care services, growth statistics India's current health spend is as low as 4% compared to its peers. In addition, government spending is comparatively less hence there is a huge scope to improve this health care spending by government as well as private players. This will eventually add up to the overall health care expenditure.
4. **Infrastructure Development:** Infrastructure in Indian healthcare is not up to the mark as per standards all over the world. Best quality infrastructure, facilities, amenities as well as skilled personnel are necessity in current healthcare industry.

FUTURE RESEARCH DIRECTIONS

Technological Innovations

Technologies are already introduced in healthcare organizations and there is a huge scope to upgrade these technologies further. Healthcare research in hospitals and information technology companies supporting technological innovations. Example: Disease coding in USA based hospitals has moved from ICD-9 to ICD-10 with tremendous changes in health system and operations. Technologies are evolving and have positive effect on cost as well as quality that is required for efficiency.

CONCLUSION

Healthcare is a service industry, which constitutes major part in India's economy. Hospitals are bigger and most important segment of healthcare. A hospital as an organization offers an intangible service called healthcare, which is characterized by intangibility, variability, inseparability and perishability. Key strategies for health care service industry's consistent growth are expansion of hospital network, diversification and acquisitions. Players in the health care have diverse geographical presence and range of services from single specialty, super-specialty or multispecialty health care.

Though demand is growing for quality healthcare, sector is facing challenges being underserved and under consumed. Good infrastructure and talent are in short supply, and there are significant regional variations in health care delivery. Health insurance covers very small part of population. Investments in primary care and public health have long been inadequate.

Recently various important trends are being observed in healthcare like changing disease patterns, people becoming aware and health conscious, health care insurance penetration, prevalence of technology through telemedicine, digital health etc. Main objective of the healthcare is to gain government support and grow at rapid pace hence there is a need for creating public-private partnerships and inter-sectoral partnerships in various sectors of healthcare industry such as technology, pharmaceutical and insurance.

Other immediate priorities include:

- Investment in IT and other relevant technologies is necessary to deal with healthcare challenges and improve quality, accessibility and patient engagement.
- Establishing standards in health care products and services in terms of quality, outcomes etc.
- Encouraging medical tourism through cost leadership, efficient treatments etc.
- Developing healthcare research areas to gain advantage in R&D, clinical trials and outsourcing through contract research.
- Increasing penetration and coverage of healthcare insurance in India.

REFERENCES

Ahmad, Al-Qarni, Alsharqi, Qalai, & Kadi. (2013). The Impact of Marketing Mix Strategy on Hospitals Performance Measured by Patient Satisfaction: An Empirical Investigation on Jeddah Private Sector Hospital Senior Managers Perspective. *International Journal of Marketing Studies*.

Alrubaiee. (2011). The Mediating Effect of Patient Satisfaction in the Patients' Perceptions of Health Care Quality – Patient Trust Relationship. *International Journal of Marketing Studies, 3*(1).

Apollo Hospitals Enterprises Limited. (2015). *Investor Presentation*. New Delhi: AHEL.

ASA Associates LLP. (2015). *Health care, Telemedicine and Medical Tourism*. Gurgaon: ASA Associates. Retrieved from www.asa.in/pdfs/surveys-reports/Health care-Telemedicine-and-Medical-Tourism-in-India.pdf

Badri, M., Attia, S., & Ustadi, A. (2009). Healthcare quality and moderators of patient satisfaction testing for causality. *International Journal of Health Care Quality Assurance*, *22*(4), No-4. doi:10.1108/09526860910964843 PMID:19725210

Bain and Company & NATHEALTH. (2015). *Aarogya Bharat – India Healthcare Roadmap for 2025*. Author.

Bertolini, C., Liu, Z., Schäf, M., & Stolz, V. (2011). *Towards a formal integrated model of collaborative healthcare workflows*. Tech. Rep. 450, IIST. United Nations University.

Bitner, M. J., Booms, B. H., & Tetreault, M. S. (1990). The Service Encounter: Diagnosing Favorable and Unfavorable Incidents. *Journal of Marketing*, *54*(1), 71–84. doi:10.2307/1252174

Blankson, C., & Kalafatis, S. P. (1999). Issues and challenges in the positioning of service brands: A review. *Journal of Product and Brand Management*, *8*(2), 106–118. doi:10.1108/10610429910266968

Canadian Center of Science and Education. (2013). *International Journal of Marketing Studies, 5*(6).

Cowell, D. (1984). *The Marketing of Services*. London: Heinemann Professional Publishing Ltd.

Day, G. S., & Wensley, R. (1988). Assessing advantage: A framework for diagnosing competitive superiority. *Journal of Marketing*, *52*(2), 1–20. doi:10.2307/1251261

Deloitte. (2015b). *2015 Global healthcare outlook: common goals, competing priorities*. Available at www.deloitte.com/content/dam/Deloitte/global/Documents/Life-Sciences-Health-Care/gxlshc-2015-health-care-otlook-global.pdf

Ernst and Young. (2015). *Megatrends 2015: Making Sense of a World in Motion*. Available from www.ey.com/Publication/vwLUAssets/ ey-megatrends-report-2015/$FILE/ey-megatrends-report-2015.pdf

Essiam. (n.d.). *Service Quality and Patients Satisfaction with Health care Delivery: Empirical Evidence from Patients of the Out Patient Department of a Public University Hospital in Ghana*. Academic Press.

Farid. (2008). *Development of a Model for Healthcare Service Quality*. Academic Press.

Fitzsimmons, J. A., & Sullivan, R. S. (1982). The service sector: Challenges and imperatives for research in operations management. *Journal of Operations Management*, *2*(4).

Friesner, D., Neufelder, D., Raisor, J., & Bozman, C. S. (2009). How to Improve Patient Satisfaction When Patients Are Already Satisfied: A Continuous Process Improvement Approach. *Hospital Topics*, (Winter): 2009. PMID:19103585

Furrer, O., Liu, B. S. C., & Sudharshan, D. (2000). The relationships between culture and service quality perceptions. *Journal of Service Research*, *2*(4), 355–371. doi:10.1177/109467050024004

Ginzberg, E. (Ed.). (1991). *Health services research: key to health policy*. Cambridge, MA: Harvard University Press.

India Brand Equity Foundation (IBEF). (2015). *IBEF Health care*. IBEF Website.

Institute of Medicine. (1991). *Improving information services for health services researchers: a report to the National Library of Medicine*. Washington, DC: National Academy Press.

Kalra, A. (2015). *Exclusive: Modi govt puts brakes on India's universal health plan*. Retrieved from http://in.reuters.com/article/india-health-idINKBN0MM2UT20150327

Kazandjian, V. A., & Terry, R. (2001). Lied, Small hospitals and performance measurement: Implications and strategies. *International Journal of Health Care Quality Assurance*.

Kemp, E., Jillapalli, R., & Becerra, E. (2014). Healthcare branding: Developing emotionally based consumer brand relationships. *Journal of Services Marketing*, *28*(2), 126–137. doi:10.1108/JSM-08-2012-0157

Kotler, P. (2011, July). Reinventing Marketing to Manage the Environmental Imperative. *Journal of Marketing*, *75*(4), 132–135. doi:10.1509/jmkg.75.4.132

Kotler, P., Armstrong, G., & Cunningham, M. H. (2005). Principles of marketing. Toronto: Pearson Prentice Hall.

KPMG. (2015). *Health care: The neglected GDP driver*. Academic Press.

Lohr, K. N., & Steinwachs, D. M. (2002, February). Health services research: An evolving definition of the field. *Health Services Research*, *37*(1), 7–9. PMID:11949927

Naidu, A. (2009). Factors Affecting Patient Satisfaction and Healthcare Quality. *International Journal of Health Care Quality Assurance*, *22*(4), 366–381. doi:10.1108/09526860910964834 PMID:19725209

Nargundkar, R. (2008). *Marketing Research- Text and Cases* (3rd ed.). New Delhi: The McGraw-Hill.

Overseas Indian Facilitation Centre. (2013). *Booming Indian health care industry*. Retrieved from www.oifc.in/booming-indian-healthcare-industry

Planning Commission (India). (n.d.). In *Wikipedia*. Retrieved September 2016. From https://en.wikipedia.org/wiki/Planning_Commission_(India)

Rao, K. D., Peters, D. H., & Bandeen-Roche, K. (2006, December). Towards patient-centered health services in Indiaa scale to measure patient perceptions of quality. *International Journal for Quality in Health Care*, *18*(6), 414–421. doi:10.1093/intqhc/mzl049 PMID:17012306

Sasser, et al. (1978). Management of Service Operations. Allyn and Bacon.

Sciulli, L. M., & Missien, T. L. (2015). Hospital service-line positioning and brand image: Influences on service quality, patient satisfaction, and desired performance. *Innovative Marketing*, *11*(2), 2015.

Smith, R. D. (2004). Foreign direct investment and trade in health services: A review of the literature. *Social Science & Medicine*, *59*(11), 2313–2323. doi:10.1016/j.socscimed.2004.03.021 PMID:15450706

Smith, R. D., Rupa, C., & Viroj, T. (2009b). Trade in health-related services. *Lancet*, *373*(9663), 593–601. doi:10.1016/S0140-6736(08)61778-X PMID:19167053

Steinwachs & Hughes. (2008). *Health Services Research: Scope and Significance*. Retrieved from www.ncbi.nlm.nih.gov/books/NBK2660/

Talib F., et al. (2011a). Total Quality Management and Service Quality: An Exploratory Study of Management Practices and Barriers in Service Industries. *International Journal of Services and Operations Management*.

The Guardian. (2012). *Health care spending around the world, country by country*. Retrieved from www.theguardian.com/news/datablog/2012/jun/30/healthcare-spending-world-country.

WHO. (2015). *Global Health Observatory data repository*. Retrieved from WHO Website: http://apps.who.int/gho/data

WHO Regional Office for Europe's Health Evidence Network (HEN). (2003). *What are the best strategies for ensuring quality in hospitals?* Retrieved from www.euro.who.int/__data/assets/pdf_file/0006/74706/E82995.pdf

Woodside, A. G., Frey, L. L., & Daly, R. T. (1989). Linking service quality, customer satisfaction, and behavioural intention. *Journal of Health Care Marketing*, *9*(4). PMID:10304174

Zeithaml, V. A., & Bitner, M. J. (2000). *Services Marketing: Integrating Customer Focus across the Firm* (2nd ed.). Boston, MA: McGraw-Hill.

This research was previously published in Strategic Marketing Management and Tactics in the Service Industry edited by Tulika Sood, pages 280-303, copyright year 2017 by Business Science Reference (an imprint of IGI Global).

APPENDIX: ABBREVIATIONS

CAGR: Compound Annual Growth Rate
EMR: Electronic Medical Records
FDI: Foreign Direct Investment
FY: Indian Financial Year (April to March)
GDP: Gross Domestic Produc
GOI: Government of India
HIS: Healthcare Information System
ICT: Information and Communications Technology
INR: Indian Rupee
IT: Information Technology
mHealth: Mobile Health
PE: Private Equity
PPP: Public Private Partnerships
R&D: Research and Development
US: United States
USD: United States Dollar
WHO: World Health Organization

Chapter 9
Telemedicine and Telehealth:
Academics Engaging the Community in a Call to Action

Kim L. Brown-Jackson
The National Graduate School of Quality Management, USA

ABSTRACT

Despite identifying the importance of telemedicine and telehealth education in scholarly environments, researchers rarely explore the dynamics of taking some of this learning to the community to engage in prevention. Medical professionals are consistently receiving education to enhance their knowledge, skills, and capabilities. Telemedicine and Telehealth have a new role in the community and is akin to house calls from the past. Engaged in this text is the action for medical professionals, government officials, and civic leaders to work together to move prevention health study to the community. This movement promotes the sharing of knowledge and understanding between the scholarly world and the communities they serve. The researcher concludes with the discussion of the responsibility required in the learning process at all levels. This text will provide a guideline for such an engaged and shared approach to healthcare prevention, as well as implications for future research and practice.

INTRODUCTION

Telemedicine is a groundbreaking method of bridging the health care delivery gap by expanding access to services for medically underserved communities (Bashshur et al., 2014). Telehealth is a propelling force for speaking to existing challenges in the healthcare environment and enhancing the quality of healthcare effectively. Telemedicine and Telehealth has been promoted as a disruptive innovative (Grady, 2014) approach to bridging the health care delivery gap by increasing access to services for medically underserved communities. Other terms associated with telemedicine and telehealth include but are not limited to telecare, telenursing, online health, eHealth, connected health (Cason, 2014; Maheu, Whitten, & Allen, 2002) or virtual care (West & Mehrotra, 2016). For the purpose of this text, telemedicine and telehealth are used interchangeably, although they do have slightly different definitions.

DOI: 10.4018/978-1-5225-6198-9.ch009

Civic responsibility connects healthcare to the community, working with the community to solve the problem. Healthcare in the United States is encountering an upsetting predicament (World Health Organization, 2012). Such predicament includes a change in thought from the traditional provider-centered, disease-oriented methodology to a patient-centered, health management model (Bauer & Ringel, 2009; Haughton, 2000; Sacristán, 2013; Truog, 2012). According to World Health Organization, through this shift in thought patterns, the data shows an extraordinary void of valuable resources (World Health Organization [WHO], 2012, The World Health Report). The void continues to be a contrasting force not in support of the commencement of positive occurrences in healthcare, and the associated research, and expansions. Craig, Eby, and Whittington, (2011), posits that such forces include but are not limited to overpriced generic, medical, technological, and pharmaceutical progress. In determining a path to engaged scholarship and civic responsibility, the researcher provides analysis that contributes to the literature mentioned in addition to the data that is generally discussed. The text is divided into five sections that will be separately reviewed, and then blended to show how scholarship and community can work together for the common good.

BACKGROUND

Telemedicine refers to medical data transmitted between sites through electronic communications to enhance patients' health status (Ajami & Lamoochi, 2014; American Telemedicine Association [ATA], 2015; Majerowicz & Tracy, 2010; Mokdad, Marks, Stroup, & Gerberding, 2004). Another way to refer to telemedicine is that telemedicine refers to the remote delivery of medical care. Telemedicine is the use of electronic information and communications technologies (ICT) to provide clinical services when participants are at different locations. Telemedicine does not represent a separate medical specialty; rather it is a tool that can be used by health providers to extend the traditional practice of medicine outside the walls of the typical medical practice (Bashshur & Shannon, 2009). Telehealth is a term used to encompass a broader application of technologies to distance education, consumer outreach, and other applications wherein ICT are used to support healthcare services. Physicians to physicians and physicians to patience have communicated over distance using technology since the early 1900s (Bashshur & Shannon, 2009). Current technological progress and a shifting health care landscape have altered telemedicine from a novelty into a flourishing industry.

The business of treating patients via telehealth in the U.S. will dramatically increase to nearly $2 billion in revenue within five years from $240 million today, an annual growth rate of 56 percent as reported in Forbes (Jaspen, 2013). To date, there are over 200 telemedicine programs in hospitals in the U. S. with at least one telehealth program and those hospitals that do not use telemedicine are already behind the leaders in health care delivery (ATA, 2014) and are losing market share or paying fines due to 30-day hospital admissions (Centers for Medicare and Medicaid Services [CMS], 2014). Traditional models center on videoconferencing, telephone, and email. These modalities remain relevant, but the field has rapidly added capabilities and indications (Di Cerbo, Morales-Medina, Palmieri, & Iannitti, 2015). Examples include videoconferencing, transmission of still images, e-health including patient portals, remote monitoring of vital signs, continuing medical education and nursing call centers are all considered part of telemedicine and telehealth. Telemedicine includes diagnostics, treatment, monitoring, consultation, and education among other domains. Figure 1 show Telemedicine Typology and Figure 2 Telehealth Typology. Telemedicine encompasses different types of programs and services provided for

Figure 1. Telemedicine Typology. Source: Adapted from Welsh, 2002

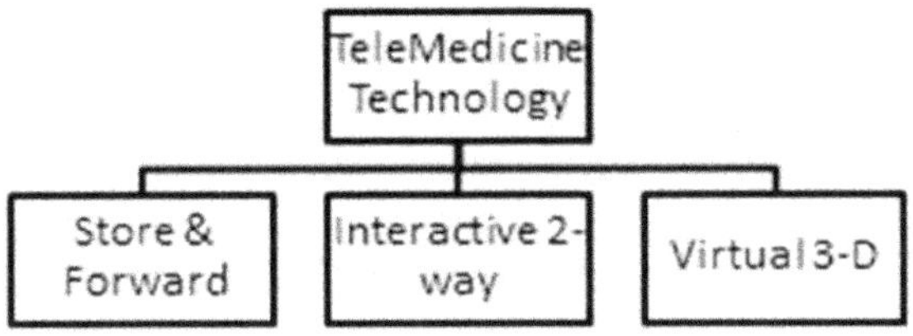

Figure 2. Telehealth Typology. Source: Adapted from Welsh, 2002

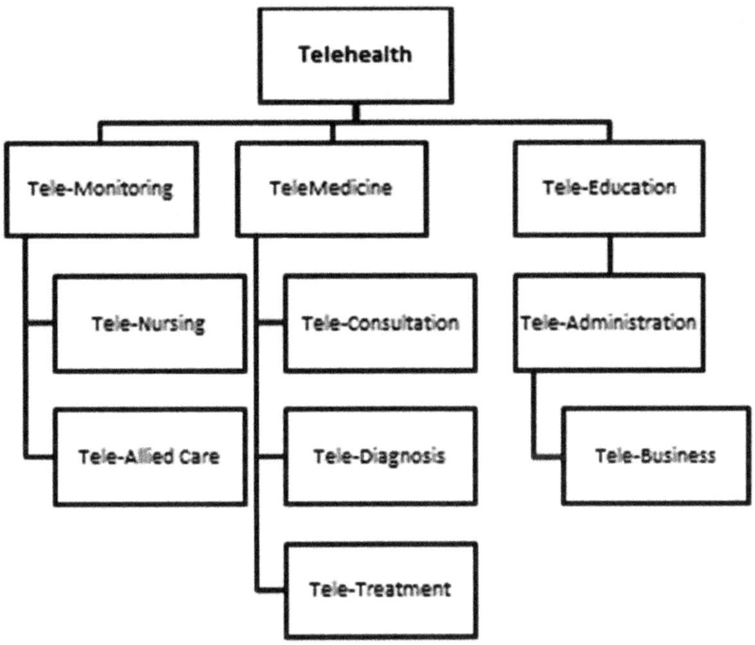

the patient. This typology model of telehealth and telemedicine depicts the types of general health care services offered by nurses and allied health care workers, primary care and specialty physicians, and health care educators and administrators.

Telehealth complements traditional ambulatory and hospital-based practices, which tends to be provider-centric (Schwamm, 2014). Telehealth creates delivery systems that are more patient centered and that use technology to increase access and quality, decrease cost, and help providers manage an ever-increasing volume of information and relationships (Bashshur, Reardon, & Shannon, 2000; Schwamm, 2014).

Telehealth has become a fundamental piece of American health care delivery because it helps address issues of both health care costs and access. Moving forward, digital health capabilities will only continue to grow. In order to most effectively leverage these tools, we must ensure providers use them effectively and appropriately. Today's medical trainees are well versed in technology, but the practice of telemedicine is not necessarily intuitive. Therefore, we advocate the introduction of telemedicine training into medical schools. Telemedicine offers a means to help transform healthcare itself by encouraging greater consumer involvement and civic responsibility in decision making and providing new approaches to maintaining a healthy lifestyle.

WHAT IS THE CONNECTION BETWEEN HEALTH CARE AND COMMUNITY CONCERN?

Chronic Disease Management

Health care and its delivery in the United States has been experiencing an explosive conversion during the early decades of the 21st century with the impactful implementation of the Patient Protection and Affordable Care Act (ACA), an onward movement towards value-based care, and a focused national Triple Aim for health care of enhancing the patient experience, improving population health, and controlling costs (Institute for Healthcare Improvement [IHI], 2015). At the center of these phenomena is telemedicine and telehealth. As a dimension of a model of care, an effective telehealth model of care can improve clarity of purpose, improve quality and outcomes, and enhanced use of resources (Thaker, Monypenny, Olver & Sabesan, 2013). Figure X, The Chronic Care Model of Wagner (1998) describes the essential elements for quality of care. The purpose of the Chronic Care Model of Wagner model is to optimize the chronic care management (1998).

The scholarship of engagement means connecting the rich resources of the university to our most pressing social, civic and ethical problems. Leaders from academic institutions, professional associations, community-based organizations, philanthropy, and government must come together to take a leadership role in creating a more supportive culture for community-engaged solutions in health care.

Figure 3. The Chronic Care Model. Source: Adapted from Wagner, 1998.

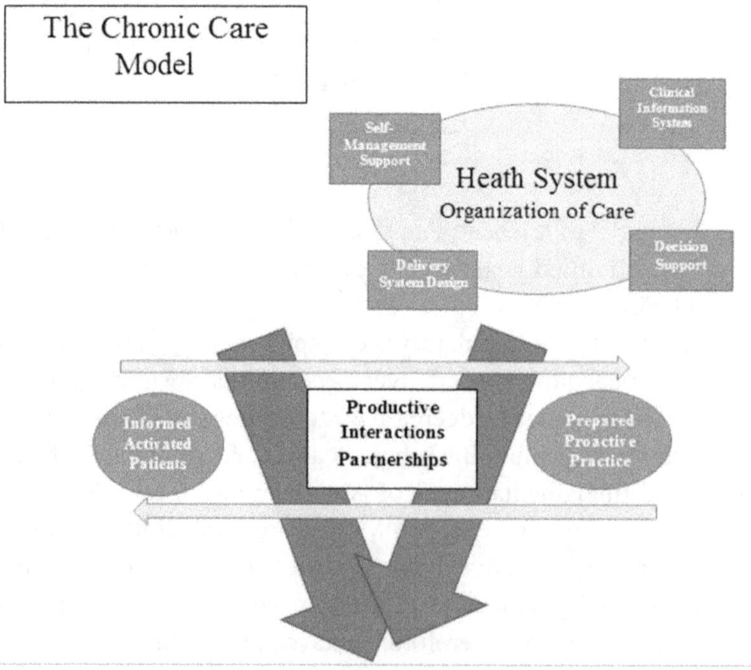

Ensuring High-Quality Care

In health care delivery, one should refrain from assuming that digital native physicians will deliver high-quality virtual care without formal and systematic training. Current research suggests that telemedicine has a tremendous promise, but successful studies are typically carried out in academic medical centers by a limited number of well-trained doctors (Flodgren, Rachas, Farmer, Inzitari, Shepperd, 2015). Most residency programs are sponsored by and take place in large teaching hospitals and academic health centers. However, as health care services are increasingly provided in ambulatory and community-based settings, residency training is beginning to expand to non-hospital sites (Chen, Chen, & Mullan, 2012). Based on the rapid evolution in health system delivery involving an increasing emphasis on non-hospital-based care, many experts recommend an acceleration of this transition (Barr, 2016; Fuchs, 2011). Other studies have shown that telemedicine can lead to mixed-quality care (Mehrotra, Paone, Martich, Albert, & Shevchik, 2013). Schoenfeld et al., (2016) revealed substantial variation in the quality of care provided by commercial telemedicine companies. As remote patient visits increase and indications for telemedicine become more complex, the need to train physicians to offer virtual care equivalent with in-person consultation is paramount.

Medical education must recognize the intrinsic differences between the practice of traditional medicine and that of telemedicine. For instance, it is difficult to remotely carry out a physical exam, which fundamentally changes the diagnostic process. Technological limitations may cause marked variation in data quality between clinic and remote visits. A patient's self-reported blood pressure from home may differ from that measured by a nurse in clinic. Providers need to be able to judge those differences.

Telemedicine has its limitations in other dimensions as well. Pain management is difficult to gauge from afar. Complex diagnoses and the initial phases of patient education may be better done in person. The nature of the doctor-patient relationship is different. Given these limitations, practitioners must be able to determine when telemedicine is appropriate and how to optimally process information when they see patients remotely. They must also understand how to navigate the many medical and legal issues that remain in telemedicine, including the role of Health Insurance Portability and Accountability Act regulations, restrictions due to licensing laws, and issues regarding malpractice. Telemedicine is a rapidly evolving field with many stakeholders and murky regulation; providers must learn how to interact with such a system.

Healthcare Workforce Shortage

This is a time of tremendous change and uncertainty in U.S. healthcare. Key provisions of the Patient Protection and Affordable Care Act (ACA) are not yet implemented. Many health providers and policy makers worry that the Act's expansion of health insurance coverage to millions of Americans—combined with the aging of the population—will overwhelm the workforce we have. Some analysts have projected dramatic workforce shortages—especially for physicians—that could prevent many people from getting needed health services (Association of American Medical Colleges [AAMC], 2013; Petterson et al., 2012). There are also widespread concerns that the nation is not training the right specialty mix of physicians to meet society's needs (Fraher, Knapton, & Holmes, 2017; Gaynor, Mostashari, & Ginsburg, 2017), and that these physicians are not geographically well distributed (Iglehart, 2011; Rafieri, Mohebbifar, Hashemi, Ezzatabadi, & Farziampour, 2016). At the same time, current economic pressures dictate placing full attention to this issue.

Numerous goals established by the U.S. Department of Health and Human Services (HHS) and the Institute of Medicine (IOM) directly influence the health care workforce. One of HHS Office of Disease Prevention and Health Promotion (ODPHP) goals for HHS Healthy People 2020 is to integrate more information technology in health care (HHS Office of Disease Prevention and Health Promotion [ODPHP], 2011). IOM made recommendations about education for health professionals in order to address their ability to keep pace with and be responsive to shifting patient demographics and desires, changing health system expectations, evolving practice requirements and staffing arrangements, new information, a focus on improving quality, and new technologies (2014). The magnitude, alignment, distribution, and expertise of the health care workforce will determine the success of health care reform in the U.S. (Fraher, Knapton, & Holmes, 2017). Despite the size of the workforce required in the future to meet society's needs, education of health professionals' merit attention (AAMC, 2013).

WHY ARE TELEMEDICINE AND TELEHEALTH A SCHOLARLY AND COMMUNITY CONCERN?

Patient-Centered Healthcare

As health care moves rapidly toward a value-based delivery model, a greater emphasis will be placed on care coordination (Darkins et al., 2009; Kvedar et al., 2014; Psek, Stametz, Bailey-Davis, & Davis, 2015). Patients not only need to receive the right care at the right time and in the right setting, but also every part of the delivery system is connected and understands that patients' needs will be critical going forward (Darkins et al., 2009; Nundy & Oswald, 2014). Providers and payers must understand the market and the three channels of patient engagement as the first step for cost effective sustainable telehealth strategy. Information technology will be instructional in making sure that cost effective sustainable telehealth strategy occurs and in providing clinicians with valuable new decision support tools (Li-Chen et at., 2012; Wildevuur & Simonse, 2015). By electronically connecting the continuum of clinicians to clinicians, patients to clinicians, and patients to other patients, the promise of telehealth can be realized closer to or in the home. The community-based telehealth program is at the heart of connecting the continuum (Craig, Eby, & Whittington, 2011).

Despite recognizing the significance of telehealth and telemedicine in the community, researchers have consistently explored the state of health care quality and its dynamics in a silo (Agency for Healthcare Research and Quality [AHRQ], 2015). Resolving these contradictions, the researcher proposes that scholars connect with civic leaders to responsibly engage in research and learning in a holistic manner. The holistic manner is for scholars to engage in scholarship through civic responsibility and the likely outcomes that follow these paths.

Engaged Scholarship involves academic resources from institutions of higher learning, as well as resources from governments and businesses to collaborate with communities to define and then work for the betterment of community-determined needs (Ellaway, 2016). The outcomes and measurements are focused on civic engagements that better communities and augment the research and learning of academics, businesses, and governments. Thus, the three salient focal points are academic service learning, community-based research, and community engagement. These focal points integrate civic and hands-on experiential education into the frameworks of academic and business (Giardino, 2016). Figure 4. Focal points for Patient-Centered Telehealth Civic Engagement depicts these focal points. Further, engaged scholarship entails utilizing technologies to improve engagement with communities, institutions of higher learning, businesses, and governments.

Telemedicine Education: Training Digital Natives

Telemedicine has an important role in the delivery of health care, and is projected to have a major role in the delivery of quality care. Current medical students are among the first generation of "digital natives" who are well versed in the incorporation of technology into social interaction. These students are well positioned to apply advances in communications to patient care. Providers require training to effectively leverage these opportunities. Therefore, we recommend introducing telemedicine training into medical school curricula and propose a model for incorporation (Moore, 2016; Pathipati, Azad, & Jethwani, 2016).

Formal training is the best way to teach providers how to approach the challenges and opportunities inherent in telemedicine (Moore, 2016; Pathipati et al., 2016). The researcher proposes that this training should begin in medical school. Today's medical trainees are the first generation of digital natives—individuals who grew up surrounded by digital technology and are therefore comfortable processing information in an electronic world. This fact is not enough to guarantee high-quality telemedicine care. Formal training can extend and amplify the impact that telemedicine brings to health care. Current medi-

Figure 4. Focal Points for Patient-Centered Telehealth Civic Engagement

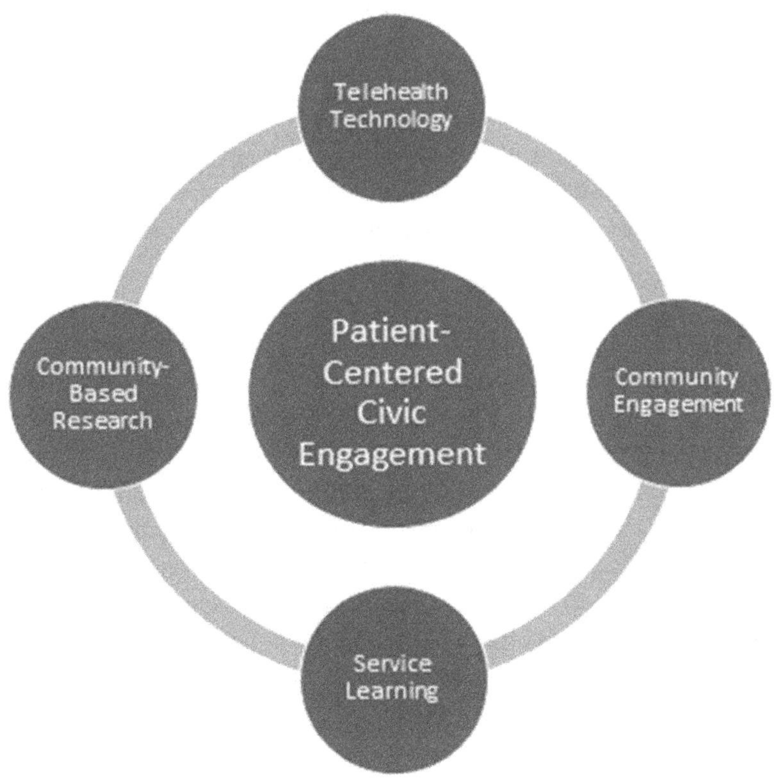

cal students' inherent comfort with technology should be nurtured through structured training. Without this, providers will be ill prepared to take advantage of innovations in telemedicine.

With this in mind, the researcher proposes incorporating telemedicine training into the standardized medical school curriculum (Chacko & Supe, 2015). We have an opportunity to translate students' familiarity with technology into superior medical care. Creating a formal training program will allow students to directly compare and contrast telemedicine with traditional medicine, recognize when to use it, and learn best practices. Placing the training program in medical schools would ensure that all new doctors have that ability. To ensure high-quality telemedicine care, we must train students to practice telemedicine with the same level of skill they demonstrate delivering traditional care.

STRENGTHENS, WEAKNESSES, OPPORTUNITIES, AND THREATS FOR COMMUNAL COLLABORATION IN UNDERSTANDING?

Prevention and Wellness

Starting a telemedicine and telehealth program in a health care organization with little prior experience with the concepts and technologies of telemedicine is challenging (Mid-Atlantic Telehealth Resource Center, 2016, Resources). Telemedicine care is one of the solutions to the problem of health care delivery in rural areas by improving communication capabilities and providing convenient access to current health care interventions on-line, consultations, and other forms of multimedia support (Lilly et al., 2014).

Ageing Population and Chronic Disease Management

An ageing population and an increase in chronic diseases are challenges for the future (Bui & Fonarow, 2012; Willemse, Adriaenssens, Dilles, & Remmen, 2014). Since home health care is the fastest growing area of health care in the U.S. today, it is important to address effectively and efficiently the needs of the chronically ill and aging population should be addressed effectively and efficiently (CMS, 2015a; Hartgerink et al., 2014). It is an ever-aging population that uses the majority of health care resources (Bui & Fonarow, 2012).

As a result of the chronically ill and aging population shift, a need exists for new organizational structures and new partnerships (Stanhope & Lancaster, 2013). The Chronic Care Model of Wagner (1998) model summarizes the basic components for improving care in health systems at the community, organization, practice, and patient levels. The dimensions of the Chronic Care Model are the community, the health system, self-management support, delivery system design, decision support, and clinical information systems (Wagner, 2013). There are inherently numerous deficiencies in the management of chronic care to include the lack of coordination and patients inadequately trained to manage their illnesses (Parekh, Goodman, Gordon, & Koh, 2011; Wagner, 2013). Overcoming these deficiencies will require a transformation of health care, from a system that is in essence reactive to one that is proactive and focused on keeping a person as healthy as possible (Wagner, 2013). When health care emphases are on social and societal aspects, the advancement of the harmonization of care and multidisciplinary alliance is at the juncture (Adeleye & Ofili, 2010).

Patient Engagement

Collectively, in communities, there needs to be a shift from being volume conscious to value conscious and focused on outcomes. One way to do that is to actively have a paradigm shift in the way professionals approach health care. To successfully achieve patient engagement in the community health center or in a health care practice, five elements should be considered: (1) define your organization's vision for patient engagement, (b) create a culture of engagement, (c) employ the right technology and services, (d) empower patients to become collaborators in their care, and (e) chart progress and be ready to change and adapt. (Brown-Jackson, 2017). These points are what are noted as a best practice. According to Berwick et al, (2008) and Stiefel and Nolan (2012), leveraging telehealth technologies, while engaging with patients, lowers costs, improves quality outcomes, and enhances the patient experience which can serve as telehealth strategy and a shared value proposition of the Triple Aim. This message is very important for those in the health care field to understand and for higher education to impart to its students who will engage patients in a wide variety of settings to improve outcomes. The advances in technology, the delivery of health care and support to patients in rural, remote, and underserved locations has become feasible through telehealth; therefore, with telehealth technology, distance and travel issues are no longer a barrier for the underserved in rural or isolated populations (Bailey, 2009; Buzza et al., 2011).

Better Care Coordination

Effective inter-professional communication is key in care coordination and impacts patients, saves costs for the facility/medical practice and the community they serve (IOM, 2014). The threat inherit in the lack of inter-professional communication is that the patient can suffer, i.e., 30-day readmissions due to lack of understanding discharge instructions, lack of full-circle communication among providers, disjointed priorities that neither benefits patient outcomes or management of chronic diseases, and workflows impacted which lead to unproductive staff and physicians (IOM, 2014). A semi-structured, grounded theory qualitative focus group research study conducted with experts who are directly involved in the strategy, policy, management, technology, teaching, and delivery of telehealth and telemedicine and all members of the American Telemedicine Association, offered several salient points about better care coordination. "When we conduct health fairs on the weekend, we use our telehealth kiosks to help us see as many patients as possible. We have learned over time how vital it is to have care coordination." Another participant stated, "The doctors we work with in rural and underserved areas have had their workflows impacted in positive ways to help these patients. It is with the use of telehealth interventions that we are able to get these patients linked up with primary care to help them on the journey to managing their health." "What we have seen workflow wise is that when we coordinate with a telehealth service, a multitude of things happen from administrative, to employing more physician assistants, monitoring our patients care in a more personalized manner a lot sooner than waiting for an acute care event to happen. Acute care only sets our patients back and fills up my waiting room." (Brown-Jackson, 2016).

Many changes in workflow were also highlighted from the perspective of convenience. All of these examples are attributes that higher education cannot provide alone; it is through academic engagement with health care providers and practices that are using technology to aid in better care delivery and through opportunities while still in school that can provide invaluable knowledge. Knowledge can be shared, experiential learning can be incorporated, and bridges can begin to be built as a scaffolding to help better the lives of caregivers and those who are the most vulnerable.

WHAT ARE THE STEPS FORWARD FOR EDUCATION AND STUDY BETWEEN ACADEMIA AND THE COMMUNITY?

Importance of the Framework of Scholarship

Health care education has an abundant amount of literature that praises the value and importance of quality research that underpins evidence-based practice (Doolan-Grimes, 2013). Higher education more recent focus on the research excellence framework in order to secure financial subsistence and gain credibility as a research focused center of excellence. Despite this emphasis on research, there appears to be a gap in developing a culture of scholarship (Fitzpatrick & McCarthy, 2010). Taking into consideration the fundamental principles of professionals in health care, Fitzpatrick and McCarthy (2010), posited that academic professional leaders must ensure a broad vision of scholarship is incorporated within their educational offerings and embolden a significantly wider breadth of examination. The seminal work of Boyer (1990) proposed and furthered the depth of study that led to a progressive framework. Boyer's framework presented an avenue for academicians to use order to encourage a holistic, energetic, methodical approach to learning.

Historical Context

From a historical perspective, Rolfe (2009) posited the definition of scholarship as the various approaches to learning pursued in academics, including critical dialogue, philosophical examination, and empirical research. The resulting tangible output created and provided for critique by academic peers. As argued by Boyer (1990) and Boyer, Moser, Ream, and Braxton, (2015), the framework was key to the conception of scholarship; however, research had become the dominant factor preceding all other types of scholarship.

Boyer

Boyer's framework includes four inter-related and overlapping domains (Kelsey, 2016) and has been increasingly influential in helping to frame scholarship with higher education institutes (Boyd, 2013). In Boyer's new model, the tenants of academic scholarship which encourages freedom to integrate new knowledge, to re-evaluate your thoughts, to generate new connections, to build bridges (that enable the application of novel and innovative ways of operationalizing and producing), and to disseminate this new knowledge in a way that is value and purpose. In order to achieve this successfully, health care professionals must embrace their own distinctive role, as educationalists, and mentors and work collaboratively through mutual respect to develop practitioners who are adequately skilled and knowledgeable to safeguard the better wellness and care (Kelsey, 2016).

It is the discovery of knowledge through research, integration of theories into clinical practice, application of intellect to professional dilemmas, and the sharing of knowledge through teaching, a process by which fresh ideas are gained and new learning advocated. It could be posited that this process encourages the concept of lifelong learning and supports the values of a profession committed to both the art and science of health care advancement. A salient point about Boyer's framework is that it has the potential to help bridge the theory-practice gap between research and teaching, as well as integration of learning and application of that learning into professional practice (Boyd, 2013). Utilizing this framework has the potential to improve quality of care so that the discovery and generation of new knowledge continues to transcend throughout clinical care, creating a cyclical approach that keeps on flowing (Smith and

Crookes, 2011). The researcher argues that Boyer's scholarship framework used effectively and with fervent dedication and focus, inspire a new class of practitioners to function as thought leaders engaged in social matters of true significance and become mentors, future leaders and educationalists.

Reciprocal Partnership

There is a great opportunity for research to be conducted using health care providers working through engaged scholarship and civic responsibility. Engaged scholarship, from application of academic expertise to community engaged scholarship, is a reciprocal partnership with the community; engaged scholarship is interdisciplinary. The infusion of this type of scholarship in integrates roles of teaching, research, and service. Engaged scholarship is demarcated by the partnership among academics and individuals outside of a regular college or university setting for the equally valuable exchange of knowledge, skills, and resources in a setting of partnership and exchange. These setting could be community centers, health centers within rural, remote, underserved communities. Particular types of engagement could be informational meetings, focus groups to gather data, as well as classes to provide information on the management of chronic disease, in addition to gain information on what class participants do to manage chronic disease. Particular to future research, engaged scholarship and civic responsibility academics refer to the telemedicine and telehealth professionals. These professionals would work with rural, remote, and underserved communities.

The scholarship of engagement comprises self-governing amounts of inspiring the participation of non-academics in matters that improve and widen engagement and conversation about major social issues (i.e., stroke, PTSD, and specialty shortages) inside and outside the college/university setting. Telehealth professionals participating in engaged scholarship and civic responsibility would lead diverse issues individually to facilitate as much dialog and to gain as much information as possible. Specific to future studies engaged scholarship and civic responsibility relate to telehealth. Tele-professionals would facilitate a more vigorous and engaged democracy by getting the community and those concerned with chronic disease, into the understanding of problem-solving work through avenues that improve the public good with and not merely for the public, but for learning purposes, as well.

Curriculum Development

To date, telemedicine training has been limited to small research settings (ATA, 2015). It is posited that telemedicine training should become a more prominent part of the medical school and allied health curriculums moving forward (AAMC, 2015; Hawkins, 2012). For example, specific health care providers participate in engaged scholarship and civic responsibility includes, but not limited to, tele-counselors and college and university tele-mental health servicers (Bashshur et al., 2009). Topics of study could include clinical outcomes, satisfaction outcomes, operational outcomes, and financial outcomes. The frequency of the practitioners within the community engaging in these topics and the method of knowledge delivery within the community, educating the community are the topics and activities that would support civic responsibility.

Telehealth studies specific to practitioner's training and learning communities should include training and learning specific to health issues of the community, the total of multi-cultural and diverse health care specialty understanding to include, innumerable learning, the means in that information is offered to practitioners, and the volume of health care practice opportunities available to support knowledge development of all learning participants, health care professionals and the community. Additionally, directing further examinations on the way training and learning communities can influence practitioners' intellectual capital and empower these practitioners to become more educated, experienced, and informed health care practitioners, would add to the body of knowledge for telehealth and telemedicine in the disruptive environment, and to specifically support engaged scholarship and civic responsibility.

A quality service learning experience provides meaningful service that addresses community-defined needs and course relevant learning to the students. The challenge for faculty is to integrate, at the core of the course, civic responsibility concepts and practices that contribute to a quality service learning experience (Ander & Love, 2017b).

SOLUTIONS AND RECOMMENDATIONS

Civic responsibility means active participation in the public life of a community in an informed, committed, and constructive manner, with a focus on the common good. Given the working definition, how can faculty make civic responsibility an integral part of their curriculum and potentially affect student-learning outcomes.

Service Learning and Community Engagement

What is often missing from these experiences is structured reflection that leads to critical thinking about how the service experience is related to the life of the volunteer as an individual and as a community member. One avenue for providing students with such structured support is service learning (Rafiei, Mohebbifar, Hashemi, Ezzatabadi, & Farzianpour, 2016). While some service is performed through involvement in extracurricular activities such as student government, clubs, religious groups, sports teams, or honor societies, a growing number of students today are introduced to service through service learning in their college classrooms. Service learning can be a way for students to connect or re-connect with civic society. By engaging in these activities, they may develop a set of attitudes and behaviors that is consistent with the expectations of citizenship.

Studies have shown students who performed community service were more aware of the need to become involved in the policy process, felt a greater connection to the community, and were better able to view situations from others' perspectives (Ander & Love, 2017; Eyler, Giles, & Braxton 1997). It is important to note, however, that unless civic responsibility is intentionally integrated into the academic curriculum, this potential is not likely to be realized. It is argued that education is the key to civic engagement. Therefore, institutions of learning must adequately prepare students for such activity and should be viewed as beacon of society that should model community behavior (Ehrlich 2000, 1999). Using service learning to develop civic responsibility allows community colleges to fulfill their basic mission of providing a quality educational experience and serving the needs of the community.

Community-Based Research

Telehealth Resource Centers (TRC), the American Telemedicine Association, the Association of American Medical Colleges, the Centers for Medicare and Medicaid Services, state and local health centers, and rural research centers among other organizations all conduct community-based research. Telehealth Resource Centers are funded by the U. S. Department of Health and Human Services' (HHS), Health

Resources and Services Administration (HRSA), Office for the Advancement of Telehealth (OAT), a part of the Office of Rural Health Policy. Nationally, 14 TRCs exist, with 12 Regional Centers, all with different strengths and regional expertise, and 2 National Centers that focus on areas of technology assessment and telehealth policy (U. S. Department of Health and Human Services', Health Resources and Services Administration [HHS HRSA], 2014). These Centers are astute at providing advice, examples, resources, best practices, and case studies that link what students in healthcare can do to make the connection of engaging academics with the community and fulfilling the call to action of civic responsibility that telehealth is best suited to provide.

PRACTICAL IMPLICATIONS

Leaders from academic institutions, professional associations, community-based organizations, philanthropy, and government must come together to take a leadership role in creating a more supportive culture for community-engaged solutions in health care.

Research is needed for telemedicine and telehealth that encapsulates a community of practice (CoP). Communities of practice (CoP) are collections of individuals with common concerns who meet regularly for communal interests whether they be for activities or learning engagements to enhance knowledge, skill, or ability (Barnett, Jones, Caton, Iverson, Bennett, & Robinson, 2014). CoP theory is an applicable architype for clarifying medical information sharing and for overcoming professional isolation (Barnett et al., 2014). CoP groups should be formed by telehealth professionals such as nurses, doctors, specialists, or therapist in a distance learning format (Allen, Seaman, & Sloan, 2011). According to Burton (2016), these groups could share in a persistent process of sincere peripheral participation for greater knowledge, skills, or abilities. As groups of people associated together by a need to resolve business concerns and perfect skills by sharing common practices and experiences, these CoPs would pull health care information from individual members that are inseparable as both members and participants; therefore, organization would occur around common needs and interest (e.g., telehealth and chronic illness data), thus would provide members a sense of collaborative enterprise. Specifically, through CoPs, medical professionals could share a domain of medical knowledge, with diverse intensities of expertise. Through this shared practice of medical knowledge enhanced application could occur. Through CoPs, users can become stronger in understanding how to better function in their professional areas (Burton, 2016). Further, the CoP becomes an avenue for mentoring and coaching, which can lead to learning and training.

Facilitation of knowledge and skills in CoPs, as well as, the honing of abilities for telehealth practitioners in training, learning, and development happens as it relates to mentoring, coaching, and experience. Training and learning in health care communities propose a procedure wherein numerous individuals work in partnership across diverse health care communities and specialties (i.e., telestoke, tele-mental health, and dermatology) to unite disjointed data (Leonard, 2013; Rodriguez, 2013). Performing numerous telehealth training and learning CoPs would foster practitioners and researchers to examine diversity in learner engagement, as well as, knowledge transfer. Knowledge transfer through CoPs can address major medical concerns that could lead to successful application within health care organizations. Through CoPs knowledge is managed and disseminated instead of remaining confined within the heads of its individual knowledge holders. The key advantage of CoPs is that they can be used to leverage and manage new and existing knowledge through the use of existing information technology.

FUTURE RESEARCH DIRECTIONS

Future studies could include homogenous groups of medical providers with at least 10 years in in the field with direct patient care in rural and underserved populations who are engaged to provide their insights using a modified Adelphi study. An Adelphi study would enable the group to build off of each other's experiences in higher education and how it would be beneficial to incorporate civic responsibility and engaged scholarship in order to provide new offering into curriculums. The best way to add value into higher education experiences is to have academics engaging the community, the actual providers who can share their real life experiences with those who are yet to follow in their footsteps.

Future studies that involve the aforementioned development of CoPs could look for ways to design cohort learning and sharing that could be researched for its potential impact on communities that have high incidences of chronic diseases to determine success factors on patient outcomes, practitioner's happiness and well-being, and total community involvement. It is through both giving and receiving through actions such as mentoring, coaching, and teaching, that all participants can evolve and better networks can be developed.

CONCLUSION

The responsibility required in the learning process at all levels can be greatly improved when the patient is at the center and the main goal is empowerment. When the paradigm shift is moved just a few dials more toward patient-centered, disease management, it is the community that benefits. The researcher posits that telehealth is a civic responsibility and this means the betterment of society. We all benefit when everyone heeds the call to action and the pendulum swings towards academic engagement earlier in the preparation for service in health care which can impart better care coordination, better patient engagement, and health care providers working through engaged scholarship and civic responsibility that can lead to prevention and wellness-patient centered care.

REFERENCES

Adeleye, O. A., & Ofili, A. N. (2010). Strengthening intersectoral collaboration for primary health care in developing countries: Can the health sector play broader roles? *Journal of Environmental and Public Health*, *2010*, 1–6. doi:10.1155/2010/272896 PMID:20454703

Agency for Healthcare Research and Quality (AHRQ). (2015). *The state of health care quality: How good is care?* Retrieved from http://archive.ahrq.gov/hcqual/meetings/mar12/chap01.html

Ajami, S., & Lamoochi, P. (2014). Use of telemedicine in disaster and remote places. *Journal of Education and Health Promotions*, *3*, 26. doi:10.4103/2277-9531.131886 PMID:25013819

Allen, I., Seaman, J., & Sloan, C. (2011). *Going the distance: Online education in the United States, 2011*. Sloan Consortium. Retrieved from http://files.eric.ed.gov/fulltext /ED529948.pdf

American Telemedicine Association. (2014). Telemedicine frequently asked questions (FAQs). Retrieved from http://www.americantelemed.org/about-telemedicine/faqs

American Telemedicine Association (ATA). (2012a). *What are the benefits of telemedicine?* Retrieved from www.americantelemed.org

American Telemedicine Association (ATA). (2012b). *Telemedicine practice guidelines*. Retrieved from www.americantelemed.org

Ander, D. S., & Love, J. (2017). Educational scholarship. *Western Journal of Emergency Medicine, 18*(1).

Ander, D. S., & Love, J. N. (2017b). The evolving definition of education scholarship: What the clinician educator needs to know. *Western Journal of Emergency Medicine*, *18*(1), 1–3. doi:10.5811/westjem.2016.12.33326 PMID:28115998

Association of American Medical Colleges (AAMC). (2012). *Recent studies and reports on physician shortages in the US.* Association of American Medical Colleges (AAMC) Center for Workforce Studies. Retrieved from https://www.aamc.org/download/100598/data/

Association of American Medical Colleges (AAMC). (2013). *Data and analysis*. Retrieved from www.aamc.org/data

Association of American Medical Colleges (AAMC). (2015). *The complexities of physician supply and demand: Projections from 2013 to 2025.* Retrieved from https://www.aamc.org/download/426260/data/physiciansupplyanddemandthrough2025keyfindings.pdf

Bailey, J. M. (2009). The top 10 rural issues for health care reform. *Center for Rural Affairs*, *2*(1), 1–8.

Barnett, S., Jones, S. C., Caton, T., Iverson, D., Bennett, S., & Robinson, L. (2014). Implementing a virtual community of practice for family physician training: A mixed-methods case study. *Journal of Medical Internet Research*, *16*(3), e83. doi:10.2196/jmir.3083 PMID:24622292

Barr, D. A. (2016). *Introduction to US health policy: The organization, financing, and delivery of health care in America*. Baltimore, Maryland: Johns Hopkins University Press.

Bashshur, R. L., Reardon, T. G., & Shannon, G. W. (2000). Telemedicine: A new health care delivery system. *Annual Review of Public Health*, *21*(1), 613–637. doi:10.1146/annurev.publhealth.21.1.613 PMID:10884967

Bashshur, R. L., & Shannon, G. W. (2009). *History of telemedicine. Evolution, context, and transformation*. New Rochelle: Mary Ann Liebert, Inc. Publishers.

Bashshur, R. L., Shannon, G. W., Krupinski, E. A., Grigsby, J., Kvedar, J. C., Weinstein, R. S., ... Tracy, J. (2009). National telemedicine initiatives: Essential to healthcare reform. *Telemedicine Journal and e-Health*, *15*(6), 600–610. doi:10.1089/tmj.2009.9960 PMID:19534591

Bashshur, R. L., Shannon, G. W., Smith, B. R., Alverson, D. C., Antoniotti, N., Barsan, W. G., ... Ferguson, S. (2014). The empirical foundations of telemedicine interventions for chronic disease management. *Telemedicine Journal and e-Health*, *20*(9), 769–800. doi:10.1089/tmj.2014.9981 PMID:24968105

Bauer, J., & Ringel, M. (2009). *Telemedicine and the reinvention of healthcare* (pp. 59–74). New York: McGraw-Hill.

Berwick, D., Nolan, T., & Whittington, J. (2008). The triple aim: Care, cost, and quality. *Health Affairs*, *27*(3), 759–769. Retrieved from http://content.healthaffairs.org/content/27/3/759.full.pdf doi:10.1377/hlthaff.27.3.759 PMID:18474969

Boyd, W. E. (2013). Does Boyers integrated scholarships model work on the ground? An adaption of Boyers model for scholarly professional development. *International Journal for the Scholarship of Teaching and Learning*, *7*(2), 1. doi:10.20429/ijsotl.2013.070225

Boyer, E. L. (1990). *Scholarship reconsidered. Priorities of the professoriate*. California: The Carnegie Foundation for the Advancement of Teaching.

Boyer, E. L. (1997). Enlarging the perspective. In *Scholarship reconsidered: Priorities of the Professoriate* (pp. 15–25). Princeton, NJ: The Carnegie Foundation.

Boyer, E. L., Moser, D., Ream, T. C., & Braxton, J. M. (2015). *Scholarship reconsidered: priorities of the professoriate* (2nd ed.). San Francisco, CA: John Wiley & Sons.

Brown-Jackson, K. L. (2017). *Disrupting and retooling: A model for a community-based telehealth program.* Unpublished doctoral dissertation, The National Graduate School of Quality Management, Falmouth, MA.

Bui, A. L., & Fonarow, G. C. (2012). Home monitoring for heart failure management. *Journal of the American College of Cardiology*, *59*(2), 97–104. Retrieved from http://www.ncbi.nlm.nih.gov/pubmed/22222071 doi:10.1016/j.jacc.2011.09.044 PMID:22222071

Bulik, R. J., & Shokar, G. S. (2010). Integrating telemedicine instruction into the curriculum: Expanding student perspectives of the scope of clinical practice. *Journal of Telemedicine and Telecare*, *16*(7), 355–358. doi:10.1258/jtt.2010.090910 PMID:20643847

Burton, S. L. (2016). *Best practices for faculty development through andragogy in online distance education.* Manuscript submitted for publication.

Buzza, C., Ono, S. S., Turvey, C., Wittrock, S., Noble, M., Reddy, G., & Reisinger, H. S. (2011). Distance is relative: Unpacking a principal barrier in rural healthcare. *Journal of General Internal Medicine*, *26*(Suppl. 2), 648–654. doi:10.100711606-011-1762-1 PMID:21989617

Cason, J. (2014). Telehealth: A rapidly developing service delivery model for occupational therapy. *International Journal of Telerehabilitation*, *6*(1), 29–36. Retrieved from http://telerehab.pitt.edu/ojs/index.php/Telerehab/article/view/6148 doi:10.5195/IJT.2014.6148 PMID:25945220

Centers for Medicare and Medicaid Services (CMS). (2015a). *Home health quality initiative*. Retrieved from https://www.cms.gov/Medicare/Quality-Initiatives-Patient-Assessment-Instruments/HomeHealthQualityInits/index.html

Centers for Medicare and Medicaid Services (CMS). (2015c). *National health expenditure.* Retrieved from https://www.cms.gov/Research-Statistics-Data-and-Systems/Statistics-Trends-and-Reports/NationalHealthExpendData/NHE-Fact-Sheet.html

Chacko, T. V., & Supe, A. (2015). Scholarship in medical education. *The Art of Teaching Medical Students*, 448. [https://books.google.com/books?hl=en&lr=&id=LTqLCgAAQBAJ&oi=fnd&pg=PA448&dq=glassick+scholarship&ots=K5hxzvyCTA&sig=zk8cQkmBRtC3KAX-218NtdNJ3Sw#v=onepage&q=glassick%20scholarship&f=false]

Chen, C., Chen, F., & Mullan, F. (2012). Teaching health centers: A new paradigm in graduate medical education. *Academic Medicine: Journal of the Association of American Medical Colleges*, *87*(12), 1752–1756. doi:10.1097/ACM.0b013e3182720f4d PMID:23095929

Craig, C., Eby, D., & Whittington, J. (2011). *Care coordination model: Better care at lower cost for people with multiple health and social needs. IHI Innovation Series white paper.* Cambridge, Massachusetts: Institute for Healthcare Improvement. Retrieved from www.IHI.org

Darkins, A., Ryan, P., Kobb, R., Foster, L., Edmonson, E., Wakefield, B., & Lancaster, A. E. (2009). Care coordination/home health: The systematic implementation of health informatics, home health, disease management to support the care of veteran patients with chronic conditions. *Telemedicine Journal and e-Health*, *14*(10), 1118–1126. doi:10.1089/tmj.2008.0021 PMID:19119835

Di Cerbo, A., Morales-Medina, J. C., Palmieri, B., & Iannitti, T. (2015). Narrative review of telemedicine consultation in medical practice. *Patient Prefer Adherence*, *9*, 65–75. doi:10.2147/PPA.S61617 PMID:25609928

Doolan-Grimes, M. (2013). *Clinical nurse managers understanding of use of evidence based practice.* Dublin City University: School of Education Studies.

Ellaway, R. H. (2016). Ideology and health professional education scholarship. *Advances in Health Sciences Education: Theory and Practice*, *21*(3), 501–503. doi:10.100710459-016-9684-7 PMID:27138693

Eyler, J., Giles, D. E. Jr, & Braxton, J. (1997). The impact of service-learning on college students. *Michigan Journal of Community Service Learning*, *4*, 5–15.

Fitzpatrick, J. J., & McCarthy, G. (2010). Boyer and Scholarship Revisited. *Applied Nursing Research*, *23*(3), 121. doi:10.1016/j.apnr.2010.04.001 PMID:20643320

Flodgren, G., Rachas, A., Farmer, A. J., Inzitari, M., & Shepperd, S. (2015). Interactive telemedicine: Effects on professional practice and health care outcomes. *Cochrane Database of Systematic Reviews*, *9*, CD002098. doi:10.1002/14651858.CD002098 PMID:26343551

Fraher, E. P., Knapton, A., & Holmes, G. M. (2017). A methodology for using workforce data to decide which specialties and states to target for graduate medical education expansion. *Health Services Research*, *52*(Suppl. 1), 508–528. doi:10.1111/1475-6773.12649 PMID:28127769

Fuchs, V. R. (2011). *Who shall live?: Health, economics and social choice.* Hackensack, New Jersey: World Scientific Publishing Co Inc. doi:10.1142/8167

Gaynor, M., Mostashari, F., & Ginsburg, P. B. (2017). Making health care markets work: Competition policy for health care. *Journal of the American Medical Association, 317*(13), 1313–1314. doi:10.1001/jama.2017.1173 PMID:28253376

Giardino, A. P. (2016). Quality improvement and safety in healthcare: Reflections on essential frameworks to guide applied scholarship that promotes transformation and innovation. *Journal of Nursing & Interprofessional Leadership in Quality & Safety, 1*(1). Retrieved from http://digitalcommons.library.tmc.edu/uthoustonjqualsafe/vol1/iss1/6

Grady, J. (2014). CE: Telehealth: A case study in disruptive innovation. *The American Journal of Nursing, 114*(4), 38–45. doi:10.1097/01.NAJ.0000445682.52553.89 PMID:24632888

Grady, J. L. (2011). The virtual clinical practicum: An innovative telehealth model for clinical nursing education. *Nursing Education Perspectives, 32*(3), 189–194. doi:10.5480/1536-5026-32.3.189 PMID:21834382

Hartgerink, J. M., Cramm, J. M., Bakker, T. J. E. M., van Eijsden, R. A. M., Mackenbach, J. P., & Nieboer, A. P. (2014). The importance of relational coordination for integrated care delivery to older patients in the hospital. *Journal of Nursing Management, 22*(2), 248–256. doi:10.1111/j.1365-2834.2012.01481.x PMID:24576117

Haughton, J. (2000). A paradigm shift in healthcare: From disease management to patient-centered systems. *M.D. Computing, 17*(4), 34–38. Retrieved from www.mdcomputing.com/issues/v17n4/haughton.htm PMID:11014076

Hawkins, S. Y. (2012). Telehealth nurse practitioner student clinical experiences: An essential educational component for todays health care setting. *Nurse Education Today, 32*(8), 842–845. doi:10.1016/j.nedt.2012.03.008 PMID:22503296

Institute for Healthcare Improvement (IHI). (2015). *The IHI triple aim iInitiative*. Retrieved from http://www.ihi.org/engage/initiatives/tripleaim/pages/measuresresults.aspx

Institute of Medicine. (IOM). (2012). *The role of telehealth in an evolving health care environment: Workshop summary*, (p. 159). Washington, DC: National Academies Press. Retrieved from https://www.nap.edu/catalog/13466/the-role-of-telehealth-in-an-evolving-health-care-environment

Institute of Medicine (IOM). (2005). *Quality through collaboration: Future of rural health*. Washington, DC: National Academies Press. Retrieved from https://www.nap.edu/catalog/11140/quality-through-collaboration-the-future-of-rural-health-care

Institute of Medicine (IOM). (2008). *Retooling for an aging America, building the health care workforce*. Washington (DC): National Academies Press. Retrieved from http://www.nationalacademies.org/hmd/reports/2008/retooling-for-an-aging-america-building-the-health-care-workforce.aspx

Institute of Medicine (IOM). (2013). *Interprofessional education for collaboration: Learning how to improve health from interprofessional models across the continuum of education to practice: Workshop summary*. Washington, DC: The National Academies Press. Retrieved from http://www.nationalacademies.org/hmd/Reports/2013/Interprofessional-Education-for-Collaboration.aspx

Institute of Medicine (IOM). (2014). *Building health workforce capacity through community-based health professional education: Workshop summary*. Washington, DC: The National Academies Press. Retrieved from http://iom.nationalacademies.org/Reports/2014/BuildingHealthWorkforceCapacity.aspx

Kelsey, C. (2016). Rethinking scholarship in nurse education. In *Papers from the Education Doctoral Research Conference 2015*, Birmingham, University of Birmingham (pp. 53-59).

Kvedar, J., Coye, M. J., & Everett, W. (2014). Connected health: A review of technologies and strategies to improve patient care with telemedicine and telehealth. *Health Affairs*, *22*(2), 194–199. doi:10.1377/hlthaff.2013.0992 PMID:24493760

Leonard, K. G. (2013). *Online communities of practice: Factors that affect usage from a multiple stakeholder perspective*.

Li-Chen, C., Chi-Wen, C., Yung-Ching, W., Rung-Ji, S., Hui-chu, Y., Yufang, C., & Feipei, L. (2012). An information technology framework for strengthening telehealthcare service delivery. *Telemedicine Journal and e-Health*, *18*(8), 596–603. doi:10.1089/tmj.2011.0267 PMID:23061641

Lilly, C. M., Zubrow, M. T., Kempner, K. M., Reynolds, H. N., Subramanian, S., Eriksson, E. A., ... Cowboy, E. R. (2014). Critical care telemedicine: Evolution and state of the art. *Critical Care Medicine*, *42*(11), 2429–2436. doi:10.1097/CCM.0000000000000539 PMID:25080052

Maheu, M., Whitten, P., & Allen, A. (2002). *E-Health, telehealth, and telemedicine: A guide to startup and success. CA*. San Francisco: John Wiley & Sons.

Majerowicz, A., & Tracy, S. (2010). Telemedicine: Bridging gaps in healthcare delivery. *Journal of American Health Information Management Association*, *81*(5), 52–53, 56. PMID:20496638

Mehrotra, A., Paone, S., Martich, G. D., Albert, S. M., & Shevchik, G. J. (2013). A comparison of care at e-visits and physician office visits for sinusitis and urinary tract infection. *JAMA Internal Medicine*, *173*(1), 72–74. doi:10.1001/2013.jamainternmed.305 PMID:23403816

Mid-Atlantic Telehealth Resource Center (MATRC). (2016). *Resources*. Retrieved from www.matrc.org

Mokdad, A. H., Marks, J. S., Stroup, D. F., & Gerberding, J. L. (2004). Actual causes of death in the United States, 2000. *Journal of the American Medical Association*, *291*(10), 1238–1245. doi:10.1001/jama.291.10.1238 PMID:15010446

Moore, J. D., Jr. (2016, April 27). *Medical schools transform teaching methods to strengthen, sharpen health care workforce*. California Health Care Foundation. Retrieved from http://www.chcf.org/articles/2016/04/medical-schools-transform

Nundy, S., & Oswald, J. (2014). Relationship-centered care: A new paradigm for population health management. [Autralia: Elsevier.]. *Health Care*, *2*(4), 216–219. PMID:26250627

Parekh, A. K., Goodman, R. A., Gordon, C., & Koh, H. K. (2011). Managing multiple chronic conditions: A strategic framework for improving health outcomes and quality of life. *Public Health Reports*, 126(4), 460–471. PMID:21800741

Pathipati, A. S., Azad, T. D., & Jethwani, K. (2016). Telemedical education: Training digital natives in telemedicine. *Journal of Medical Internet Research, 18*(7), e193. doi:10.2196/jmir.5534 PMID:27405323

Petterson, S. M., Law, W. R., Phillips, R. L., Rabin, D. L., Meyers, D. S., & Bazemore, A. W. (2012). Projecting US primary care physician workforce needs: 20102025. *Annals of Family Medicine, 6*(10), 503–509. doi:10.1370/afm.1431 PMID:23149526

Psek, W. A., Stametz, R. A., Bailey-Davis, L. D., Davis, D., Darer, J., Faucett, W. A., … Gerrity, G. (2015). Operationalizing the Learning Health Care System in an Integrated Delivery System. *eGEMs, 3*(1), 1122. doi:10.13063/2327-9214.1122

Rafiei, S., Mohebbifar, R., Hashemi, F., Ezzatabadi, M. R., & Farzianpour, F. (2016). Approaches in health human forecasting: A roadmap for improvement. *Electronic Physician, 8*(9), 2911–2917. doi:10.19082/2911 PMID:27790343

Rodriguez, G. (2013). *Perceptions within a virtual community of practice: A Q-method study.*

Sacristán, J. A. (2013). Patient-Centered Medicine and Patient-Oriented Research: Improving health outcomes for individual patients. *BMC Medical Informatics and Decision Making, 13*(1), 6. doi:10.1186/1472-6947-13-6 PMID:23294526

Schoenfeld, A. J., Davies, J. M., Marafino, B. J., Dean, M., DeJong, C., Bardach, N. S., ... Dudley, R. A. (2016). Variation in quality of urgent health care provided during commercial virtual visits. *JAMA Internal Medicine, 176*(5), 635–642. doi:10.1001/jamainternmed.2015.8248 PMID:27042813

Schwamm, L. H. (2014). Telehealth: Seven strategies to successfully implement disruptive technology and transform health care. *Health Affairs, 33*(2), 200–206. Retrieved from http://search.proquest.com/docview/1498231606?accountid=62746 doi:10.1377/hlthaff.2013.1021 PMID:24493761

Stanhope, M., & Lancaster, J. (2013). *Public health nursing-revised reprint: Population-centered health care in the community* (8th ed.). Maryland Heights, Missouri: Elsevier Mosby.

Stiefel, M., & Nolan, K. (2012). *A guide to measuring the triple aim: population health, experience of care, and per capita cost. IHI Innovation Series white paper.* Cambridge, Massachusetts: Institute for Healthcare Improvement. Retrieved from www.IHI.org

Thaker, D. A., Monypenny, R., Olver, I., & Sabesan, S. (2013). Cost savings from a telemedicine model of care in northern Queensland, Australia. *The Medical Journal of Australia, 199*(6), 414–417. doi:10.5694/mja12.11781 PMID:24033216

Truog, R. D. (2012). Patients and doctors evolution of a relationship. *The New England Journal of Medicine, 366*(7), 581–585. doi:10.1056/NEJMp1110848 PMID:22335734

U.S. Department of Health and Human Services (HHS). Office of Disease Prevention and Health Promotion. (2011). *Healthy people 2020.* Washington, DC: Government Printing Office. Retrieved from https://www.healthypeople.gov

Wagner, E. H. (1998). Chronic disease management: What will it take to improve care for chronic illness? *Effective Clinical Practice, 1*(1), 2–4. PMID:10345255

Wagner, E. H. (2013). The chronic care model. Group health research institute.

Welsh, T. S. (2002). *Organizational structure of telehealth care: An examination of four types of telemedicine systems.*

West, J. & Mehrotra, A. (2016). The future ecology of care. *Annals of Internal Medicine*. doi:10.7326/M15-1978

Wildevuur, S. E., & Simonse, L. W. (2015). Information and communication technology-enabled person-centered care for the big five chronic conditions: Scoping review. *Journal of Medical Internet Research*, *19*(3), e77. doi:10.2196/jmir.3687 PMID:25831199

Willemse, E., Adriaenssens, J., Dilles, T., & Remmen, R. (2014). Do telemonitoring projects of heart failure fit the Chronic Care Model? *International Journal of Integrated Care*, *14*(3). Retrieved from http://www.ijic.org/index.php/ijic/article/view/URN%3ANBN%3ANL%3AUI%3A10-1-114790/2421 doi:10.5334/ijic.1178 PMID:25114664

World Health Organization (WHO). (2012). *The World Health Report 2008 - primary health care (now more than ever).* Retrieved from http://www.who.int/whr/2008/en/

KEY TERMS AND DEFINITIONS

American Telemedicine Association (ATA): A non-profit association based in Washington, DC with a membership network of more than 10,000 industry leaders and healthcare professionals. ATA is the largest telehealth-focused organization focused on helping to transform healthcare by improving the quality, equity, and affordability of healthcare throughout the world.

Chronic Disease Management: A patient-centered model of care that encompasses integrated and coordinated primary medical care and specialty care, patient and clinician education, and evidence-based care plans and dependent on expert care accessibility.

Community of Practice (CoP): A collection(s) of individuals with common concerns for communal interests, who meet on a consistent basis whether in person and/or virtually, to enhance their knowledge, skill, or ability or focus on the betterment of the community members.

Disruptive Innovation: Disruption means the departure from the traditional patient–clinician face-to-face relationship.

Medical Education: Education related to the practice of being a medical practitioner; either the initial training to become a physician in medical school and through internships, and typically includes additional training in the form of residency and fellowships.

Patient-Centered Healthcare: Replaces our current physician centered system with one that revolves around the patient. It is a methodology of providing health care that depend on upon effective communication, empathy, and a feeling of partnership between physician and patient to improve patient care outcomes, patient satisfaction, to manage effectively patient symptoms through prevention, and to reduce unnecessary costs.

Rural: A rural area is an area that is unincorporated or towns with a population of under 10,000.

Teleconferencing: Interactive electronic communication between multiple users at two or more sites that facilitates voice, video, and/or data transmission systems: audio, graphics, computer, and video systems.

Virtual Care: A term that is associated with telemedicine and telehealth include but are not limited to telecare, telenursing, online health, eHealth, or connected health. Virtual care is based on rapid accessibility and online collaboration with fellow colleagues and patients. It is also platform agnostic and can be accomplished using any hardware or software platforms.

This research was previously published in Engaged Scholarship and Civic Responsibility in Higher Education edited by Sharon L. Burton, pages 166-193, copyright year 2018 by Information Science Reference (an imprint of IGI Global).

Chapter 10
Modelling Factors Affecting Patient–Doctor–Computer Communication in Primary Care

Shiri Assis-Hassid
Harvard University, USA

Tsipi Heart
Ono Academic College, Israel

Iris Reychav
Ariel University, Israel

Joseph S. Pliskin
Ben-Gurion University of the Negev, Israel

ABSTRACT

This work presents a conceptual model aimed at explaining factors affecting the formation of effective patient-doctor-computer communication at the primary care clinic. The authors define a new construct – patient-doctor-computer communication (PDCC), aimed to replace the traditional concept of dyad patient-doctor communication (PDC). PDC has been characterized as one of the most significant factors affecting healthcare outcomes. To better understand PDCC and its antecedents, the authors integrate theories from the patient-centered care and the Information Systems domains and suggest that the characteristics of the EMR, the user (doctor) and the task determine the doctor's perception of fit between the EMR and the medical task, which in turn positively affects PDCC. The suggested conceptual model contributes to both theory and practice. On the theoretical side, it opens several new research trajectories. For practice, the model implies that there is a need for a tighter collaboration between experts from both the information systems and medicine domains in designing EMR systems that are aligned with and support the medical task at hand.

DOI: 10.4018/978-1-5225-6198-9.ch010

INTRODUCTION

The goal of this paper is to develop a conceptual model that explains how the doctor's perception of fit between the Electronic Medical Record (EMR) system and the medical task affects the formation of effective patient-doctor communication (PDC) at the primary care clinic.

To this end, we first explore the role of the physician-patient relationship in patient care and highlight findings of how this relationship is affected by the introduction of the EMR into the doctor's workflow. While it may appear that the EMR provides an opportunity to improve patient-doctor communication (PDC) as part of the clinical practice (Haluza & Jungwirth, 2014), and hence medical outcomes, there is evidence that the introduction of ICT into the clinic significantly affects the physician-patient relationship (Haluza & Jungwirth, 2014; Pearce, Arnold, Phillips, Trumble, & Dwan, 2009). Surprisingly, there is scant literature suggesting how to address adverse implications of EMR use during the medical encounter on PDC (Shachak & Reis, 2009). This existing gap is particularly unsettling since communication has been characterized as one of the most powerful, encompassing, and versatile instruments available to the physician (Weiner, 2012). Furthermore, patient-centered care, which is nowadays the leading paradigm in primary care, emphasizes the need to build rapport between doctor and patient through undivided attention and communication (Assis-Hassid, Heart, Reychav, Pliskin, & Reis, 2013), yet studies imply negative effects of EMR use on PDC and patient-centered aspects (Shachak & Reis, 2009). As will be shown, the patient-centered care approach assumes that good communication can improve healthcare outcomes from better treatment adherence and fewer interactions leading to malpractice suits (Frankel et al., 2005; Lorig, 2012).

This work focuses on the effects of using the EMR system on patient-doctor communication at the primary care clinic. We explore the behavioral aspects of the EMR's introduction in the clinic as well as physical aspects such as room configuration. Regarding both aspects, we show the benefits and shortcomings of EMR use and their influence on the patient-doctor communication. Finally, we develop a model of factors affecting PDCC in the primary care clinic based on theoretical frameworks from the information systems (IS) field, and from the patient-centered concept, which hypothesize that the characteristics of the user (doctor), the EMR system and the task affect the doctor's perception of the fit between the EMR and the task at hand. Based on IS theories, it is conjectured that this perception of fit affects the doctor's ability to develop effective PDC.

The main contribution of this work is in extending the current understanding of the physician-patient relationship in a computerized environment. Moreover, we believe that the application of IS theoretical frameworks and the concepts highlighted in this work can be generalized to other Healthcare Information Systems (HIS), implying what needs to be done in order to minimize current obstacles in HIS utilization while improving the patient-doctor communication in the computerized medical environment.

BACKGROUND AND LITERATURE SURVEY

The following section provides an overview of the physician-patient relationship and imperative role of communication in healthcare which have been recognized prior to the introduction of the EMR system. Namely, we discuss elements of patient-centered care, physicians' communication behavior, and characteristics of the medical task at the GP's clinic.

Physician-Patient Communication and Patient-Centered Care

Primary medical care requires effective physician-patient communication (Lorig, 2012; Mead, Bower, & Hann, 2002). Thousands of medical interactions have been studied to elucidate the key 'ingredients' of good consultations (Lorig, 2012). Increasingly, researchers have adopted the concept of 'patient-centered care' as an indicator of good quality consulting. The concept of patient-centeredness has received numerous definitions: Balint (1961) defines patient-centered medicine in cognitive terms, as understanding the patient as a unique human being; McWhinney (1989) narrows the concept from understanding the patient to understanding the patient's experience of the illness. Hall and Dornan (1988) provide a framework linking physicians' consulting behaviors with patient outcomes. The concept of patient-centeredness has evolved from the bio-psychosocial model (G. Engel, 1977; G. L. Engel, 1981) which places suffering, disease, and illness in the broad context of biological, psychological and social dimensions. In practice, according to the model, physicians need to process information that is provided to them by patients in both a biomedical and a psychosocial context. According to the model suggested by Engel (1977), physicians need to "listen with both ears," that is, symbolically assigning one ear to receive biomedical and the other ear to receive psychosocial information, for example by being attentive to the stories patients tell, nuances of the patient's body position, facial expressions, etc. (Epstein, Campbell, Cohen-Cole, McWhinney, & Smilkstein, 1993).

In the following sections, we describe different types of physicians' communication behavior during the medical interview as well as the required skills for conducting an effective medical interview.

Physicians' Communication Behaviors at the Clinic

Hall et al. (1988) distinguish two main types of physician communication behaviors: The first behavior is recognized as 'instrumental' and the second as 'socio-emotional.' Instrumental behaviors are concerned with the medical tasks of the consultation, such as problem diagnosis and management. Such behaviors include question-asking and providing the patient with information and advice. Socio-emotional behaviors facilitate interpersonal aspects of the interaction. These may be verbal or non-verbal; for example, greetings, expressions of empathy, reassurance, partnership-building statements, eye contact, jokes or social talk.

Well before the introduction of the Electronic Medical Record (EMR) and computerization into the medical environment, researchers addressed the importance of non-verbal communication in the medical encounter. For example, it has been suggested that physicians' non-verbal skills are associated with outcome variables such as patient satisfaction, patient recall of medical information, compliance with keeping appointments, and compliance with medical regimens (DiMatteo, Taranta, Friedman, & Prince, 1980; Hall, Harrigan, & Rosenthal, 1995; Hall, Roter, & Rand, 1981; Larsen & Smith, 1981; Ong, de Haes, Hoos, & Lammes, 1995). Physicians' indirect or broken eye contact and indirect facial orientation have been associated with less patient disclosure (Duggan & Parrott, 2001).

The Medical Interview Process

Effective Physician-patient communication during the medical interview is based on the patient-centeredness concept and the *three function model* developed by Bird and Cohen-Cole, (1990), and Epstein et al., (1993). The model highlights three core functions of the interaction between physician and patient and stresses that the conduct of the medical interview is critical to optimizing biological outcomes as well as psychological and social well-being. The three functions are (1) Gathering data to understand the patient; (2) Development of rapport and responding to the patient's emotions; (3) Patient education and behavioral management.

Each of the above functions includes supportive skills required by the physician (see Table 1). For example, in order to successfully conduct data gathering (function 1), the physician must be able to apply open-ended questioning; in order to develop a rapport with the patient (function 2) the physician must be able to show support and respect towards the patient. Finally, in order to carry out patient education (function 3), the physician must be able, for example, to elicit basic understanding with the patient and/or deliver information in a convenient manner.

The Impact of Using the EMR on Patient-Doctor Communication

So far we have discussed the importance of physician-patient communication for healthcare quality and more specifically, the patient-centered approach which focuses on the psychological and social aspects of the medical encounter in addition to the obvious biomedical aspects. Furthermore, we have discussed the different skills required from physicians during the medical encounter, in general, regardless of EMR use. Next, we turn to find out how the medical encounter and relationship between physician and patient is affected by the introduction of the EMR into healthcare in general and the clinic in particular.

Dyadic Relationships Transform Into Triadic

Most of the research and teaching of patient-centered communication has assumed that the relationship between physician and patient is purely dyadic. However, the computer may be playing a more significant and visible role that has turned this relationship into a triadic one (Pearce, Walker, & O'Shea, 2008). As a result, it is important to discuss and obtain a better understanding of this triadic relationship and how

Table 1. The three function model of the medical interview - required skills

Function	Objective	Required Skills
1) Data gathering and patient understanding	Gather data accurately and efficiently, and understand the patient and his problem.	- Use of an open-to-closed cone of questioning to progressively narrow the focus of the narrative - Facilitation - Clarification - Checking - Surveying for new problems
2) Rapport developing and emotional response to the patient	Increasing the efficiency of data gathering Increasing patient satisfaction Recognition of psychiatric distress Humanization of the doctor-patient encounter Improving physical outcomes Increasing physician satisfaction	Five basic skills: - Reflection - Legitimation - Support - Partnership - Respect
3) Patient education and behavioral management	Providing diagnostic information and therapeutic recommendations to patients in a manner that they can understand and put into action. When this function of the interview is successful, physician and patient will agree on a course of action and patient adherence will be maximized	- Eliciting baseline understanding - Eliciting the patient's preferences and obtaining a statement of commitment - Eliciting the patient's preferences and ideas for changes

computers and information systems can be incorporated in healthcare-related encounters so that their benefits to healthcare quality are maximized.

A computerized exam room has the potential to shift the physician's attention and involvement away from the patient to the keyboard and monitor. Since attention to the patient may be associated with positive outcomes of care (Lorig, 2012), it is important to examine how computer use enhances or interferes with the physician's attention to the patient and the effects on physician-patient communication.

The Electronic Medical Record (EMR)

The most apparent manifestation of ICT in healthcare is the continuingly growing use of Electronic Medical Records (EMRs) in the exam room.

The EMR is a longitudinal electronic record of patient health information generated by one or more encounters in a care delivery setting. Records include patient demographics, progress notes, problems, medications, vital signs, past medical history, immunizations, laboratory data and radiology reports. The EMR is destined to automate and streamline the clinician's workflow (Hayrinen, Saranto, & Nykanen, 2008; HIMSS). A full function EMR should include the following functions: 1) patient charting, 2) order communication system or computerized physician order entry (CPOE), 3) clinical decision-making support system 4) document and image management, 5) patient portal, 6) management of internal and external clinical documents and notes, 7) statistics and reporting (HealthTechnologyReview.com)

Benefits of EMR Use in Healthcare

The motivation to use computerized information systems in healthcare is driven by the expectation that such systems will improve the quality of care, patient safety and lower medical costs (Bates, Ebell, Gotlieb, Zapp, & Mullins, 2003). The potential benefits of EMR include: the ability to comprehensively document the patient's medical history, easy access to medical data, improvements in accuracy and completeness of medical data, clinical decision support, minimization of human and medical errors as a result of using printed prescriptions instead of handwritten ones, on-screen reminders and alerts, electronic sharing of medical information among physicians and more (Chaudhry et al., 2006; McGrath, Arar, & Pugh, 2007; Shachak, Hadas-Dayagi, Ziv, & Reis, 2009).

It is becoming more difficult to imagine healthcare without EMRs (Margalit, Roter, Dunevant, Larson, & Reis, 2006) as their use is continually increasing around the world. In 2003, nearly 5 million patient records were entered by nearly 5000 physicians (in the areas of family medicine, internal medicine, and pediatrics) in the four major Israeli Health Maintenance Organizations (HMOs) (Margalit et al., 2006). Europe has shown high rates of EMR use as well. By 1995, 80% of primary care physicians in the UK worked in a computerized environment - over 60% of them were using the EMR during the consultation; 70% of Danish primary care physicians used EMRs, 60% in Sweden and 40% in the Netherlands (Als, 1997) and Australia, 93% of physicians used computers as part of their clinical routine (Pearce et al., 2009). In contrast, until recently only 5-15% of general practitioners in the US use EMR (Bates, Ebell, et al., 2003; DesRoches et al., 2008). This however has changed as a result of the "Meaningful Use" Act initiate by the Obama regime in 2009, proposing financial incentives to practices adopting and using an EMR system (Blumenthal & Tavenner, 2010). Since then, implementation rates have risen from 18% in 2001 to 78% in 2013 in the US (Hsiao, Hing, Socey, & Cai, 2011). The healthcare community has slowly accepted health IT as a tool assisting medical staff.

Adverse Implications of EMR Use in Healthcare

Despite the apparent benefits of computerization in healthcare, several adverse implications can be recognized as well. The structured format of EMRs (e.g., drop-down lists, etc.) is very different from the narrative structure of the traditional patient record and may result in loss of information (V. L. Patel, Kushniruk, Yang, & Yale, 2000). Moreover, experienced users perform selections from lists very quickly, in an automatic manner making it easy to select the wrong item. Two commonly reported errors with EMRs are selecting the wrong medication and adding to the wrong patient's chart (Shachak et al., 2009). Other examples are the use of copy-paste and templates that may raise concerns regarding data quality (Thielke, Hammond, & Helbig, 2007). In terms of physician-patient communication, EMRs were often found to have a negative effect on patient centeredness (Shachak & Reis, 2009). The computer often caused physicians to lose rapport with their patients, for example by screen gazing while the patient is talking or while talking to the patient (Booth, Robinson, & Kohannejad, 2004; G. Makoul, Curry, & Tang, 2001; Margalit et al., 2006).

Barriers to EMR Adoption

Several potential negative effects of computer use in primary care may have slowed the EMR dissemination processes. Concerns about increased costs, lengthened visit time, additional training needs and treatment inflexibility have been raised (Boonstra & Broekhuis, 2010; Jha et al., 2009; Miller & Sim, 2004) as well as potential privacy and confidentiality breaches that may cause lower rapport between physician and patient (Jha et al., 2009; Vimla L. Patel, Arocha, & Kushniruk, 2002). Bates et al.(2003) argue that physician resistance to using EMRs as part of their routine may be an additional barrier. This resistance may derive from physicians' perception that the EMR use will negatively affect their workflow. For example, data entry may take extra time. Financial barriers and questionable return on investment have also been reported as adoption inhibitors (DesRoches et al., 2008). Another concern that has been raised is the possibility of a negative effect on physician-patient communication. This latter concern will be discussed in detail in the sequel.

The Effects of EMR Use on Physician-Patient Relationship

Empirical studies of EMR use through direct observation have provided insight into the EMR's effect on the physician-patient communication dynamics. An early study conducted by Greatbatch et al. (1995) videotaped physicians' early adapting to EMR use. Physicians' behavior was described as 'pre-occupied' with attention largely focused on the computer monitor and only occasionally on the patient. In addition, they found that the visits were characterized by long periods of silence and minimal verbal engagement with patients. Patel et al. (2002) observed physicians with different expertise levels of EMR use and found that less experienced physicians were strongly influenced by the order and organization of the information presented by the EMR on the screen when asking the patient questions and entering data. The more experienced physicians showed greater flexibility in their interviewing style by moving back and forth between sections. According to Shachak and Reis (2009), the EMR had a positive impact on information related tasks and information exchange (the first function of the three-function model medical interview). However, the EMR was found to have a negative impact on the second function of the medical interview, establishing rapport with patients. There is an indication that the EMR organizes the encounter around data gathering demands rather than on the patient's narrative. Shachakand Reis (2009) indicate that physicians rarely used the computer for the third function of the medical interview - patient education and behavioral management.

Physician-Patient Communication Styles During EMR Use

An observational study conducted by Booth et al., (2004) describes three distinct styles of EMR use by physicians: the first, the *controlling style* – the physician directs the patient not to interrupt during computer use; the second, the *responsive style* – the physician uses gaps in the conversation to glance at the computer screen and the third, *the ignoring style* – the physician is so occupied with using the computer that s/heignores the patient's comments. Patel et al. (2002) argue that without focused training, the demands of the computer and patient interaction may result in cognitive overload with consequences for both computer use and attentiveness to patients.

Makoul et al. (2001) directly compared the communication patterns of physicians in the same clinic that used either EMRs or paper charts. As was later corroborated by Shachak and Reis (2009), Makoul et al. found that EMR use strengthened the physician's ability to complete information tasks but reduced the physician's attention to the patient-centered aspects of patient communication. Margalit et al.,(2006) extended the previous studies by capturing patterns of verbal communication and non-verbal communication such as eye contact, head nods, smiling and leaning forward. They also documented indicators of computer use, for example the number of seconds the physicians gazed at the computer screen and levels of active keyboarding. The research results show a significant relationship between computer use and physician's behavior of data gathering, patient education, and counseling functions. Patient disclosure of biomedical information to the physician was positively related to levels of physician keyboarding. Interestingly, it was found that screen gaze was inversely related to the physician's use of psychosocial questions. Emotional exchange (empathy, concern, reassurance) appeared lessened with screen gaze.

A Unified Classification of Physician Communication Styles During EMR Use

Another classification, based on Booth et al.(Booth et al.) and Ventres et al. (2005) is suggested by Shachak and Reis (2009). They identify three major styles: (1) informational-ignoring, (2) controlling-managerial, (3) interpersonal.

The informational-ignoring style is similar to the ignoring style described by Booth et al. (2004) and is characterized by focusing on details of information and extensive information gathering, which involves much focus on the computer. Information-ignoring physicians tended to lose rapport with their patients while engaged with the EMR. They frequently spoke while gazing at the computer screen, hardly faced the patient while entering or retrieving data from the computer and did not usually use the computer for information sharing purposes with the patient.

The controlling-managerial style is similar to the controlling style described by Booth et al. (2004) and is characterized by separating communication with the patient from computer use. Such physicians altered their attention between the patient and computer in defined stages of the encounter. They usually used gestures such as turning their body or gazing in order to indicate their switch of attention.

The interpersonal style is similar to the responsive style described by Booth et al. (2004) and is characterized by focusing on the patient. Such physicians did not tend to speak while using the computer. They faced the patient with their body even while using the EMR and used the computer in order to share and review information together with the patient. In addition, physicians demonstrating this style did not use computer-guided questions, spent less time entering data and refrained from using the computer at the beginning of the encounter.

Effective Computer Use During the Medical Encounter

The previous section mapped the different behavioral styles exemplified by physicians during the medical encounter and while using the EMR. The question that still remains unanswered is whether or not computers enhance or interfere with the medical encounter, and more specifically, which type of behavior style is recommended while using an EMR?

Frankel et al. (2005) tried to assert whether the computer enhanced or interfered with the physician's attention to the patient. They identified three ways by which physicians maintained communication with patients during computer use: (1) Verbally, the physician maintained conversation while looking at the screen or typing; (2) Visually, the physician formed occasional eye contact with the patient during computer use; and (3) Posturally, the physician was facing the patient rather than turning away from the patient during computer use. They also identify four domains in which exam-room computing affected physician-patient communication: (1) visit organization;(2) verbal and nonverbal behavior;(3) computer navigation and mastery; and (4) spatial organization of the exam room.

In the domain of visit organization, during the visit the physician's tasks include gathering and documenting medical data, physically examining the patient, delivering diagnostic and prognostic information, providing treatment recommendations, and educating the patient. Exam-room computers have added complexity to the visit's flow by increasing the amount of clinical information or introducing additional tasks such as typing information onto the computer. Therefore, visit organization refers to managing the cognitive, physical and socio-emotional tasks that constitute the medical encounter.

Regarding verbal and nonverbal behaviors, good and bad verbal and non-verbal skills can facilitate or constrain communication. The introduction of the computer was found to amplify the physician's baseline skills and thereafter affect the communication with patients. Physician's verbal and non-verbal behaviors that include empathy, support posture, gesture, and tone of voice, have been related to outcomes of care such as patient satisfaction and adherence to the physician's medical recommendations.

Computer navigation and mastery was related to the physician's ability to navigate on the computer, influencing whether its use facilitated or impeded on communication during the visit. Various technical mastery issues, such as typing information efficiently, were also found to influence the physician-patient communication.

The spatial configuration which is the physical placement of the computer, monitor, exam table and physician's chair, was found to influence physician-patient communication. This aspect will be discussed in detail in the next section.

As can be learned from the above, in order to maintain an effective communication level with patients, physicians need to maintain communication with the patient verbally, visually and posturally. Nevertheless, the introduction of the EMR into the exam room may require updating the traditional clinical encounter paradigms. Further research should, therefore, find out how to improve visit organization, verbal and non-verbal behaviors, computer navigation and mastery and provide a supportive spatial configuration. All these factors combined offer the opportunity to maximize the benefits of EMR use in the exam room and decrease potential shortcomings.

Physical Aspects of Exam-Room Computers

As discussed shortly in the previous section, the effects of EMR use on physician-patient communication may vary in different spatial configurations. That is why in addition to technical aspects of EMRs on physician-patient communication, it is equally important to address the physical aspects of computerization in primary care, namely the spatial configuration or physical configuration of the exam room. Physicians' exam rooms have been designed well before the introduction of computers and as such, it is interesting to find out whether and how the exam room spatial configuration facilitates or interferes with the communication process.

Pearce et al. (2008) examined how Australian general practitioners (GPs) have integrated computers into a physical environment that was not designed around them, and how the physician, patient, and computer interact within this workspace. Rather than focusing on physician-patient communication, Pearce et al. (2008) focused on the physical environment and configuration, namely, on the physician's desk and surrounding chairs. Current teaching of Australian trainee GPs is based on the principles of patient-centeredness and the importance of making the patient feel more equal in the interaction. Physicians are taught that the optimal arrangement for a seated consultation is with the patient seated beside the desk with minimal obstacles between patient and physician. Pearce et al. took still photos of twenty physician rooms and analyzed them. They identified two general settings: patient inclusive and patient exclusive. In the inclusive setting, physicians were able to share the computer screen with the patient, for example by using a flat screen on a moveable arm. Such screens facilitated better physician-patient communication as well as supported patient education. In the exclusive setting, physicians limited the patient's exposure to the screen and therefore only active patients could share the screen. The patient was required to focus mostly on the physician who controlled access to the computer screen. Inclusive and exclusive settings resulted in different behaviors in both physicians and patients. For patients who desired to see the screen, exclusive settings produced challenges, and they tended to adjust their seating position or move their chair in order to be able to see the screen. Inclusive settings were much more likely to develop a triangle of physician and patient sharing screen time.

In addition, the spatial configuration of the computerized clinic was found to affect physicians' ability to utilize the EMR effectively. For example, in some configurations the computer screen interfered with eye contact or with the ability to face the patient directly which were both found to be a significant component of an effective medical encounter (Pearce et al., 2008).

The clinical encounter is a form of social engagement. One of the critical components of social engagement is eye contact and therefore, should be considered as a critical part of the clinical encounter as well (MacDonald, 2009). For example: eye contact is an implicit component of rapport building (Dyche, 2007; Mauksch, Dugdale, Dodson, & Epstein, 2008); eye contact during clinical encounters may improve memory of important clinical messages (Senju & Johnson, 2009); engaging with deliberate eye contact during critical periods results in a feeling of shared purpose and hope(Larson & Yao, 2005). Moreover, changes in eye contact may signal unspoken emotional responses (such as anxiety or shame). Being able to recognize these cues is part of the basic empathic skills of a physician (Dyche, 2007; Suchman, Markakis, Beckman, & Frankel, 1997).

DEVELOPING A CONCEPTUAL MODEL

As illustrated above, using the EMR during the medical encounter may bear significant implications due to the effects on patient-doctor communication. The question, therefore,, is: how can we model and measure this effect? This is important because results can highlight misalignment between the traditional medical encounter and the modern one that includes EMR use. Only when well empirically identified, these obstacles can be remedied. To this end, in this part, we propose a research model that shows factors affecting the physician's ability to develop effective PDCC.

The above literature review highlights several behaviors that drive effective patient-doctor communication (PDC), place the patient in the center of the medical encounter and result in positive medical outcomes. The literature also describes attempts to discuss how the computer can be used to minimize disturbance to effective PDC, yet these efforts are still in their infancy.

To advance understanding of the new communication task at the clinic, where the computer, doctor and patient form a triadic relationship, we define a new construct, patient-doctor-computer communication (PDCC) and suggest a theoretical model of factors affecting this construct. The model is based on theories borrowed from the IS discipline, as well as form the patient-centeredness concept.

The IS theories describe factors affecting user acceptance of IS, user satisfaction and ultimately – effective use. An overview of the relevant IS theories discussed below as well as possible application in healthcare is summarized in Table 2.

Patient-Doctor-Computer Communication (PDCC) and Its Antecedents

PDCC is defined as *the physician's ability to maintain effective communication with the patient while using the EMR system during the medical encounter at the primary care clinic (Assis-Hassid et al., 2013).* Effective communication refers to the concept of patient-centeredness (Mead & Bower, 2000)

Table 2. IS theoretical framework and application in healthcare

Theory	Researchers	Antecedents and Dependent Constructs	Potential Research Applications in Healthcare
Utilization/Technology Acceptance Model (TAM)	Davis et al. 1989	The role of perceived usefulness and perceived ease in determining IS utilization	Examining the effects of perceived usefulness and perceived ease of use on EMR use in healthcare.
Technology to Performance Chain (TPC)	Goodhue and Thompson, 1995	Task, technology, and individual characteristics affect use, user satisfaction and user performance	Exploring individual characteristics' moderating effect on the "fit" between physicians' task and technology at use. The effect of the fit described above on physician's performance in terms of communicating with patients following the patient-centeredness principles. Effects of perceived fit on physician satisfaction
Post Acceptance Model (PAM)	Bhattacherjee, 2001	PAM is the first model to address the importance of continued IS use PAM emphasizes user satisfaction as a variable that effects IS use	Physician's long-term satisfaction level where EMR continuous use is demonstrated.
Fit between Individual Technology and Task (FITT)	Ammenwerthet al., 2006	Adds focus on the fit between individual and task which has not been examined in TTF and TPC.	Examine physician motivation to use EMRs affected by the "fit" between physician's characteristics, EMR perceptions, and task habits.

and the biopsychosocial model (G. Engel, 1977) that have been discussed above. Assis-Hassid et al. (2013) found PDCC to be an inherently different construct then PDC.

As of the late 1980s, researchers noticed that the mere introduction of IS in organizations did not necessarily imply acceptance by the users and effective use. It has been shown that over 50% of projects, including clinical IS, fail to deliver the expected benefits (Paré, Sicotte, Jaana, & Girouard, 2008). Thus, users rejection of IS resulted in sub-optimal use, dissatisfaction and often even productivity loss (Brynjolfsson, 1993). Several theories illustrated factors affecting IS success in terms of user satisfaction, use and user performance, among them are task-technology fit (TTF) or Technology-to-Performance Chain (TPC), the Technology Acceptance Model (TAM), the Post Acceptance Model (PAM) and the FITT Framework. Although developed in the late 1980s, these theories are still valid and are the foundation for further theoretical developments in the IS research field (Cady & Finkelstein, 2014; Chiasson, Kelley, & Downey, 2015; Pai & Huang, 2011).

The TTF and the later TPC models "highlight the role of fit between technologies and users' tasks in achieving individual performance impacts from information technology" (Goodhue & Thompson, 1995, p. 213). Figure 1 illustrated the constructs associated with the TPC theory.

The 'Technology Acceptance Model' (or TAM) is the most cited theory in IS research. It asserts that beliefs about usefulness and ease of use predict the actual use of information systems (Davis, Bagozzi, & Warshaw, 1989; Venkatesh, Morris, Davis, & Davis, 2003). TAM represents a family of theories hypothesizing that increased use will lead to positive performance impacts. For example, in the context of the medical encounter, EMR used to exchange information with the patient may increase patient satisfaction and compliance, a positive performance outcome under the patient-centeredness concept.

Ammenwerth et al., (2006) suggest the FITT framework for ICT adoption that extends the TTF model (See Figure 2) by adding the relationship between the three factors affecting the 'fit' construct. The basic concept is that ICT adoption depends not only directly on the three antecedents, but also on cross-impact among the individual users' attributes (such as computer anxiety, motivation), technology attributes (such as: usability, functionality) and task attributes (such as: task complexity, organization).

Figure 1. The Technology-to-Performance (TPC) Chain (Goodhue & Thompson, 1995)

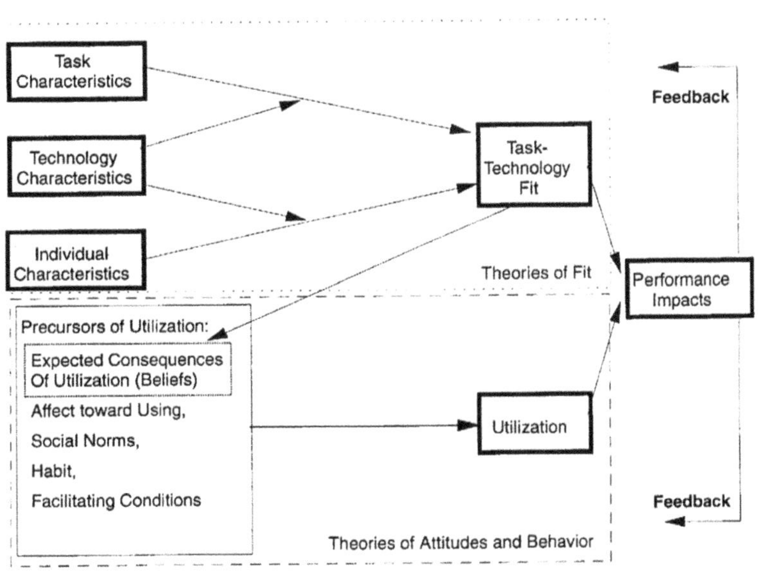

Figure 2. The FITT framework (©Ammenwerth 2006, Used with permission)

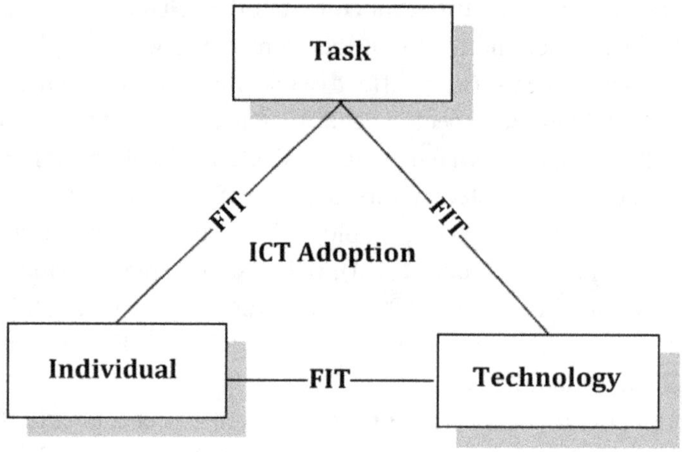

Initial acceptance of information systems has been widely discussed (Legris, Ingham, & Collerette, 2003). However, long-term success of information systems depends on their *continued* use (Bhattacherjee, 2001). The Post Acceptance Model (PAM)is based on Oliver's (1980) expectation confirmation theory (ECT) which has been widely used in the IS users behavior literature (Lee, 2010). The PAM model asserts that after initially accepting and using the IS, users form an opinion on the extent to which their expectations are confirmed. Alongside, they develop their opinion regarding the benefits / perceived usefulness of the IS. Both of these factors influence the user's satisfaction with the IS, which explains the user's willingness to continue with IS use. Bhattacherjee found that satisfaction with IS use was the strongest predictor of users' continuous intention, followed by perceived usefulness. Bhattacherjee's study shows that ignoring post-acceptance user satisfaction might result in disastrous consequences for continuous use (Figure 3).

Table 2 summarizes the constructs affecting user IS acceptance, satisfaction, use and performance as conjectured by the above theories.

The Proposed Research Model

Based on the above theories we propose a model describing factors affecting the physician's performance of PDCC, as previously defined (Figure 4). Thus, PDCC is the model's dependent variable, which is hypothesized to be affected by the perceived fit between the EMR and the medical encounter task. The

Figure 3. A Post-Acceptance Model of IS Continuance (© Bhattacherjee, 2001, MISQ, Used with permission)

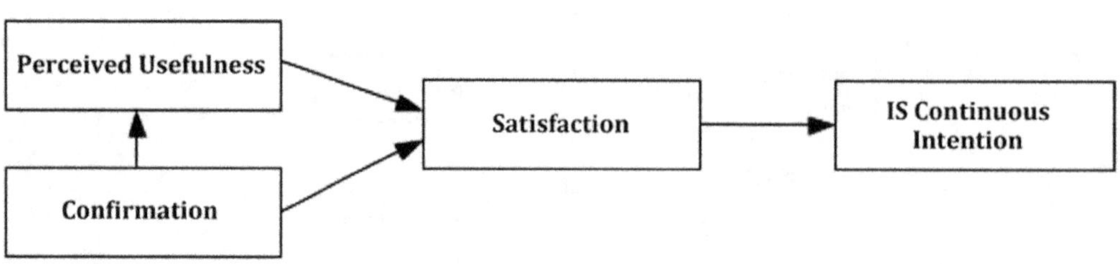

Figure 4. Proposed conceptual model

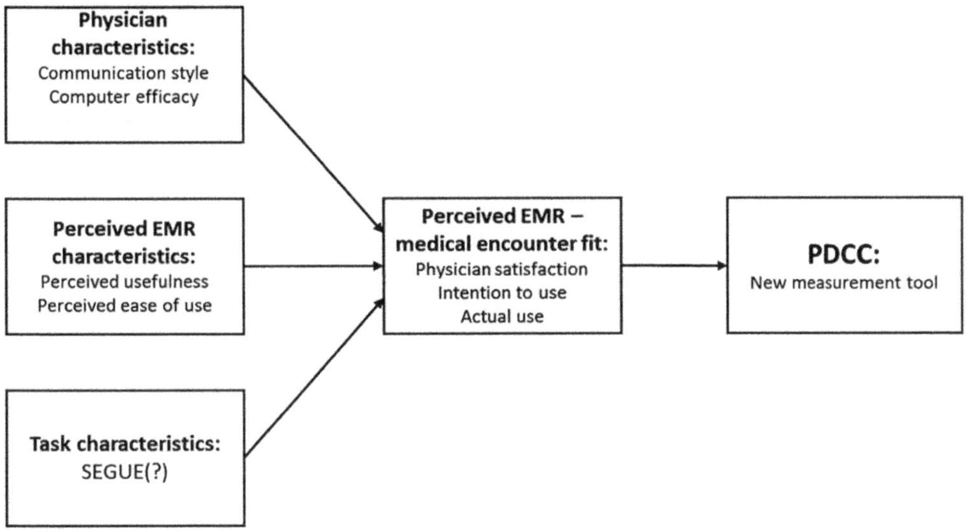

"perceived fit" construct is defined as *the degree to which the physician believes that the EMR design and structure is aligned (fits) with the demands of the medical encounter at the primary care clinic.* This construct can be measured by the physician's intention to use the EMR, satisfaction, and actual use, as predicted by TAM and PAM. This perception, however, is affected by the characteristics of the physician (user), the task and the EMR. The physicians characteristics can be determined by their communication style as defined by Shachak and Reis (2009), by their computer efficacy (the degree to which they perceive themselves as able to smoothly operate the computer), and possibly by several demographic attributes. The perceived EMR characteristics can be measured by the physician's beliefs about usefulness and ease of using the EMR during the medical encounter, as hypothesized by TAM. Task characteristics measurement is more complex and is left for further research. We suggest considering Makoul's (2001) five stage SEGUE model as a plausible framework.

Finally, we suggest developing a PDCC measurement tool based on principles of EMR use during the medical encounter that is aligned with patient-centeredness behaviors. This call is timely since existing tools measuring PDC based on patient centeredness principles disregard behavioral changes required to adapt the communication practices to the new, computerized era (Assis-Hassid et al., 2013). Future research may consider Makoul's SEGUE tool as a framework underlying the development of the new instrument (Assis-Hassid, Reychav, Heart, Pliskin, & Reis, 2015).

SUMMARY AND CONCLUSION

As shown in the previous sections, PDC is one of the most important tools available to a physician and holds the ability to improve health outcomes. Moreover, as shown in Engel's biopsychosocial model (1977) and the three function model of the medical interview (Bird & Cohen-Cole, 1990), good patient-doctor communication constitutes the heart of primary care.

Our literature review shows that previous research has explored the effects of EMR implementation on the patient-doctor relationship and that it implies possible adverse implications. Yet, there is only scant inter-disciplinary research that synthesizes knowledge from the IS and Medicine disciplines in order to explain this phenomenon and improve EMR acceptance by medical staff. It is believed that better acceptance and use will enhance PDCC. It seems that as EMRs' use around the world is continuously expanding; further research in this area must be conducted.

While EMR use during the medical encounter is becoming prevalent in most primary care clinics, patient centeredness principles crafted prior to the EMR era should be adapted and updated to include the computer. Hence, it is necessary to provide guidelines for physicians' communication behaviors at the clinic to assure patient-centeredness is maintained while using the EMR. Furthermore, it seems that in order to obtain the potential benefits of incorporating the EMR into healthcare, models offering successful IS implementation and long-term use cannot be ignored, especially those that describe factors affecting IS users performance for example in terms of PDCC.

In this paper, we defined a new construct – PDCC, regarded a physician's performance indicator, which prior research has shown is affected by EMR use. Based on theories explaining IS success in terms of use and user performance, we propose that perceived fit between the EMR and the medical task will positively affect PDCC as it is associated with user satisfaction and continuous use. The perceived fit is conjectured to be affected by antecedents defined by the Technology-to-Performance Chain model, as depicted in Figure 4.

Contribution and Future Research

This work contributes to both theory and practice. First, we defined the PDCC as a different construct than PDC (Assis-Hassid et al., 2013). This new construct should replace PDC wherever EMRs have been implemented and where physicians are required to extensively use it during the medical encounter in primary care while still maintain effective communication with their patients.

Second, we synthesized theories from the IS and Medicine domains to explain and describe what should be considered in order to encourage EMR use during the medical encounter in a manner that promotes, rather than hinders, PDCC. Finally, we suggested how the model's constructs can be measured in the context of EMR use at the clinic and highlighted the need to formulate a new PDCC measurement instrument.

Regarding contribution to practice, to our knowledge, most of the theories discussed in the previous section have not yet been applied or researched in healthcare in the context of patient centeredness and PDC. We find that the healthcare industry can significantly benefit by learning from IS implementation experience in other industries that went through productivity loss issues as a result of IS adoption.

In this paper, we provided an overview of relevant theoretical frameworks from the IS field. These theories focus on the variables that need to be considered when designing and implementing information systems in general. The field of healthcare should consider applying these theories when designing and implementing healthcare information systems as well.

Future research should further refine the model's constructs and measures, paying special attention to measuring the task characteristics and PDCC. These can be based on existing frameworks enhanced and adapted to include EMR use in a manner that contributes to effective patient-doctor communication.

A PDCC measurement instrument is particularly important, as it can serve as an educational, benchmarking and guidelines tool used by physicians and educators. An initial step in this direction was attempted by Assis-Hassid et al. (2015).

A better understanding of the effect of EMR use on PDCC is essential to elicit its full potential and benefits.

REFERENCES

Als, A. B. (1997). The desk-top computer as a magic box: Patterns of behaviour connected with the desk-top computer; GPs' and patients' perceptions. *Family Practice*, *14*(1), 17–23. doi:10.1093/fampra/14.1.17 PMID:9061339

Ammenwerth, E., Iller, C., & Mahler, C. (2006). IT-adoption and the interaction of task, technology and individuals: A fit framework and a case study. *BMC Medical Informatics and Decision Making*, *6*(1), 3. doi:10.1186/1472-6947-6-3 PMID:16401336

Assis-Hassid, S., Heart, T., Reychav, I., Pliskin, J. S., & Reis, S. (2013). Existing instruments for assessing physician communication skills: Are they valid in a computerized setting? *Patient Education and Counseling*, *93*(3), 363–366. doi:10.1016/j.pec.2013.03.017 PMID:23623463

Assis-Hassid, S., Reychav, I., Heart, T., Pliskin, J. S., & Reis, S. (2015). Enhancing patient-doctor-computer communication in primary care: towards measurement construction. *Israel journal of health policy research, 4*(1), 4.

Balint, M. (1961). The doctor, his patient and the illness. *Cerebral Palsy Bulletin*, *3*(4), 395–411.

Bates, D. W., Ebell, M., Gotlieb, E., Zapp, J., & Mullins, H. C. (2003). A Proposal for Electronic Medical Records in U.S. Primary Care. *Journal of the American Medical Informatics Association*, *10*(1), 1–10. doi:10.1197/jamia.M1097 PMID:12509352

Bates, D. W., Evans, R. S., Murff, H., Stetson, P. D., Pizziferri, L., & Hripcsak, G. (2003). Detecting Adverse Events Using Information Technology. *Journal of the American Medical Informatics Association*, *10*(2), 115–128. doi:10.1197/jamia.M1074 PMID:12595401

Bhattacherjee, A. (2001). Understanding Information Systems Continuance: An Expectation-Confirmation Model. *Management Information Systems Quarterly*, *25*(3), 351–370. doi:10.2307/3250921

Bird, J., & Cohen-Cole, S. A. (1990). The three-function model of the medical interview. In H. MS (Ed.), Methods in Teaching Consultation-Liaison Psychiatry: Adv Psychosom Med. (Vol. 20, pp. 65-88). Karger. doi:10.1159/000418260

Blumenthal, D., & Tavenner, M. (2010). The "Meaningful Use" Regulation for Electronic Health Records. *The New England Journal of Medicine*, *363*(6), 501–504. doi:10.1056/NEJMp1006114 PMID:20647183

Boonstra, A., & Broekhuis, M. (2010). Barriers to the acceptance of electronic medical records by physicians from systematic review to taxonomy and interventions. *BMC Health Services Research, 10*(1), 231–247. doi:10.1186/1472-6963-10-231 PMID:20691097

Booth, N., Robinson, P., & Kohannejad, J. (2004). Identification of high - quality consultation practice in primary care: The effects of computer use on doctor - patient rapport. *Informatics in Primary Care, 12*(2), 75–83. PMID:15319059

Brynjolfsson, E. (1993). The productivity paradox of information technology. *Communications of the ACM, 36*(12), 66–77. doi:10.1145/163298.163309

Cady, R. G., & Finkelstein, S. M. (2014). Task–Technology Fit of Video Telehealth for Nurses in an Outpatient Clinic Setting. *Telemedicine Journal and e-Health, 20*(7), 633–639. doi:10.1089/tmj.2013.0242 PMID:24841219

Chaudhry, B., Wang, J., Wu, S., Maglione, M., Mojica, W., Roth, E., ... Shekelle, P. G. (2006). Systematic Review: Impact of Health Information Technology on Quality, Efficiency, and Costs of Medical Care. *Annals of Internal Medicine, 144*(10), 742–752. doi:10.7326/0003-4819-144-10-200605160-00125 PMID:16702590

Chiasson, M., Kelley, H., & Downey, A. (2015). Understanding Task-Performance Chain Feed-Forward and Feedback Relationships in E-health. *AIS Transactions on Human-Computer Interaction, 7*(3), 167–190.

Davis, F. D., Bagozzi, R. P., & Warshaw, P. R. (1989). User Acceptance of Computer Technology: A Comparison of Two Theoretical Models. *Management Science, 35*(8), 982–1003. doi:10.1287/mnsc.35.8.982

DesRoches, C. M., Campbell, E. G., Rao, S. R., Donelan, K., Ferris, T. G., Jha, A., ... Blumenthal, D. (2008). Electronic Health Records in Ambulatory Care -- A National Survey of Physicians. *The New England Journal of Medicine, 359*(1), 50–60. doi:10.1056/NEJMsa0802005 PMID:18565855

DiMatteo, M. R., Taranta, A., Friedman, H. S., & Prince, L. M. (1980). Predicting Patient Satisfaction from Physicians' Nonverbal Communication Skills. *Medical Care, 18*(4), 376–387. doi:10.1097/00005650-198004000-00003 PMID:7401698

Duggan, P., & Parrott, L. (2001). Physicians' nonverbal rapport building and patients' talk about the subjective component of illness. *Human Communication Research, 27*(2), 299–311. doi:10.1111/j.1468-2958.2001.tb00783.x

Dyche, L. (2007). Interpersonal Skill in Medicine: The Essential Partner of Verbal Communication. *Journal of General Internal Medicine, 22*(7), 1035–1039. doi:10.100711606-007-0153-0 PMID:17437144

Engel, G. (1977). The need for a new medical model: A challenge for biomedicine. *Science, 196*(4286), 129–136. doi:10.1126cience.847460 PMID:847460

Engel, G. L. (1981). The Clinical Application of the Biopsychosocial Model. *The Journal of Medicine and Philosophy, 6*(2), 101–124. doi:10.1093/jmp/6.2.101 PMID:7264472

Epstein, R. M., Campbell, T. L., Cohen-Cole, S. A., McWhinney, I. R., & Smilkstein, G. (1993). Perspectives on patient-doctor communication. *The Journal of Family Practice, 37*(4), 377–388. PMID:8409892

Frankel, R., Altschuler, A., George, S., Kinsman, J., Jimison, H., Robertson, N. R., & Hsu, J. (2005). Effects of Exam-Room Computing on Clinician–Patient Communication. *Journal of General Internal Medicine*, *20*(8), 677–682. doi:10.1111/j.1525-1497.2005.0163.x PMID:16050873

Goodhue, D. L., & Thompson, R. L. (1995). Task-Technology Fit and Individual Performance. *Management Information Systems Quarterly*, *19*(2), 213–236. doi:10.2307/249689

Greatbatch, D., Heath, C., Campion, P., & Luff, P. (1995). How do desk-top computers affect the doctor-patient interaction. *Family Practice*, *12*(1), 32–36. doi:10.1093/fampra/12.1.32 PMID:7665038

Hall, J. A., & Dornan, M. C. (1988). What patients like about their medical care and how often they are asked: A meta-analysis of the satisfaction literature. *Social Science & Medicine*, *27*(9), 935–939. doi:10.1016/0277-9536(88)90284-5 PMID:3067368

Hall, J. A., Harrigan, J. A., & Rosenthal, R. (1995). Nonverbal behavior in clinician—patient interaction. *Applied & Preventive Psychology*, *4*(1), 21–37. doi:10.1016/S0962-1849(05)80049-6

Hall, J. A., Roter, D. L., & Katz, N. R. (1988). Meta-Analysis of Correlates of Provider Behavior in Medical Encounters. *Medical Care*, *26*(7), 657–675. doi:10.1097/00005650-198807000-00002 PMID:3292851

Hall, J. A., Roter, D. L., & Rand, C. S. (1981). Communication of Affect between Patient and Physician. *Journal of Health and Social Behavior*, *22*(1), 18–30. doi:10.2307/2136365 PMID:7240703

Haluza, D., & Jungwirth, D. (2014). ICT and the future of health care: Aspects of doctor-patient communication. *International Journal of Technology Assessment in Health Care*, *30*(3), 298–305. doi:10.1017/S0266462314000294 PMID:25308693

Hayrinen, K., Saranto, K., & Nykanen, P. (2008). Definition, structure, content, use and impacts of electronic health records: A review of the research literature. *International Journal of Medical Informatics*, *77*(5), 291–304. doi:10.1016/j.ijmedinf.2007.09.001 PMID:17951106

What is an EMR. (n. d.). HealthTechnologyReview.com. Retrieved from http://www.healthtechnology-review.com/emr-functions.php

HIMSS Electronic Health Records. (n. d.). Retrieved from http://www.himss.org/library/ehr/

Hsiao, C., Hing, E., Socey, T., & Cai, B. (2011). Electronic health record systems and intent to apply for meaningful use incentives among office-based physician practices: United States, 2001–2011. In N. C. H. Statistics (Ed.), *NCHS data brief, no 79*. Hyattsville: MD National Center for Health Statistics.

Jha, A. K., Bates, D. W., Jenter, C., Orav, E. J., Zheng, J., Cleary, P., & Simon, S. R. (2009). Electronic health records: Use, barriers and satisfaction among physicians who care for black and Hispanic patients. *Journal of Evaluation in Clinical Practice*, *15*(1), 158–163. doi:10.1111/j.1365-2753.2008.00975.x PMID:18759752

Larsen, K., & Smith, C. K. (1981). Assessment of nonverbal communication in the patient-physician interview. *The Journal of Family Practice*, *12*(3), 481–488. PMID:7462949

Larson, E. B., & Yao, X. (2005). Clinical empathy as emotional labor in the patient-physician relationship. *Journal of the American Medical Association*, *293*(9), 1100–1106. doi:10.1001/jama.293.9.1100 PMID:15741532

Lee, M.-C. (2010). Explaining and predicting users' continuance intention toward e-learning: An extension of the expectation–confirmation model. *Computers & Education*, *54*(2), 506–516. doi:10.1016/j.compedu.2009.09.002

Legris, P., Ingham, J., & Collerette, P. (2003). Why do people use information technology? A critical review of the technology acceptance model. *Information & Management*, *40*(3), 191–204. doi:10.1016/S0378-7206(01)00143-4

Lorig, K. (2012). Patient-Centered Care: Depends on the Point of View. *Health Education & Behavior*, *39*(5), 523–525. doi:10.1177/1090198112455175 PMID:22991049

MacDonald, K. (2009). Patient-Clinician Eye Contact: Social Neuroscience and Art of Clinical Engagement. *Postgraduate Medicine*, *121*(4), 136–144. doi:10.3810/pgm.2009.07.2039 PMID:19641279

Makoul, G. (2001). The SEGUE Framework for teaching and assessing communication skills. *Patient Education and Counseling*, *45*(1), 23–34. doi:10.1016/S0738-3991(01)00136-7 PMID:11602365

Makoul, G., Curry, R. H., & Tang, P. C. (2001). The Use of Electronic Medical Records: Communication Patterns in Outpatient Encounters. *Journal of the American Medical Informatics Association: JAMIA*, *8*(6), 610–615. doi:10.1136/jamia.2001.0080610 PMID:11687567

Margalit, R. S., Roter, D., Dunevant, M. A., Larson, S., & Reis, S. (2006). Electronic medical record use and physician-patient communication: An observational study of Israeli primary care encounters. *Patient Education and Counseling*, *61*(1), 134–141. doi:10.1016/j.pec.2005.03.004 PMID:16533682

Mauksch, L. B., Dugdale, D. C., Dodson, S., & Epstein, R. (2008). Relationship, communication, and efficiency in the medical encounter: Creating a clinical model from a literature review. *Archives of Internal Medicine*, *168*(13), 1387–1395. doi:10.1001/archinte.168.13.1387 PMID:18625918

McGrath, J. M., Arar, N. H., & Pugh, J. A. (2007). The influence of electronic medical record usage on nonverbal communication in the medical interview. *Health Informatics Journal*, *13*(2), 105–118. doi:10.1177/1460458207076466 PMID:17510223

McWhinney, I. (1989). The need for a transformed clinical method. In M. Stewart & D. Roter (Eds.), Communicating with medical patients (pp. 25-40). London: Sage.

Mead, N., & Bower, P. (2000). Patient-centredness: A conceptual framework and review of the empirical literature. *Social Science & Medicine*, *51*(7), 1087–1110. doi:10.1016/S0277-9536(00)00098-8 PMID:11005395

Mead, N., Bower, P., & Hann, M. (2002). The impact of general practitioners' patient-centredness on patients' post-consultation satisfaction and enablement. *Social Science & Medicine*, *55*(2), 283–299. doi:10.1016/S0277-9536(01)00171-X PMID:12144142

Miller, R. H., & Sim, I. (2004). Physicians' Use of Electronic Medical Records: Barriers and Solutions. *Health Affairs*, *23*(2), 116–126. doi:10.1377/hlthaff.23.2.116 PMID:15046136

Oliver, R. L. (1980). A Cognitive Model of the Antecedents and Consequences of Satisfaction Decisions. *JMR, Journal of Marketing Research*, *17*(4), 460–469. doi:10.2307/3150499

Ong, L. M. L., de Haes, J. C. J. M., Hoos, A. M., & Lammes, F. B. (1995). Doctor-patient communication: A review of the literature. *Social Science & Medicine*, *40*(7), 903–918. doi:10.1016/0277-9536(94)00155-M PMID:7792630

Pai, F.-Y., & Huang, K.-I. (2011). Applying the Technology Acceptance Model to the introduction of healthcare information systems. *Technological Forecasting and Social Change*, *78*(4), 650–660. doi:10.1016/j.techfore.2010.11.007

Paré, G., Sicotte, C., Jaana, M., & Girouard, D. (2008). Prioritizing the Risk Factors Influencing the Success of Clinical Information System Projects - A Delphi Study in Canada. *Methods of Information in Medicine*, *47*(3), 251–259. doi:10.3414/ME0512 PMID:18473092

Patel, V. L., Arocha, J., & Kushniruk, A. (2002). Patients' and physicians' understanding of health and biomedical concepts: relationship to the design of EMR systems. *Journal of Biomedical Informatics, 35*(1), 8-16.

Patel, V. L., Kushniruk, A. W., Yang, S., & Yale, J. F. (2000). Impact of a Computer - based Patient Record System on Data Collection, Knowledge Organization, and Reasoning. *Journal of the American Medical Informatics Association: JAMIA*, *7*(6), 569–585. doi:10.1136/jamia.2000.0070569 PMID:11062231

Pearce, C., Arnold, M., Phillips, C., Trumble, S., & Dwan, K. (2011). The patient and the computer in the primary care consultation. *Journal of the American Medical Informatics Association*, *18*(2), 138–142. doi:10.1136/jamia.2010.006486 PMID:21262923

Pearce, C., Walker, H., & O'Shea, C. (2008). A visual study of computers on doctors' desks. *Informatics in Primary Care*, *16*(2), 111–117. PMID:18713527

Senju, A., & Johnson, M. H. (2009). The eye contact effect: Mechanisms and development. *Trends in Cognitive Sciences*, *13*(3), 127–134. doi:10.1016/j.tics.2008.11.009 PMID:19217822

Shachak, A., Hadas-Dayagi, M., Ziv, A., & Reis, S. (2009). Primary Care Physicians' Use of an Electronic Medical Record System: A Cognitive Task Analysis. *Journal of General Internal Medicine*, *24*(3), 341–348. doi:10.100711606-008-0892-6 PMID:19130148

Shachak, A., & Reis, S. (2009). The impact of electronic medical records on patient–doctor communication during consultation: A narrative literature review. *Journal of Evaluation in Clinical Practice*, *15*(4), 641–649. doi:10.1111/j.1365-2753.2008.01065.x PMID:19522722

Suchman, A. L., Markakis, K., Beckman, H. B., & Frankel, R. (1997). A model of empathic communication in the medical interview. *Journal of the American Medical Association*, *277*(8), 678–682. doi:10.1001/jama.1997.03540320082047 PMID:9039890

Thielke, S., Hammond, K., & Helbig, S. (2007). Copying and pasting of examinations within the electronic medical record. *International Journal of Medical Informatics*, *76*, S122–S128. doi:10.1016/j.ijmedinf.2006.06.004 PMID:16899403

Venkatesh, V., Morris, M. G., Davis, G. B., & Davis, F. D. (2003). User Acceptance of Information Technology: Toward a Unified View. *Management Information Systems Quarterly*, *27*(3), 425–478.

Ventres, W., Kooienga, S., Marlin, R., Vuckovic, N., & Stewart, V. (2005). Clinician style and examination room computers: A video ethnography. *Family Medicine*, *37*(4), 276–281. PMID:15812698

Weiner, J. P. (2012). Doctor-patient communication in the e-health era. *Isr J Health Policy Res*, *33*(1), 1–7. PMID:22929000

This research was previously published in the International Journal of Reliable and Quality E-Healthcare (IJRQEH), 5(1); edited by Anastasius Moumtzoglou, pages 1-17, copyright year 2016 by IGI Publishing (an imprint of IGI Global).

Chapter 11
Ethical Challenges in Online Health Games

Matthieu J. Guitton
Laval University, Canada

ABSTRACT

Using heath-oriented applications supported by and delivered through the Internet, eHealth and mHealth strategies are part of the modern therapeutic arsenal. Among them, online health games cumulate the advantages of online support groups and telerehabilitation therapies in a playful environment. Furthermore, compared to health games not delivered online, they abolish geographical constraints and make it possible to simultaneously reach large numbers of individuals – health professionals and patients alike. However, online health games also raise several ethical questions which may hinder their practical efficiency and their expansion. Ethical challenges related to online health games echo some of the concerns already identified for online games and online spaces, operationalized in the particular context of health applications. This Chapter will summarise and address these challenges, ranging from the "out of the game" ethical challenges to the "in game" ethical challenges, and suggest practical recommendations in order to implement efficient, safe, and ethical online health games.

INTRODUCTION

With the rise of new technologies of information, the recent years have witnessed the emergence of new health-related strategies. Using heath-oriented applications supported by and delivered through the Internet, eHealth and mHealth strategies are rapidly becoming part of the modern therapeutic arsenal. Among these new emerging tools, online health games represent a weapon of choice.

Among online games, massively multiplayer online games, such as the virtual platform of Second Life or the famous massively multiplayer online role-playing game World of Warcraft, are extremely popular. Massively multiplayer online games are able to support the simultaneous presence in the virtual environment of considerable numbers of human-controlled avatars, thus leading to the possibility of extremely important interactions for players immersed in the game through their avatars. In the context of healthcare applications, this type of online games cumulate several advantages, including those of

DOI: 10.4018/978-1-5225-6198-9.ch011

online support groups and telerehabilitation therapies, in a playful and pleasant environment. However, although online health games bear promises of fascinating developments, they also raise several ethical questions which may hinder their practical efficiency as well as their expansion. As for any medical innovation, identifying, understanding, and adequately answering the ethical challenges related to this new way to take care of patients will be one of the first required steps before the development and operative implementation of large scale online health games.

The nature of ethical challenges related to online health games is complex and multiform. Indeed, arising at the intersection of both the medical and the information technology fields, online health games present ethical challenges specific to each of these two fields, and ethical challenges resulting from their interaction. Therefore, ethical challenges related to online health games echo some of the concerns already identified for online games and online virtual spaces, operationalized in the particular context of health applications, as well as challenges linked to democratization of eHealth applications such as the problems related to the interactions between online health games and electronic health record (EHR).

This Chapter will address a few of these challenges, ranging from issues surrounding games (e.g., the problem of the game access related to patient's self-disclosure) to issues related to the games themselves, in order to help serious game developers and eHealth practitioners to optimize these tools for a broad population.

1. HEALTH GAMES, ONLINE HEALTH GAMES, AND MMO HEALTH GAMES

Whatever the therapeutic strategy envisioned, patient engagement is one of the keys to insure the success of the therapy. Reaching satisfactory levels of engagement has thus been a critical issue in developing eHealth and mHealth applications. By using game aesthetic and thinking and group-based mechanics, gamification processes help to engage people more strongly (Kapp, 2012). Indeed, computer games promote interactivity, and ultimately engagement from the player by promoting several aspects such as intrinsic motivation, cognitive apprenticeship, and flow (Dede, 2009).

Compared to health games not delivered online, online health games abolish geographical constraints, and, in a context of limited resources, allow therapists to reach patients whatever the distance. They also allow patients to use the game in the comfort of their home, and in the safety of relative anonymity.

Multiplayer online games – and particularly massively multiplayer online (MMO) games – add a social dimension to online health games. People using a MMO game are not just playing a game, they become part of a community (Guitton, 2012). Furthermore, MMO games make it possible to simultaneously reach large numbers of individuals – health professionals and patients alike, which is an important point when considering the potential benefits of a health game. Indeed, not only MMO games can promote peer-interactions, they can as well offer the possibility for health professionals to be part of the virtual world as agents. By allowing health professional to directly monitor the progresses of the patients, or to provide instantaneous support or advice in the virtual space of the game, this possibility considerably enhances the value of the therapeutic potential of the experience. Hence, MMO games cumulate the advantages of health games with those of online support groups and telemedicine-based distance therapies, in a playful and pleasant environment.

Interestingly enough, such health-oriented applications – and the related communities of users – can spontaneously emerge even in MMO settings not primarily designed for health purpose, such as the virtual world of Second Life (Lomanowska and Guitton, 2014). Indeed, the possibility for Second Life

Table 1. Summary of the principal ethical challenges of online health games

"Out of the Game" Ethical Challenges
Risk to develop excessive use (game addiction)
Mechanisms of registration in the game (entering the game)
Risk of developing a stigma for the patients
Interactions between the game record and the electronic health record
Risk of break of privacy due to online security issues
"In Game" Ethical Challenges
Risk for the patient of shifting the interest from the therapeutic effects to the game mechanics (meta-gaming)
Risk of isolation from the external world within a patient community
Risk of problematic social comparisons (including risk of bullying)
Risk of language and/or cultural issues
Risk of break of privacy due to unnecessary disclosure from the patients

avatars to become "virtually pregnant" led to the spontaneous emergence of a vast community, where "role-play" interests are largely masked by motivations such as curiosity (the wish to experiment it virtually first) and other health related concerns (Lomanowska and Guitton, 2014).

2. "OUT OF THE GAME" ETHICAL CHALLENGES

While online health games offer fascinating possibilities, their optimal use is still challenged by several ethical concerns – the first category of them arising from the different aspects surrounding the game rather than the game itself (Table 1).

The first ethical issue of using online health games is the potential risk for the patients using them to develop an excessive use, which could ultimately lead to a form of addiction (Lortie and Guitton, 2013). Indeed, notably due to their highly immersive features and strong social dimension, MMO games have been consistently described to be highly addictive (Caplan, 2003; Griffiths *et al.*, 2004; van Rooij *et al.*, 2010; Block, 2008; Gentile *et al.*, 2011). From a clinical perspective, increasing the features of a game in such a way that the patients will be more likely to play would maximize the effectiveness of the therapeutic intervention. Empowerment in a game can result in various motivational states. However, used on a vulnerable population (the patients), in a therapeutically endorsed process (leading the patients to be psychologically empowered for using the game), the addictive potential of MMO games might well be even strengthened. Online health game developers are thus facing a classical ethical vs. practical risk/benefit challenge. Would the game offer a possibility to help against a given pathology while taking the risk to develop another one (in this case, a form of Internet addiction)? Alternatively, would it be safer to make sure the features of the health game are not addictive, in other words not optimal, potentially compromising the therapeutic impact of the game?

A second major set of major ethical issues of online health game is related to the mechanism supporting the process of entering the game. Entering an online game specifically designed as a health game would by default carry the risk of adding some stigma to the users. More specifically, questions may arise regarding the exact mechanisms supporting joining the game. A few mechanisms could be considered, but all of them bear some issues. A first approach is to go without *a priori* information. No personal data would be provided before registering into the game. While this solution would be the simplest one – as well as the less invasive in terms of sharing the patients' personal information, it is far from being perfect either. Such issues are not merely theoretical: one could indeed recall the famous case of Experience Project – a website which was initially developed as an anonymous patient support group, but which became within a rather short time one of the widest non-controlled adult forum. In order to avoid such risks, a next level could easily be implemented by having the users provide limited health-related information when registering. That would however not solve the problem, as the validity of the provided information would not be controlled, and the stigma-related aspects would grow even larger. Another alternative could thus be to directly link the game registration to the electronic health record of the patient – this being done either directly by health care providers in the case of a health game that would be part of the actual treatment of the patient, or by the patient himself in the case of an open type of health game. In both cases, the potential cross-talk – and the risk of resulting overlap – between the electronic health record and the online health game raises numerous ethical and legal questions.

The issue of the interactions between online health games and electronic health records is probably one of the most problematic ethical concerns related to online health games. Even if they became a recognized therapeutic tool, online health games are not primarily intended to document patients' health status, but rather to contribute to provide a form of cure. Nonetheless, in order to fulfil its role of therapeutic tool, it is legitimate to assume that information from the health game could be gathered to be analysed by health professionals, if only to assess the evolution of the patient and the effects of the game on the patient's condition. Ethically speaking, this diffuse border is obviously extremely problematic. Which types of information could "leak" from online health game to enrich the electronic health record of a patient? As for any distributed online information system, the question of the database access is thus of a central importance for online health games. In fact, the very same questionings that those happening regarding data management in electronic health records would thus apply for online health games (Anderson, 2007; Blumenthal and Tavenner, 2010; Hayrinen *et al.*, 2008). Who could have access to the patients' data? How would the data be stored? For how long could and should information related to the game be stored? Knowing that such possibilities of leak would exist from the online in-game actions to the offline health record would also be likely to make numerous patients refuse to join the health game – or at the very least to alter their in-game behavior (Powell *et al.*, 2006).

These concerns related to the interactions between online health games and electronic health records also obviously raise technical questions. Indeed, the storage of health-related sensitive information is effected in the cloud. While at first glance these issues might appear purely related to security standards, they do have important ethical consequences. Indeed, a break of data privacy in the context of online health games actually entails important risks to do harm to the patient. In addition to the disclosure of sensitive information regarding the patient's privacy, the absolute irreversibility of such a breach is to be noted here – once health-related personal information have been publicly revealed, there is no going back.

3. "IN GAME" ETHICAL CHALLENGES

The main objective of a health game is to trigger behavioral or physical changes during or after the game which will positively modify health behavior or health status in the real life of the user (Lieberman, 2012; Primack *et al.*, 2012). Issues related to transfer – how the game experience can transfer into real life effects – is therefore of a central importance while considering health game outcomes. The fact that experiences gained in virtual spaces (ideally positive game experiences in the case of health games) can transfer into real life effects is supported by a set of psychological mechanisms often collectively referred to as the Proteus effect (Yee and Bailenson, 2007). Due to this premium on transfer, more than for a purely leisure-oriented game, the appropriate balance needs to be found between the Proteus effect and in-game immersion in online health games (Guitton, 2012). Health games should be designed in such a way that users can undergo the necessary experiences required to impact their real life behavior without taking the risk of becoming so immersed that they could get absorbed in the process (Table 1).

In MMO games, power and prestige are earned and not bought (Guitton, 2012). On the one hand, this favors player motivation (the "drive" to get involved in the game), and reinforces player and community engagement. On the other hand, game achievements can be optimized by a better understanding of the interface, i.e. the game mechanics. This self reflexive attitude toward the rules and mechanics of the game ("playing" vs. "gaming") is what is referred to as meta-gaming. The ratio between playing and meta-gaming should always favor the first, to avoid the risk for the patient of shifting the interest from the therapeutic effects to the game mechanics. Patient focus should remain centered on the different aspects of the game experience identified as relevant for the therapeutic effect and not meta-gaming (the game itself). Unfortunately, the very high potential of MMOs to elicit high engagement from the players comes with a very high potential to elicit meta-gaming too. While this is not such an issue in the context of regular leisure-oriented games, it becomes a concern for online health games. In this case, ethical questions get embedded with issues related to game development, both from purely technical (where information is located in the game?) and theoretical perspectives (what is in the game and what is out of the game?).

When engaged in a massively multiplayer online game, users have been shown to display an extremely strong – and often purely unconscious – form of homophily (Lortie and Guitton, 2011; 2012). In other words, when in 3D immersive virtual spaces, people tend to gather with people (or avatars) who look similar to themselves. Obviously, this phenomenon of homophily reinforces group cohesion and the positive effects of a classical peer support group. Nonetheless, it also bears the risk of having a too strong emphasis put on the fact of being similar to each others. The final aim of an online health game is not for patients to be willing to stay among themselves, but rather to help them fight and win against their disease, or at the very least improve their health status. Online health games, in contrast to peer support forums, attempt to help the patients to get open to others – others in this case referring to the non-patient population.

Whether the premium is on collaboration or on competition, massively multiplayer online games are a place for social comparison. While lateral comparison (comparison with a person one considers as an equal) may be valuable for the patients in the case of online health games, upward comparison (comparison with someone perceived as better off or unattainable) is clearly not suitable for patients. Similarly, downward comparison (comparison with someone perceived as worse off) would clearly not be desirable among an online community of sufferers, as "feeling better at looking how worse others are" would easily lead to disdain and risks of bullying.

In the context of MMO health games, a legitimate worry could spring from the possibility to have patients coming from different countries in the same virtual setting, with possibly major divergences in terms of language, social, or cultural origin. However, this might not be an immediate ethical concern. Indeed, while online health games abolish geographical constraints, they still do not abolish the specificity of each national health system. Online health games differ from standard online games in the sense that they are part of a therapeutic process. Hence, their suggested use by patients depends on the way these games are perceived by a given health system. Due to both legal and technical reasons, it is unlikely that the first online health game applications will be massively trans-national. Thus, language issues or major differences of cultural background are unlikely to represent an immediate worry. However, with the further development of these new tools, these issues may become more important, and gain their status of ethical concern in their own right.

Finally, echoing the concerns regarding data security already mentioned in the "out of the game" section, in-game situations might also lead to the development of important breaches in terms of safety and privacy. Disclosure is likely to happen much more frequently in peer-to-peer online discussions and interactions – both of them being highly favored in MMO games. While such a risk exists in any situation in which patients would be interacting with each other, the perceived distance that the patient may have with the avatars might increase the development of dangerous behaviors in terms of personal disclosure. In addition to proper education and warning for patients, a solution could be to implement specific regulations and "codes of conducts" in the game. It should however be noted that, in the context of the extreme space of liberty that online games represent, implementing such rules might not be that easy. Constant and appropriate control of health supervisors is thus a requisite to insure that no misbehavior occur which could potentially endanger the privacy of the patients.

CONCLUSION

To keep their promises as new weapons in the therapeutic arsenal of the 21st century, online health games will have to be the target of intense reflections. This short survey of some of the prominent ethical questions that the increase of online health games weight in online therapies will elicit clearly demonstrates that implementing successful eHealth strategies including online health games, while of major interest, will bear a few significant challenges as well.

In order to meet the ethical challenges related to the game mechanisms themselves, to how people would behave when within the game, and to the game surroundings, intense collaborations with common ground will have to emerge between game developers and health specialists. As it occurs within a massively multiplayer online game, a premium on collaboration will have to be put to optimize health game development (Table 2).

Game design has always been centered on a communication perspective: designing a computer game means answering questions such as how a digital space can be structured to help users in their interactions. Defining game rules involves in fact specifying the appropriate interactions (whether being moves or behavior) as well as clarifying the roles that the players can adopt. Maintaining a careful cost/benefit ratio while keeping in mind ethical concerns will be central in the future of online health games. By questioning how to combine ethical aspects with efficiency, the development of online health games is

Table 2. Practical recommendations to implement ethical online health game

Take into consideration the different ethical challenges related to online health games
Evaluate the benefit/risk ratio of the online game as a treatment option
Keep the highest security standards regarding data protection and privacy
Comply to the highest regulatory standards
Try to establish independent evaluation

likely to enrich in the coming years the process of game thinking, as well as serious games development strategies. In conclusion, in addition to the obvious implementation of technically sound security protocols, taking into account the multi-disciplinary nature of the ethical challenges raised by online health games is likely to contribute to the development and optimization of new tools to support patients with online virtual settings and games through eHealth or mHealth approaches.

REFERENCES

Anderson, J. G. (2007). Social, ethical and legal barriers to e-health. *International Journal of Medical Informatics*, *76*(5-6), 480–483. doi:10.1016/j.ijmedinf.2006.09.016 PMID:17064955

Block, J. J. (2008). Issues for DSM-V: Internet addiction. *The American Journal of Psychiatry*, *165*(3), 306–307. doi:10.1176/appi.ajp.2007.07101556 PMID:18316427

Blumenthal, D., & Tavenner, M. (2010). The "meaningful use" regulation for electronic health records. *The New England Journal of Medicine*, *363*(6), 501–504. doi:10.1056/NEJMp1006114 PMID:20647183

Caplan, S. E. (2003). Preference for online social interaction: A theory of problematic Internet use and psychosocial well-being. *Communication Research*, *30*(6), 625–648. doi:10.1177/0093650203257842

Dede, C. (2009). Immersive interfaces for engagement and learning. *Science*, *323*(5910), 66–69. doi:10.1126cience.1167311 PMID:19119219

Gentile, D. A., Choo, H., Liau, S., Sim, T., Li, D., Fung, D., & Khoo, A. (2011). Pathological video game use among youths: A two-year longitudinal study. *Pediatrics*, *127*(2), e319–e329. doi:10.1542/peds.2010-1353 PMID:21242221

Griffiths, M. D., Davies, M. N. O., & Chappell, D. (2004). Online computer gaming: A comparison of adolescent and adult gamers. *Journal of Adolescence*, *27*(1), 87–96. doi:10.1016/j.adolescence.2003.10.007 PMID:15013262

Guitton, M. J. (2012). Living in the Hutt Space: Immersive Process in the Star Wars Role-Play community of Second Life. *Computers in Human Behavior*, *28*(5), 1681–1691. doi:10.1016/j.chb.2012.04.006

Häyrinen, K., Saranto, K., & Nykanen, P. (2008). Definition, structure, content, use and impacts of electronic health records: A review of the research literature. *International Journal of Medical Informatics*, *77*(5), 291–304. doi:10.1016/j.ijmedinf.2007.09.001 PMID:17951106

Kapp, K. M. (2012). *The gamification of learning and instruction: Game-based methods and strategies for training and education*. Hoboken, NJ: John Wiley & Sons.

Lieberman, D. A. (2012). Digital games for health behavior change: Research, design, and future directions. In S. M. Noar & N. G. Harrington (Eds.), *eHealth applications: Promising strategies for behavior change* (pp. 110–127). New York, NY: Routledge.

Lomanowska, A. M., & Guitton, M. J. (2014). My avatar is pregnant! Representation of pregnancy, birth, and maternity in a virtual world. *Computers in Human Behavior*, *31*, 322–331. doi:10.1016/j.chb.2013.10.058

Lortie, C. L., & Guitton, M. J. (2011). Social organization in virtual settings depends on proximity to human visual aspect. *Computers in Human Behavior*, *27*(3), 1258–1261. doi:10.1016/j.chb.2011.01.006

Lortie, C. L., & Guitton, M. J. (2012). Looking similar promotes group stability in a game-based virtual community. *Games for Health Journal*, *1*(4), 274–278. doi:10.1089/g4h.2012.0016 PMID:26191630

Lortie, C. L., & Guitton, M. J. (2013). Internet addiction assessment tools: Dimensional structure and methodological status. *Addiction (Abingdon, England)*, *108*(7), 1207–1216. doi:10.1111/add.12202 PMID:23651255

Powell, J., Fitton, R., & Fitton, C. (2006). Sharing electronic health records: The patient view. *Informatics in Primary Care*, *14*(1), 55–57. PMID:16848967

Primack, B. A., Carroll, M. V., McNamara, M., Klem, M. L., King, B., Rich, M., & ... (2012). Role of video games in improving health-related outcomes: A systematic review. *American Journal of Preventive Medicine*, *42*(6), 630–638. doi:10.1016/j.amepre.2012.02.023 PMID:22608382

Van Rooij, A. J., Schoenmakers, T. M., Vermulst, A. A., van den Eijnden, R. J. J. M., & van de Mheen, D. (2010). Online video game addiction: Identification of addicted adolescent gamers. *Addiction (Abingdon, England)*, *106*(1), 205–212. doi:10.1111/j.1360-0443.2010.03104.x PMID:20840209

Yee, N., & Bailenson, J. (2007). The Proteus Effect: The effect of transformed Self-representation on behavior. *Human Communication Research*, *33*(3), 271–290. doi:10.1111/j.1468-2958.2007.00299.x

ADDITIONAL READING

Alamri, A., Hassan, M. M., Hossain, M. A., Al-Qurishi, M., Aldukhayyil, Y., & Hossain, M. S. (2014). Evaluating the impact of a cloud-based serious game on obese people. *Computers in Human Behavior*, *30*, 468–475. doi:10.1016/j.chb.2013.06.021

Bouchard, S., Bernier, F., Boivin, E., Morin, B., & Robillard, G. (2012). Using biofeedback while immersed in a stressful videogame increases the effectiveness of stress management skills in soldiers. *PLoS ONE*, *7*(4), e36169. doi:10.1371/journal.pone.0036169 PMID:22558370

Girard, B., Turcotte, V., Bouchard, S., & Girard, B. (2009). Crushing virtual cigarettes reduces tobacco addiction and treatment discontinuation. *Cyberpsychology & Behavior*, *12*(5), 477–483. doi:10.1089/cpb.2009.0118 PMID:19817561

Grenier, S., Forget, H., Bouchard, S., Isere, S., Belleville, S., Potvin, O., & ... (2014). Using virtual reality to improve the efficacy of cognitive-behavioral therapy (CBT) in the treatment of late-life anxiety: Preliminary recommendations for future research. *International Psychogeriatrics*, *10*, 1217–1225. doi:10.1017/S104161021400074X PMID:25381697

Guitton, M. J. (2012). The immersive impact of meta-media in a virtual world. *Computers in Human Behavior*, *28*(2), 450–455. doi:10.1016/j.chb.2011.10.016

Guitton, M. J. (2014). The importance of studying the dark side of social networks. *Computers in Human Behavior*, *31*, 335. doi:10.1016/j.chb.2013.10.055

Guitton, M. J., & Cristofari, C. (2014). Does surviving the zombie apocalypse represent a good model of human behavior in response to pandemics? *Journal of Public Health Management and Practice*, *20*(4), 375–377. doi:10.1097/PHH.0000000000000043 PMID:24368409

Jerabeck, J. M., & Ferguson, C. J. (2013). The influence of solitary and cooperative violent video game play on aggressive and prosocial behavior. *Computers in Human Behavior*, *29*(6), 2573–2578. doi:10.1016/j.chb.2013.06.034

Kim, S. Y., Prestopnik, N., & Biocca, F. A. (2014). Body in the interactive game: How interface embodiment affects physical activity and health behavior change. *Computers in Human Behavior*, *36*, 376–384. doi:10.1016/j.chb.2014.03.067

Kowert, K., Domahidi, E., Festl, R., & Quandt, T. (2014). Social gaming, lonely life? The impact of digital game play on adolescents' social circles. *Computers in Human Behavior*, *36*, 385–390. doi:10.1016/j.chb.2014.04.003

Lomanowska, A. M., & Guitton, M. J. (2012). Virtually Naked: Virtual environment reveals sex-dependent nature of skin disclosure. *PLoS ONE*, *7*(12), e51921. doi:10.1371/journal.pone.0051921 PMID:23300580

Marinho Pessoa, T., Sousa Coutinho, D., Martinho Pereira, V., Pinho de Oliveira Ribeiro, N., & Nardi, A. E. (2014). The Nintendo Wii as a tool for neurocognitive rehabilitation training and health promotion. *Computers in Human Behavior*, *31*, 384–392. doi:10.1016/j.chb.2013.10.025

McKernan, B., Martey, R. M., Stromer-Galley, J., Kenski, K., Clegg, B. A., Folkestad, J. E., & ... (2015). We don't need no stinkin' badges: The impact of reward features and feeling rewarded in educational games. *Computers in Human Behavior*, *45*, 299–306. doi:10.1016/j.chb.2014.12.028

Tear, M. J., & Nielsen, M. (2014). Video games and prosocial behavior: A study of the effects of non-violent and ultra-violent gameplay. *Computers in Human Behavior*, *41*, 8–13. doi:10.1016/j.chb.2014.09.002

KEY TERMS AND DEFINITIONS

Immersion: Cognitive processes leading to the embodiment of an avatar in a virtual space.

Massively Multiplayer Online (MMO) Games: Online games which can support the simultaneous interactions of large to very large numbers of players.

MMORPGs: MMORGPs is the acronym of massively multiplayer online role-playing games, a particular type of massively multiplayer online games in which players assume the role of a specific character in a role-playing context.

Online Disclosure: Fact of disclosing personal information online usually with a feeling of safety related to apparent anonymity.

Proteus Effect: Psychological phenomenon supporting the fact that experiences gained in virtual spaces can transfer into real-life effects.

Social Comparison: Comparison that individuals make with others, can be lateral (comparison with a person one considers as equal), upward (comparison with a person perceived as better), or downward (comparison with a person perceived as worse off).

This research was previously published in the Handbook of Research on Holistic Perspectives in Gamification for Clinical Practice edited by Daniel Novák, Bengisu Tulu, and Håvar Brendryen, pages 1-9, copyright year 2016 by Medical Information Science Reference (an imprint of IGI Global).

Chapter 12
Personal Health Systems for Diabetes Management, Early Diagnosis and Prevention

Konstantia Zarkogianni
National Technical University of Athens, Greece

Konstantina Nikita
National Technical University of Athens, Greece

ABSTRACT

This chapter aims at the presentation and comparative assessment of tools and methodologies used for the development of Personal Health Systems (PHSs) for diabetes management, early diagnosis and prevention. Medical decision support systems such as glucose prediction models, risk assessment models for long-term diabetes complications, models for early diagnosis of diabetes and closed-loop glucose controllers along with integrated systems for diabetes management are described. The outcomes of a wide range of research studies demonstrate the feasibility of providing safe, reliable and cost-effective solutions towards improving patients' quality of life through the application of PHSs. Specific limitations that prevent these systems from being fully adopted in clinical practice are highlighted, while challenges and future research directions are summarized.

INTRODUCTION

Diabetes Mellitus (DM) is a group of chronic metabolic diseases characterized by elevated blood glucose levels for a prolonged period. The deregulation of glucose metabolism is due to either the insufficient insulin secretion from the pancreatic cells or impaired response of the body cells to insulin. DM is broadly classified into three main categories:

1. **Type 1 Diabetes Mellitus (T1DM)**: T1DM is an autoimmune disease caused by the destruction of insulin-producing beta cells of the pancreas resulting in the absence of insulin secretion. T1DM is usually diagnosed in children and young adults and accounts for only 5% of patients with diabetes.

DOI: 10.4018/978-1-5225-6198-9.ch012

2. **Type 2 Diabetes Mellitus (T2DM)**: T2DM is characterized by, either or both, insulin resistance and relative insulin deficiency. It is the most common form of diabetes accounting for at least 90% of all cases of diabetes.
3. **Gestational Diabetes Mellitus (GDM)**: GDM is characterized by high blood glucose levels during pregnancy. GDM accounts for one per 25 pregnancies worldwide. GDM usually disappears after pregnancy but it is a risk factor, for both the mother and the child, to develop T2DM in the future. Approximately half of women with a history of GDM develop T2DM within five to ten years after delivery.

DM has severe short-term and long-term complications. In particular, diabetic ketoacidosis, and hyperglycemia hyperosmolar state, are acute episodes, which may lead to diabetic coma if not treated promptly and properly. Moreover, severe hypoglycemic episodes which are caused by overdoses of administered insulin, may lead to the lost of consciousness. The excess glucose circulating through the body in the blood stream over time leads to damage of blood vessels and severe long term mortality related complications such as cardiovascular disease, diabetic neuropathy, and diabetic retinopathy.

According to the International Diabetes Federation (IDF) (2013), 382 million people (8.3% of adults) suffer from DM worldwide, while 175 million people with DM are undiagnosed. By 2035, it is estimated that 592 million people will have DM. In 2013, 5.1 million of deaths were attributed to DM while at least USD 548 billion dollars of health expenditure were caused from DM, which corresponded to a percentage of 11% of the total health spending on adults. According to the outcomes of the Diabetes Control Complications Trial, intensive glycemic control, reduces the long-term diabetes complications in T1DM (The Diabetes Control and Complications Trial Research Group, 2003). Moreover, several studies have investigated the importance of tight glycemic control for protection against the incidence of microvasclular and cardiovascular disease in T2DM (Giorgino, Leonardini, & Laviola, 2013). Intensive glycemic control involves regular glucose measurements and exogenous insulin administration, in case of insulin treated patients. To this end, latest technological advances have led to the development of Continuous Glucose Measurement Systems (CGMS) able to provide the information of glucose levels every 1 *min* or 5 *min*, and subcutaneous insulin infusion pumps (Klonoff, 2005).

The high prevalence of DM, and the rapidly growing number of patients with DM, along with the rising costs of care, the predictable number of deaths and medical errors, poses the need to move from a health system that focuses on the disease to a health system that focuses on personalized care. Optimal management of diabetes requires deep understanding of the risk factors associated with the disease, early diagnosis and treatment of the disease before the occurrence of complications, and tight glycemic control. Toward this direction, several technological advances in the fields of sensors for physiological parameters measurement and drug delivery systems coupled with advanced Information and Communications Technology (ICT) that enable efficient monitoring, data presentation, decision support, and social networking, make feasible the development of personal health systems for the remote monitoring and management of patients with DM. Intelligent systems for multi-parametric biomedical data analysis (such as physiological measurements, genetic data, medical images, laboratory examinations, activity, lifestyle and data from the surrounding environment) play a crucial role in personal health systems, focusing on the processing and interpretation of the data for accurate and timely alerting, signalling of risks, supporting clinical decisions and empowering patients for self-managing their disease.

The development of Personal Health Systems (PHS) for DM management, early diagnosis and prevention, attracts great interest from the international research community leading to a wide range of studies and collaborative national and international research projects. However, these systems suffer from several limitations in terms of reliability, accuracy, performance, and user acceptance, thus creating important challenges.

In view of all the above, the present chapter focuses on the introduction and comparative assessment of Medical Decision Support Systems and Integrated Systems, for diabetes management, early diagnosis and prevention of long-term complications, focusing mainly on T1DM and T2DM since these two categories consist the most frequent target groups for applying PHS in the field of diabetes management. Moreover, the limitations of these systems are highlighted while challenges and future research directions are presented.

BACKGROUND

State of the Art in Medical Decision Support Systems for Diabetes Management

In recent years, many efforts have been reported in the literature towards the development of medical decision support systems for the management of DM, focusing on: i) glucose prediction models for patients with T1DM, ii) Artificial Pancreas (AP), iii) risk assessment models for long-term complications of T1DM and T2DM, and iii) models for early diagnosis of T2DM. Each of the above is further detailed in the following.

Glucose Prediction Models for Patients with Type 1 Diabetes

Tight glycemic control is difficult to be achieved in patients with T1DM, since several environmental factors such as nutrition, physical activity, patient's psychological status and overall lifestyle along with endogenous processes, such as circadian rhythms, strongly affect glucose metabolism. Furthermore, intra- and inter- patient variability in response to therapy, makes the regulation of glucose levels very difficult. The aforementioned difficulties can be addressed through the development of computational models able to produce accurate and reliable estimations of future glucose profile in response to various stimuli. Predicted glucose profile is mainly used for producing early warnings of the upcoming hypoglycemic/hyperglycemic episodes or for adjusting insulin injections and insulin infusion rates in insulin treated patients.

Many research approaches have been reported for the development of predictive glucose/insulin metabolism models for patients with T1DM, based on Compartmental Models (CMs), which represent fundamental glucoregulatory processes such as glucose distribution, production and utilization along with the effect of insulin action. The most widespread model is the minimal model, which quantifies pancreatic response and insulin sensitivity by incorporating three compartments representing plasma insulin, remote insulin, and plasma glucose, respectively (Bergman, Phillips, & Cobelli, 1981). Aiming at enhancing the minimal model, the use of two compartments for describing glucose kinetics have been proposed by Caumo and Cobelli (1993), Vicini et al. (1997), and Cobelli et al. (1999), while the parameters' identification has been based either on the usage of glucose tracers or the application of Bayesian techniques. Moreover, noteworthy attempts have been made in order to estimate the simultaneous effect of insulin on glucose distribution/transport, glucose disposal, and endogenous production during an intravenous glucose tolerance test (IVGTT) (Ferrannini et al., 1985; Hovorka et al., 2002).

Hence, insulin action has been modeled by postulating three compartments representing the effect of insulin in glucose distribution/transport, disposal, and production (Hovorka et al., 2002).

Glucose-insulin metabolism models like those mentioned above, have been used as the core models towards the development of simulators of T1DM aiming at facilitating the conduction of *in silico* experiments for testing the performance of closed-loop glucose controllers. For the same purpose, several CMs for the simulation of subcutaneous insulin kinetics (Nucci & Cobelli, 2000), subcutaneous glucose kinetics (Freeland & Bonnecaze, 1999; Schmidtke, Freeland, Heller & Bonnecaze, 1998; Wilinska 2004) and glucose absorption from the gut (Arleth, Andreassen, Orsini-Federici, Timi, & Massi Benedetti, 2000; Dalla Man, Raimondo, Rizza, & Cobelli, 2007; Dalla Man, Camilleri, & Cobelli, 2006) have been developed. Limited efforts have been reported for quantifying the effect of physical activity in glucose metabolism (Derouich, & Boutayeb, 2002; Chassin, Wilinska, & Hovorka, 2007). All these CMs have been integrated towards the development of T1DM simulators with the ability to simulate cohorts of *in silico* subjects. The Hovorka's model is equipped with 18 *in silico* patients with T1DM, while it simulates the intra-patient variability through applying sinusoidal oscillations in specific parameters (Hovorka et al., 2004a). The UVa T1DM simulator includes a total of 300 T1DM children, adolescents and adults. It has been validated against actual clinical data and it has been accepted by FDA as a substitute for pre-clinical animal trials of glucose control strategies for T1DM (Kovatchev, Breton, Dalla Man, & Cobelli, 2009).

Although, the CM - based models provide a virtual environment for the simulation of clinical trials, their acceptance in forecasting glucose levels has been limited because these systems take into account only a confined number of factors associated with glucose metabolism, and they are not easily adapted to accurately simulate metabolic processes for a specific patient with DM. Moreover, the identification of CMs' parameters requires clinical measurements, which are not typically available in the clinical settings. In order to overcome these limitations, the use of data-driven modeling techniques has been proposed which disregard physiological insights and use pattern recognition techniques to simulate glucose metabolism. Volterra series models, Time Series Analysis and Machine Learning Methods are the most widespread data-driven techniques towards the development of glucose prediction models. In particular, nonlinear Volterra models of glucose-insulin dynamics have been shown to provide accurate predictions in the absence of noise (Florian, & Parker, 2005; Mitsis, Markakis, & Marmarelis, 2009). Autoregressive exogenous input (ARX) and Box-Jenkins (BJ) models with constant parameters and various model orders (high and low) have also been applied to simulate glucose-insulin dynamics (Finan, Zisser, Jovanovic, Bevier, & Seborg, 2006). Several types of Artificial Neural Networks (ANNs) such as multilayer perceptron (MLP) Neural Networks (NN) (Pappada, Cameron, & Rosman, 2008; Perez-Gandia et al., 2010), Radial Basis Function (RBF) NNs (Baghdadi, & Nasrabadi, 2007), wavelet NNs (Zainuddin, Pauline, & Ardil, 2009), neurofuzzy systems applying wavelets as activation functions (Zarkogianni et al., 2014), and Recurrent Neural Networks (RNNs) (Mougiakakou et al., 2008), have been deployed for the simulation of glucose metabolism. Furthermore, glucose prediction models based on Gaussian processes have been developed (Valleta, Chipperfield, & Byrne, 2009). Additionally, hybrid glucose-insulin metabolism models based on the combined use of CMs and data-driven modeling techniques, such as RNN (Mougiakakou et al., 2008), SVR (Georga et al., 2013) and Self-Organizing Maps (SOMs) (Zarkogianni, Litsa, Vazeou, & Nikita, 2013), have produced promising results. In Table 1, the aforementioned glucose prediction models, which are based on Artificial Intelligence, are summarized while additional information regarding their evaluation assessment results is presented. Although a direct and fair comparison between the models' performance is not feasible due to different datasets, input spaces and evaluation methodologies, substantial differences can be obtained. As it is expected, the input space plays a crucial role in the models' predictive performance, since higher input space dimension results in improved performance. Moreover, the hybrid glucose-insulin metabolism models have produced the best results, demonstrating, thus, increased capabilities of accurately and reliably capturing the metabolic behavior of a specific patient with T1DM.

A personalized glucose-insulin metabolism model developed at the Biomedical Simulations and Imaging (BIOSIM) Laboratory, School of Electrical and Computer Engineering, National Technical University of Athens is presented in (Zarkogianni, Litsa, Vazeou, & Nikita, 2013). The glucose-insulin metabolism model has been based on the combined use of a mathematical model (MM) module and an ANN. The MM module consists of two CMs, which simulate subcutaneous (sc) insulin kinetics and glucose absorption into the blood from the gut, respectively, while the ANN module incorporates a SOM,

Table 1. Glucose prediction models based on Artificial Intelligence

Model	Input Space	No. of T1DM Patients (Monitoring Period)	Evaluation Results	Reference
Multilayer FNN	CGM data, Blood glucose readings, Insulin dosages, Carbohydrate intake, Hyperglycemic and hypoglycemic symptoms, Lifestyle (activities and events), Emotional states	18 (3-9 days)	PH[a] (*min*)/ MAD[b] (%): 50/6.7, 120/14.5%, 180/18.9%	(Pappada, Cameron, & Rosman, 2008)
FNN with 2 hidden layers	CGM data	9 (12 days), 6 (2 days)	PH (*min*)/ RMSE[c] (*mg/dl*): 15/10, 30/18, 45/27	(Perez-Gandia et al., 2010)
RBF NN	Blood glucose readings, Insulin dosages, Food intake, Stress, Level of exercise	1 (77 days)	Interval / RMSE (*mg/dl*): morning/1.49, afternoon/0.92, evening/0.67, night/0.21	(Baghdadi, & Nasrabadi, 2007)
Wavelet NN	Blood glucose readings, Insulin dosages, Food intake, Stress, Level of exercise	1 (77 days)	Interval / RMSE (*mg/dl*): morning/0.81 afternoon/0.63, evening/0.60, night/0.30	(Zainuddin, Pauline, & Ardil, 2009)
Neurofuzzy (applying wavelets as activation functions)	CGM data, Physical activity data from sensor	6 (7-15 days)	PH (*min*)/ RMSE (*mg/dl*): 15/14.42, 30/20.20, 45/24.79, 60/28.49	(Zarkogianni et al., 2014)
Hybrid model based on the combined use of CMs and RNN	CGM data, Insulin infusion rates, Carbohydrates ingested	9 (10 days)	PH (*min*)/ RMSE (*mg/dl*): 30/18.34	(Mougiakakou et al., 2008)
Hybrid model based on the combined use of CMs and SVR	CGM data, Insulin dosages, Carbohydrates ingested, Physical Activity data from sensor, Time	15 (5 - 22 days)	PH (*min*)/ RMSE (*mg/dl*): 15/5.21, 30/6.03, 60/7.14, 120/7.62	(Georga et al., 2013)
Hybrid model based on the combined use of CMs and SOM	CGM data, Insulin Infusion rates, Carbohydrates ingested	12 (10 days)	PH (*min*)/ RMSE (*mg/dl*): 30/14.10, 60/23.19	(Zarkogianni, Litsa, Vazeou, & Nikita, 2013)

[a]Prediction Horizon
[b]Mean Absolute Difference
[c]Root Mean Square Error

Figure 1. Outline of the Personalized Glucose-Insulin Metabolism Model presented in (Zarkogianni, Litsa, Vazeou, & Nikita, 2013)

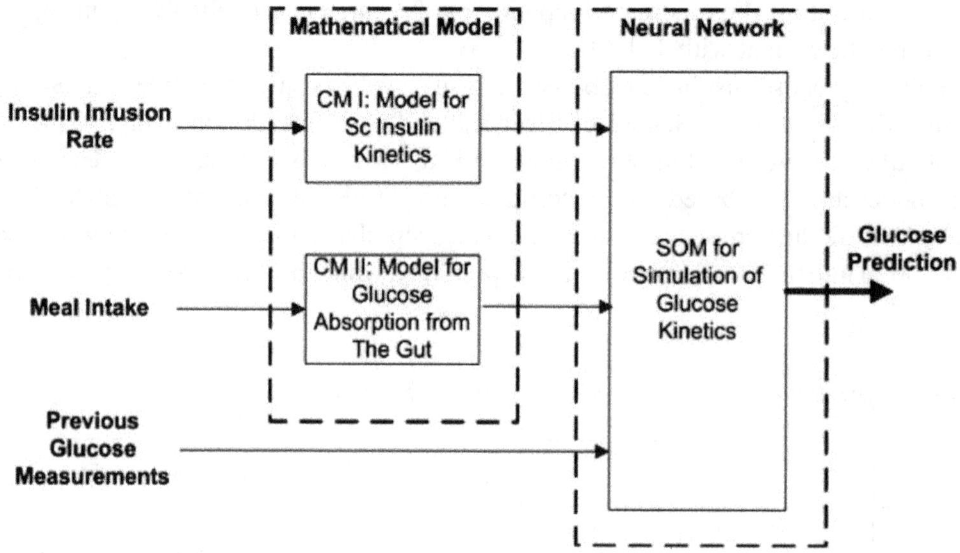

which models the patient's glucose kinetics (Figure 1). Information regarding recent sc insulin infusion rate and meal intake are fed to the MM module. CMs' outputs along with the most recent sc glucose measurement are applied to the SOM in order for the latter to provide glucose predictions.

The personalized glucose-insulin metabolism model has been evaluated using data from the medical records of 12 patients with T1DM (Mougiakakou et al., 2008). The patients' characteristics are shown in Table 2. For each patient, data corresponding to the 60% of the monitored days were used for training purposes (model development), while the remaining 40% for testing (model evaluation). The predictive performance of the proposed model was evaluated considering a prediction horizon (PH) equal to 30 min and 60 min with a 5-min resolution. Root-mean-squared error (RMSE) and correlation coefficient (CC) corresponding to the testing dataset were calculated to evaluate the performance of the model in terms of matching the predicted glucose with the original one. Continuous Glucose- Error Grid Analysis (CG-EGA) can be used to evaluate the clinical accuracy of the glucose predictions and their effect

Table 2. T1DM patients' characteristics used in (Zarkogianni, Litsa, Vazeou, & Nikita, 2013; Zarkogianni et al., 2014)

	Population for the Development and the Evaluation of the Glucose-Insulin Metabolism Hybrid Model	**Population for the Development and the Evaluation of the Neuro-Fuzzy Based Glucose Prediction Model**
	Mean ± Standard Deviation	**Mean ± Standard Deviation**
Age	19.83 ± 12.28	48.00 ± 9.69
Diabetes Duration	12.67 ± 7.74	27.20 ± 12.85
BMI	22.00 ± 4.88	22.64 ± 1.79
HbA1c	6.78 ± 0.94	7.70 ± 0.72

on decisions to avoid hypo- and hyperglycemic events (Kovatchev, Gonder-Frederick, Cox, & Clarke, 2004). The estimates of point and rate precision are combined in a single accuracy assessment for each of the blood glucose ranges: hypoglycemia, euglycemia, and hyperglycemia. To this end, the point error grid analysis (P-EGA) and the rate error grid analysis (R-EGA) are combined in these three clinically relevant regions. Clinically accurate glucose predictions are considered to be within the zones A and B on both P-EGA and R- EGA. Clinically benign errors correspond to acceptable point accuracy (i.e., A or B P-EGA zones) and significant errors in rate accuracy (i.e., C, D, or E R-EGA zones), which are unlikely to lead to severe clinical consequences. Clinically significant errors are those that could lead to inappropriate clinical actions.

From both the RMSE (mean ± standard deviation (SD): 14.10 ± 4.57) and CC (mean ± SD: 0.94 ± 0.02), it is obvious that the predicted glucose profile follows the original one for 30 min PH. Even for the case of 60 min PH, the RMSE (mean ± SD: 23.19 ± 6.40) and the CC (mean ± SD: 0.84 ± 0.05) indicate that the glucose predictions are close to the original values. Moreover, from the CG-EGA presented in Table 3, it is observed that for 30 min PH, high percentage of glucose predictions are accurate readings. Regarding the 60 min PH, most of the glucose predictions are accurate readings in the ranges of normal glycemia and hyperglycemia while most erroneous errors are observed in the range of hypoglycemia. The obtained results demonstrate the ability of the personalized glucose-insulin model to accurately predict the glucose profile.

Most of the aforementioned models are fed with information related to glucose records, insulin injections/infusion and meal intakes. Only a limited number of studies take into account physical activity, recorded by means of appropriate sensors, as input to the models. An exemplar personalized glucose prediction model based on neuro-fuzzy techniques applying wavelets as activation functions (FWNN) is presented in (Zarkogianni et al., 2014). The model receives as input, data from two body sensors recording glucose and physical activity, respectively and produces estimations of the future glucose profile. The reason for selecting this particular input space, is twofold: i) to develop a glucose prediction model, which does not require information inserted manually from the patient and shortcomings, and ii) to avoid feeding the model with subjective information which suffers from large inaccuracies.

The data used for the development and the evaluation of the model has been collected within the framework of the EU research project METABO (Georga et al., 2009). In particular, data from the medical records of 6 T1DM patients for an observation period ranging from 7 to 15 days (mean ± standard deviation: 10.83 ± 3.86) have been used. Patients' characteristics are presented in Table 2. In order to evaluate the predictive performance and the generalization abilities of the FWNN the ten-fold cross validation has been applied individually for each patient. A prediction horizon (PH) up to 15 min, 30 min, 45 min and 60 min, respectively, has been considered. The RMSE and CC values, which

Table 3. Continuous Glucose Error Grid Analysis for the clinical evaluation of the hybrid personalized glucose-insulin metabolism model (Zarkogianni, Litsa, Vazeou, & Nikita, 2013)

	Hypoglycemia (%)		Normal Glycemia (%)		Hyperglycemia (%)	
PH	30 *min*	60 *min*	30 *min*	60 *min*	30 *min*	60 *min*
Accurate readings	81.06	63.22	92.18	91.71	88.27	87.19
Benign Errors	3.48	2.83	7.59	7.97	2.96	3.45
Erroneous Readings	15.46	33.95	0.23	0.32	8.77	9.36

are presented in Table 4, demonstrate the ability of the model to produce accurate glucose predictions. Moreover, from the EGA presented in Table 5, it is observed that most of the glucose predictions are in the clinically accepted zones of A and B. The MARD (Table 4) is below 7.5% for PH equal to 15 min while it increases for larger PHs.

Artificial Pancreas

One of the most important research directions concerns the development of AP, which applies particularly in patients with T1DM since this target group needs exogenous insulin administration in order to regulate their blood glucose levels. AP is a closed loop blood glucose control system that combines a glucose sensor, a control algorithm and an insulin infusion pump. Information from the glucose sensor is fed to the control algorithm, which determines the parameters of the insulin infusion pump in order for the latter to release optimal amounts of insulin. The problem of maintaining blood glucose levels within the physiological range is particularly complex in patients with diabetes and it is strongly related with the control of various parameters that affect glucose metabolism. These parameters are exogenous, such as the administered insulin, the diet, the exercise, the stress and the concomitant treatments, as well as intrinsic, such as the function of the liver, the brain, the muscle tissue, the kidney and the intestine. Controlling all these parameters is particularly difficult because of their complex interaction mechanism, and the ever-changing and unpredictable nature of glucose metabolism resulting in wide inter- and intra-patient variability. Therefore, glucose control algorithms should be able to provide with individualized treatment recommendations.

Table 4. Evaluation of the FWNN's predictive in terms of RMSE, CC, MARD (Zarkogianni et al., 2014)

PH (*min*)	**RMSE (Mean ± Standard Deviation (%))**	**CC (Mean ± Standard Deviation (%))**	**MARD (Mean ± Standard Deviation (%))**
15	14.42 ±4.26	96.74 ± 1.63	6.68 ± 2.18
30	20.20 ± 4.20	93.77 ± 3.65	10.24 ± 1.48
45	24.79 ± 1.33	89.73 ± 5.72	13.45 ± 2.56
60	28.49 ± 2.04	86.43 ± 7.03	15.92 ± 3.29

Table 5. EGA for the evaluation of the FWNN's predictive performance (Zarkogianni et al., 2014)

Zones	**PH (15** *min***)**	**PH (30** *min***)**	**PH (45** *min***)**	**PH (60** *min***)**
	Mean ± Standard Deviation (%)			
A	94.35 ± 5.66	86.70 ± 3.76	78.08 ± 7.56	71.89 ± 9.33
B	4.56 ± 4.98	10.89 ± 3.41	17.73± 5.22	22.03 ± 7.17
C	0.09 ± 0.13	0.10 ± 0.19	0.22 ± 0.12	0.13 ± 0.12
D	0.97 ± 0.76	2.29 ± 1.27	3.95 ± 2.68	5.90 ± 3.90
E	0.00 ± 0.00	0.00 ± 0.00	0.00 ± 0.00	0.00 ± 0.00

Closed-loop glucose control systems can be categorized according to the degree of user input required. In "fully closed-loop" mode, insulin is delivered without information about the time or size of the meal. In "semiclosed-loop" control, the controller is fed with information regarding the meal size and generates advice on prandial insulin. Meal announcement provides significant benefit to the glucose controller since it helps overcoming the delays associated with the appearance of glucose in the bloodstream due to meal ingestions. Although a wide range of algorithms have been proposed (Cobelli et al., 2009), the most common approaches are based on proportional integral derivative controller (Steil, Panteleon, & Rebrin, 2004; Marchetti, Barolo, Jovanovic, Zisser, & Seborg, 2008), and model-predictive controller (MPC) (Abu-Rmileh, & Garcia-Garbin, 2010; Wang, Dassau, & Doyle, 2010; Hovorka et al., 2011; Hyunjin, & Bequette, 2009; Magni et al., 2007, Chase et al., 2007; Magni et al., 2009; Hovorka et al., 2004b; Schaller et al., 2006). MPC seems to be the most appropriate for the development of AP, since it is able to handle problems related to 1) high nonlinearity of the glucose–insulin metabolism, caused by saturation and inhibition effects evidenced by chemical substrates and hormones involved in enzyme dynamics and hormonal control effects, 2) time delays in sc–sc route due to the delayed effect of infused sc insulin and food intake to the blood and, consequently, of glucose diffusion from the blood to the sc space, and the lag time between sc glucose value and glucose sensor (in the case of sensors based on microdialysis or microperfusion), and 3) noise to the sc glucose measurements. The models used to develop glucose controllers based on linear MPC are usually discrete linearized state-space models obtained from the average original nonlinear patient's model, which serves as the *in silico* T1DM patient for the evaluation of the glucose controllers (Magni et al., 2007; Abu-Rmileh, & Garcia-Garbin, 2010). However, such an approach would suffer from the lack of personalization and from dependencies between the predictive model integrated in the glucose controller and the *in silico* patient model, thus limiting the reliability of the *in silico* evaluation of the controller (Cobelli et al., 2009). In order to overcome these problems, the use of data-driven models has been proposed. In particular, a model- predictive iterative learning based on a data-driven linear autoregressive exogenous model (ARX) is presented in (Wang, Dassau, & Doyle, 2010). Although this model cannot accurately simulate the glucose-insulin metabolism of a specific patient with T1DM, the proposed control law performed well, especially in the case of repetitive diets. Meal detection and meal size estimation algorithms have been developed to improve meal glucose disturbance rejection when incoming meals are not announced (Hyunjin, & Bequette, 2009). Furthermore, several attempts have been made toward the development of glucose controllers based on nonlinear model-predictive control (NMPC), and the effectiveness of the NMPC over the linear MPC has been studied (Chase et al., 2007; Hovorka et al., 2011; Zarkogianni et al., 2011).

Several clinical studies have been conducted in recent years, in order to test the performance of the closed loop glucose controllers and compare their performance with that obtained by applying conventional therapies (Kumareswaran, Evans, & Hovorka, 2012; Hovorka et al., 2010; Hovorka et al., 2011). Overnight closed loop experiments using different MPC controllers have demonstrated the superiority of the closed loop control over conventional pump treatment (Kumareswaran, Evans, & Hovorka, 2012). The same conclusion has been obtained from the outcomes of closed loop clinical studies lasting more than 24 hours (Kumareswaran, Evans, & Hovorka, 2012).

An exemplar adaptive glucose control algorithm (Insulin Infusion Advisory System - IIAS) developed at BIOSIM laboratory is presented in (Zarkogianni et al., 2011). The IIAS is based on auto-tuning model predictive control and operates in "semi-closed loop" control, meaning that the controller is provided with meal information. The IIAS has been evaluated using 10 virtual adults with T1DM provided in the training version of the UVa T1DM simulator. Several experiments have been carried out to test the ability of the IIAS to handle meal disturbances, fasting conditions, inter-patient variability, robustness against erroneous estimation of carbohydrates' amount in ingested meals (up to 40% Overestimation (OEE) and Underestimation (UEE)), and intraday variation in physiological parameters. The simulation scenarios consider a two-day testing period with varying meal timings and amounts. The target glucose range has been set to 70 mg/dl-180 mg/dl. The blood glucose trace along with the min/max envelop presented in Figure 2, when the IIAS is applied to 10 adults, shows that the blood glucose levels even for the hard-to-control patients are within the acceptable range 70–236 mg/dl, managing to avoid hypoglycemic episodes and severe hyperglycemic episodes (above 280 mg/dl). To provide more details about the IIAS's performance for all simulation scenarios, the numerical metrics of average glycemia, percentages above and below the target range, risk associated with extreme glucose deviations (low blood glucose index (LBGI), high blood glucose index (HBGI), and total risk index), are presented in Table 6. It can be observed that most of the time, blood glucose levels are kept within the target range, while the risk indices (LBGI, HBGI, total risk index) have low values, showing that tight glycemic control is achieved for all simulation scenarios. Furthermore, the performance of the IIAS has been compared to the one obtained by applying an open loop pre-adjusted treatment, which is supported by the UVa T1DM Simulator. The results presented in Table 6 demonstrate the superiority of the IIAS over the open loop pre-adjusted treatment. The results from the *in silico* evaluation of the IIAS encourage further evaluation through clinical trials.

Since the discovery of new insulin analogues, which have improved the subcutaneous insulin absorption, great progress has been made to enhance the safety and the accuracy of automated insulin delivery systems. However, the risk of hypoglycemia, caused by overestimated insulin infusion rates, is not completely eliminated. In order to prevent and treat hypoglycemia, latest research directions focus on the administration of both the insulin and the insulin-counteracting hormone glucagon. A representative example of a Bihormonal Closed-Loop Artificial Pancreas is the so-called "Bionic Pancreas", a

Figure 2. Blood Glucose trace for the 10 adults of the UVa T1DM simulator when the IIAS is applied. Mean response (solid curve), standard deviation (dashed-dotted curve), min/max envelop (dashed curve) (Zarkogianni et al., 2011)

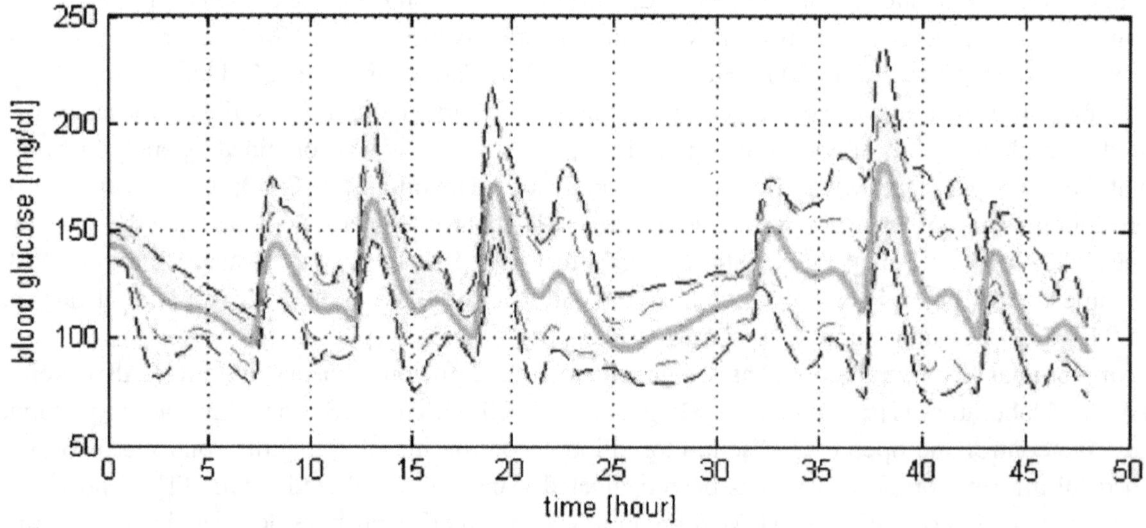

Table 6. Control performance of the IIAS (Zarkogianni et al., 2011)

Simulation Scenario	Pre meal BG Mean (±SD)	% below Target Mean (±SD)	% above Target Mean (±SD)	LBGI Mean (±SD)	HBGI Mean (±SD)	Risk Index Mean (±SD)
Accurate meal announcement (IIAS)	114.27 (12.31)	0 (0)	2.51 (2.76)	0.35 (0.35)	1.09 (0.64)	1.45 (0.66)
40% OEE (IIAS)	105.57 (8.99)	1.01 (1.51)	2.40 (3.13)	0.72 (0.34)	0.92 (0.62)	1.64 (0.66)
40% UEE (IIAS)	110.11 (11.13)	5.15 (5.07)	4.36 (4.43)	0.99 (0.71)	1.27 (0.86)	2.26 (1.10)
Intraday variation in physiological parameters (IIAS)	128.25 (15.79)	0.66 (0.87)	3.41 (3.25)	0.4 (0.27)	1.2 (0.65)	1.6 (0.71)
Open loop preadjusted treatment	124.4 (6.65)	1.14 (3.61)	3.68 (3.78)	0.30 (0.48)	1.65 (0.69)	1.96 (0.91)

device that mimics the behavior of beta cells and alpha cells of the pancreas, which has been developed at Imperial College London (Herrero, Georgiou, Oliver, Johnston, & Toumazou, 2012). It consists of an electrochemical glucose sensor, which penetrates the skin, and two small portable infusion pumps, insulin and glucagon, respectively. Every 5 minutes, the sensor detects glucose levels. If the glucose concentration is increased, the artificial beta cells produce a signal which drives a motor that pushes the syringe to release insulin. If the sensor detects low concentrations of glucose, artificial alpha-cells activate the glucagon infusion pump.

Several clinical studies have been carried out in order to test the feasibility of achieving safe and good glycemic control by applying bi-hormonal closed loop glucose controller (Bakhtiani, Zhao, Youssef, Castle, & Ward, 2013). Within the framework of a crossover study to compare the performance of insulin-only closed glucose controllers versus insulin and glucagon closed loop glucose controllers, 15 T1DM patients have been recruited in the Oregon Health and Science University (OHSU) outpatient clinics and underwent one closed-loop study with insulin plus placebo and one study with insulin plus glucagon, given at times of impending hypoglycemia (Castle et al., 2010). The Fading Memory Proportional Derivative (FMPD) algorithm has been used to determine the insulin and glucagon delivery rates. According to the obtained results, the application of the bi-hormonal closed loop glucose controller has reduced the rate of hypoglycemia. A group from Boston has developed a bi-hormonal system using MPC and PID for the estimation of insulin and glucagon delivery rates, respectively, while the pharmacokinetics models parameters for the lispro insulin have been adjusted accordingly in order to account for the inter-patient variability (El-Khatib, Russell, Nathan, Sutherlin, & Damiano, 2010). The system has been tested on 11 T1DM patients for 27 hours and has achieved good glycemic control. In order to test whether safe and effective glycemic control could be achieved in T1DM using a bihormonal bionic endocrine pancreas, a clinical study has been carried out, in which 6 T1DM patients have participated in two experiments lasting 51 *hours* and including six high-carbohydrate meals and exercise (Russell et al., 2012). Insulin dosing has been controlled by a customized model predictive control algorithm incorporating a pharmacokinetic model for insulin lispro, while a customized PID has been used to control glucagon dosing. The obtained results have shown that good glycemic control can be achieved with minimal hypoglycemia.

Risk Assessment Models for Long-term Type 1 and Type 2 Diabetes Complications

Patients with T1DM or T2DM are at high risk of developing serious and mortality related macrovascular and microvascular diseases in the long run. In order to delay or even prevent such complications, appropriate prevention treatment should be initiated. Within this context, reliable risk score calculators provide valuable support for medical decision making. Several studies have been devoted to the development of diabetes risk engines able to provide estimations of the probability of a diabetic patient to develop long-term diabetes complications and the effect of different interventions in reducing this probability. These diabetes risk engines are used to facilitate patients' stratification according to their probability of developing complications. To estimate the risk of a diabetic patient to develop long-term complications, multi-parametric processing of medical history data, clinical measurements and environmental data is applied. The most common predictions are related to Cardiovascular Diseases (CVD) and Diabetic Retinopathy.

Towards this direction, large scale clinical studies with a duration of more than 5 years, have been conducted (Lagani, Koumakis, Chiarugi, Lakasing, & Tsamardinos, 2013). A well-known multicenter, randomized clinical study is the Diabetes Control and Complications Trial (DCCT), conducted by the United States National Institute of Diabetes and Digestive and Kidney Diseases (The Diabetes Control and Complications Trial Research Group, 2003). The study has been designed to compare the effects of intensive versus conventional diabetes therapy on the incidence of vascular and neurological complications. A total of 1441 patients with T1DM across 29 medical centers in the United States and Canada have been recruited and followed up from 1983 through 1989. The Epidemiology of Diabetes Interventions and Complications (EDIC) study has started after the DCCT with the aim to describe the development and progression of CVD and investigate the effects of intensive control on cost-effectiveness and quality of life (The Epidemiology of Diabetes Interventions and Complications (EDIC) Research Group, 1999). The EDIC study has followed up more than 90% of the participants from the DCCT. The outcomes of both studies have been merged in order to explore the complex relationship between diabetes and CVD. Such analysis has resulted in the development of three risk prediction models. In particular, Cox proportional hazard models have been applied to assess the effect of time dependent covariates with respect to the incidence of CVD, while the Kaplan Meier method have been used to produce the cumulative incidence of a CVD taking as input clinical variables (age, smoking habits), biochemical markers (cholesterol level, albuminuria, microalbuminuria, HbA1c) and the presence of renal impairment (The Diabetes Control and Complications Trial/Epidemiology of Diabetes Intervention and Complications (DCCT/EDIC) Study Research Group, 2005). Furthermore, cox regression models have been used to identify biological risk factors, which are strongly related with atherosclerotic occlusion in peripheral vascular disease (Carter et al., 2007). Tobit survival regression models have been also applied in order to identify relevant risk factors for the development of coronary artery calcification (CAC) events in T1DM, resulting in age, sex, smoking, waist-to-hip ratio, diabetes duration, HbA1c, hypercholesterolemia, hypertension and albumin excretion rate as significant predictors of CAC (Cleary et al., 2006).

QRisk is another large-scale study aiming at the development of a CVD risk calculator (Hippisley-Cox et al., 2008). In this study, 2.3 million patients aged 35–74 with 140,000 cardiovascular events have been recruited. Of the participants, 2.22 million people were either white or whose ethnic group was not recorded, 22.013 south Asian, 11.595 black African, 10.402 black Caribbean, and 19.792 from Chinese or other Asian ethnic groups. The current version of the calculator, the Q-Risk 2, produces estimations of the risk of having a heart attack or stroke over the next 10 years receiving as inputs epidemiological data (age, sex, ethnicity) and clinical information (smoking status, diabetes status, angina or heart attack in a 1st degree relative < 60, chronic kidney disease, atrial fibrillation, on blood pressure treatment, cholesterol/HDL ratio, systolic blood pressure, height and weight). The development of QRisk-2 calculator has been based on both Cox proportional hazard models for the estimation of coefficients and fractional polynomials for modeling non-linear risk relations with continuous variables.

The UKPDS study is a widely known randomized controlled trial which has demonstrated that intensive treatment of both blood glucose and blood pressure in newly diagnosed patients with T2DM has the potential to decrease the risk of the incidence of diabetes-related complications (Stevens, Kothari, Adler, & Stratton, 2001). The UKPDS cohort has included 5.102 T2DM patients, who have been followed up for a median of 10,7 years, and enrolled over a period of 14 years (1977 to 1991). Multivariate logistic regression has been applied using data corresponding to subsets of the UKPDS patients in order to investigate the relevance of the selected risk factors with respect to the incidence of fatal and non-fatal myocardial infarction (MI) and stroke, respectively (Stevens et al., 2004). Advanced age at diagnosis of diabetes, diabetes duration, HbA1c, SBP and urinary albumin, have been identified as the most significant risk factors for MI fatality. Moreover, sex, HbA1c, SBP, white blood cell count, and presence of previous stroke have been identified as significant stroke risk factors. Survival analysis has been performed on the UKPDS cohort, in order to predict the risk of the incidence of stroke in patients with T2DM (Kothari et al., 2002). Another work based on the UKPDS cohort is the development of a model able to estimate the probability of the occurrence of death, MI, other ischemic heart diseases, stroke, heart failure, amputation, renal failure and diabetic eye disease. To this end, Weibull proportional hazard regression models have been applied for each outcome (Clarke, 2004). The UKPDS cohort has also been used towards the development of the CDC/RTI Type 2 Diabetes Progression Model, which is based on Markov modeling (The CDC Diabetes Cost-effectiveness Group, 2002). The model incorporates five different disease paths in order to simulate the complications of diabetes (nephropathy, neuropathy, retinopathy, CHD, and stroke).

Another study is the EuroDiab prospective complications study, in which 3.250 T1DM patients from 16 different European countries have been recruited (Porta et al., 2001). A subset (1115) of this cohort has been used for the development and the validation of a model able to estimate the risk of microalbuminuria (Vergouwe et al., 2010). The risk factors which are fed to the model include age, sex, duration of diabetes, HbA1c, AER, fasting triglycerides, non-HDL (total cholesterol minus the HDL level) and LDL cholesterol, waist–hip ratio (WHR), BMI, pulse pressure, hypertension and smoking. The model has been based on logistic regression while a transformation in a risk chart for making the risk calculation easier in real medical settings has been applied. Table 7 summarizes information regarding the data, the type of diabetes and the methodology adopted, towards the development of the aforementioned diabetes risk engines while the target complications are also presented.

Retinopathy constitutes one of the most important complications of T1DM, representing the leading cause of blindness in young adults. It affects up to 80 percent of all patients who have had diabetes for 10 years or more (Kertes, & Johnson, 2007). Different methodologies based on Artificial Intelligence have been utilized towards the development of personalized decision support systems for the risk prediction of diabetic retinopathy development in patients with T1DM (Skevofilakas, Zarkogianni, Karamanos, & Nikita, 2010). In particular, a Feedforward Neural Network (FNN), a Classification and Regression Tree (CART), and an improved Hybrid Wavelet Neural Network (iHWNN) have been developed and evaluated using data from the medical records of 55 Type 1 DM patients, who were recalled for 5 years follow up. The data set was acquired by the Athens Hippokration Hospital in close collaboration with

Table 7. Diabetes risk engines

Reference	Risk Assessment Model	Data From Reference Study	Type and no. of Patients	Target Complications
(The Diabetes Control and Complications Trial/ Epidemiology of Diabetes Intervention and Complications (DCCT/EDIC) Study Research Group, 2005)	Cox regression model	DCCT/EDIC	1441 patients with T1DM	CVD
(Carter et al., 2007)	Cox regression models	DCCT/EDIC	1441 patients with T1DM	Atherosclerotic occlusion in peripheral vascular disease
(Cleary et al., 2006)	Tobit survival regression models	DCCT/EDIC	1441 patients with T1DM	CAC
(Hippisley-Cox et al., 2008)	Cox proportional hazard models and fractional polynomials	QRisk	1.280.000 patients with T2DM	CVD
(Stevens et al., 2004)	Multivariate logistic regression	UKPDS	5.102 T2DM patients	Fatal and non-fatal MI and stroke
(Kothari et al., 2002)	Survival analysis	UKPDS	5.102 T2DM patients	Stroke
(Clarke, 2004)	Weibull proportional hazard regression models	UKPDS	5.102 T2DM patients	Death, MI, stroke, heart failure, amputation, renal failure, diabetic eye disease
(The CDC Diabetes Cost-effectiveness Group, 2002)	Markov modeling	UKPDS	5.102 T2DM patients	Nephropathy, neuropathy, retinopathy, CHD, and stroke
(Vergouwe et al., 2010)	Logistic regression	EuroDiab	1.115 T1DM patients	Microalbuminuria

the EURODIAB research team. In this dataset, 17 out of the 55 Type 1 DM patients (30.9%) developed retinopathy during their treatment period (Porta et al., 2009). The following risk factors have been found as strongly correlated with the retinopathy complication by the EURODIAB Prospective Complications Study Group: i) age, ii) diabetes duration, iii) glycosylated hemoglobin (HbA1c), iv) total cholesterol, v) triglycerides, vi) incidents of hypertension, vii) treatment duration. Moreover, since the diabetic retinopathy is a condition that develops slowly over time, the risk of the disease is not only related to the current values of the predictors, but also to their temporal evolution over the past years. To this regard, the risk assessment model should take into account the evolution of the predisposing factors over time. Toward this direction additional inputs were applied to the model, corresponding to the average values calculated over the years from the baseline until the year of prediction. This procedure was applied in the measurements of HbA1c, total cholesterol and triglycerides. Receiving Operating Characteristics (ROC) and 10-fold cross validation have been used to evaluate the performance of the developed risk assessment models. The results from the ROC analysis, presented in Table 8, indicate that the best performance among the models is achieved by the iHWNN providing an accuracy of 97%. Moreover, iHWNN is far superior to the other models in estimating the high risk of a diabetic patient to develop retinopathy, achieving a high value of sensitivity (94%) and specificity (98%) while the False Positive Rate (FPR) is low (1%). The application of the FNN and CART has resulted in low sensitivity and high specificity. The above analysis provides the basis to enhance and extend the risk assessment models to other long-term diabetes complications such as cardiovascular disease, stroke etc.

Table 8. ROC Analysis for the evaluation of the performance of the risk prediction models for the incidence of diabetic retinopathy (Skevofilakas, Zarkogianni, Karamanos, & Nikita 2010)

	FNN	CART	iHWNN
Sensitivity (%)	58	58	94
False Positive Rate (FPR) (%)	7	9	1
Accuracy (%)	88	86	97
Specificity (%)	92	90	98
Positive Predictive Value (PPV) (%)	55	50	88
Negative Predictive Value (NPV)(%)	93	93	99
False Discovery Rate (FDR)(%)	44	50	11

Models for early diagnosis of Type 2 Diabetes

T2DM is frequently asymptomatic in the early stages and although the detection of the disease has been improved in the recent years, the delay from disease onset to diagnosis may exceed 10 years and it is usually done after the incidence of diabetes related complications (Shankaracharya, Odedra, Samanta, & Vidyarthi, 2010). Hence there is a great need for early detection of diabetes, which requires multifactorial analysis. Towards this direction, various risk scores have been proposed while the Finnish Diabetes Risk Score has been judged as the most convenient risk score for early diagnosis of DM (Schwarz, Li, Lindstorm, & Tuomilehto, 2009). However, this method is sensitive to human errors since it requires human intervention in deciding criteria and score. In order to overcome this problem, statistic pattern recognition analysis and machine learning techniques have been proposed aiming at enabling early detection of T2DM.

Several risk factors such as age, gender, body mass index, waist-to-hip ratio, waist circumference, random blood sugar test results, fasting blood sugar test results, post plasma blood, sugar tests, history of hyperlipidemia, and history of hypertension, and race/ethnicity, occupation, blood pressure medication, cholesterol medication, gestational diabetes, high blood pressure, high cholesterol, parental history of diabetes and exercise, have been identified as strongly related to the incidence of DM (Shankaracharya, Odedra, Samanta, & Vidyarthi, 2010). Logistic regression analysis has been applied for detecting DM and has resulted to a classification accuracy of the order of 80 (Shanker, 1996; Heikes, Eddy, Arondekar, & Schlessinger, 2008). According to an extensive literature review analysis, which has been devoted to computational intelligence for early diagnosis of DM, the application of clustering techniques that make use of k-means, mixture-of-Gaussians, SOM and neural gas (NG) have been investigated for the diagnosis of DM (Shankaracharya, Odedra, Samanta, & Vidyarthi, 2010). Furthermore, SVM and several types of Neural Networks, such as multilayer, back-propagated, radial basis function (RBF), general regression Neural Networks (GNN) and neuro-fuzzy inference systems have been used for classifying subjects in diabetics and non-diabetics

Although good performance has been achieved by the application of a single classifier, more accurate and reliable predictions can be obtained by using a mixture of experts (ME), which combines the outputs of single experts for providing the final decision. Furthermore, a modified mixture of experts (MME), which includes an assembly of expert networks and a gate-bank, has the potential to further increase the classification prediction performance. To this end, MEs and MMEs have been applied for early diagnosis of DM. In Table 9 the classification performance of the selected models for early DM diagnosis are summarized. It should be noted that all the performance of all the models presented in Table 6, apart from the SVM, has been assessed on the Pima Indian diabetes dataset (Shankaracharya, Odedra, Samanta, & Vidyarthi, 2010). The best performance has been achieved by applying the MME.

Integrated Systems for Diabetes Management

Medical decision support systems like those presented in the above section, constitute key modules of integrated systems for diabetes management. The overall architecture of an integrated system for diabetes management (Figure 3) includes: i) the personal loop, where information from portable devices for measuring physiological parameters is utilized locally, and ii) the remote loop, which is achieved through remote connection between the patient and health care professional. The personal loop incorporates sensors for measuring physiological parameters such as CGMS, and devices for measuring physical activity (such as pedometers, accelerometers, heart rate sensors). The collection and sending of the data is achieved through a Personal Smart Assistant (PSA), which integrates applications for data entry, processing and presentation.

The central unit, which closes the loop between the patient and the health care professional, incorporates the clinical information systems which are dedicated to collect, store and manage medical data. Specifically, clinical information systems:

- provide health care professionals and patients with proper reminders
- identify relevant subpopulations of patients for preventive care
- facilitate personalized care
- share information between patients and health care professionals
- monitor the performance of the team and the system.

It is worth to mention that the latest trends in PSA's technology include the development of advanced mobile applications such as the quantification of the ingested meal content in carbohydrates, proteins

Table 9. Classification performance of computational intelligence-based models for DM diagnosis (Shankaracharya, Odedra, Samanta, & Vidyarthi, 2010)

Model	Accuracy	Sensitivity	Specificity	Reference
Modified FNN	80.07	84.38	74.00	(Kamruzzaman, Hasan, Siddiquee, & Mazumder, 2004)
Adaptive neuro-fuzzy inference system	98.14	98.58	96.97	(Ubeyli, 2010)
SVM	94.00	94.00	93.00	(Barakat, Bradley, & Barakat, 2010)
Linear discriminant analysis and adaptive network based fuzzy inference system.	84.61	85.18	83.33	(Polat, Gunes, & Arslan, 2008)
Multilayer FNN	91.53	91.19	92.42	(Ubeyli, 2009)
ME	97.93	98.01	97.73	(Ubeyli, 2009)
MME	99.17	99.43	98.48	(Ubeyli, 2009)

Figure 3. Overall architecture of integrated systems for the management of DM

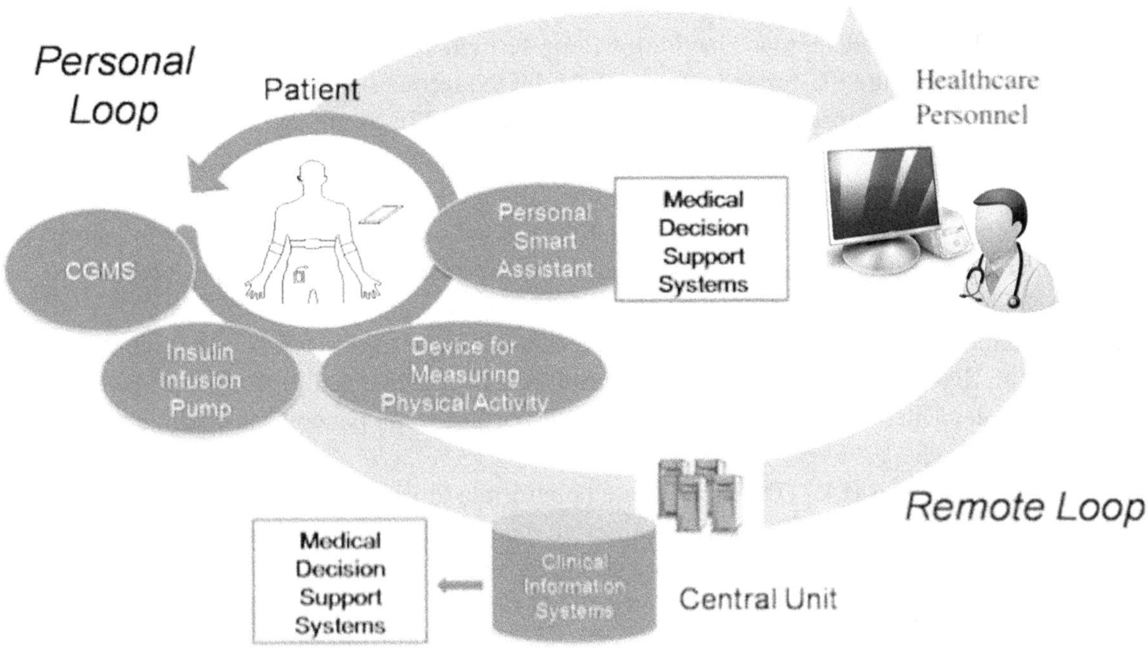

and fats, based on a snapshot of the meal (Kong, & Tan, 2012). This application allows individuals to take a snapshot of the meal, which is automatically quantified based on its contents so as to extract its nutritional value. This application incorporates advanced image processing techniques to quantify the content of the ingested meal in calories, carbohydrates, proteins and fats based on snapshots of the meal. The whole procedure usually includes four steps. In the first step, image segmentation techniques are deployed in order to distinguish and extract the relevant areas corresponding to different food contents. The second step concerns the classification of every food item. Following the identification of the food items as particular types of food, the third step focuses on the estimation of the volume of each food item, by applying methods for 3D reconstruction. Once the type and the volume of the food items are determined, the nutrient content is estimated in the fourth step, using standardized food databases.

Several projects have been devoted to the development of integrated systems for diabetes management. In particular, within the framework of the METABO project (Controlling Chronic Diseases related to metabolic disorders), a comprehensive platform with the ability to relate the clinical dimension of the patients with their daily life and reality has been developed (Georga et al., 2009). This has been achieved through the development of a multi-parametric monitoring system that gathers information of the metabolic status of patients and their lifestyle. The physicians can access all these data through a web-based application.

Within the framework of the EU-funded project INCA (Intelligent Control Assistant for Diabetes), the challenge to create a personal smart assistant, able to provide patients with closed-loop control strategies (personal and remote loop) has been addressed (Gomez et al., 2008). It is based on a real-time continuous glucose sensor (Guardian RT, Medtronic), an insulin pump (D-TRON, Disetronic Medical Systems), and a mobile general packet radio service (GPRS)-based telemedicine system. Moreover, a multi-access telemedicine central server provides to the patients and the doctors a web-based access to continuous glucose monitoring and insulin infusion data. Hence, patient therapeutic decision making can be supervised by the doctors.

An integrated ICT-based approach facilitating long-term management of T1DM has been developed within the framework of the EU-funded project REACTION (Remote Accessibility to Diabetes Management and Therapy in Operational Healthcare Networks) (Spanakis et al., 2012). An intelligent service platform facilitating professional, remote monitoring and therapy management to T1DM patients in different healthcare regimes across Europe has been implemented. The REACTION system features an interoperable peer-to-peer communication platform based on a service oriented architecture, which executes several applications for continuous glucose monitoring, context awareness, integrative risk assessment, event and alarm handling as well as automated closed loop insulin delivery.

Artificial Pancreas at home (AP@home) is an EU-funded project, with the vision to improve treatment of patients with diabetes at home, through the development of enhanced closed loop glucose controllers and the implementation and validation of a single-port artificial pancreas providing remote alarms (Heinemann, Benesch, & DeVries, 2011).

In the same field, the SMARTDIAB integrated platform addresses the challenge of remote monitoring and management of patients with T1DM, through the integration of state-of-the-art technologies in the fields of biosensors, telecommunications, simulation algorithms, and data mining (Mougiakakou et al., 2010). The system consists of: i) the Patient Unit (PU), which incorporates a continuous glucose measurement system, an insulin pump for subcutaneous insulin infusion and a personal smart assistant, and ii) the Patient Management Unit (PMU), which resides on a central web server and incorporates a diabetes data management system (DDMS), a decision support system (DSS) that provides risk assessment for long-term diabetes complications, and an insulin infusion advisory system (IIAS). The PU and PMU communicate with each other for data exchange.

Summarizing the application of PHSs towards the management, and early diagnosis of the DM has attracted great interest from the international research community in the recent years, leading to the development of a wide range of tools, methods and services. PHSs assist in the provision of continuous, quality controlled, cost-effective and personalized health services. However, due to the lack of extensive and reliable clinical and technical validation, the acceptance of these systems in clinical practice has been limited. Such extensive research, though, provides the fundamental basis for future research activities.

FUTURE RESEARCH DIRECTIONS

It is widely known that ICT-based solutions for the management of diabetes, such as those described in the previous section, have the potential to empower patients towards the self management of their disease, to facilitate remote monitoring and management, to handle acute episodes, to initiate appropriate interventions and consequently to reduce the health cost expenditures and to enhance the patients' quality of life. Although great progress has been made in this field in the recent years, there are still several limitations that prevent those systems to be fully adopted in clinical practice (Bellazzi, 2008). In particular, the PHS for diabetes management, suffer from biased data analysis since many studies have applied different criteria for selecting patients and controls and different approaches for the treatment of control groups. Another issue is the lack of substantial economic analysis for justifying the cost–effectiveness of the telemedicine systems. Moreover, there are still concerns regarding the privacy and security issues. Taking also into account that PHS exhibit varying speed performance, user interfaces, and communication standards, an in-depth technical and usability validation is needed in order to measure the acceptability of these tools from both the physicians and the patients.

User-centered approaches, mobile communication, context awareness and wearable systems remain the most promising areas for the next generation of PHS for diabetes management. Future research directions focus on empowering citizens and patients towards the self-management of their own health and disease outside institutions, improving, thus, health outcomes in terms of both quality of life and health expenditures. This requires a holistic patient-centered approach, including healthy lifestyle interlinked with disease management, and focusing on health education, secondary prevention and self-management of the disease. Moreover, research into socio-economic, environmental, clinical and biological factors along with cultural values, behavioral and social models is needed. Within this context, computer-based predictive models, providing personalization capabilities and combined with heterogeneous sources of data, have a great potential to raise individual awareness, promote behavioral lifestyle changes, support treatment, and monitor the disease.

Digital representation of health data to improve DM diagnosis and treatment represents another important challenge. Although several digital personalized models, tools and standards are currently available, there is a need for greater integration of patient information. Thus, multi-scale and multi-level physiological models with current and historical patient-specific and population-specific data should be used to generate new clinical information for patient management. New DSSs coupled with heterogeneous data sources and patient specific computer models can provide valuable tools to the healthcare professionals for personalized prediction and decision in prevention, diagnosis or treatment. The sustainability of these systems is an important feature and should be supported by taking advantage of the personal medical data accumulated over time, the standardization of data formats, and the integration of data provided by new technologies. Furthermore, the usability and acceptability of these systems should be ensured through highly visual data representation and the creation of user-friendly interactive exploratory interfaces. Gender and ethical issues should be duly considered while issues related to data security and protection should be adequately addressed. In this direction, the EU-funded project MOSAIC (Models and simulation techniques for discovering diabetes influence factors) aims at the development of mathematical models and algorithms that can enhance the current tools and standards for the diagnosis of T2DM, Impaired Glucose Tolerance and Impaired Fasting Glucose, that can improve the characterization of patients suffering those metabolic disorders and that can help evaluating the risk of developing T2DM and their related complications (Dagliati et al., 2014). These models will be integrated into a platform for diabetes management and remote monitoring, in order to facilitate the interpretation and visualization of the data and to enable a comprehensive understanding of the information by the health care professionals.

Increased emphasis should be given towards the development of DSSs in order to improve interactions between patients and health professionals within the context of co-decision making. Furthermore, the creation of ICT based ecosystems for DM management, involving multiple stakeholders such as patients, families, diabetologists, general practitioners (GPs), case managers, who undertake activities related to the coordination of services (assessment, planning, facilitation, evaluation, monitoring the patient's progress and promoting cost-effective care) on behalf of an individual patient, and health care policy makers is particularly challenging.

Although many attempts have been reported focusing on the development and evaluation of AP, these systems still suffer from limitations in terms of reliability, usability and safety. Although inpatient and short-term outpatient clinical studies for testing closed loop glucose controllers have shown promising results, the ultimate goal is the chronic use in the outpatient setting. In particular, inpatient clinical studies using closed loop systems have lasted less than one week while outpatient testing has lasted even less. In order to justify the reliability and safety of these systems, much longer-term outpatient studies are needed, which will enable assessment of metabolic control and quality of life. Furthermore, although sensor technology for continuous glucose recording has advanced significantly in recent years, there are still occasional inaccuracies in glucose readings. This problem is usually addressed through the usage of more than one sensor, which may enhance the glucose readings' accuracy. Another important limitation refers to the technical usability of the AP systems. Currently, multiple handheld devices are being used for receiving sensor signals, calculating hormone delivery rates, and controlling hormone delivery pumps. The complexity of these systems causes sensor signal dropouts and failure of pumps to receive commands. These overall system failures require professional intervention during the clinical trials, which is not acceptable especially in the outpatient settings. Hence, further integration of the devices and optimization of the communication infrastructure is needed. Furthermore, the administration of insulin in the subcutaneous space, causes significant delays in insulin action which makes the problem of estimating insulin infusion rates very difficult. On going research focuses on the development of more effective insulin analogs. Bi-hormonal closed loop systems seem to be very promising in achieving optimal glycemic control. However, the instability of glucagon remains a major concern, particularly for outpatient trials. More stable glucagon preparations are needed in order for the glucagon to remain in a wearable pump for at least 3-7 days.

CONCLUSION

In the present chapter, the use of PHSs for diabetes management and prevention of complications is described. PHSs have the potential to provide cost-effective, safe and reliable solutions for the management of the disease in terms of treatment, prevention, and enhancement of patients' quality of life. In particular, MDSSs for glucose prediction, closed loop glucose control, risk assessment of the incidence of complications, and early diagnosis along with integrated systems which incorporate these MDSSs are the most popular PHSs for diabetes management. Many research studies and funded research projects have been devoted to the design, the development and the evaluation of such PHSs. The outcomes of these studies/projects are very promising and justify the feasibility of improving patients' quality of life by applying cost-effective personalized ICT-based services and tools. Early diagnosis, prevention and proper management of acute episodes, patient risk stratification, lifestyle behavioral changes, adherence to treatment, promotion of socializing and the implementation of artificial pancreas, are currently the main goals of the PHSs.

Although remarkable advances have been made in the field in recent years, there are still several limitations that should be taken into consideration in order to put these systems in clinical practice. The most important issue refers to the lack of long-term and reliable validation. Moreover, aspects related to data protection and security, are not adequately addressed. Substantial economic studies need to be carried out in order to provide evidence regarding the cost-effectiveness of the proposed systems.

REFERENCES

Abu-Rmileh, A., & Garcia-Garbin, W. (2010). A gain-scheduling model predictive controller for blood glucose control in type 1 diabetes. *IEEE Transactions on Bio-Medical Engineering*, *57*(10), 2478–2484. doi:10.1109/TBME.2009.2033663 PMID:19846371

Arleth T., Andreassen S., Orsini-Federici M., Timi A., & Massi Benedetti M. (2000). A model of glucose absorption from mixed meal. Model. *Control Biomed. Syst.*, 307-312.

Baghdadi, G., & Nasrabadi, A. M. (2007). Controlling blood glucose levels in diabetics by neural network predictor. In *Proceedings of 29th Annual International Conference of the IEEE Engineering in Medicine and Biology Society* (3216-3219). 10.1109/IEMBS.2007.4353014

Bakhtiani, P. A., Zhao, L. M., Youssef, J. E., Castle, J. R., & Ward, W. K. (2013). A Review of Artificial Pancreas Technologies with an Emphasis on Bi-hormonal Therapy. *Diabetes, Obesity & Metabolism*, *15*(12), 1065–1070. doi:10.1111/dom.12107 PMID:23602044

Barakat, N. H., Bradley, A. P., & Barakat, M. B. (2010). Intelligible sup- port vector machines for diagnosis of diabetes mellitus. *Trans. Inf. Technol. Biomed.*, *14*(4), 1114–1120. doi:10.1109/TITB.2009.2039485 PMID:20071261

Bellazzi, R. (2008). Telemedicine and Diabetes Management: Current Challenges and Future Research Directions. *Journal of Diabetes Science and Technology*, *2*(1), 98–104. doi:10.1177/193229680800200114 PMID:19885183

Bergman, R. N., Phillips, L. S., & Cobelli, C. (1981). Physiologic evaluation of factors controlling glucose tolerance in man: Measurement of insulin sensitivity and b-cell glucose sensitivity from the response to intravenous glucose. *The Journal of Clinical Investigation*, *68*(6), 1456–1467. doi:10.1172/JCI110398 PMID:7033284

Carter, R. E., Lackland, D. T., Cleary, P. A., Yim, E., Lopes-Virella, M. F., Gilbert, G. E., & Orchard, T. J. (2007). Intensive treatment of diabetes is associated with a reduced rate of peripheral arterial calcification in the diabetes control and complications trial. *Diabetes Care*, *30*(10), 2646–2648. doi:10.2337/dc07-0517 PMID:17623823

Castle, J. R., Engle, J. M., Youssef, J., Massoud, R. G., Yuen, K. C., Kagan, R., & Ward, W. K. (2010). Novel use of glucagon in a closed-loop system for prevention of hypoglycemia in type 1 diabetes. *Diabetes Care*, *33*(6), 1282–1287. doi:10.2337/dc09-2254 PMID:20332355

Caumo, A., & Cobelli, C. (1993). Hepatic glucose production during the labeled IVGTT: Estimation by deconvolution with a new minimal model. *The American Journal of Physiology*, *264*, 829–841. PMID:8498505

Chase, J. G., Hann, C. E., Shaw, G. M., Wong, J., Lin, J., Lotz, T., ... Lonergan, T. (2007). Overview of glycemic control in critical care: Relating performance and clinical results. *Journal of Diabetes Science and Technology*, *1*(1), 82–91. doi:10.1177/193229680700100113 PMID:19888384

Chassin, L. J., Wilinska, M. E., & Hovorka, R. (2007). Intense exercise in type 1 diabetes: Exploring the role of continuous glucose monitoring. *Journal of Diabetes Science and Technology*, *1*(4), 164–167. doi:10.1177/193229680700100415 PMID:19885120

Clarke, P. M., Gray, A. M., Briggs, A., Farmer, A. J., Fenn, P., Stevens, R. J., ... Holman, R. R. (2004). A model to estimate the lifetime health outcomes of patients with type 2 diabetes: The United Kingdom Prospective Diabetes Study (UKPDS) outcomes model. *Diabetologia*, *47*(10), 1747–1759. doi:10.100700125-004-1527-z PMID:15517152

Cleary, P. A., Orchard, T. J., Genuth, S., Wong, N. D., Detrano, R., Backlund, J. Y., ... Nathan, D. M. (2006). The effect of intensive glycemic treatment on coronary artery calcification in type 1 diabetic participants of the Diabetes Control and Complications Trial/Epidemiology of Diabetes Interventions and Complications (DCCT/EDIC) Study. *Diabetes*, *55*(12), 3556–3565. doi:10.2337/db06-0653 PMID:17130504

Cobelli, C., Caumo, A., & Omenetto, M. (1999). Minimal model S-G overestimation and S-I underestimation: Improved accuracy by a Bayesian two-compartment model. *The American Journal of Physiology*, *277*, 481–488. PMID:10484360

Cobelli, C., Dalla Man, C., Sparacino, G., Magni, L., De Nicolao, G., & Kovathev, B. (2009). Diabetes: Models, signals, and control. *IEEE Rev. Biomed. Eng.*, *2*, 54–96. doi:10.1109/RBME.2009.2036073 PMID:20936056

Dagliati, A., Sacchi, L., Bucalo, M., Segagni, D., Zarkogianni, K., Martinez Millana, A., ... Bellazzi, R. (2014). A data gathering framework to collect Type 2 diabetes patients data. In *Proceedings of IEEE-EMBS International Conference on Biomedical and Health Informatics* (pp. 244-247). 10.1109/BHI.2014.6864349

Dalla Man, C., Camilleri, M., & Cobelli, C. (2006). A system model of oral glucose absorption: Validation on gold standard data. *IEEE Transactions on Bio-Medical Engineering*, *53*(2), 2472–2478. PMID:17153204

Dalla Man, C., Raimondo, D. M., Rizza, R. A., & Cobelli, C. (2007). GIM, simulation software of meal glucose-insulin model. *Journal of Diabetes Science and Technology*, *1*(3), 323–330. doi:10.1177/193229680700100303 PMID:19885087

Derouich, M., & Boutayeb, A. (2002). The effect of physical exercise on the dynamics of glucose and insulin. *Journal of Biomechanics*, *35*(7), 911–917. doi:10.1016/S0021-9290(02)00055-6 PMID:12052393

El-Khatib, F. H., Russell, S. J., Nathan, D. M., Sutherlin, R. G., & Damiano, E. R. (2010). Bi-hormonal Closed-Loop Blood Glucose Control for Type 1 Diabetes. *Science Translational Medicine*, *2*(27), 27ra27. doi:10.1126citranslmed.3000619 PMID:20393188

Ferrannini, E., Smith, J. D., Cobelli, C., Toffolo, G., Pilo, A., & De Fronzo, R. A. (1985). Effect of Insulin on the Distribution and Disposition of Glucose in Man. *The Journal of Clinical Investigation*, *76*(1), 357–364. doi:10.1172/JCI111969 PMID:3894421

Finan, D. A., Zisser, H., Jovanovic, L., Bevier, W. C., & Seborg, D. E. (2006). Identification of linear dynamic models for type 1 diabetes: a simulation study. In *Proceedings of 14th IFAC ADCHEM Symposium* (pp. 503-508).

Florian, J. A., & Parker, R. S. (2005). Empirical modeling for glucose control in diabetes and critical care. *Eur. J. Control*, *11*, 344–350. doi:10.3166/ejc.11.601-618

Freeland, A. C., & Bonnecaze, R. T. (1999). Inference of blood glucose concentrations from subcutaneous glucose concentrations: Applications to glucose biosensors. *Annals of Biomedical Engineering*, *27*(4), 525–537. doi:10.1114/1.196 PMID:10468237

Georga, E., Protopappas, V., Guillen, A., Fico, G., Ardigo, D., Arredondo, M. T., ... Fotiadis, D. I. (2009). Data mining for blood glucose prediction and knowledge discovery in diabetic patients: the METABO diabetes modeling and management system. In *Proceedings of 31st Annu. Int. Conf. IEEE Engineering in Medicine and Biology Society* (pp. 5633-5636). 10.1109/IEMBS.2009.5333635

Georga, E., Protopappas, V. C., Ardigo, D., Marina, M., Zavaroni, I., Polyzos, D., & Fotiadis, D. I. (2013). Multivariate prediction of subcutaneous glucose concentration in Type 1 diabetes patients based on support vector regression. *IEEE Journal of Biomedical and Health Informatics*, *17*(1), 71–81. doi:10.1109/TITB.2012.2219876 PMID:23008265

Giorgino, F., Leonardini, A., & Laviola, L. (2013). Cardiovascular disease and glycemic control in type 2 diabetes: Now that the dust is settling from large clinical trials. *Annals of the New York Academy of Sciences*, *1281*(1), 36–50. doi:10.1111/nyas.12044 PMID:23387439

Gomez, E. J., Perez, M., Vering, T., Rigla Cros, M., Bott, O., Garcia-Saez, G., ... de Leiva, A. (2008). The INCA System: A Further Step Towards a Telemedical Artificial Pancreas. *IEEE Transactions on Information Technology in Biomedicine*, *12*(4), 470–479. doi:10.1109/TITB.2007.902162 PMID:18632327

Heikes, K. E., Eddy, D. M., Arondekar, B., & Schlessinger, L. (2008). Diabetes risk calculator: A simple tool for detecting undiag- nosed diabetes and pre-diabetes. *Diabetes Care*, *31*(5), 1040–1045. doi:10.2337/dc07-1150 PMID:18070993

Heinemann, L., Benesch, C., & DeVries, J. H. (2011). AP@home: A Novel European Approach to Bring the Artificial Pancreas Home. *Journal of Diabetes Science and Technology*, *5*(6), 1363–1372. doi:10.1177/193229681100500607 PMID:22226253

Herrero, P., Georgiou, P., Oliver, N., Johnston, D. G., & Toumazou, C. (2012). A Bio-Inspired Glucose Controller Based on Pancreatic β-Cell Physiology. *Journal of Diabetes Science and Technology*, *6*(3), 606–616. doi:10.1177/193229681200600316 PMID:22768892

Hippisley-Cox, J., Coupland, C., Vinogradova, Y., Robson, J., Minhas, R., Sheikh, A., & Bindle, P. (2008). Predicting cardiovascular risk in England and Wales: Prospective derivation and validation of QRISK2. *BMJ (Clinical Research Ed.)*, *336*(7659), 1475–1482. doi:10.1136/bmj.39609.449676.25 PMID:18573856

Hovorka, R., Allen, J. M., Elleri, D., Chassin, L. J., Harris, J., Xing, D., ... Dunger, D. B. (2010). Manual closed-loop insulin delivery in children and adolescents with type 1 diabetes: A phase 2 randomised crossover trial. *Lancet*, *375*(9716), 743–751. doi:10.1016/S0140-6736(09)61998-X PMID:20138357

Hovorka, R., Canonico, V., Chassin, L. J., Haueter, U., Massi-Benedetti, M., Federici, M. O., ... Wilinska, M. E. (2004). Nonlinear model predictive control of glucose concentration in subjects with type 1 diabetes. *Physiological Measurement*, *25*(4), 905–920. doi:10.1088/0967-3334/25/4/010 PMID:15382830

Hovorka, R., Chassin, L. J., Wilinska, M. E., Canonico, V., Akwi, J. A., Federici, O., ... Pieber, T. R. (2004). Closing the loop: The adicol experience. *Diabetes Technology & Therapeutics*, *8*(3), 307–318. doi:10.1089/152091504774197990 PMID:15198833

Hovorka, R., Kumareswaran, K., Harris, J., Allen, M. J., Elleri, D., Xing, D., ... Evans, M. L. (2011). Overnight closed loop insulin delivery (artificial pancreas) in adults with type 1 diabetes: Cross over randomized controlled studies. *BioMed J.*, *342*, 1855.

Hovorka, R., Shojaee-Moradie, F., Carroll, P. V., Chassin, L. J., Gowrie, I. J., Jackson, N. C., ... Jones, R. H. (2002). Partitioning glucose distribution/transport, disposal, and endogenous production during IVGTT. *American Journal of Physiology. Endocrinology and Metabolism*, *282*(5), 992–1007. doi:10.1152/ajpendo.00304.2001 PMID:11934663

Hyunjin, L., & Bequette, B. W. (2009). A closed-loop artificial pancreas based on model predictive control: Human-friendly identification and automatic meal disturbance rejection. *Biomedical Signal Processing and Control*, *4*(4), 347–354. doi:10.1016/j.bspc.2009.03.002

International Diabetes Federation. (2013). *IDF Diabetes Atlas (6th Edition)*. Retrieved December 1, 2014 from http://www.idf.org/diabetesatlas/download-book

Kamruzzaman, S. M., Hasan, A. R., Siddiquee, A. B., & Mazumder, E. H. (2004). Medical diagnosis using neural network. In *Proceedings of 3rd International Conference on Electrical and Computer Engineering* (pp. 537-540).

Kertes, P. J., & Johnson, T. M. (2007). *Evidence Based Eye Care*. Philadelphia, PA: Lippincott Williams & Wilkins.

Klonoff, D. C. (2005). Continuous glucose monitoring: Roadmap for 21st century diabetes therapy. *Diabetes Care*, *28*(5), 1231–1239. doi:10.2337/diacare.28.5.1231 PMID:15855600

Kong, F., & Tan, J. (2012). DietCam: Automatic dietary assessment with mobile camera phones. *Pervasive and Mobile Computing*, *8*(1), 147–163. doi:10.1016/j.pmcj.2011.07.003

Kothari, V., Stevens, R. J., Adler, A. I., Stratton, I. M., Manley, S. E., Neil, H. A., & Holman, R. R. (2002). UKPDS 60: Risk of stroke in type 2 diabetes estimated by the UK Prospective Diabetes Study risk engine. *Stroke*, *33*(7), 1776–1781. doi:10.1161/01.STR.0000020091.07144.C7 PMID:12105351

Kovatchev, B. P., Breton, M., Dalla Man, C., & Cobelli, C. (2009). In Silico Preclinical Trials A Proof of Concept in Closed-Loop Control of Type 1 Diabetes. *Journal of Diabetes Science and Technology*, *3*(1), 44–55. doi:10.1177/193229680900300106 PMID:19444330

Kovatchev, B. P., Gonder-Frederick, L. A., Cox, D. J., & Clarke, W. L. (2004). Evaluating the accuracy of continuous glucose-monitoring sensors: Continuous glucose error grid analysis illustrated by TheraSense Freestyle Navigator data. *Diabetes Care*, *27*(8), 1922–1928. doi:10.2337/diacare.27.8.1922 PMID:15277418

Kumareswaran, K., Evans, M. L., & Hovorka, R. (2012). Closed-loop Insulin Delivery: Towards Improved Diabetes Care. *Discovery Medicine*, *13*(69), 159–170. PMID:22369975

Lagani, V., Koumakis, L., Chiarugi, F., Lakasing, E., & Tsamardinos, I. (2013). A systematic review of predictive risk models for diabetes complications based on large scale clinical studies. *Journal of Diabetes and Its Complications*, *27*(4), 407–413. doi:10.1016/j.jdiacomp.2012.11.003 PMID:23273850

Magni, L., Raimondo, D. M., Bossi, L., Dalla Man, C., De Nicolao, G., Kovatchev, B., & Cobelli, C. (2007). Model predictive control of type 1 diabetes: An in silico trial. *Journal of Diabetes Science and Technology*, *1*(6), 804–812. doi:10.1177/193229680700100603 PMID:19885152

Magni, L., Raimondo, D. M., Dalla Man, C., De Nicolao, G., Kovatchev, B., & Cobelli, C. (2009). Model predictive control of glucose concentration in type 1 diabetic patient: An in silico trial. *Biomedical Signal Processing and Control*, *4*(4), 338–346. doi:10.1016/j.bspc.2009.04.003

Marchetti, G., Barolo, M., Jovanovic, L., Zisser, H., & Seborg, D. E. (2008). An improved PID switching control strategy for type 1 diabetes. *IEEE Transactions on Bio-Medical Engineering*, *55*(3), 857–865. doi:10.1109/TBME.2008.915665 PMID:18334377

Mitsis, G. D., Markakis, M. G., & Marmarelis, V. Z. (2009). Nonlinear modeling of the dynamic effects of infused insulin on glucose: Comparison of compartmental with Volterra models. *IEEE Transactions on Bio-Medical Engineering*, *56*(10), 2347–2358. doi:10.1109/TBME.2009.2024209 PMID:19497805

Mougiakakou, S. G., Bartsocas, C., Bozas, E., Chaniotakis, N., Iliopoulou, D., Kouris, I., ... Nikita, K. S. (2010). SMARTDIAB: A Communication and Information Technology Approach for the Intelligent Monitoring, Management and Follow-up of Type 1 Diabetes Patients. *IEEE Transactions on Information Technology in Biomedicine*, *14*(3), 622–633. doi:10.1109/TITB.2009.2039711 PMID:20123578

Mougiakakou, S. G., Prountzou, A., Zarkogianni, K., Bartsocas, C., Nikita, K. S., & Gerasimidi-Vazeou, A. (2008). Prediction of glucose profile in children with type 1 diabetes mellitus using continuous glucose monitors and insulin pumps. *Hormone Research*, *70*, 22–23. PMID:18493146

Nucci, G., & Cobelli, C. (2000). Models of subcutaneous insulin kinetics. A critical review. *Computer Methods and Programs in Biomedicine*, *62*(3), 249–257. doi:10.1016/S0169-2607(00)00071-7 PMID:10837910

Pappada, S. M., Cameron, B. D., & Rosman, P. M. (2008). Development of a neural network for prediction of glucose concentration in Type 1 diabetes patients. *Journal of Diabetes Science and Technology*, *2*(5), 792–801. doi:10.1177/193229680800200507 PMID:19885262

Perez-Gandia, C., Facchinetti, A., Sparacino, G., Cobelli, C., Gómez, E. J., Rigla, M., ... Hernando, M. E. (2010). Artificial neural network algorithm for online glucose prediction from continuous glucose monitoring. *Diabetes Technology & Therapeutics*, *12*(1), 81–88. doi:10.1089/dia.2009.0076 PMID:20082589

Polat, K., Gunes, S., & Arslan, A. (2008). A cascade learning system for classification of diabetes disease: Generalized discriminant analysis and least square support vector machine. *Expert Systems with Applications*, *34*(1), 482–487. doi:10.1016/j.eswa.2006.09.012

Porta, M., Sjoelie, A. K., Chaturvedi, N., Stevens, L., Rottiers, R., Veglio, M., & Fuller, J. H.EURODIAB Prospective Complications Study Group. (2001). Risk factors for progression to proliferative diabetic retinopathy in the EURODIAB Prospective Complications Study. *Journal of Diabetologia*, *44*(12), 2203–2209. doi:10.1007001250100030 PMID:11793022

Russell, S. J., El-Khatib, F. H., Nathan, D. M., Magyar, K. L., Jiang, J., & Damiano, E. R. (2012). Blood glucose control in Type 1 Diabetes with a bihormonal bionic endocrine pancreas. *Diabetes Care, 35*(11), 2148–2155. doi:10.2337/dc12-0071 PMID:22923666

Schaller, H. C., Schaupp, L., Bodenlenz, M., Wilinska, M. E., Chassin, L. J., Wach, P., ... Pieber, T. R. (2006). On-line adaptive algorithm with glucose prediction capacity for subcutaneous closed loop control of glucose: Evaluation under fasting conditions in patients with type 1 diabetes. *Diabetic Medicine, 23*(1), 90–93. doi:10.1111/j.1464-5491.2006.01695.x PMID:16409572

Schmidtke, D. W., Freeland, A. C., Heller, A., & Bonnecaze, R. T. (1998). Measurement and modeling of the transient difference between blood and subcutaneous glucose concentrations in the rat after injection of insulin. *Proceedings of the National Academy of Sciences of the United States of America, 95*(1), 294–299. doi:10.1073/pnas.95.1.294 PMID:9419369

Schwarz, P. E., Li, J., Lindstorm, J., & Tuomilehto, J. (2009). Tools for predicting the risk of type 2 diabetes in daily practice. *Hormone and Metabolic Research. Hormon- und Stoffwechselforschung. Hormones et Metabolisme, 41*(2), 86–97. doi:10.1055-0028-1087203 PMID:19021089

Shankaracharya, D. O., Odedra, D., Samanta, S., & Vidyarthi, A. S. (2010). Computational intelligence in early diabetes diagnosis: A review. *The Review of Diabetic Studies; RDS, 7*(4), 252–262. doi:10.1900/RDS.2010.7.252 PMID:21713313

Shanker, M. S. (1996). Using neural networks to predict the onset of diabetes mellitus. *Journal of Chemical Information and Computer Sciences, 36*(1), 35–41. doi:10.1021/ci950063e PMID:8576289

Skevofilakas, M., Zarkogianni, K., Karamanos, B. G., & Nikita, K. S. (2010). A hybrid decision support system for the risk assessment of retinopathy development as a long term complication of type 1 diabetes mellitus. In *Proceedings of Annual International Conference of the IEEE Engineering in Medicine and Biology Society* (vol.1, pp. 6713-6716). 10.1109/IEMBS.2010.5626245

Spanakis, E., Chiarugi, F., Kouroubali, A., Spat, S., Beck, P., Asanin, S., ... Thestrup, J. (2012). Diabetes management using modern information and communication technologies and new care models. *Interact. J. Med. Res., 1*(2), 1–12. PMID:23612026

Steil, G. M., Panteleon, A. E., & Rebrin, K. (2004). Closed-loop insulin delivery- the path to physiological glucose control. *Advanced Drug Delivery Reviews, 56*(2), 125–144. doi:10.1016/j.addr.2003.08.011 PMID:14741112

Stevens, R. J., Coleman, R. L., Adler, A. I., Stratton, I. M., Matthews, D. R., & Holman, R. R. (2004). Risk factors for myocardial infarction case fatality and stroke case fatality in type 2 diabetes. *Diabetes Care, 27*(1), 201–207. doi:10.2337/diacare.27.1.201 PMID:14693990

Stevens, R. J., Kothari, V., Adler, A. I., & Stratton, I. M. (2001). The UKPDS risk engine: A model for the risk of coronary heart disease in type II diabetes. *Clinical Science, 101*(6), 671–679. doi:10.1042/CS20000335 PMID:11724655

The CDC Diabetes Cost-effectiveness Group. (2002). Cost-effectiveness of intensive glycemic control, intensified hypertension control, and serum cholesterol level reduction for type 2 diabetes. *Journal of the American Medical Association, 287*(19), 2542–2551. doi:10.1001/jama.287.19.2542 PMID:12020335

The Diabetes Control and Complications Trial Research Group. (2003). The effect of intensive treatment of diabetes on the development and progression of long-term complications in insulin-dependent diabetes mellitus. *The New England Journal of Medicine*, *329*(14), 977–986. PMID:8366922

The Diabetes Control and Complications Trial/Epidemiology of Diabetes Intervention and Complications (DCCT/EDIC) Study Research Group. (2005). Intensive diabetes treatment and cardiovascular disease in patients with type 1 diabetes. *The New England Journal of Medicine*, *53*, 2643–2653. PMID:16371630

The Epidemiology of Diabetes Interventions and Complications (EDIC) Research Group. (1999). Design, implementation, and preliminary results of a long-term follow-up of the Diabetes Control and Complications Trial cohort. *Diabetes Care*, *22*(1), 99–111. doi:10.2337/diacare.22.1.99 PMID:10333910

Ubeyli, E. D. (2009). Modified mixture of experts for diabetes diagnosis. *Journal of Medical Systems*, *33*(4), 299–305. doi:10.100710916-008-9191-3 PMID:19697696

Ubeyli, E. D. (2010). Automatic diagnosis of diabetes using adaptive neuro-fuzzy inference systems. *Expert Systems: International Journal of Knowledge Engineering and Neural Networks*, *27*(4), 259–266. doi:10.1111/j.1468-0394.2010.00527.x

Valleta, J., Chipperfield, A., & Byrne, C. (2009). Gaussian process modelling of blood glucose response to free-living physical activity data in people with type 1 diabetes. In *Proceedings of 1st Annual International Conference of the IEEE Engineering in Medicine and Biology Society* (pp. 4913-4916). 10.1109/IEMBS.2009.5332466

Vergouwe, Y., Soedamah-Muthu, S. S., Zgibor, J., Chaturvedi, N., Forsblom, C., & Snell-Bereon, J. K. (2010). Progression to microalbuminuria in type 1 diabetes: Development and validation of a prediction rule. *Diabetologia*, *53*(2), 254–262. doi:10.100700125-009-1585-3 PMID:19908023

Vicini, P., Caumo, A., & Cobelli, C. (1997). The hot IVGTT two-compartment minimal model: Indexes of glucose effectiveness and insulin sensitivity. *The American Journal of Physiology*, *273*, 1024–1032. PMID:9374690

Wang, Y., Dassau, E., & Doyle, F. J. (2010). Closed-loop control of artificial pancreatic β-cell in type 1 diabetes mellitus using model predictive iterative learning control. *IEEE Transactions on Bio-Medical Engineering*, *57*(2), 211–219. doi:10.1109/TBME.2009.2024409 PMID:19527957

Wilinska, M. E., Bodenlenz, M., Chassin, L. J., Schaller, H. C., Schaupp, L. A., Pieber, T. R., & Hovorka, R. (2004). Interstitial glucose kinetics in subjects with type 1 diabetes under physiologic conditions. *Metabolism: Clinical and Experimental*, *53*(11), 1484–1491. doi:10.1016/j.metabol.2004.05.014 PMID:15536606

Zainuddin, Z., Pauline, O., & Ardil, C. (2009). A neural network approach in predicting the blood glucose level for diabetic patients. *International Journal of Computational Intelligence*, *5*(1), 72–79.

Zarkogianni, K., Litsa, E., Vazeou, A., & Nikita, K. S. (2013). Personalized glucose-insulin metabolism model based on self-organizing maps for patients with type 1 diabetes mellitus. In *Proceedings of 13th IEEE International Conference on BioInformatics and BioEngineering* (pp. 1-4). 10.1109/BIBE.2013.6701604

Zarkogianni, K., Mitsis, K., & Arredondo, M.-T., Fico G., Fioravanti A., & Nikita K. S. (2014). Neuro-Fuzzy based Glucose Prediction Model for Patients with Type 1 Diabetes Mellitus. In *Proceedings of IEEE-EMBS International Conferences on Biomedical and Health Informatics* (pp. 1-4). 10.1109/BHI.2014.6864351

Zarkogianni, K., Vazeou, A., Mougiakakou, S. G., Prountzou, A., & Nikita, K. S. (2011). An insulin infusion advisory system based on autotuning nonlinear model-predictive control. *IEEE Transactions on Bio-Medical Engineering*, *58*(9), 2467–2477. doi:10.1109/TBME.2011.2157823 PMID:21622071

KEY TERMS AND DEFINITIONS

Artificial Pancreas: Closed loop controller for the regulation of blood glucose levels in diabetic patients, through the integration of the following modules: i) insulin infusion pump, ii) algorithm for the estimation of insulin delivery rates/bolus, iii) Continous Glucose Measurement System.

Compartmental Model: Type of mathematical model with particular application in describing the transmission of materials and energies among the compartments.

Data-Driven Modeling: Method used for either static or dynamic modeling, which is based on learning from the input and output data while disregarding explicit knowledge of the system's physical behavior.

Diabetes: Chronic metabolic disease characterized by the deregulation of glucose metabolism

Machine Learning: Scientific discipline which is oriented to the investigation and construction of algorithms that can learn from the data.

Medical Decision Support Systems: Applications that provide to the users support/assistance within the context of decision making.

Patient Empowerment: Approach that puts the patient at the center of the services and provides him/her with the appropriate tools and services towards the self-management of his/her disease.

Personal Smart Assistant: Smart mobile software agent, with the ability to perform tasks, or services, on behalf of an individual based on a combination of user input, location awareness, and the ability to communicate remotely with other devices.

Telemedicine: The combined use of telecommunication and information technologies for providing remote clinical health care.

This research was previously published in the Handbook of Research on Trends in the Diagnosis and Treatment of Chronic Conditions edited by Dimitrios I. Fotiadis, pages 465-492, copyright year 2016 by Medical Information Science Reference (an imprint of IGI Global).

Section 3
Privacy and Security

Chapter 13
Secure Storage and Transmission of Healthcare Records

Grasha Jacob
Rani Anna Govt. College, Tirunelveli

Murugan Annamalai
Dr. Ambedkar Govt. Arts College, India

ABSTRACT

Telemedicine has become a common method for transmission of medical images and patient data across long distances. With the growth of computer networks and the latest advances in digital technologies, large amount of digital data gets exchanged over various types of insecure networks - wired or wireless. Modern Healthcare Management Systems need to change to accommodate these new advances. There is an urgent need to protect the confidentiality of health care records that are stored in common databases and transmitted over public insecure channels. This chapter outlines DNA sequence based cryptography which is easy to implement and is robust against cryptanalytic attack as there is insignificant correlation between the original record and the encrypted image for the secure storage and transmission of health records.

INTRODUCTION

India is providing quality health care of international standards at a relatively low cost and has attracted the patients from across the globe. India is now one of the favorite destinations for the health care services. With the advances in technology that is witnessed each passing day, there is not a dimension of development that can sustain unless technology is embraced by it. Information Technology receives a benevolent face when it delivers value addition to medical field. It ranges from processing of patient data to computer aided drug discovery. One of the greatest challenges facing mankind in the 21st century is to make high-quality health care available to all. Such a vision has been expressed by the World Health Organization (WHO) in its health-for-all strategy in the 21st century. A Health Telematics Policy, a document from World Health Organization states telemedicine motivation as - "…integrate the appro-

DOI: 10.4018/978-1-5225-6198-9.ch013

priate use of health telematics in the overall policy and strategy for the attainment of health for all in the 21st century, thus fulfilling the vision of the world in which benefits of science, technology and public health development are made equitable available to all people everywhere". Telemedicine is defined as, "the delivery of health care and the exchange of health information across distances including all medical activities: making diagnosis, treatment, prevention, education and research" (Wootton & Craig, 1999). Telemedicine is connecting remote locations and helps in addressing the inadequacies associated with health care. Telemedicine can improve equity of access to health care in the rural and urban areas. Widespread adoption of telemedicine would permit decentralization and could potentially have the greatest effect, allowing underserved people to benefit from a greatly improved standard of health care. In remote or rural areas, telemedicine could have a great impact, permitting better diagnostic and therapeutic services, faster and easier access to medical knowledge, obviating the need for patients and health-care workers to travel. Even in urban areas, however, telemedicine can improve access to health services and to information. A health record may contain patient information along with a scanned image. According to Norcen et al (2003), "The organization of today's health systems often suffers from the fact that different doctors do not have access to each other's patient data. The enormous waste of resources for multiple examinations, analyses, and medical check-ups is an immediate consequence. In particular, multiple acquisitions of almost identical medical image data and loss of former data of this type have to be avoided to save resources and to provide a time-contiguous medical report for each patient. A solution to these problems is to create a distributed database infrastructure where each doctor has electronic access to all existing medical data related to a patient, in particular to all medical image data acquired over the years. When such an infrastructure is established, it is possible to give precise, advanced, and entire information about patients at the time of care, permit speedy access to patient records for more synchronized and well organized care and distribute digital information to patients and other clinicians Additionally, many medical professionals are convinced that the future of health care will be shaped by tele-radiology, tele-cardiology and technologies such as telemedicine in general. Telemedicine is of most use to the remote and disconnected areas which do not have access to best of medical care, both infrastructural and human resource. With advancement in mobile technologies, mobile apps can be developed which can work in synchronization with the ERP system at the hospital. Thus, Telemedicine reaches wider audience through the outreach of smartphones. Remote prescription, drug administration, oversight etc. can be managed remotely and thus reducing the travel, nursing and hospital admitting costs. In the fields like psychiatry, telemedicine is the most effective way, as videoconferencing is required. Post-surgical monitoring can also be done remotely as patient may be allowed to recover in congenial homely environment. Shortage of physicians and paramedical persons can be addressed with initial investment in telemedicine technology. Though there are several advantages of Tele-medicine, several threats like accidental disclosure, insider curiosity, insider subordination, uncontrolled secondary usage and outsider intrusion exist to patient information confidentiality. Accidental disclosure occurs when medical personnel make innocent mistakes during multiple electronic transfers of data to various entities, medical personnel and cause disclosure of data. Insider curiosity is the result of misusing their access rights to patient information out of curiosity. Insider subordination results when medical personnel leak out personal medical information for spite, profit, revenge, or other purposes. Uncontrolled secondary usage occurs when those who are granted access to patient information solely for the purpose of supporting primary care can exploit that permission for reasons not listed in the contract, such as research. Outsider intrusion occurs when former employees, network intruders, hackers, or others may access information, damage systems or disrupt operations. All electronic health information must be encrypted and decrypted as necessary according to user defined preferences in accordance with the best available encryption key strength. During data exchange all electronic health information must be suitably encrypted and decrypted when exchanged in accordance with an encrypted and integrity protected link. All actions related to electronic health information must be recorded with the date, time, patient identification, and user identification whenever any electronic health information is created, modified, deleted, or printed; and an indication of which action(s) took place must also be recorded. As health care records contain health information which is protected under legislation, image encryption schemes for health care records have been increasingly studied to meet the demand for real-time secure storage and transmission over the Internet.

BACKGROUND

In the past, the medical record was a warehouse of information in papers that was reviewed or used for clinical, research, administrative, and financial purposes. It was strictly restricted in terms of accessibility, available to only one user at a time. The paper-based record sometimes had bad handwritten information written by physicians which could not be read or understood and was updated manually, resulting in delays for record completion that persisted from 1 to 6 months or more. Most medical record departments were housed in institutions' basements because the weight of the paper precluded other locations. The physician was in control of the care and documentation processes and authorized the release of information. Patients rarely viewed their medical records. Another limitation of the paper-based medical record was the lack of security. Right to use was controlled by doors, locks, identification cards, and tedious sign-out procedures for authorized users. Unauthorized access to patient information triggered no warnings, nor was it known what information had been observed. It is "the right of individuals to keep information about themselves from being disclosed to others; the claim of individuals to be let alone, from surveillance or interference from other individuals, organizations or the government". The information that is shared as a result of a clinical relationship is considered confidential and must be secure. The information can take several forms (including identification data, diagnoses, treatment and progress notes, and laboratory results) and can be stored in numerous ways. Patient information should be released to others only with the patient's permission or as allowed by law. This is not, however, to say that physicians cannot gain access to patient information. Information can be released for treatment, payment, or administrative purposes without a patient's authorization. The patient, too, has federal, state, and legal rights to view, obtain a copy of, and amend information in his or her health record. The key to preserving confidentiality is making sure that only authorized individuals have the right to use the information. The process of controlling the access - restricting who can see what—begins with authorizing users. In the case of health record access, the practice administrator identifies the users, determines what level of information is needed by whom and for what purpose it is needed, and assigns usernames and passwords. Basic standards for passwords include requiring that they be changed at set intervals, setting a minimum number of characters, and prohibiting the reuse of passwords. The user's access is based on pre-established, role-based privileges. A physician, a nurse and a receptionist have very different tasks and responsibilities; therefore, they do not have access to the same information. Hence, designating user privileges is a critical aspect of medical record security: all users who have access to the information have to realize their roles and responsibilities, and must know that they are accountable for use or misuse of the information they view and change.

SECURE STORAGE OF HEALTH RECORDS

Telemedicine is good for every nation for the betterment of Medicare services and for developed nations, it is even more necessary to setup and generate immediate and expert medical care. Government and healthcare people are converting the paper-based health records into electronic health records in order to reduce the healthcare cost, improving the patient care and decreasing the medical faults. A health care record may include medical history, notes, and other information about health including the symptoms, diagnoses, medications, lab results, vital signs, immunizations, and reports from diagnostic tests such as x-rays and scans. Health records aim to store a patients' health information in one place, providing quick access when required. Telemedicine improves efficiency and coordination of the administration of hospitals. Telemedicine reduces the communication distance between experts and consultants, as the sharing of reports of patient and discussion over the treatment becomes online. But there are lots of research issues in storing and transmitting electronic health records which require an optimal solution. These facts show very clearly that there is an urgent need to provide and protect the confidentiality of patient related medical image data when stored in databases or cloud and transmitted over networks of any kind" (Norcen et. al, 2003). Measures that can be built in to such systems may include: "Access control" tools like passwords and PIN numbers, to help limit access to your information to authorized individuals, "Encrypting" the stored information - health information is unintelligible except by those using a system that can "decrypt" it with a "key" and an "audit trail" feature, which records the details such as who accessed your information, what changes were made and when. When a message is generated by the user and transmitted by the application software, until it is received by an authorised recipient the data is subject to a range of security risks. These risks are inclusive of standard security protections relating to hardware, software, human interventions, natural disasters, network issues and logical problems. Clinical information must be transferred securely and like other types of secure transmission must include identification, authorization, authentication, confidentiality, integrity and nonrepudiation. Protecting the data in transit is subject to the same security threats as required for other sensitive data; data may be subject to loss, late delivery, damage, or attack. All information in a patient's health care record is confidential and subject to the prevailing privacy laws and policies. Like all computerized systems, electronic records are vulnerable to crashes. Back-ups should be taken now and then and should be stored in two or more databases or multi-cloud to prevent loss of information.

Secure Transmission of Health Records

An electronic health record (EHR) is a record of a patient's medical details (including history, physical examination, investigations and treatment) in digital format. Physicians and hospitals are implementing EHRs because they offer several advantages over paper records. They increase access to health care, improve the quality of care and decrease costs. However, ethical issues related to EHRs confront health personnel. Telemedicine can make health care more efficient and less expensive, and improve the quality of care by making patients' medical history easily accessible to all who treat them. But as health care providers adopt electronic records, the challenges have proved daunting, with a potential for mix-ups and confusion that can be frustrating, costly and even dangerous. The medical record, either paper-based or electronic, is a communication tool that supports clinical decision making, coordination of services, evaluation of the quality and efficacy of care, research, legal protection, education, and accreditation and regulatory processes. The way to keep the information in these exchanges secure is a major concern. There is no way to control what information is being transmitted, the level of detail, whether communications are being intercepted by others, what images are being shared, or whether the mobile device is encrypted or secure. Mobile devices are largely designed for individual use and were not intended for centralized management by an information technology (IT) department. Computer workstations are rarely lost, but mobile devices can easily be misplaced, damaged, or stolen. Encrypting mobile devices that are used to transmit confidential information is of the utmost importance. Another potential threat is that data can be hacked, manipulated, or destroyed by internal or external users, so security measures and on-going educational programs must include all users. Some security measures that protect data integrity include firewalls, antivirus software, and intrusion detection software. Regardless of the type of measure used, a full security program must be in place to maintain the integrity of the data, and a system of audit trails must be operational. Providers and organizations must formally designate a security officer to work with a team of health information technology experts who can inventory the system's users, and technologies; identify the security weaknesses and threats; assign a risk or likelihood of security concerns in the organization; and address them. The responsibilities for privacy and security can be assigned to a member of the physician office staff or can be outsourced. With the advent of audit trail programs, organizations can precisely monitor who has had access to patient information. Audit trails track all system activity, generating date and time stamps for entries; detailed listings of what was viewed, for how long, and by whom; and logs of all modifications to electronic health records. Administrators can even detail what reports were printed, the number of screen shots taken, or the exact location and computer used to submit a request. Alerts are often set to flag suspicious or unusual activity, such as reviewing information on a patient one is not treating or attempting to access information one is not authorized to view, and administrators have the ability to pull reports on specific users or user groups to review and chronicle their activity. Software companies are developing programs that automate this process. End users should be mindful that, unlike paper record activity, all EHR activity can be traced based on the login credentials. Audit trails do not prevent unintentional access or disclosure of information but can be used as a deterrent to ward off would-be violators. To prevent intentional and unauthorized access, encryption renders a helping hand.

MAIN FOCUS OF THE CHAPTER

From time immemorial, the art of communicating secretly has been imperative. German's Enigma machine and American Indian Language Navajo used during World War II are concrete examples to explain everything about cryptographic techniques. A cryptosystem is a system that allows two parties to communicate secretly by transforming an intelligible message into unintelligible form and then retransforming that back to its original form. The main objective of cryptography is confidentiality. The usefulness of developing techniques of DNA computing and ultimately developing working DNA computers can be described as falling into one of the following three general categories:

1. Applications making use of "classic" DNA computing schemes where the use of massive parallelism holds an advantage over traditional computing schemes, including potential polynomial time solutions to hard computational problems;
2. Applications making use of the "natural" capabilities of DNA, including those that make use of informational storage abilities and those that interact with existing and emerging biotechnology;

3. Contributions to fundamental research within both computer science and the physical sciences especially concerned with exploring the limitations of computability and understanding and manipulating biomolecular chemistry.

DNA Computing is an exciting fast developing interdisciplinary area concerned with the use of DNA molecules and associated quaternary coding for the implementation of computational processes. Adleman's pioneering experimental work in 1994 opened a new world of possibilities in computing. He established the use of nucleic-acid strand interactions for computing and solved the travelling-salesman problem (Hamiltonian Path Problem - to find the shortest route that visits each city exactly once in a given network of connected cities), in a test tube using specially sequenced DNA molecules and standard molecular-biology procedures in a biological laboratory. Although solving such a specific task is a far cry from building a general-purpose computer, this ground-breaking work revealed that information could indeed be processed using interactions between the strands of DNA. Adleman's work urged other researchers to develop DNA-based logic circuits using a variety of approaches. The resulting circuits were able to perform simple mathematical and logical operations, recognize patterns based on incomplete data and play simple games. Molecular circuits even detect and respond to a disease signature inside a living cell and open up the possibility of medical treatments based on man-made molecular software. Lipton extended the work of Adleman and investigated the solution of Satisfiability of Propositional Formula pointing to new opportunities of DNA computing. Research work is being done on DNA Computing either using test tubes (biologically) or simulating the operations of DNA using computers (Pseudo or Virtual DNA computing). L. Kari (1997) gave an insight of the various biological operations concerning DNA.

Taylor et al. (1997) proposed a substitution cipher for plaintext encoding where base triplet was assigned to each letter of the alphabet, numeral and special characters and demonstrated a steganographic approach by hiding secret messages encoded as DNA strands among multitude of random DNA. Decryption was difficult with the use of sub-cloning, sequencing and there was a need of an additional triplet coding table. Gehani et al.(2000) introduced a trial of DNA based Cryptography and proposed two methods: i) a substitution method using libraries of distinct one time pads, each of which defines a specific, randomly generated, pair-wise mapping and ii) an XOR scheme utilizing molecular computation and indexed random key strings were used for encryption. They used the natural DNA sequences to encode the information and encrypted an image by using the XOR logic operation. Such experiments could be done only in a well-equipped lab using modern technology, and it would involve high cost. Leier et al. (2000) also presented two different cryptographic approaches based on DNA binary strands with the idea that a potential interceptor cannot distinguish between dummies and message strand. The first approach hid information in DNA binary strands and the second designed a molecular checksum. Decryption was done easily using PCR and subsequent gel electrophoresis without the use of sub-cloning, sequencing and additional triplet coding table. Although the approach of generating bit strands shown here had advantages such as rapid readout, it also had practical limitations. One of the limitations was the resolution of the used agarose-gels. Chen (2003) presented a novel DNA-based cryptography technique that took advantage of the massive parallel processing capabilities of biomolecular computation. A library of one time pads in the form of DNA strands was assembled. Then, a modulo-2 addition method was employed for encryption whereby a large number of short message sequences could be encrypted using one time pads. A novel public-key system using DNA was developed by Kazuo et al. (2005) based on the one-way function. The message-encoded DNA hidden in dummies could be restored by PCR amplification, followed by sequencing.The YAEADNA algorithm (Sherif et al., 2006) used a search technique in order to locate and return the position of quadruple DNA nucleotide sequence representing the binary octets of plain text characters. Plain text character and a random binary file were given as input and the output PTR was a pointer to the location of the found quadruple DNA nucleotide sequence representing the binary octet. The encryption process was tested on images to show how random the selection of DNA octet's locations is on the encrypting sequence. Cui et al. (2008) designed an encryption scheme by using the technologies of DNA synthesis, PCR amplification, DNA digital coding and the theory of traditional cryptography. The data was first pre-processed to get completely different ciphertext to prevent attack from a possible word as PCR primers. Then, the DNA digital encoding technique was applied to the ciphertext. After coding sender synthesizes the secret-message DNA sequence which was flanked by forward and reverse PCR primers, each 20-mer oligo nucleotides long. Thus, the secret-message DNA sequence was prepared and at last sender generated a certain number of dummies and put the sequence among them. Once the data in encrypted form reached the receiver's side the reverse procedure was followed to decrypt it. Biological difficult issues and cryptography computing difficulties provided a double security safeguard for the scheme. The intended PCR two primer pairs used as the key of this scheme was designed by the complete cooperation of sender and receiver to increase the security of this encryption scheme. Ning (2009) explained the pseudo encryption methodology based upon the work of Gehani. The plain text was converted to DNA sequences and these sequences were converted to the spliced form of data and protein form of data by cutting the introns according to the specified pattern and it was translated to mRNA form of data and mRNA was converted into protein form of data. The protein form of data was sent through the secure channel. The method did not really use DNA sequences, but only the mechanisms of the DNA function; therefore, the method was a kind of pseudo DNA cryptography methods. The method only simulates the transcription, splicing, and translation process of the central dogma; thus, it was a pseudo DNA cryptography method.

Sadeg et al. (2010) proposed a symmetric key block cipher algorithm which included a step that simulated ideas from the processes of transcription (transfer from DNA to mRNA) and translation (from mRNA into amino acids). Though the encryption algorithm (OTP) proposed was theoretically unbreakable, it experienced some disadvantages in its algorithm. These drawbacks had prevented the common use of its scheme in modern cryptosystems. In 2010, Qinghai had also proposed a method to protect information, including representing information using biological alphabets to enhance the security of traditional encryption, using DNA primer for secure communication and key distribution, and using the chemical information of DNA bases for steganography. Alice and Bob share a secret DNA sequence codebook. Alice can design a sequence that can maximally match one of the sequences in the codebook and then send the designed sequence to Bob through a public channel. When Bob receives the sequence he would use the non-matching letters in the private sequence as the encryption key. Knowing the public string only, an attacker cannot decrypt the transmitted information. L.Xuejia et al. (2010) also proposed an asymmetric encryption and signature cryptosystem by combining the technologies of genetic engineering and cryptology. It was an exploratory research of biological cryptology. DNA-PKC uses two pairs of keys for encryption and signature, respectively. Using the public encryption key, everyone can send encrypted message to a specified user, only the owner of the private decryption key can decrypt the ciphertext and recover the message; in the signature scheme, the owner of the private signing key can generate a signature that can be verified by other users with the public verification key, but no else can forge the signature. DNA-PKC differs from the conventional cryptology in that the keys and the cipher texts are all biological molecules. The security of DNA-PKC relies on difficult biological problems instead of computational problems; thus DNA-PKC is immune from known attacks, especially the quantum computing based attacks. In the image encryption algorithm based on DNA sequence addition operation combined with logistic chaotic map to scramble the location and value of pixel of an image presented by Qiang et al., (2010) a DNA sequence matrix was obtained by encoding the original image and it was divided into some equal blocks and two logistic maps. DNA complementarity and DNA sequence addition operations were utilized to add these blocks. DNA sequence matrix was decoded to get the encrypted image. The experimental results and security analysis showed that the proposed algorithm had larger key space and resisted exhaustive, statistical and differential attacks. In 2012, Qiang et al. presented a novel image encryption algorithm based on DNA subsequence operations that uses the idea of DNA subsequence operations (such as elongation operation, truncation operation, deletion operation, etc.) combining with the logistic chaotic map to scramble the location and the value of pixel points from the image. The experimental results and security analysis showed that the proposed algorithm was easy to be implemented, had good encryption effect and a wide secret key's space, strong sensitivity to secret key, and had the abilities of resisting exhaustive attack and statistic attack but the defect was its weak ability of resisting differential attack. Encryption protects both data at rest and data in transit or transmission. The concept of using DNA computing in the fields of cryptography is studied in order to enhance the security of cryptographic algorithms. Adleman, with his pioneering work set the foundation for the new field of bio-computing research. His main notion was to use actual chemistry to solve problems that were either unsolvable by conventional computers, or required a massive amount of computation. Gehani et al. introduced the first trial of DNA-based cryptography. In this work, the principle ideas of the central dogma of molecular biology are used for computing instead of real biological DNA strands. The method simulates the transcription, splicing, and translation process of the central dogma; thus, it is a virtual DNA cryptographic method.

DNA Coding

An electronic computer needs only two digits, 0 and 1 for coding information. As a single strand of DNA is similar to a string consisting of a combination of four different symbols, A, C, G and T, DNA coding should reflect the biological characteristics of the four nucleotide bases- A, C, G and T along with the Watson-Crick complementary rule (A is complementary to T and C is complementary to G). Out of the twenty four combinations of the four nucleotides, only eight combinations (00 01 10 11 - C T A G, 00 01 10 11 - C A T G, 00 01 10 11 - G T A C, 00 01 10 11 - G A T C, 00 01 10 11 - T C G A, 00 01 10 11- T G C A, 00 01 10 11 - A C G T, 00 01 10 11 - A G C T) satisfy the complementary rule of the nucleotides.

In accordance with the increasing molecular weight of the four nucleotides, (C -111.1 g/mol, T - 126.1133 g/mol, A - 135.13 g/mol and G - 151.13 g/mol) C T A G, is the best coding pattern and is used as DNA Coding. According to DNA Coding Technology, C denotes the binary value 00, T denotes 01, A denotes 10 and G denotes 11 so that Watson-Crick complementary also holds good. Table 1 gives the DNA Coding used in this work. This pattern perfectly reflects the biological characteristics of the four nucleotide bases and has biological significance.

Table 1. DNA Coding

Digital value	DNA base	Molecular weight g/mol
00	C	111.1
01	T	126.11
10	A	135.13
11	G	151.13

Axiomatic Definition of DNA Algebra

DNA algebra is an algebraic structure defined on a set of elements B{C, T, A, G} together with two binary operators '∨' and '∧' provided the following Huntington postulates are satisfied.

1. a. Closure with respect to the operator ∨.
   b. Closure with respect to the operator ∧.
2. 2. a. An identity element with respect to ∨, designated by C:

$$x \vee C = x, \forall x \in B$$

   b. 2. b. An identity element with respect to ∧, designated by G:

$$x \wedge G = x, \forall x \in B$$

3. 3. a. Commutative with respect to ∨.

$$x \vee C = C \vee x, \forall x \in B$$

   b. 3. b. Commutative with respect to ∧.

$$x \wedge G = G \wedge x, \forall x \in B$$

4. 4. a. ∨ is Distributive over ∧.

$$x \vee (y \wedge z) = (x \vee y) \wedge (x \vee z), \forall x, y, z \in B$$

   b. 4. b. ∧ is Distributive over ∨.

$$x \wedge (y \vee z) = (x \wedge y) \vee (x \wedge z), \forall x, y, z \in B$$

5. For every element $x \in B$, there exists an element $x' \in B$ (called the complement of x) such that
   a. $x \vee x' = G$ and
   b. $x \wedge x' = C$
6. There exists at least two elements $x, y \in B$, such that $x \neq y$.

The following De' Morgan's laws also hold good.

$$(x \vee y)' = x' \wedge y', \forall x, y \in B$$

$$(x \wedge y)' = x' \vee y', \forall x, y \in B$$

The primitive logic operations AND, OR, XOR, and NOT can be carried out and the results will be obtained as given in the characteristic tables Table 2, Table 3, Table 4 and Table 5.

Table 2. AND characteristic table

>	**C**	**T**	**A**	**G**
C	C	C	C	C
T	C	T	C	T
A	C	C	A	A
G	C	T	A	G

Table 3. OR characteristic table

V	**C**	**T**	**A**	**G**
C	C	T	A	G
T	T	T	G	G
A	A	G	A	G
G	G	G	G	G

Table 4. XOR

⊕	**C**	**T**	**A**	**G**
C	C	T	A	G
T	T	C	G	A
A	A	G	C	T
G	G	A	T	C

Table 5. Not characteristic table

X	**C**	**T**	**A**	**G**
~X	G	A	T	C

DNA Sequence Based Image Representation

The term image refers to a two-dimensional light intensity function, denoted by f(x,y), where the value of f at spatial coordinates(x,y) gives the intensity(brightness) of the image at that point. As light is a form of energy f(x,y) must be nonzero and finite, that is f(x) must lie between zero and infinity ($0 < f(x,y) < \infty$). A digital image is an image f(x,y) that has been discretized both in spatial coordinates and brightness. A digital image can be considered a matrix (two dimensional array) whose row and column indices identify a point in the image and the corresponding matrix element value identifies the gray level at that point. The elements of such a picture array are called pixels. Each pixel of the image consists of 8 bits. Using DNA coding principle, substituting C for 00, A for 01, T for 10 and G for 11, each pixel of the DNA image is represented as a quadruple nucleotide sequence.

Arithmetic and Logic Operations on Images

Arithmetic and logic operations between pixels are used extensively in most of the branches of image processing and are generally carried out on images pixel by pixel. The principle use of image addition is for image averaging - to reduce noise. Image addition is mainly done using XOR operation for brightening the image and for image security applications. Image subtraction is a basic tool in medical imaging where it is used to remove static background information. Image multiplication or division is used to correct gray level shading resulting from non-uniformities in illumination or in the sensor used to acquire the image. Arithmetic operations can be done "in place" such that the result of performing an operation can be stored in that location in one of the existing images. With DNA computing, two DNA images can be added using parallel addition of the rows of the two images.

DEFINITIONS FOR THE PROPOSED SYSTEM

Cryptosystem

A cryptosystem is a five tuple (M, C,κ, ε, D), where the following conditions are satisfied.

1. M is a finite set of possible plain-text (images).
2. C is a finite set of possible cipher-text.
3. κ is a finite set of possible keys.
4. E is a finite set of encryption rules indexed by K, such that for each $K \in \kappa$, there is a function E_K: $M \rightarrow C$.
5. D is a finite set of encryption rules indexed by K, such that for each $K \in \kappa$, there is a function D_K: $C \rightarrow M$.
6. For each $K \in \kappa$, $D_K(E_K(x)) = x$, for every plain-text $x \in M$..

Condition (vi) enables a user to decrypt a received cipher-text, since $D_K(E_K(x)) = x$, for every plain-text $x \in M$.

For unambiguous decryption, it is required that $E_K(x1) \neq E_K(x2)$, if $x_1 \neq x_2$. On the other hand, if $E_K(x_1) = E_K(x_2)$, and $x_1 \neq x_2$ then decryption is not unique and hence it is impossible for a recipient to decide whether the intended plain text was x_1 or x_2 upon receipt of $E_K(x_1) = E_K(x_2)$.

Synthesis

In standard solid phase DNA synthesis, a desired DNA molecule is built up nucleotide by nucleotide on a support particle in sequential coupling steps. For example, the first nucleotide (monomer), say A, is bound to a glass support. A solution containing C is poured in, and the A reacts with the C to form a two-nucleotide (2-mer) chain AC. After washing the excess C solution away, one could have the C from the chain AC coupled with T to form a 3-mer chain (still attached to the surface) and so on.

Creation of DNA sequences for the image data is referred to as Synthesis. DNA sequences are made up of four bases – A, C, T and G. According to the DNA Coding Technology, C denotes 00, T – 01, A - 10 and G – 11. According to the coding method proposed for each pixel of a digital image by Qiang et al, each pixel of the digital image is converted into its corresponding DNA coded value.

Translation

When the positions of sequences are translated, the sequences are interchanged.
Translation is represented as

$$P_1P_2P_3P_4 \leftrightarrow P_5P_6P_7P_8$$

Substitution

Each quadruple nucleotide sequence is substituted by the value returned by the DNA Sequence Crypt function. Substitution is represented by the following expression.

$$V \leftarrow DNASequenceCryptfn(P_1P_2P_3P_4)$$

Re-Substitution

Each value, V in the encrypted image is replaced by the corresponding quadruple nucleotide sequence from that position in the DNA Sequence File.

$$P_1P_2P_3P_4 \leftarrow V$$

Detect

Detect searches for a quadruple nucleotide sequence of the image starting from a random position in the DNA sequence file and returns true if a match is found and false otherwise.

Re-Synthesis

Re-synthesis is the process of converting each sequence into its digital form.

KEY DEPENDENT DYNAMIC S-BOX FUNCTION

A Key Dependent Dynamic S-Box function is a mapping function,

$f:\{0,1\}^m \rightarrow \{0,1\}^n$,

that maps n bit input string X into n bit output string Y based on the codeword and has the following properties:

1. **Bijection**: A bijection (or bijective function or one-to-one correspondence) is a function between the elements of two sets, where every element of one set is paired with exactly one element of the other set, and every element of the other set is paired with exactly one element of the first set. A bijective function $f: X \rightarrow Y$ is a one to one and onto mapping of a set X to a set Y. A bijection from the set X to the set Y has an inverse function from Y to X.
2. **Strict Avalanche Criterion**: If a one bit change in the input results in at least 50 percent changes in the output bits, then there is Strict Avalanche criteria.
3. **Correlation-Immunity:** If the output bits act independently from each other, then there is Correlation-immunity.
4. **Nonlinearity**: A function with its corresponding vector is said to be highly nonlinear when the resulting vector $_{yi}$ from a function f_i has a high Hamming distance with all the linear vectors in the set of B_n.
5. **Balance**: An S-box with n input bits and m output bits, $m \leq n$, is balanced if each output occurs 2^{n-m} times. For the S-box to be balanced it should have the same number of 0's and 1's.

Hamming Distance

Hamming distance between two binary vectors of equal length is the number of places for which the corresponding entries are different.

DNA Sequence Crypt Function

DNA Sequence Crypt function is a function that returns one of the many positions of the quadruple DNA sequence in the key DNA sequence file.

A one to many DNA Sequence Crypt function is a one-to-many function d(x), which has the following three properties:

1. A pointer, h maps an input quadruple nucleotide sequence, x to one of the many positions obtained in random in the key DNA sequence file.
2. Ease of computation: Given d and an input x, d(x) is easy to compute.

3. Resistance to guess: In order to meet the requirements of a cryptographic scheme, the property of resistance to guess is required of a crypt function with input x, x_1 and outputs y, y_1.

As similar quadruple nucleotide sequence that occur in a plain text are mapped to different positions in the DNA nucleotide sequence file(one to many mapping), it is difficult for a recipient to guess the plain-text. The sender and receiver agree on a key DNA sequence file, R which can be freely and easily downloaded from the DNA GenBank®, the NIH genetic sequence database, an annotated collection of all publicly available DNA sequences. GenBank is part of the International Nucleotide Sequence Database Collaboration, which comprises the DNA DataBank of Japan (DDBJ), the European Molecular Biology Laboratory (EMBL), and GenBank at NCBI. These three organizations exchange data on a daily basis. Query sequence(s) to be used for a BLAST search should be pasted in the 'Search' text area. It accepts a number of different types of input and automatically determines the format or the input.

A sequence in FASTA format begins with a single-line description, followed by lines of sequence data. The description line is distinguished from the sequence data by a greater-than (">") symbol at the beginning. It is recommended that all lines of text be shorter than 80 characters in length. In cryptographic applications, long DNA sequences are generally used.

DUAL ENCRYPTION METHOD

The Dual Encryption method consists of two phases. Phase I is the encryption process and Phase II is the decryption process. The decryption is the reverse of the encryption process. In Phase I of the proposed Encryption scheme, a codeword is first generated based upon a 64 bit key. For simplicity, the key is denoted as Hex value in this paper. Once the codeword is generated, based upon the codeword, a Dynamic key-dependent S-Box is generated and a DNA sequence file is selected at run-time. Then the input image (health record) is transformed into a coded image based on the Dynamic key-dependent S-Box. With the key dependent Dynamic S-Box function, it is a single simple function applied over and over again to each byte of the input image, returning a byte. Each of the 256 possible byte values is transformed to another byte value with the key dependent SubBytes (S-Box) transformation, which is a full permutation, meaning that every element gets changed, and all 256 possible elements are represented as the result of a change, so that no two different bytes are changed to the same byte. Figure 1 represents the key dependent Dynamic SBox generated at runtime, when the key is A451B67290F7DE38 (hex representation of the 64 bit key 1010 0100 0101 0001 1011 0110 0111 0010 1001 0000 1111 0111 1101 1110 0011 1000). The coded image is then synthesized - transformed into DNA image. The encrypted image obtained by substituting each quadruple DNA nucleotides sequence of the translated image by one of the many positions of the quadruple nucleotides sequence (which is randomly obtained) in the gene sequence file by applying DNA Sequence Crypt function is sent to the receiver through any channel or stored in a public database in the encrypted form. The sender and the receiver/retriever should agree upon the 64 bit-key that is generated and the receiver/retriever can decrypt the encrypted image.

In the retrieval phase, the authorized receiver upon receiving the encrypted image converts it into DNA image by mapping the pointers from the encrypted image onto the key DNA sequence file and the decrypted image is then obtained by Re-Synthesis followed by Inverse S-Box substitution.

Figure 1. Dynamic SBox when the key is A451B67290F7DE38

36	C7	77	B7	2F	B6	F6	5C	03	10	76	B2	EF	7D	BA	67
AC	28	9C	D7	AF	95	74	0F	DA	4D	2A	FA	C9	4A	27	0C
7B	DF	39	62	63	F3	7F	CC	43	5A	5E	1F	17	8D	13	51
40	7C	32	3C	81	69	50	A9	70	21	08	2E	BE	72	2B	57
90	38	C2	A1	B1	E6	A5	0A	25	B3	6D	3B	92	3E	F2	48
35	1D	00	DE	02	CF	1B	B5	A6	BC	EB	93	A4	C4	85	FC
0D	FE	AA	BF	34	D4	33	58	54	9F	20	F7	05	C3	F9	8A
15	3A	04	F8	29	D9	83	5F	CB	6B	AD	12	01	FF	3F	2D
DC	C0	31	CE	F5	79	44	71	4C	7A	E7	D3	46	D5	91	37
06	18	F4	CD	22	A2	09	88	64	EE	8B	41	ED	E5	B0	BD
0E	23	A3	A0	94	60	42	C5	2C	3D	CA	26	19	59	4E	97
7E	8C	73	D6	D8	5D	E4	9A	C6	65	4F	AE	56	A7	EA	80
AB	87	52	E2	C1	6A	4B	6C	8E	DD	47	F1	B4	DB	B8	A8
07	E3	5B	66	84	30	6F	E0	16	53	75	9B	68	1C	D1	E9
1E	8F	89	11	96	9D	E8	49	B9	E1	78	9E	EC	55	82	FD
C8	1A	98	D0	FB	6E	24	86	14	99	D2	F0	0B	45	BB	61

Algorithm Dynamic_S-Box

```
Input  :     codeword cw
Output:     D_ S-Box
Begin
Z = Bin2dec(cw) + 1
Case Z of
  1 - 32      : D_ S-Box = row transformation of AES_S-Box
  33 - 64     : D_ S-Box = column transformation of AES_S-Box
  65 - 96     : D_ S-Box = transpose of results obtained in case 1- 32
  97 - 128   : D_ S-Box = transpose of results obtained in case 33- 64
 129 - 160  : D_ S-Box = nibble exchange of results obtained in case 1- 32
 161 - 192 : D_ S-Box = nibble exchange of results obtained in case 33- 64
193 - 224  : D_ S-Box = transpose of results obtained in case 129- 160
225 -256  : D_ S-Box = transpose of results obtained in case 161- 192
EndCase
Return D_ S-Box
End Dynamic_S-Box
```

Generation of Codeword

The key used for encryption is considered to be 64 bits in length. From the key, a codeword of 8 bits ($C_8C_7C_6C_5C_4C_3C_2C_1$) is generated at run-time based upon the Hamming Distance and Hamming Weight. The key-dependent S-box generated at runtime based upon the codeword is non-linear in nature. The codeword is also used to randomly select a DNA sequence file. For generating the codeword from the 64 bit key, the following two assumptions are considered.

1. Each bit of the codeword calculates the parity (Hamming weight) for certain bits in the key.
2. A parity bit is set to 1 if the total number of ones in the positions it checks is odd or 0 otherwise.
   Algorithm Codeword_Generator

```
Input  :      64 bit key K
Output:      codeword cw
Begin
C8  =  (check all the bits of the key(1-64)) ? 1 : 0
C7  =  (check 1 bit, skip 1 bit, check 1 bit  etc. (1,3,5,7,9,11,13,15,...)) ?
1 : 0
C6 =  (check 2 bits, skip 2 bits, check 2 bits etc. (2,3,6,7,10,11,14,15,...))
? 1 : 0
C5 =  (check 4 bits, skip 4 bits, check 4 bits etc. (4,5,6,7,12,13,14,15,...))
? 1 : 0
C4 =  (check 8 bits, skip 8 bits, check 8 bits etc. (8-15,24-31,40-47,...)) ? 1
: 0
C3 =  (check 16 bits, skip 16 bits, check 16 bits etc. (16-31,48-56)) ? 1 : 0
C2 =  (check 32 bits, skip 32 bits, check 32 bits etc. (32-63)) ? 1 : 0
C1 =  (check bit 1 and bit 64) ? 1 : 0
cw = C8C7C6C5C4C3C2C1
Return cw
End Codeword_Generator
```

The codeword for the key A451 B672 90F7 DE38 and A451 B672 90F7 DE39 are 10111001 and 00111000, respectively.

Algorithm CodeImage

```
Input   :  Input image X, codeword cw
Output:  Coded image  C
Begin
DS ← Dynamic_S-Box(cw)
C  ← SubByte Transformation of  X based on DS
Return C
End CodeImage
SubByte Transformation
```

With the Dynamic S-Box function, the SubByte transformation is a single simple function applied over and over again to each byte of the input image. In the SubByte transformation, each of the 256 possible byte values is treated independently and transformed to another byte value based on the Dynamic S-Box, which is a full permutation, meaning that every element gets changed, and all 256 possible elements are represented as the result of a change, so that no two different bytes are changed to the same byte.

The SubByte transformation is pictorially represented in Figure 2. To substitute a byte, the byte is interpreted as two hexadecimal digits. The left digit defines the row and the right digit defines the column of the substitution table. The two hexadecimal digits at the junction of the row and the column is the new byte.

Figure 3 is an illustration of a SubByte transformation of an image of size 4 x 4, when the key is **{F95BBAF3656EFB64}.**

Figure 2. SubByte transformation

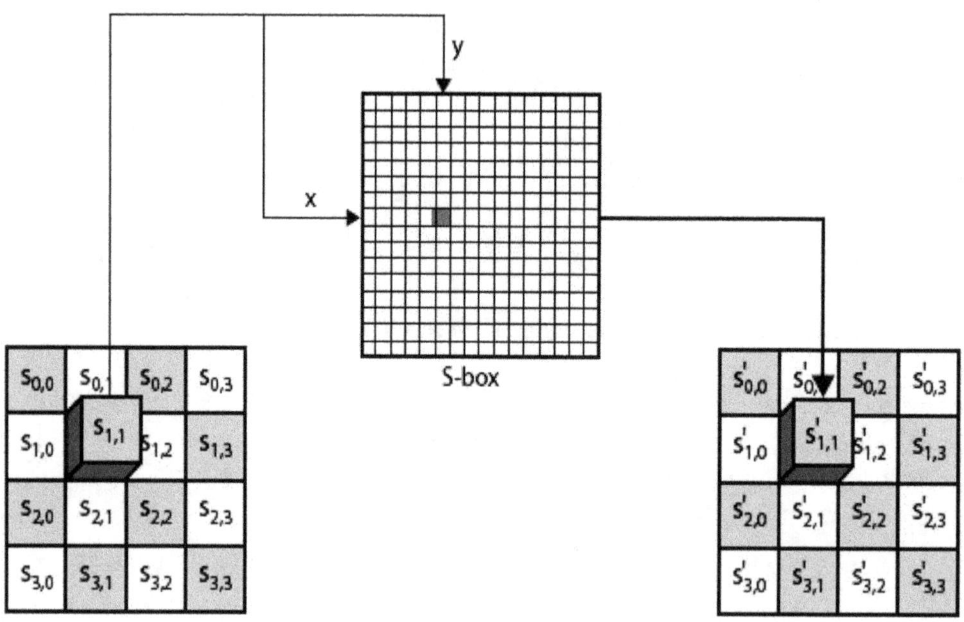

Figure 3. SubByte transformation when the key is {F95BBAF3656EFB64}

EA	04	65	85
83	45	5D	96
5C	33	98	B0
F0	2D	AD	C5

→

87	BF	4D	97
EC	6E	4C	90
4A	C3	46	E7
63	D8	95	A6

The coded image obtained in the encryption phase is then encrypted based upon the reference DNA Sequence file using the encryption algorithm. The encrypted image thus obtained is sent to the receiver. The sender and the receiver should agree upon the 64 bit-key that is used for encryption. Phase I and II of the proposed scheme can be pictorially represented as shown in Figure 4.

Algorithms EncryptImage, DNASequenceCrypt, Detect and DNA are given below.

Algorithm EncryptImage

```
Input  :    64 bit key, image to be encrypted  Y
Output:     Encrypted image E
Begin
Image X, codedimage C
DNA Sequence file R
Integer I
1.        cw ← Codeword_Generator(key)
2.        C ← CodeImage(Y, cw)
3.        Synthesis
 X ← DNA(Y)   // Convert the coded image into a DNA image
4.        R ← Select a  DNA Sequence file at runtime based on the Codeword
5.        Detection and Substitution
  For each quadruple DNA nucleotide sequence in X
      E(I)←DNASequenceCrypt(R, X(P1P2P3P4))
      I = I + 1
  End_For
6.        Return E
End EncryptImage
```

Algorithm DNASequenceCrypt

```
Input   :  DNA Sequence file R, quadruple nucleotide sequence x(P1P2P3P4)
Output:  index in DNA Sequence file h
```

Figure 4. Encryption and decryption phase

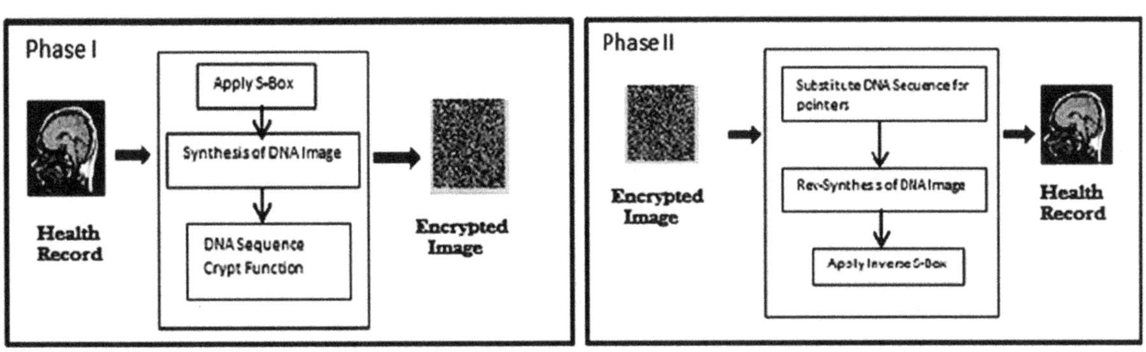

```
Begin
Integer i, J
J = rnd(30000)
repeat
   rn = rnd(J)
   for  i = rn to length(R) - 4
        if Detect(i, R, x)
              h = i
              Return h
       endif
   end for
      J = J - 100
until J > 5
End DNASequenceCrypt
Algorithm Detect
Input   :   search index rn, DNA Sequence file R,
                   quadruple nucleotide sequence x(P1P2P3P4)
Output:   Boolean T
Begin
String s, Integer i
     i = rn
     s = R(i)+R(i+1)+R(i+2)+R(i+3)
     if  (s = x)
           T ← true
     else
           T ← false
     end if
    return T
End Detect
```

Algorithm DNA

```
Input   :   image X
Output:   DNA image Y
Integer i, j, m, n, k
Begin
   m = w(Image) // width of image
   n  = h(Image) // height of image
   for i = 1 to h
     k = 1
    for j = 1 to w-1
        if (X(i, j) = 0 and (X(i, j) = 0) then Y(i,k) = 'C'
       if (X(i, j) = 0 and (X(i, j) = 1) then Y(i,k) = 'T'
       if (X(i, j) = 1 and (X(i, j) = 0) then Y(i,k) = 'A'
```

```
      if (X(i, j) = 1 and (X(i, j) = 1) then Y(i,k) = 'G'
     k = k + 1
  end  // end for i
  end // end for j
  Return Y
End DNA
```

Decryption Phase

The decryption is just the reverse of the encryption process. In the decryption phase, the received encrypted image (2D array) is actually pointers (index) to the DNA sequence file. The indices are therefore substituted by the quadruple nucleotide sequence starting from that index and the corresponding DNA image in encrypted form is obtained. The encrypted DNA image is converted to binary form and the Inverse SubByte transformation is performed on each byte of the resultant image to get the original image that was transmitted in the encrypted form. The inverse SubByte transformation is the inverse of the SubByte transformation and is obtained from Inverse Dynamic S_Box.

The algorithm for Inverse Dynamic S_Box is given as follows:

Algorithm Inv_Dynamic_S-Box

```
 Input :   codeword cw
Output:   Inv_D_SBox
Begin
         Inv_D_SBox  ←  Dynamic_S_Box(cw) -1
        Return  Inv_D_SBox
End Inv_Dynamic_S-Box
          The Decryption algorithm is given below:
```

Algorithm DecryptImage

```
Input  :  Encrypted Image E, Codeword cw
Output:  Decrypted Image X
Begin
1.        Select the DNA Sequence file based on the Codeword
2.        Re-Substitution
   Convert E into DNA sequence
             F     ←      E
3.         Re-Synthesis
    Convert F into its binary equivalent
4.        Substitution
                X  ←  Apply Inv_Dynamic_S-Box(cw) on F
5.        Return X
End DecryptImage
```

Experimental Results and Analysis

To prove the validity of the proposed scheme, experiments were performed on different images of varied sizes using Matlab 2009a on DELL Inspiron ACPIx64 based notebook PC. Table 2 gives the encoded values of a health care record of size 8 x 8. Figure 5 is an example of an original image and its corresponding encrypted and decrypted images and shows that the encrypted image reveals no information to the invader. The experimental results and security analysis reveal that the proposed algorithm has strong sensitivity to the key and DNA sequence file used. The level of security that the proposed encryption algorithm offers is double-fold – Dynamic S Box and DNA Sequence based.

Security analysis is defined as the technique of finding the weakness of a cryptographic scheme and retrieving whole or a part of the encrypted image without knowing the decryption key or the algorithm. For an encryption scheme to be good, it should be robust against statistical, brute-force and differential attacks, and therefore the proposed method was examined for these attacks.

Statistical Attack

The stability of the proposed method is examined via statistical attacks - correlation between adjacent pixels and the histogram. The encrypted image should not have any statistical likeness (similarity) with the original image to prevent the leakage of information.

Correlation Coefficient Analysis

In most of the plaintext-images, there exists high correlation among adjacent pixels, while there is little correlation between neighboring pixels in the encrypted image. It is the main task of an efficient image encryption algorithm to eliminate the correlation of pixels. Two highly uncorrelated sequences have approximately zero correlation coefficient and two strongly correlated sequences have a correlation coefficient nearly equal to one. Scatter graphs are drawn taking pixel value at location (x, y) on the x-axis and pixel value at location (x, y+1) on the y-axis for both the original and encrypted images.

Figure 5. Original, encrypted and decrypted images

The Pearson's Correlation Coefficient is determined using the formula:

$$\gamma = \frac{n\sum xy - (\sum x)(\sum y)}{\sqrt{n(\sum x^2)(\sum x)^2}\sqrt{n(\sum y^2)(\sum y)^2}} \tag{1}$$

where x and y are the grey-scale values of two neighboring pixels in the image and n is the total number of pixels selected from the image for the calculation is used to investigate the correlation between adjacent pixels.

Table 6 tabulates the correlation coefficient between adjacent pixels of the original and encrypted images. If there is insignificant correlation, correlation coefficient is close to 0. It is clear from Table 6 that there is negligible correlation between the two adjacent pixels in the encrypted image and gives no information to the invader regarding the nature of the original image that is being transmitted. However, the two adjacent pixels in the original image are highly correlated. Figure 6 reveals that the correlation of the encrypted images is unvaryingly distributed and reveals no information to the attacker.

Table 6. Encoded values of a health record of size 8 x 8

10454	57803	33952	33493	62362	70194	37655	43645
8200	1769	37656	35110	53348	42641	71593	77842
53016	18200	40047	4689	66314	4047	57830	66341
49	16522	48211	53335	26241	41520	59784	69671
3	15694	80186	1991	35188	33691	30117	30694
12	13848	3765	67898	76356	2061	15059	44617
261	22420	42508	39119	1820	1796	5334	744
883	31999	78981	43153	14211	50947	23941	14628

Figure 6. (a) to (f) represent the correlation between the adjacent pixels of the original and encrypted images column-wise, row-wise and diagonal-wise.

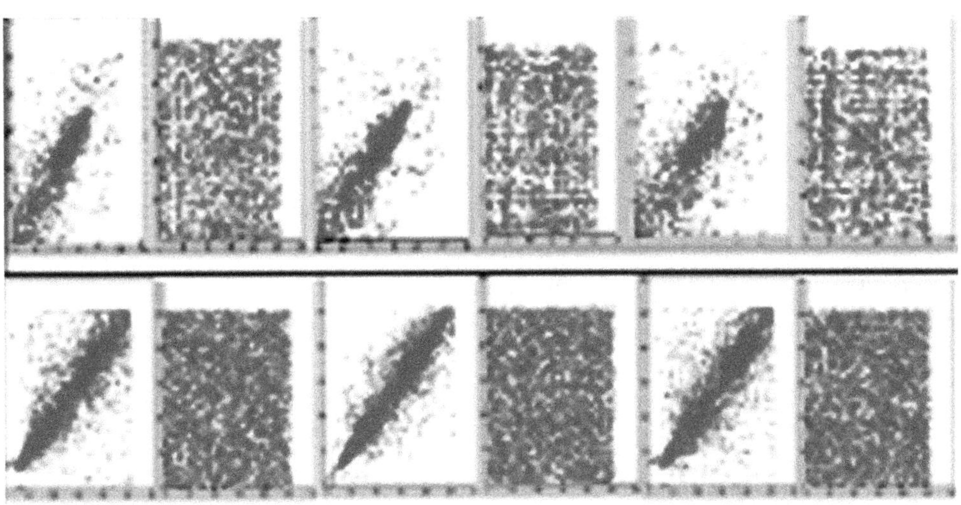

a) b) c) d) e) f)

Figure 7. Correlation coefficient graph

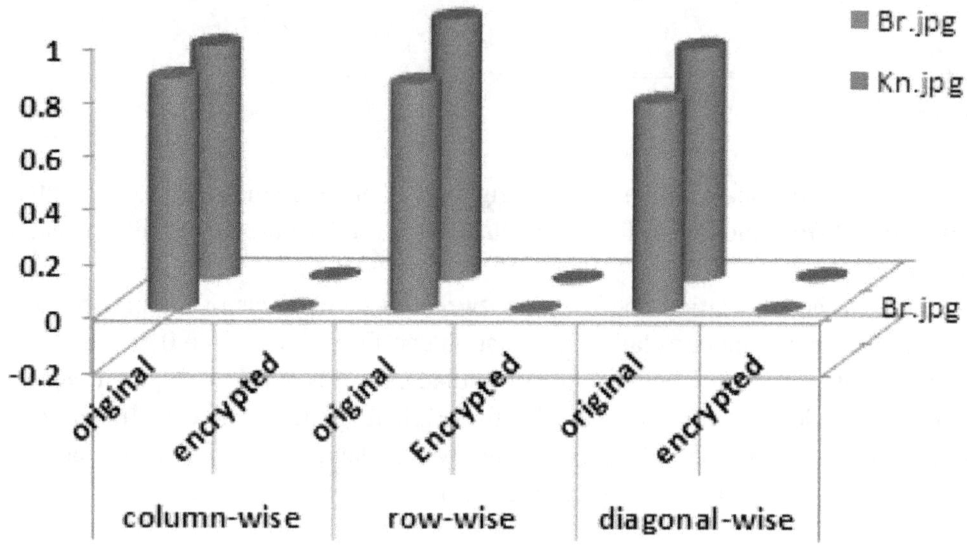

Table 7. Correlation coefficient of original and encrypted images

Image	Correlation Coefficient					
	Column-Wise		Row-Wise		Diagonal-Wise	
	Original	Encrypted	Original	Encrypted	Original	Encrypted
Br.jpg	0.8551	0.0014	0.8371	-0.0037	0.7698	0.00062
Kn.jpg	0.86	0.0062	0.959	-0.0102	0.8539	0.0108

Histogram Analysis

The histograms present the statistical characteristics of an image. If the histograms of the original image and encrypted image are different, then the encryption algorithm has good performance. An attacker cannot extract the pixels' statistical nature of the original image from the encrypted image and the algorithm can resist a chosen plain image or known plain image attack. Figure 8 reveals that the histograms of the encrypted images are fairly uniform and significantly different from the original image. It is also observed that the histograms of the encrypted images look quite similar (though the original images are different) and are completely different from that of the histograms of the original images and do not provide any information regarding the distribution of gray values to the attacker; As the encrypted image does not provide any information regarding the distribution of gray values to the attacker, the proposed algorithm can resist any type of histogram based attacks and strengthens the security of the encrypted images significantly.

Figure 8. . Histogram of brain and bone image – original and encrypted

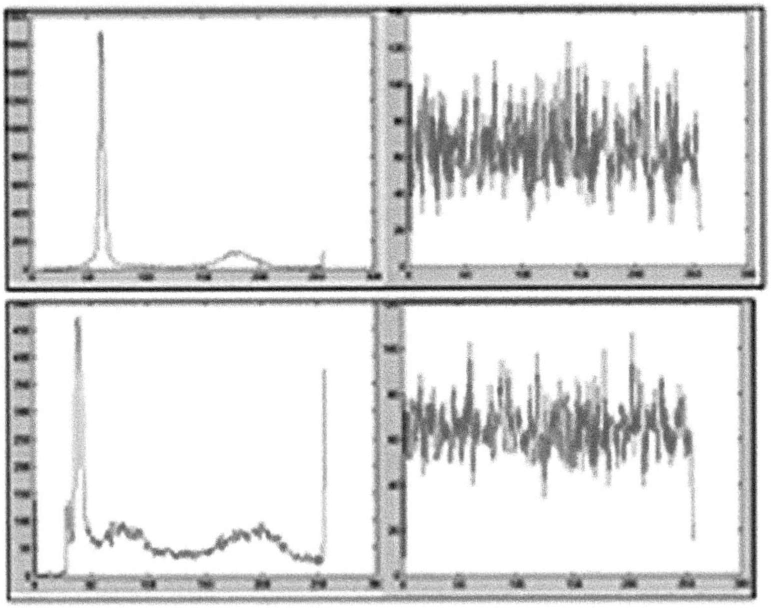

Brute Force Attack

A Brute Force Attack is an approach that can be used against any encrypted data by an intruder who is incapable to take benefit of any weakness in an encryption system that would otherwise make his task at ease. It involves methodically examining all probable keys until the right key is established. The encrypted file is principally arbitrarily generated pointers to the DNA sequence file and rarely there is a prospect of more than one quadruple nucleotide sequence pointing to the same position in the DNA sequence file. Moreover, the feature of bio-molecular setting is tougher to access as it is extremely difficult to recover the DNA digital code without knowing the correct coding technology used. An incorrect coding will cause biological pollution, which would lead to a corrupted image. Since there are many web-sites and roughly 55 million publicly accessible DNA sequences, it is practically impossible to guess the key sequence.

Differential Attack

The aim of differential attack analysis is to determine the sensitivity of encryption algorithm to slight changes. If an attack is made to create a small change in the plain image to observe the results, this influence causes a noteworthy change in the encrypted image and the antagonist will not be able to find a meaningful association between the original and encrypted image with respect to diffusion and confusion. A different sequence used or slight change made to the original plain image will result in a completely different encrypted image proving that the algorithm is highly intricate to small changes.

FUTURE RESEARCH DIRECTIONS

The two major drivers of contemporary telemedicine development are a high volume demand for clinical service, and a high criticality of need for security of health care records. These areas offer promise for further study and enhancement of security in health care management system and have the potential for large-scale deployments internationally, which would contribute significantly to the advancement of healthcare.

CONCLUSION

Electronic health record (EHR) is gradually being implemented in many countries. It is the need of the hour because it improves the quality of health care and is also cost-effective. The convolution and haphazardness of DNA based encryption of health record provides a great ambiguity which makes it better than other mechanism of cryptography and guarantees secure storage and transmission of health records. Integrating DNA based encryption along with Dynamic S-Box substitution helps in a double fold security. The proposed Encryption Scheme is easy to implement and can resist brute-force, statistical and differential attack and is suitable for the secure storage and transmission of health-care records.

REFERENCES

Adleman, L. (1994). Molecular Computation of Solutions to Combinatorial Problems. *Science*, *266*(5187), 1021–1024. doi:10.1126cience.7973651 PMID:7973651

Alvarez, G., Li, S., & Hernandez, L. (2007). Analysis of security problems in a medical image encryption system. *Computers in Biology and Medicine*, *37*(3), 424–427. doi:10.1016/j.compbiomed.2006.04.002 PMID:16872592

Amos, M., Paun, G., Rozenberg, G., & Salomaa, A. (2002). Topics in the theory of DNA computing. *Theoretical Computer Science*, *287*(1), 3–38. doi:10.1016/S0304-3975(02)00134-2

Chen, J. (2003) A DNA-based biomolecular cryptography design. *IEEE International Symposium on Circuits and Systems*, (pp. 822–825).

Craig, J., & Patterson, V. (2005). Introduction to the practice of Telemedicine. *Journal of Telemedicine and Telecare*, *11*(1), 3–9. doi:10.1258/1357633053430494 PMID:15829036

Cui, G. Z., Qin, L. M., Wang, Y. F., & Zhang, X. (2007). Information Security Technology Based on DNA Computing. *IEEE International Workshop on Anti-counterfeiting, Security, Identification*, (pp. 288–291). 10.1109/IWASID.2007.373746

Cui, G. Z., Qin, L. M., Wang, Y. F., & Zhang, X. (2008). An Encryption Scheme using DNA Technology. *International Conference on Bio-Inspired Computing: Theories and Applications*, (pp. 37-42).

Gehan, A., LaBean, T., & Reif, J. (2000). DNA Based Cryptography. *DIMACS Series in Discrete Mathematics and Theoretical Computer Science.*, *54*, 233–249.

Joseph, A. A., Lehmanny, C. U., Green, M. D., Pagano, M. W., Zachary, N. J. P., & Rubin, A. D. (2010). *Self-Protecting Electronic Medical Records Using Attribute-Based Encryption*. Retrieved from: https://eprint.iacr.org/2010/565.pdf

Josh, B., Melissa, C., Horvitz, E., & Lauter, K. (n.d.). *Patient Controlled Encryption: Ensuring Privacy of Electronic Medical Records*. Microsoft Research.

Ning, K. (2009). A pseudo DNA cryptography Method. *CoRR*. Retrieved from http://arxiv.org/abs/0903.2693

Norcen, R., Podesser, M., Pommer, A., Schmidt, H. P., & Uhl, A. (2003). Confidential storage and transmission of medical image data. *Computers in Biology and Medicine*, *33*(3), 273–292. doi:10.1016/S0010-4825(02)00094-X PMID:12726806

Qiang, Z., Xue, X., & Wei, X. (2012). A Novel Image Encryption Algorithm based on DNA Subsequence Operation. *TheScientificWorldJournal*.

Reif, J. (1997). Local Parallel Biomolecular Computation. *3rd DIMACS workshop on DNA based computers*, (pp. 243-258).

Richa, H. R., & Phulpagar, B. D. (2013). Review on Multi-Cloud DNA Encryption Model for Cloud Security. *Int. Journal of Engineering Research and Applications*, *3*(6), 1625–1628.

Richter, C., Leier, A., Banzhaf, W., & Rauhe, H. (2000). Private and Public Key DNA steganography. *6th DIMACS Workshop on DNA Based Computers*, (pp. 1-10).

Risca, V. I. (2001). DNA-based Steganography. *Cryptologia, Taylor and Francis*, *25*(1), 37–49. doi:10.1080/0161-110191889761

Rothemund P W K. (1996). A DNA and restriction enzyme implementation of Turing machines. *DNA Based Computers, 6*, 75-120.

Rothemund, P. W. K., Papadakis, N., & Winfree, E. (2004). Algorithmic self-assembly of DNA Sierpinski triangles. *PLoS Biology*, *2*(12), e424. doi:10.1371/journal.pbio.0020424 PMID:15583715

Rozenberg, G., Bäck, T., & Kok, J. (2012). *Handbook of Natural Computing*. Springer. doi:10.1007/978-3-540-92910-9

Rozenberg, G., & Salomaa, A. (2006). DNA computing: New ideas and paradigms. Lecture Notes in Computer Science, 7, 188-200.

Sadeg, S. (2010). An Encryption algorithm inspired from DNA. *IEEE International Conference on Machine and Web Intelligence*, (pp. 344 – 349).

Sakakibara, Y. (2005). Development of a bacteria computer: From in silico finite automata to *in vitro* and *invivo. Bulletin of EATCS*, *87*, 165–178.

Sánchez, R., Grau, R., & Morgado, E. (2006). A Novel Lie Algebra of the Genetic Code over the Galois Field of Four DNA Bases. *Mathematical Biosciences*, *202*(1), 156–174. doi:10.1016/j.mbs.2006.03.017 PMID:16780898

Shannon, C. E. (1949). Communication theory of secrecy system. *Journal of Bell System Technology, 28*(4), 656–715. doi:10.1002/j.1538-7305.1949.tb00928.x

Sherif, T. A., Magdy, S., & El-Gindi, S. (2006). A DNA based Implementation of YAEA Encryption Algorithm. *IASTED International Conference on Computational Intelligence*.

Siromoney, R., & Bireswar, D. (2003). DNA algorithm for breaking a propositional logic based cryptosystem. *Bulletin of the European Association for Theoretical Computer Science, 79*, 170–177.

Stallings W. (n.d.). *Cryptography and Network Security – Principles and Practice* (5th ed.). Prentice Hall.

Susan, K. L., & Hellman, M. E. (1994). Differential-Linear Cryptanalysis. *LNCS, 839*, 17–25.

Tausif, A., Sanchita, P., & Kumar, S. (2014). Message Transmission Based on DNA Cryptography [Review]. *International Journal of Bio-Science and Bio-Technology, 6*(5), 215–222. doi:10.14257/ijbsbt.2014.6.5.22

Taylor, C., Risca, V., & Bancroft, C. (1999). Hiding messages in DNA Microdots. *Nature, 399*(6736), 533–534. doi:10.1038/21092 PMID:10376592

Terec, R. (2011). DNA Security using Symmetric and Asymmetric Cryptography. *International Journal of New Computer Architectures and Their Applications*, 34-51.

Udo, F., Saghaf, S., Wolfgang, B., & Rauhe, H. (2002). DNA Sequence Generator: A program for the construction of DNA sequences. *DNA Computing,* 23-32.

Vazirani, V. (2004). *Approximation Algorithms*. Berlin: Springer.

Volos, C. K., Kyprianidis, I., & Stouboulos, I. (2013). Image encryption process based on chaotic synchronization phenomena. *Signal Processing, 93*(5), 1328–1340. doi:10.1016/j.sigpro.2012.11.008

Wang, X., & Wang, Q. (2014). A novel Image Encryption Algorithm based on Dynamic S-Boxes constructed by chaos. *Nonlinear Dynamics, 75*(3), 567–576. doi:10.100711071-013-1086-2

Wang, Z., & Yu, Z. (2011). Index-based symmetric DNA encryption algorithm. *Fourth International Congress on Image and Signal Processing*, (pp. 15-17).

Watson, J. D., & Crick, F. H. C. (1953). Molecular structure of nucleic acids: A structure for De-oxy ribose nucleic acid. *Nature, 25*(4356), 737–738. doi:10.1038/171737a0 PMID:13054692

Weichang, C., & Zhihua, C. (2000). Digital Coding of the Genetic Codons and DNA Sequences in High Dimension Space. *Acta Biophysica Sinica, 16*(4), 760–768.

Westlund, B. H. (2002). NIST reports measurable success of Advanced Encryption Standard. *Journal of Research of the National Institute of Standards and Technology*.

Winfree, A. (1980). *The Geometry of Biological Time*. Academic Press.

Winfree, E. (1996) On the Computational Power of DNA Annealing and Ligation. *1st DIMACS Workshop on DNA Based Computers*.

Wootton, R., & Craig, J. (1999). *History of Telemedicine. Introduction to telemedicine*. London: Royal Society of Medical Press.

World Health Organization. (1997). *Health-for-all Policy for the 21st Century, HQ (document EB101/8)*. Geneva: WHO.

Wu, Y., Noonan, J. P., & Agaian, S. (2011). NPCR and UACI Randomness Tests for Image Encryption. *Journal of Selected Areas in Telecommunications*, *4*, 31–38.

Wu, Y., Zhou, Y., Noonan, J. P., & Agaian, S. (2013). Design of Image Cipher Using Latin Squares. *Information Sciences*, *00*, 1–30.

Xiao, G., Lu, M., Qin, L., & Lai, X. (2006). New field of cryptography: DNA cryptography. *Chinese Science Bulletin*, *51*, 1139–1144.

Xiao, J. H., Zhang, X. Y., & Xu, J. (2012). A membrane evolutionary algorithm for DNA sequence design in DNA computing. *Chinese Science Bulletin*, *57*(2), 698–706. doi:10.100711434-011-4928-7

XueJia, L., MingXin, L., Lei, Q., JunSong, H., & XiWen, F. (2010). Asymmetric Encryption and signature method with DNA technology. *Information Sciences*, *53*, 506–514.

Youssef, M. I., Emam, A. E., Saafan, S. M., & Abd Elghany, M. (2013). Secured Image Encryption Scheme Using both Residue Number System and DNA Sequence. *The Online Journal on Electronics and Electrical Engineering*, *6*(3), 656–664.

Yunpeng, Z. (2012). *Research on DNA Cryptography*. InTech Press.

Yunpeng, Z., Fu, B., & Zhang, X. (2012). DNA cryptography based on DNA Fragment assembly. *IEEE International Conference Information Science and Digital Content Technology*.

KEY TERMS AND DEFINITIONS

DNA Sequence Based Cryptography: A cryptographic technique based on DNA sequences.

Dynamic S-Box: A S-Box generated at run-time to enhance security.

Health Care Record: A record of a patient's medical details (including history, physical examination, investigations and treatment) in digital format.

Healthcare Management System: A system designed and developed for better understanding of health care information and management.

S-Box: A substitution box used in cryptography to obscure the relationship between the key and the cipher text.

SubByte Transformation: Each byte is transformed to another byte based on Dynamic SBox.

Telemedicine: The use of telecommunication and information technology to provide health care eliminating the barriers of distance.

This research was previously published in Pattern and Data Analysis in Healthcare Settings edited by Vivek Tiwari, Basant Tiwari, Ramjeevan Singh Thakur, and Shailendra Gupta, pages 7-34, copyright year 2017 by Medical Information Science Reference (an imprint of IGI Global).

Chapter 14
An Intelligent and Secure Framework for Wireless Information Technology in Healthcare for User and Data Classification in Hospitals

Masoud Mohammadian
University of Canberra, Australia

Dimitrios Hatzinakos
University of Toronto, Canada

Petros Spachos
University of Toronto, Canada

Ric Jentzsh
University of Canberra, Australia

ABSTRACT

Real time data acquisition and evaluation are required to save lives. Such data with utilization and application of the latest technologies in hospitals around the world can improve patient treatments and well beings. The delivery of patient's medical data needs to be secure. Secure and accurate real time data acquisition and analysis of patient data and the ability to update such data will assist in reducing cost while improving patient's care. A wireless framework based on radio frequency identification (RFID) can integrate wireless networks for fast data acquisition and transmission, while maintaining the privacy issue. This chapter discusses the development of a framework that can be considered for secure patient data collection and communications in a hospital environment. A new method for data classification and access authorization has also been developed, which will assist in preserving privacy and security of data. Several Case studies are offered to show the effectiveness of this framework.

DOI: 10.4018/978-1-5225-6198-9.ch014

1. INTRODUCTION

Radio Frequency Identification (RFID) has been shown to be a viable and promising technology in the health care industry (Finkenzeller, 1999); Glover, & Bhatt, 2006; Mohammadian & Jentzsch, 2008; Schuster, Allen, & Brock, 2007; Shepard, 2005; Angeles, 2007; Pramatari, Doukidis, & Kourouthanassis, 2005; Mickey, 2004) Whiting, 2004). RFIDs have the capability to penetrate and add value to many areas of health care. RFIDs can lower the cost of some services as well as improve services to individuals and health care providers. The real value of RFID is achieved in conjunction with the use of intelligent software systems such as intelligent multi-agent systems. The integration of these two technologies can benefit and assist health care services.

Radio Frequency Identifiers (RFID) have been around for many years. Their current and projected use has only begun to be researched in hospitals (Fuhrer, & Guinard, 2007). This research study considers the use of RFIDs and their potential in hospitals and similar environments. RFIDs can effectively be used to collect data at the source thereby providing the data for monitoring patient's well-being in order to provide a higher level of patient health care. There are four areas where using RFIDs in their data collection role can have significant positive benefits in hospitals. These four areas are:

- **Care Tracking:** This is getting the right care to the right patient at the right time;
- **Quality of Care:** Improving the services given to the right patient at the right time in a timely manner;
- **Cost of Care:** Finding ways to be effective in the use of available resources such that the cost per patient per incident does not adversely increase to the cost of the resources; and
- **Service of Care:** More timely information to enable a more informed decision by providing more knowledge about an individual's need for care. (Mohammadian & Jentzsch, 2007,2008)

RFID tags and readers are commonly associated with inventory and tracking goods in such places as manufacturing and warehousing, but hospitals are starting to apply the technology to new purposes. Unlike bar codes, RFID technology does not require contact or line of sight for communication. RFID data can be obtained wirelessly through clothing and other non-metallic materials. This capability makes RFIDs an appropriate technology to fit into the health care environment.

Both research and practical application of RFIDs in hospitals continues to be of increasing importance. For hospitals this has meant managing inventories in a more efficient manner. Such inventories take on a variety of different roles than those found in manufacturing. The nature of the inventory and assets can include a plethora of expensive and varied equipment, drugs, beds, chairs, as well as patients themselves and staff.

People tracking can be looked at from three perspectives:

1. **Continuous and Full Time:** Often referred to as human chipping. People are tagged as an integral part of the person process 24/7.
2. **Part Time:** In which people acquire some type of tag as part of their work and / or task environment. RFIDs are well suited for this type of tagging by staff in the health care industry.
3. **Casual:** With tags used on and as needed basis.

The percentage of worldwide RFID projects related to people tagging has increased from eight percent to 11 percent since 2005 (Tindal, 2008). But the healthcare industry has yet to quantify or provide evidence of the benefit to people-tagging. Human chipping continuous full time tagging) is not new but does raise many ethical questions, subject for further study and beyond the scope of this paper (Angeles, 2007).

RFIDs are used in hospitals for tracking high-value assets and setting up automated maintenance routines to improve operational efficiencies. However the use of RFIDs in tracking beds and tracking mobile equipment is still in its infancy. RFIDs are used to monitor equipment use. The technology can be used, for example for how long a bed was used at a particular location to determine a sterilization schedule as well as bed location tracking. RFID technology is already being deployed across the pharmaceutical industry to combat drug counterfeiting and shelf life tracking. The management of patients and their condition is paramount in a hospital. RFIDs can assist in asset and in personnel tracking, patient care, and billing where unnecessary expenses can be cut, and the average length of stay of a patient is reduced. Conceivably, more patient lives can be saved due to more efficient services, and where patient records are actively and continuously updated to provide better patient care.

The health sector is already using people-tagging to enhance safety measures in hospitals. It allows a nurse to radio his or her location if they are under assault, reduce mother-baby mismatches and can prevent baby theft. RFIDs can help insure that severe diabetics receive correct treatment. The technology can monitor disoriented elderly patients without the need for a dedicated staff (Tindal, 2008).

The purpose of is to keep tracking staff is not to expand command and control, but rather to be able to locate staff with the particular skills that are needed at the right time and place. Staff wearing badges with embedded RFIDs can easily be found to help provide needed and timely patient care. A major concern with any tracking device is of course individual privacy. This is an issue that will require further study and remains beyond the scope of this paper.

There is a need for more research into applications and innovative architectures for secure access, retrieval and update of data in healthcare systems (Finkenzeller, 1999; Glover, & Bhatt, 2006; Mohammadian, 2008; Schuster, Allen, & Brock, 2007; Shepard, 2005; Angeles, 2007; Pramatari, Doukidis, & Kourouthanassis, 2005; Mickey, 2004; Whiting, 2004). Although many organizations are developing and testing the possible use of RFIDs the real value of RFIDs is achieved in conjunction with the use of intelligent software agents for processing and monitoring data obtained via RFIDs. Thus the issue becomes the integration of these two great technologies. RFIDs and controlling software - for the benefit of improving health care services.

This research study considers a framework using RFID and Intelligent Software Agents for managing patients' health care data in a hospital environment. A fuzzy data classification system has also been developed to improve the application of regulatory data requirements for security and privacy of data exchange (Mohammadian & Hatzinakos, 2009, 2013).

This manuscript is divided into six sections:

1. Introduction;
2. RFID Patient Scenario;
3. Intelligent Multi-Agent Systems;
4. RFID Description;
5. Classification of User and Data for Operation/Transaction Control; and
6. Secure RFID Data Transmission in Hospitals.

Section two considers issues that relate to RFID and data collection and profiling. Section three studies is the patient to doctor profiling and intelligent software agents. This section discusses RFIDs creation and attempts to provide a thorough description of RFIDs and their components (Mohammadian & Jentzsch, 2007 and 2008). Section three discusses several practical cases of RFID technology in and around hospitals from Mohammadian et al. 2008 (Mohammadian & Jentzsch, 2008) and lists three possible applicable cases to assist in managing patients' medical data.

Section four discusses the important issue of maintaining patients' data security and integrity.

Section five concludes the discussion and suggests directions for further study. Part of the research study presented in this paper has already been published in journals and presented in conferences and book chapters over the past few years (Mohammadian, & Jentzsch, 2007, 2008, 2012). The reference to these publications is provided in this manuscript when appropriate.

2. RFID PATIENT SCENARIO

Upon arrival to a hospital patients can be issued RFID tags embedded in their wrist bands, which can contain such patient information as:

- Their patient number;
- Their surname;
- First name;
- Reason admitted to the hospital;
- Date and time of admission;
- Their doctor's name; and
- Ongoing monitoring data.

Monitoring could include such things as: heart rate, blood pressure, patient temperature, and other vital signs that are specific to the patient's case. Monitoring would be calibrated to the needs of each individual patient. For example an hourly alert might be sufficient for most patients, but for others an every 15 minutes might be required. This example demonstrates how a particular section of a hospital might be configured according to the needs of that section. Each patient would have a patient-tag and each patient's bed assigned a bed-tag. Receivers are would be distributed within rooms, hall ways, and hospital staff stations. Every 15 minutes, for example, a given receiver would interrogate the bed and patient tags. The patient's vital signs would be sent to the patient database where the patient's condition would be recorded. The patient care profile would then be updated with this information. If any element of the patient's condition is out of range or an exception identified, the nearest nurse station would be contacted.

2.1 Data Collection

Large amounts of health care data (e.g., such as patient's condition, assigned doctors, and nurses, identification of the institution itself, prescriptions and diagnose) are collected and stored in hospitals. It is not feasible or effective in this paper to discuss all potential uses of RFIDs to collect and retrieve data. This chapter concentrates on a subset of RFID hospital based uses with the understanding that all areas could, directly or indirectly, benefit from the use of RFID and intelligent software agents in a health care environment (Mohammadian & Jentzsch, 2007, 2008, 2012).

The RFID (Bhuptani & Moradpour, 2005) provides the passive vehicle to obtain the data via its monitoring capabilities. The intelligent software agent provides the active vehicle in the interpretation profiling of the data and reporting capacity. By investigating and analyzing patient data the patient's condition can be monitored and abnormal situations can be reported in a timely fashion. Using this information an evolving profile of each patient can be constructed and analysed to assist in deciding the kind of care a patient requires, the effects of ongoing care, and the identification of appropriate hospital resources (doctors, nurses, beds, etc…). The intelligent software agent builds a profile of patients as they are admitted to the healthcare institution by analyzing recorded and stored patient data. A profile for each attending physician can also be developed based on stored data about each doctor. Patients' and doctors' profiles can be correlated to obtain the specialization and availability of the doctors to suit the patients.

3. INTELLIGENT AND MULTI-AGENT SYSTEMS

A Multi-agent System (MAS) consists of a set of intelligent agents that are connected to perform tasks in an environment to achieve a common goal. To achieve a common goal successfully the agents must cooperate and collaborate with each other. Such cooperation and collaboration is performed by such tasks as sharing knowledge or competition. Multi-agent systems are shown to be useful in diverse applications. Due to the benefits offered by multi-agent systems. These benefits include increased efficiency, reliability and scalability. In a Multi-agent system each agent is required to perform its task and collaborate and/or compete with other agents by modifying the environment in which agents operate. The agents are distributed in the agent environment and agents communicate with each other by passing messages/ requests. An agent's actions affects other agent's actions, environments and decisions. Multi-Agent systems provide several benefits, some of which are:

- **Parallel Computation:** Several agents can perform task simultaneously in parallel fashion. This in turn can increase the efficiency of the system in which they operate.
- **Reliability:** It is possible that one or more agents can fail to finalize their operations. In such a case it may be possible to delegate the process of such agents to other agents.
- **Scalability:** It is possible to increase the number of agents with the increase in the size or complexity and workload of the system in which the agents operate.

3.1 Patient Profiling

Profiling is combined with personalization, and user modeling (Wooldridge & Jennings 1995). The use of profiles in hospitals and healthcare so far has been limited. The tracking of information about consumers' interests by monitoring their movements online is considered profiling or user modeling in e-commerce systems. By analyzing the content, URL's, and other information about a user's browsing path/click-stream a profile of a user behavior is constructed. However patient profiling differs from the commercial variety. Patient profiling is useful in a variety of healthcare situations by providing a personalized service based on the patient and not on symptoms or illnesses and assisting in the identifying the requirements to help eliminate the need for the patient to return to the hospital any sooner than necessary. Patient profiling also assists in matching a doctor's specialization to the right patient. A patient profile can also assist in providing patient information on a continuous basis to help tailored and appropriate care program.

3.2 Patient to Doctor Profiling

A patient or doctor profile is a collection of information that can be used in a decision analysis among the patient, doctor and domain environment. A static profile is kept in pre-fixed-determined, longitudinally based data fields. Such profile data periods can bas as long as months or years. Data can be continuously updated for each patient. The updates may automated or performed manually with the former is especially useful in real time decision-making systems. The profiling within the patient-doctor model is based on patient / doctor information. These components include:

- The categories and subcategories of doctor specialization and categorization. That assists in information processing and patient/doctor matching.
- Use of the profile based on patient symptoms (past history problems, dietary restrictions, etc.) can assist in predicting specific patients' needs.
- Patient profiles can be matched with the available doctor profiles to quickly facilitate appropriate care.

A value denoting the degree of association can be created form the above evaluation of the doctor to patient's profile. The intelligent agent based on the denoting degrees and appropriate, available doctors can be identified and be allocated to the patient (Mohammadian & Jentzsch 2008).

In patient / doctor profiling the intelligent agent software will make distinctions in attribute values of the profiles and match the profiles with highest value. It should be noted that the intelligent agents create the patient and doctor profiles based on data obtained from both doctors and patient namely:

- Creation of accurate profiles based on the data entered by hospital staff.
- Implicit data profiling can fill profile gaps for them missing data by acquiring knowledge about the patient from his or her past visits or other relevant databases, if any then combining these data to fill the profile. Using such legacy data for complementing and updating a user profile seems to be a better choice than implicit profiling. This approach capitalizes on user's personal history (previous data from previous visit to doctor or hospital).

The proposed intelligent agent architecture allows user profiling and matching in a time intensive important application. The architecture of the agent profiling systems using RFIDs is given in Figure 1.

Profile matching (Doan, Lee, & Han, 2003) performed is based on a vector of weighted attributes using an intelligent agent system. To get this vector, the intelligent agent uses a rule-based system (or algorithms) to match the patient's attributes (stored in patient's profile) against doctor's attributes (stored in doctor's profile). If there is a partial or full match between them then the doctor will be informed (based on their availability from the hospital doctor database). Such a rules based system is built based on the knowledge of domain experts. This expert system is scalable as new domain knowledge can be added to its knowledge base as rules. Large amounts of research in the area of profiling in e-commerce,

Figure 1. Agent profiling model using RFID
(Mohammadian & Jentzsch 2008).

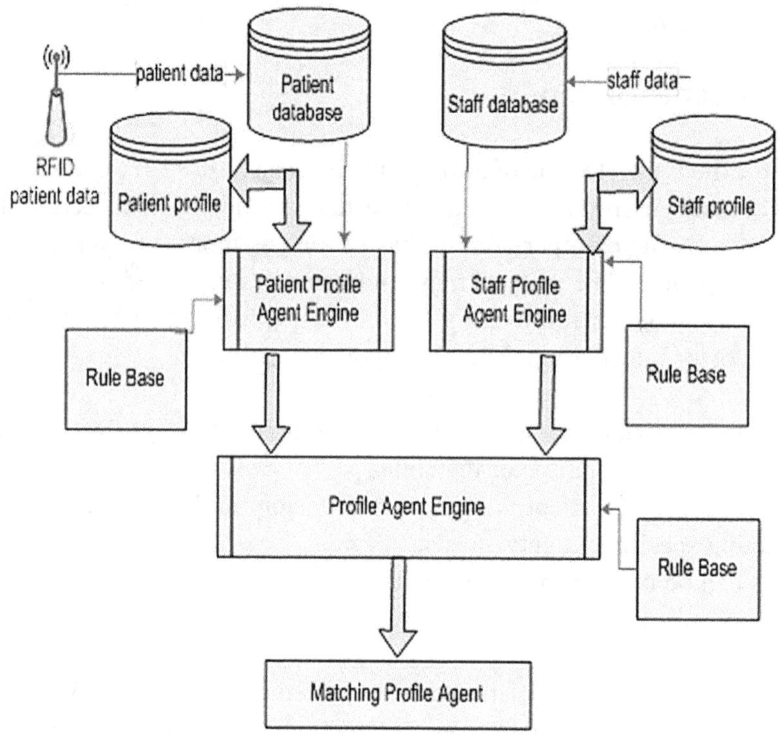

schema matching, information extraction and retrieval has shown promising results (Do & Rahin, 2002, Doan, Lee, & Han, 2003). However profiling in healthcare is still new and innovative.

Staff and patient/doctor profiling and profile matching could be the missing link in providing more tailored healthcare to patients in a hospital environment. Profile matching may consist of:

- Determining the matching algorithms required for matching patient / doctor profile,
- Determining the availability of staff and facilities required for a given patient.
- Understanding of government policies related to patient healthcare,

Another issue related to patient/doctor profiling is defining the level of matching of the patient and doctor profiles. It is not always possible to provide an exact match for a patient-healthcare because the matching doctor may be unavailable or unreachable. Some guidelines include issues such as the critical nature of the patient illness, its level of sensitivity and regulatory rules.

The rules that govern patient/doctor profile matching can be expressed in human linguistic terms which can be vague and difficult to represent formally. Fuzzy Logic (Zadeh, 1965) has been found to be useful in its ability to handle vagueness. In this research study the profiling patient / doctor and matching is based on fuzzy logic. The profiling matching system consists of a fuzzy rule based system and an inference engine between a patient profile and doctor/s profile.

The matching between a patient profile and doctor/s is then divided into the following classes: "total match", "medium match", "low match" and "no match". Based on these class categories a weighted match of patient / doctor profile can be identified. The doctors then can be categorized and ranked based on the matching profile value. The doctors can be classified into classes based on their matching profile as well as their availability such as "highly available", "more and less available" and "not available". Of course the data about availability of the doctors are obtained and updated and the profiling agent continuously checks such information from the staff database. The matched doctors then is ranked as: "high", "medium", "low" and "zero". A fuzzy rule for the profile matching system then may look like:

IF*patient_doctor_profile* is **total match**

and *doctor_availability* is **highly available**

Then*doctor_ranking* = **high**

The integration of RFID capabilities and intelligent agent techniques provides promising development in the areas of performance improvements in RFID data collection, inference and knowledge acquisition and profiling operations.

Patient_doctor_profile_match

Due to the important role of intelligent agents in this system, it is recognized that there is a need for a framework to coordinate intelligent agents so that they can perform their tasks efficiently.

Intelligent agent coordination (Wooldridge & Jennings, 1995. Bigus & Bigus, 1998. Shaalana, El-Badryb, & Rafeac, 2004) has shown to be promising. The Agent Language Mediated Activity Model (ALMA) agent architecture currently being researched is based on the mediated activity framework. We believe that such a framework is able to provide RFIDs with the necessary framework to profile a range of internal and external medical/patient profiling communication activities performed by a multi-agent system.

4. RFID DESCRIPTION

RFID or Radio Frequency Identification is a progressive technology that is easy to use and well suited for collaboration with intelligent software agents. Basically an RFID can be read-only, volatile read/ write; or write once / read many times. RFID are non-contact; and non-line-of-sight operations. Being non-contact and non-line-of-sight make RFIDs able to function under a variety of environmental con-

Table 1. Fuzzy rule base for patient_doctor_profile_match

Availability	Total Match	Medium Match	Low Match	No Match
Highly available	High	High	Low	Zero
More and less available	Medium	Medium	Low	Zero
Not available	Zero	Zero	Zero	Zero

Based on Mohammadian, Jentzsch, (2008).

ditions while still providing a high level of data integrity (Finkenzeller, 1999; Glover & Bhatt, 2006; Mohammadian & Jentzsch, 2008; Schuster, Allen, & Brock, 2007; Shepard 2005). A basic RFID system consists of four components, the RFID tag (sometimes referred to as the transponder), a coiled antenna, a radio frequency transceiver and some type of reader for the data collection.

The reader emits radio waves in ranges of anywhere from 2.54 centimeters to 33 meters. Depending upon the reader's power output and the radio frequency used and if a booster is added that distance can be increased.

The coiled antenna is used to emit radio signals to activate the tag and read or write data to it.

RFID tags can be categorized as active, semi-active, or passive. Each has and is being used in a variety of inventory management and data collection applications today. The condition of the application, place and use determines the required tag type.

The transceivers / interrogators can differ considerably in complexity, depending upon the type of tags being supported and the applications used. The overall function of the application provides the means of communicating with the tags and facilitating data transfer.

4.1 Hospital Environment

In order to manage patient medical data in a hospital environment both types; fixed and handheld transceivers and needed. Transceivers can be positioned in ceilings, walls, or doorframes to collect and disseminate data. Hospitals have become large complex environments. In a hospital nurses and physicians can retrieve the patient's medical data stored in transponders (RFID tags) before they stand beside a patient's bed or as they are entering a ward. Given the descriptions of the two types and their potential use in hospital patient data management we suggest that:

- It would be most useful to embed a passive RFID transponder into a patient's hospital wrist band;
- Also it would be most useful to embed a passive RFID transponder into a patient's medical file;

Doctors should have Netbooks or tablets equipped with RFIDs or some type of personal area network device. Either would enable them to retrieve some patient's data whenever they are near the patient, instead of waiting until the data is provided through the hospital server.

Hospital patient data management deals with sensitive and critical information. *Hands Down polling* techniques in conjunction with multiple transceivers that are multiplexed with each other, form a wireless network. The reason behind this choice is the need for high speed transference of medical data from medical equipment to or from the RFID wristband tag to the nearest RFID reader then through a wireless network or a network of RFID transceivers or LANs to the hospital server. From there it is a short distance to transmit the data to doctor's smart phone, Netbook, tablet, a laptop, or even the desktop through a wireless LAN (WLAN) or wired LAN.

The "Hand Down Polling" techniques, provides the ability to detect all detectable RFID tags at once (i.e. in parallel), preventing any unnecessary delay in transmitting medical data corresponding to each RF tagged patient. Transponder programmers are the means, by which data is delivered to write once, read many (WORM) and read/write tags. Programming can be performed off-line or on-line. For some systems re-programming may be carried out on-line, particularly if it is being used as an interactive portable data file within a production environment, for example. Data may need to be recorded during step of the process. Removing the transponder at the end of each process to read the previous process data, and to program the new data, would naturally increase process time and would detract substantially from the intended flexibility of the application. By combining the functions of a transceiver and a programmer, data may be appended or altered in the transponder as required, without compromising the production line.

4.2 Applications of the RFID Technology in a Hospital

The following section describes steps involved in the process of using RFIDs in a hospital environment for patient information management:

1. A biomedical device equipped with an embedded RFID transceiver and programmer will detect and measure the biological state of a patient. This medical data can be an ECG, EEG, BP, sugar level, temperature or any other biomedical reading. After the acquisition of the required medical data, the biomedical device will write this data to the RFID transceiver's EEPROM using the built in RFID programmer. Then the RFID transceiver through its antenna will be used to transmit the stored medical data in the EEPROM to the EEPROM in the patient's transponder (tag) which is around his/her wrist. The data received will be updated periodically once new fresh readings are available by the biomedical device. The purpose of the data stored in the patient's tag is to make it easy for the doctor to obtain current medical information regarding the patient directly via the doctor's Netbook, tablet, smart phone, or even a small laptop.
2. Similarly, the biomedical device will also transfer the measured medical data wirelessly to the nearest WLAN access point. Since high data rate transfer rate is crucial in transferring medical data, IEEE 802.11b or g is recommended for the transmission purpose.
3. Then the wirelessly sent data will be routed to the hospitals main server; to be then sent (pushed) to:
    a. Other doctors available throughout the hospital so they can be notified of any newly received medical data.
    b. To an on-line patient monitoring unit or a nurse's workstation within the hospital.
    c. Or the acquired patients' medical data can be fed into an expert (intelligent agent) software system running on the hospital server and to be then compared with other previously stored abnormal patterns of medical data, and to raise an alarm if any abnormality is discovered.
4. Another option could be to use the embedded RFID transceiver in the biomedical device to send the acquired medical data wirelessly to the nearest RFID transceiver in the room. Then the data will travel simultaneously in a network of RFID transceivers until reaching the hospital server.
5. If a specific surgeon or physician is needed in a specific hospital department, the medical staff in the monitoring unit (e.g., nurses) can query the hospital server for the nearest available doctor to the patient's location. In this scenario an intelligent agent can perform this task. The hospital server traces all doctors' locations in the hospital through detecting the presences of their wireless mobile device; e.g. Netbook, tablet, smart phone, or small laptop in the WLAN range. Physicians may also use RFID transceivers built-in the doctor's wireless mobile device.
6. Once the appropriate physician is located, an alert message can be sent to his\her Netbook, tablet, smart phone or laptop indicating the location to be reached immediately including a brief description of the patient's case.

7. The doctor can then enter into the patient's room or ward according to the alert he/she has received. The doctor can check the medical status of a certain patient and interrogates the patient's RFID wrist tag with his RFID transceiver equipped in his\her Netbook, tablet, smart phone, or laptop.

4.3 Practical Cases using RFID Technology

This section explains in details three possible applications of the RFID technology in three applicable cases.

These cases cover issues such as acquisition of patient's medical data, locating the nearest available doctor to the patients location, and how doctors can stimulate the patient's active RFID tag using their mobile device or laptop in order to acquire the required medical data.

Case I: Acquisition of Patient's Medical Data

This process can be described as follows:

A biomedical device equipped with an embedded RFID transceiver and programmer will detect and measure the biological state of a patient. This medical data can be an ECG, EEG, BP, sugar level, temperature or any other biomedical reading.

After the acquisition of the required medical data (Figure 2), the biomedical device will write -burn this data to the RFID transceiver's EEPROM using the built in RFID programmer. Then the RFID transceiver will be used to transmit the stored medical data in the EEPROM to the EEPROM in the patient's transponder (tag) attached to his/her wrist. The data received will be updated periodically once new fresh readings are available by the biomedical device. Hence, the newly sent data by the RFID transceiver will be accumulated to the old data in the tag. The purpose of the data stored in the patient's tag is to make it easy for the doctor to obtain medical information regarding the patient directly via the doctor's mobile device.

Figure 2. Acquisition of patient data
(Mohammadian & Jentzsch 2008).

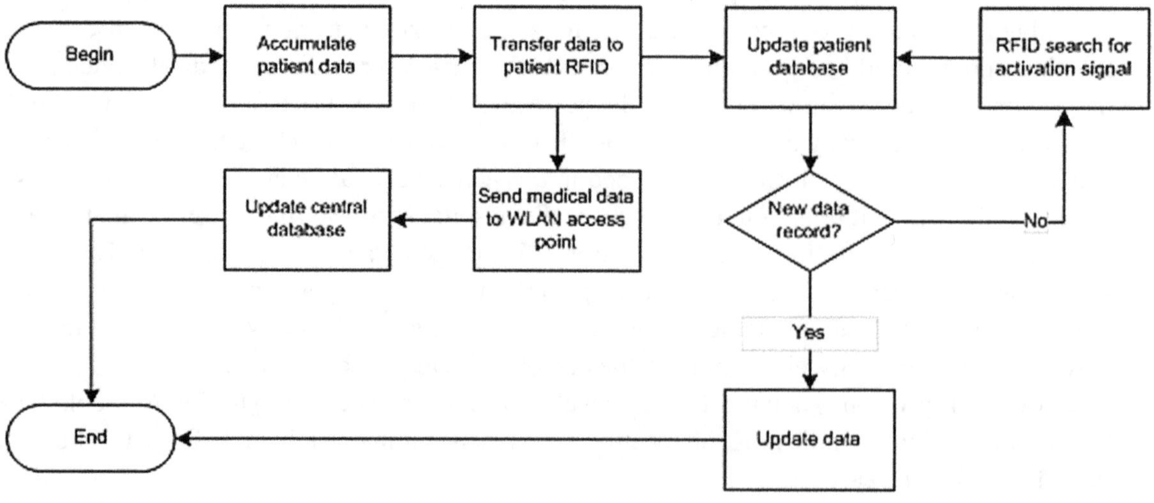

Similarly, the biomedical device will also transfer the measured medical data wirelessly to the nearest WLAN access point. Since high data rate transfer rate is crucial in transferring medical data, IEEE 802.11b, g, n (or its most recent iteration) is recommended for the transmission purpose (with n currently being optimal).

Then the wirelessly sent data will be routed to the hospitals main server; to be then sent (pushed) to:

- Other doctors available throughout the hospital so they can be notified of any newly received medical data.
- To an on-line patient monitoring unit or a nurse's workstation within the hospital.
- Or the acquired patients' medical data can be fed into an expert (intelligent) software system running on the hospital server. To be then compared with other previously stored abnormal patterns of medical data, and to raise an alarm if any abnormality is discovered.

Another option could be the use of embedded RFID transceiver in a biomedical device to send the acquired medical data wirelessly to the nearest RFID transceiver in the room. The data can then travel simultaneously through a network of RFID transceivers until reaching the hospital server.

Case II: Locating the Nearest Available Doctor to the Patients Location

If a specific surgeon or physician is needed in a specific hospital department, the medical staff in the monitoring unit (e.g., nurses) can query the hospital server for the nearest available doctor to the patient's location. In this framework an intelligent agent can perform this task.

The hospital server traces all doctors' locations in the hospital through detecting the presences of their wireless mobile devices (Figure 3).

Another method that the hospital's server can use to locate the appropriate physicians is making use of the RFID transceivers built-in the doctor's wireless mobile device. Similarly to the access points used in WLAN, RFID transceivers can assist in serving a similar role of locating doctor's location. This can be described in three steps, which are:

- The fixed RFID transceivers throughout the hospital will send a stimulation signal to detect other free RFID transceivers - which are in the doctors Netbook, tablet, smart phone, or laptop, etc…
- Then all free RFID transceivers will receive the stimulation signal and reply back with an acknowledgement signal to the nearest fixed RFID transceiver.
- Finally, each free RFID transceiver cell position would be determined by locating to which fixed RFID transceiver range it belongs to or currently operating in.

After the hospital server located positions of all available doctors, it determines the nearest requested physician (pediatrics, neurologist etc…) to the patient's location. Once the required physician is located, an alert message will be sent to his\her mobile device indicating the location to be reached immediately. This alert message could show:

The building, floor and room of the patient (e.g., 3C109);

Patient's case (e.g. heart stroke, arrhythmia, etc…) and

Figure 3. Locating nearest doctor
(Mohammadian & Jentzsch 2008).

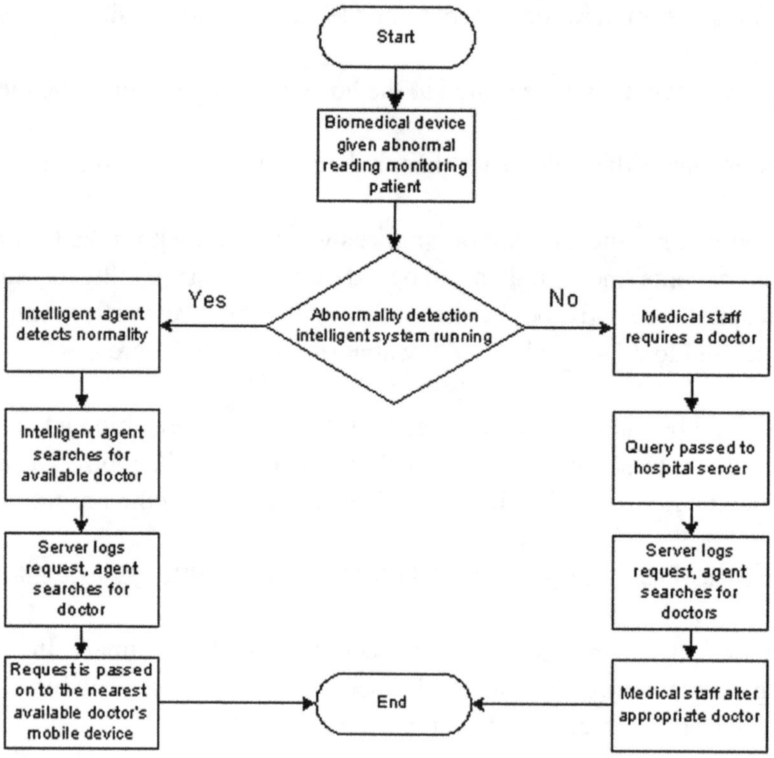

A brief description of the patient's case.

If the hospital is utilizing an intelligent agent as described in our proposed framework on its server, the process of locating and sending an alert message can be automated. This is done through comparing the collected medical data with previously stored abnormal patterns of medical data, then sending an automated message describing the situation. This system could be used to assist the staff in the patient monitoring unit or the nurse's workstation who observe and then sends an alert message manually.

Case III: Doctors Stimulates the Patient's Active RFID Tag using Their PDAs in Order to Acquire the Medical Data Stored in It

This method can be used in order to remove medical files and records placed at the front of the patient's bed. Additionally, it could help in preventing medical errors- reading the wrong file for the wrong patient and could be considered as an important step towards a paperless hospital. This case can be described in the following steps:

The doctor enters into the patient's room or ward to check the medical status of a certain patient. So instead of picking up the 'hard' paper medical file, the doctor interrogates the patient's RFID wrist tag with his RFID transceiver equipped in his\her Netbook, tablet, smart phone, or laptop.

The patient's RFID wrist tag detects the signal of the doctor's RFID transceiver coming from his\her wireless mobile device and replies back with the patient's information and medical data.

If there are multiple patients in the ward possessing RFID wrist tags, all tags can respond in parallel using Hands Down polling techniques back to the doctor's wireless mobile device. Another available option might be for the doctor to retrieve only the patient's number from the *passive* RFID wrist tag. Then through the WLAN the doctor could access the patient's medical record from the hospital's main server.

RFID technology has many potential important applications in hospitals, and the cases discussed cases provide practical examples of its uses. Two important issues can be concluded from this section: WLAN is preferred for data transfer; given that IEEE's wireless networks have much faster speed and coverage area as compared to RFID transceivers\ transponders technology. Yet, RFID technology is the best for data storage and locating positions of medical staff and patients as well.

Another point to be made is that an RFID Transceiver and programmer can be embedded in a Biomedical Device for data acquisition and dissemination, and the RFID Transceiver embedded in the doctor's wireless mobile device can be used solely for obtaining the medical data. With the continuing progress of RFID technology its use could become as standard as other wireless technologies (Bluetooth for example), and eventually manufacturers including them in all electronic biomedical devices.

5. CLASSIFICATION OF USER AND DATA FOR OPERATION/TRANSACTION CONTROL

A hospital can be impacted by the corruption, unauthorized access, or theft of its data. A data security breach can impact organization's operations as well as causing large financial, legal implications. It can impact the personal privacy and public confidence in such an organization. With the loss of manipulation of patient data in hospital human lives could be at risk by unauthorized access, corruption and modification of such a data.

An intelligent agent system has been developed to control the access from the users to the data stored in the database of a hospital. The intelligent agent system objective is to prevent the unauthorized use or observation of classified hospital data.

Discretionary Access Controls provide users with permits to allow or disallow other user's access to stored data. It is always possible however, that authorized users can pass data to unauthorized users. One way to restrict such access to data is the use of an intelligent agent capable of identifying user/s data access level. In such case each data item will contain metadata about its security and privacy (Mohammadian & Hatzinakos, 2009 and 2013). An intelligent agent can then control data access of users by checking metadata information attached to the data.

The data permission level is checked for each data item as soon as a user wishes to access a data item. If an authorized user obtains a data item and passes it to an unauthorized user, then the unauthorized user will not be able to access that data item.

Therefore each data item permits access privileges to their users based on the metadata stored with each data item. Access to a data item is left to the embedded discretion of an intelligent agent and the metadata of that data item. The intelligent agent then can provide access to authorized users based on the metadata values of a data item under their control without the intercession of other authorities such as a system administrator.

Using an intelligent agent provides a higher level of security to hospital data. As it is stated by Ferraiolo and Kuhn: "In many organizations, the end users do not ``own'' the information for which they are allowed access. For these organizations, the corporation or agency is the actual ``owner'' of system objects as well as the programs that process it (Ferraiolo & Kuhn, 1992).

Intelligent agent access control based on metadata stored with data items can provide another level of security to the already existing role-based access control (RBAC).

Data access controls in organizations are determined by user roles. In such organizations data access control decisions for user access can be based on responsibilities and duties of user/s. In a hospital situation the user role can be specified by the position of the user in the hospital such as doctor, nurse, administrator and pharmacist. Such user roles defines data access based on organization's policy and functions of user/s. Users in such environment cannot pass their access permissions on to other users but they may be able to pass the actual data they access to other users (Ferraiolo & Kuhn, 1992).

In such a case it is assumed that authorized user/s will not pass restricted data to unauthorized users and that they are aware and follow the organizations rules and governmental security and privacy laws. In hospital the data security and privacy associated with the diagnosis of ailments, treatment of disease, and the administering of medicine is of crucial importance (Ferraiolo & Kuhn, 1992).

Generally data security and privacy policies are set, controlled and maintained centrally by central security administration.

Each user or group of user in an organization is provided with access rights required for user/s to be able to access their required data and to perform their functions in the organization (Ferraiolo & Kuhn 1992). The central security administration grant/revoke access to users based on policies provided to them from government or/and organizations laws and regulations.

In such an environment data access policies provides authorization limits regarding access to sensitive data. It is possible to provide additional constraints on access and modification data by adding a metadata to support such constraint on a data item. Therefore each metadata attached to each role provides certain access and privileges to certain data items. Given the new proposed method of access control based on metadata stored the following benefits are obtained:

- *The security, privacy and confidentiality of data can be enforced further*. For example, in a hospital environment a doctor could be provided with read/write access to a prescription data whereas the pharmacist will have the read only access of prescriptions data.
- *Only authorized users can modify data items*. An unauthorized flow of information from a user with high level access to a user with a low level access is not possible.

There is a need to provide the minimum disruption when implementing this security and privacy control to an organization. Using intelligent agent technologies and metadata access control information for each data will minimize resources impact without needing to re-design databases in an organization such as a hospital. We can add extra information to each data item by adding metadata information to the attributes of each entity in relational-data bases and domains in classes in object-oriented databases.

Consider the simple relational database as shown below:

Patient (<u>Patient ID</u>, Name, Address, Tel No, Insurance ID)

Insurance (<u>Insurance ID</u>, Type, Insurance Provider ID)

Insurance Provider (Insurance Provider ID, Name, Address, Tel No, Fax No)

Doctor (Doctor ID, Name, Office No, Tel No, Pager No)

Patient Doctor (Patient Doctor ID, Patient ID, Doctor ID, Visit Date, Notes)

The meta-data information could be the value or degree of user roles and related policies for privacy and security for that data item. Metadata values can then be used for adaptation and implementation of access/operation performance identification with each data item in the above database. The meta-data values can be obtained from the knowledge workers of the organization based on organization policies, procedure and business rules as well as government requirements for data privacy and security. Table 2 shows the metadata values for table Patient attributes.

Now assume that the following domain metadata linguistic variables for the users (Doctor, Nurse, Pharmacists etc.) of data in a given hospital as: TP = "top access user", MD = "medium access user", LO = "low access user" and ZE ="no access to data".

For example Table 3 shows the metadata value related to security data access control of several kind of users based on organization's security access policy.

The values are in the range of 0 to 70, where seventy indicates the metadata for a user of the hospital data that has top (full) access and zero indicates the metadata for a user that has no access to the data in the hospital. Note that other values are also possible. For simplicity assume that the linguistic terms describing the meta-data for the attributes of entities in the above database have the values: TP = [35,..,70], MD = [25,..,37], LO = [15,..27], ZE = [17,..,0]

Based on each user ID metadata value for each user attributes the membership of that attribute to each linguistic variable can be calculated. In this case study Triangular fuzzy set was used to represent the data access classifications. The membership value of metadata for each user can be calculated for all these using the following formula:

Table 2. Metadata values for table patient attributes

Patient Attributes	Meta-Data Value of Data Security Access Control Access based on Organization Policy for Patient Data
Patient ID	70
Name	50
Address	29
Tel No	15

Based on Mohammadian & Jentzsch, 2008.

Table 3. Metadata values for different user

UserID	Meta-Data Value Given base on Organization Policy for Data Access
DoctorID	52
NurseID	29
Pharmacist ID	20

$$m_A(x) = 0, \qquad x < a_1$$

$$m_A(x) = \frac{x - a_1}{a_2 - a_1}, \quad a_1 \le x \le a_2$$

$$m_A(x) = \frac{a_3 - x}{a_3 - a_2}, \quad a_2 \le x \le a_3$$

$$m_A(x) = 0, \qquad x > a_3$$

where x is metadata value for the attribute user ID (e.g. Doctor ID, Nurse ID, Pharmacist ID etc.) and α_1, α_2 and α_3 are the lower middle and upper bound values of the fuzzy set data access security classification. Now assume that metadata value based on organization policy for users are as shown in Table 3. Based on the metadata value for each user the membership of each user to access and perform operation on data item can be calculated.

The degree of membership value of the attribute user ID based on metadata from Table 3 can then be calculated from the data shown in Table 4.

Now assume that the following access rights exist for each data item. NA = *"no access"*, RD = *"read access"*, WE = *"write access"*, RDWE =*"read and write access"*, DE = *"delete access"*, FA =*"Full access"*.

Now the data items and users of data items can be classified and categorized into fuzzy sets (with membership value), a process for determining precise actions (access rights) to be applied must be developed. This task involves writing a rule set that provides an action for any data access classification and user classification that could possibly exist. The formation of the rule set is comparable to that of an expert system, except that the rules incorporate linguistic variables with which human are comfortable. We write fuzzy rules as antecedent-consequent pairs of If-Then statements. For example:

IF*Organizational_Data_Access_Classification* is **TP** and

User_Data_Access_Classification is **TP**

Then*Level_of_Data_Access_Manipulation* is **FA**

The overall fuzzy output is derived by applying the "max" operation to the qualified fuzzy outputs each of which is equal to the minimum of the firing strength and the output membership function for each rule.

Table 4. Fuzzy membership of metadata value of users as specified in Table 3

μ(USERID)	TP	MD	LO
M DoctorID)	0.85	0	0
μ(NurseID)	0	0.66	0
μ(PharmacistID)	0	0	0.71

(Based on Mohammadian & Jentzsch, 2008).

Meta-Data Value Based on Organization Policy for Patient Data

Users' metadata and the metadata of each data item can be used to determine data access based on user security level and data security level for each data item. The precise actions that are allowed or not allowed on that data item by a given user can now be determined. The role of the intelligent agent is to perform data access authorization base on requested data by a user and to allow/disallow access to the data and the operations that can be performed on the data. The knowledge base shown in Table 5 is used by the intelligent agent in this research study for its decision making.

5.1 Framework of User Data Access for Medical Data

The previous section (Practical Cases using RFID Technology) focused on how to design a wireless framework to reflect how patient's medical data can be managed efficiently and effectively leading to the elimination of errors, delays and even paperwork. Similarly, this section will focus on the previously discussed framework from a security perspective, attempting to increase security and data integrity and user access control for:

- Acquisition of Patient's Medical Data;
- Doctors access data on the patient's active RFID tag using their wireless mobile devices in order to acquire the medical data stored in it.

The lower part of Figure 4 represents the physical (hardware) encryption layer. This part is divided into two parts. The left side demonstrates the case of a doctor acquiring patient's medical data via a passive RFID tag located in a band around the patient's wrist. The passive RFID tag contains only a very limited amount of information such as the patients name, date of admission to the hospital and above all his/her medical record number (MRN), which will grant access to the medical record containing the acquired medical data and other information regarding the patient's medical condition. This process is implemented in the following steps, and involves two pairs of encryption and decryption. The first encryption occurs after the doctor stimulates the RFID passive tag to acquire the patient's MRN, so the tag will encrypt and reply back the MRN to the doctors PDA for example. Then the doctor will decrypt the MRN and use it to access the patient's medical record from the hospital's server.

Finally the hospital server will encrypt and reply back the medical record, which will be decrypted once received by the doctor's PDA. This action is performed based on the assumption that the intelligent

Table 5. Metadata values for table patient data access

μ(DoctorID)	FA	RD	WE	RDWE	DE
PatientID	Not allowed	Allowed	Allowed	Allowed	Not Allowed
Name	Allowed	Allowed	Allowed	Allowed	Not Allowed
Address	Allowed	Allowed	Allowed	Allowed	Not Allowed
TelNo	Allowed	Allowed	Allowed	Not Allowed	Not Allowed

(Mohammadian & Jentzsch, 2008).

Figure 4. Functional flow
(Mohammadian & Jentzsch, 2008).

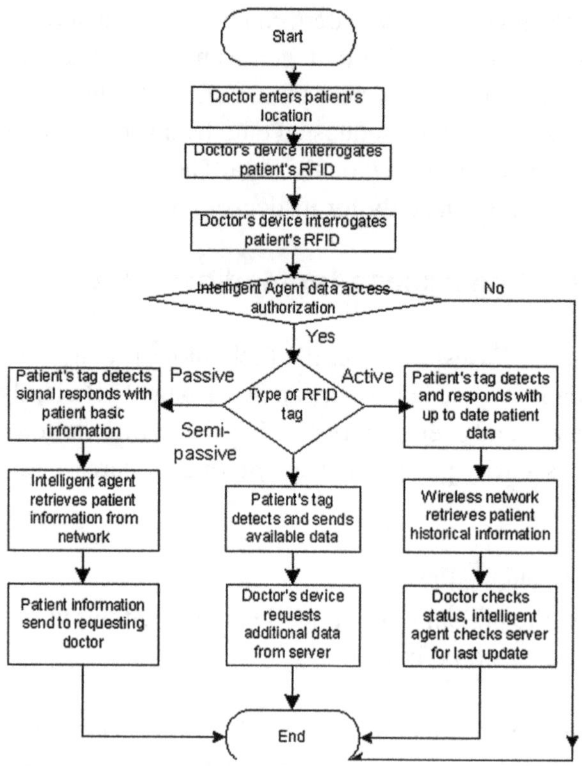

agent for data access control authorizes the doctor based metadata attached to the Doctor ID and the metadata attached to the patient data items, to access the patient data.

The right side of Figure 4 represents a similar case but this time using an active RF tag. This process involves only one encryption and decryption. The encryption occurs after the doctor stimulates the active RFID tag using her PDA which has an embedded RFID transceiver, allowing the tag to interact with the medical data encrypted. Then the received data is then decrypted through the doctor's PDA.

The upper part of Figure 4 represents the application encryption layer. Requiring the doctor to enter a password to decrypt and then access the stored medical data. So whenever the doctor wants to access patient's medical data, he\she simply enters a certain password to grant access to either wireless mobile device or a hospital server depend where the medical data actually resides. This is an extra security measure.

In conclusion securing such medial data seems to be uncomplicated, yet the main danger of compromising such data comes from the people managing it, e.g., doctors, nurses and other medical staff. For that, we have noted even though the transmitted medical data is initially encrypted from the source, doctors have to run application level encryption on their wireless mobile devices in order to protect this important data. Nevertheless, there is a compromise.

Increasing security through the use of an intelligent agent that performs the required check and authorization as described above can improve data security in this situation. It is also possible to increase length of encryption keys. However these actions will decrease the encryption\decryption speed and causes unwanted time delays. As a result, this could delay medical data sent to doctors or on-line monitoring units.

Encryption and decryption process involved in handling patient data can slow down the systems that require access / store patient data. Mohammadian et al. suggested a data classification method to identify and apply appropriate policy and security settings such as private access control and encryption requirements (Mohammadian & Hatzinakos 2009, 2013). Data classification cannot alone solve all problems related to data security and data privacy and there is a need to add security and access control privilege to applications that access data. It is envisaged that a fuzzy data classification method can be developed for data security and privacy in hospital environment using RFIDs. The process of such a data classification base on fuzzy logic can be summarized as:

- Enter security, privacy meta data values for each attributes of all table in database
- Classify data based on its security, privacy meat-data values using fuzzy logic classifier and apply the fuzzy rules for actions to be taken
- Modify data security classification based on results obtained from fuzzy classification system.

Figure 5 shows the process of decision making to classify data and level of encryption required base on a fuzzy logic classifier (Mohammadian & Hatzinakos, 2009). It is assumed that the meta-data for level of security and privacy is assigned to each data item and the fuzzy logic rules for decision making are created.

As an example consider that a hospital wishes to classify their patient data x_k, $k = 1,..,p$ which are stored in their database. There exist government and regulatory policies $G_i, i = 1,...,n$ and organizational policies $P_j, j = 1,...,m$. Assume that every data item x_k is evaluated and has been assigned meta-data values representing values or degrees of government regulatory security, privacy a_k and organizational policies b_k for that data item. Assume that a_k and b_k values are in the range of 0 to 70, where zero indicates the meta-data for a data item that is public and 70 indicates the meta-data for a data item that is top secret. Assume that the linguistic terms describing the meta-data for each data item x_k in the above database are set to be TP = top secret, SE = "secret", CO ="confidential", MC = "mission

Figure 5. Data retrieval based on proposed fuzzy logic system
(Mohammadian & Hatzinakos, 2009).

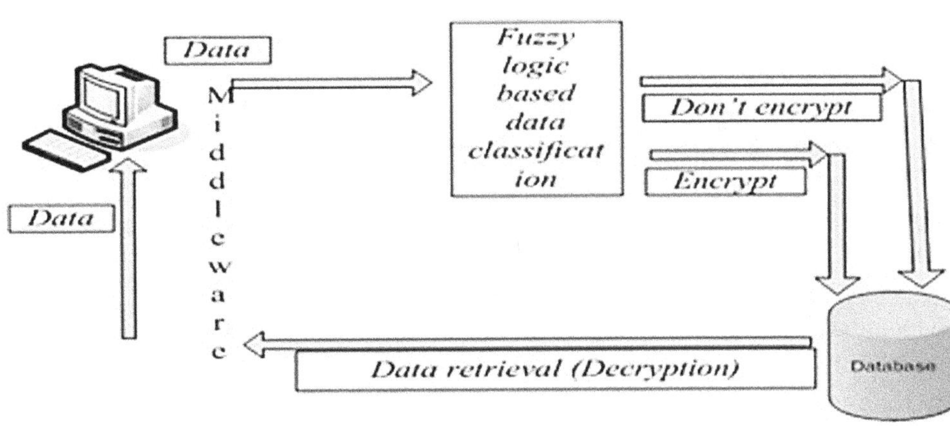

critical", NC = "not critical", PR = "private but not top secret", PU = "Public" with the following values for each linguistic variable TP = [58,..,70], SE = [48,..,60], CO =[37,..,50], MC = [28,..,40], NC = [16,..,30], PR = [8,..,20], PU = [0,..,10]. Using these meta-data values we construct discrete fuzzy sets Y_i and Z_j on the set of alternatives A_{alt} such that:

$$Y_j = \left\{\left(x_a, a_{1i}\right), \ldots, \left(x_p, a_{pi}\right)\right\}, \quad i = 1, \ldots n$$
$$Z_j = \left\{\left(x_a, b_{1j}\right), \ldots, \left(x_p, a_{pj}\right)\right\}, \quad j = 1, \ldots m$$

Then a decision formula can be constructed given by the following formula:

$$D = Y_1 \cap \ldots . Y_n \cap Z_1 \cap Z_m$$

where

$$\mu_k = \min\left(a_{k1} \ldots, a_{kn}, b_{k1}, \ldots, b_{km}\right), k = 1, \ldots, p$$

Polices with the highest membership grade among $\mu_1, \ldots, \mu_p$ will be considered the policy to be applied for the required level of encryption. Now assume that there are government and regulatory policies and organizational polices for a given data item $x_i, i = 1, .., 2$ form the set of alternatives $A_{alt} = \left\{x_1, x_2\right\}$ where the membership of x_1 in fuzzy set of government polices is $\mu_{CO} = 0.3$ and $\mu_{MC} = 0.16$. The membership of x_2 in fuzzy set of organizational polices is $\mu_{TP} = 0.8$.

Now a decision formula can created:

$$D = \left\{\left(x_1, 0.3\right), \left(x_2, 0.8\right)\right\}.$$

Now the policy x_2 has the highest membership value 0.8 hence this membership value will be applied to the consequence of the fuzzy rule base for the hospital that are triggered by the condition values of x_1 and x_2. Finally the result will be calculated by applying the decision value 0.8 (obtained from the decision formula) to the consequence of the rules which are applicable to values of values of x_1 and x_2. Such a fuzzy rule based system for encryption/ decryption can be adapted from Mohammadian and Hatzinakos (Mohammadian & Hatzinakos 2009).

6. SECURE RFID DATA TRANSMISSION IN HOSPITAL

Due to RFID's inherent broadcasting nature, wireless communications typically pose significant challenges on data security and protection, including susceptibility to unauthorized wireless data interception. Although many privacy related issues can be addressed by security mechanisms, the protection of the source location confidentiality using conventional network security methods appears untenable.

Location privacy is an important security issue, the lack of which can lead to subsequent exposure of significant traffic information on the network and the physical world entities. For instance, cardiologic data packet coming out of a hospital in a mesh network can enable an eavesdropper to analyze and find out at-risk heart patients, if the source location of those packets can be determined. Toward that goal, a number of source-location communication protocols have been proposed (Deng, Han, & Mishra, 2004, Shao, Yang, Zhu & Cao, 2008), where the main idea is a mixture of valid and fake messages. Each node transmits either a valid or a fake message, consistently. The main disadvantage of this approach is that the broadcasting of fake messages consumes significant amount of the limited energy in each sensor node. Moreover, because each node has to transmit a packet in every time slot, the effect is an increase in the number of collisions, and a decrease in the packet delivery ratio. Therefore, these approaches are not suitable especially for large scale wireless sensor networks.

Routing based protocols can also provide source-location privacy. In this paper we use the Panda-Hunter model to formalize the problem in sensor networks and propose a phantom routing technique based on both flooding and single path routing. Phantom routing involves two phases: a random walk phase, and a subsequent flooding/single path routing. Random walk is inefficient in making a fake "phantom" source far enough from the actual source. To address this problem, a direct random walk is proposed. This can be achieved by storing direction information in the header of the message. The exposure of the direction information decreases the complexity for adversaries to trace back to the true message source. One of the looming challenges that threaten the successful deployment of the proposed scheme in a WLAN is source-location privacy, especially when a network since the data that are transferred is extremely sensitive. If all the data information – data packets - that are collected in one room follow the same routing path toward the destination, it would be easy for an eavesdropper to track the direction of the transmitter of each data packet and eventually track the source and reveal the room of the data transmission. Figure 6 shows an illustrative example.

Figure 6. An illustrative example

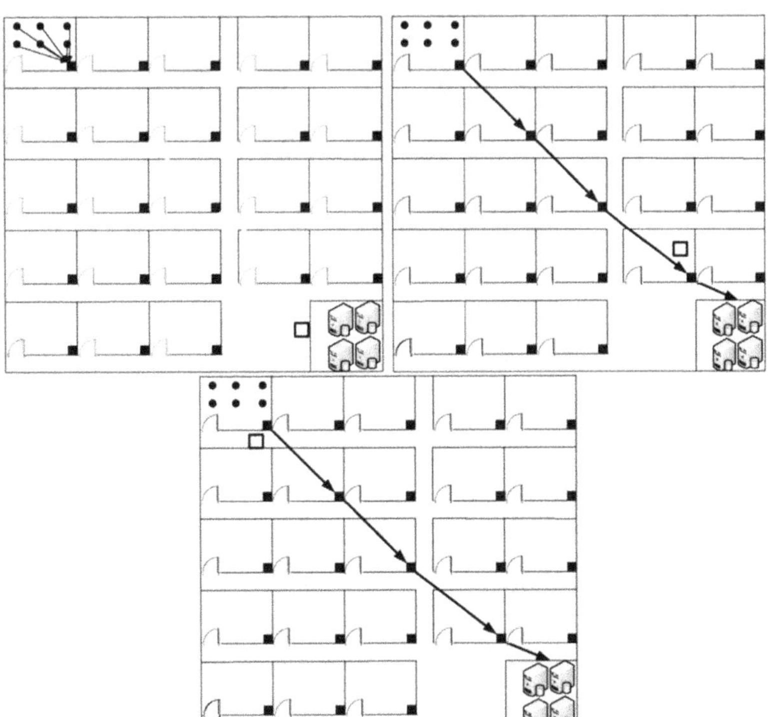

If the set V_s is empty, meaning that there is no available node, within the range of the sender node, which will make process toward the destination, the relay node is not updated. This comes from the fact that in the next time slot there will be other nodes available (because node availability is independently generated) that could provide advancement toward the destination. Finally, after a number of packet transmissions, a number of different paths toward the destination will be discovered, as illustrated in the Figure 7.

The eavesdropper is the square beside the central database room and follows the strategy as described in Algorithm 1.

Figure 7. Different paths toward the destination

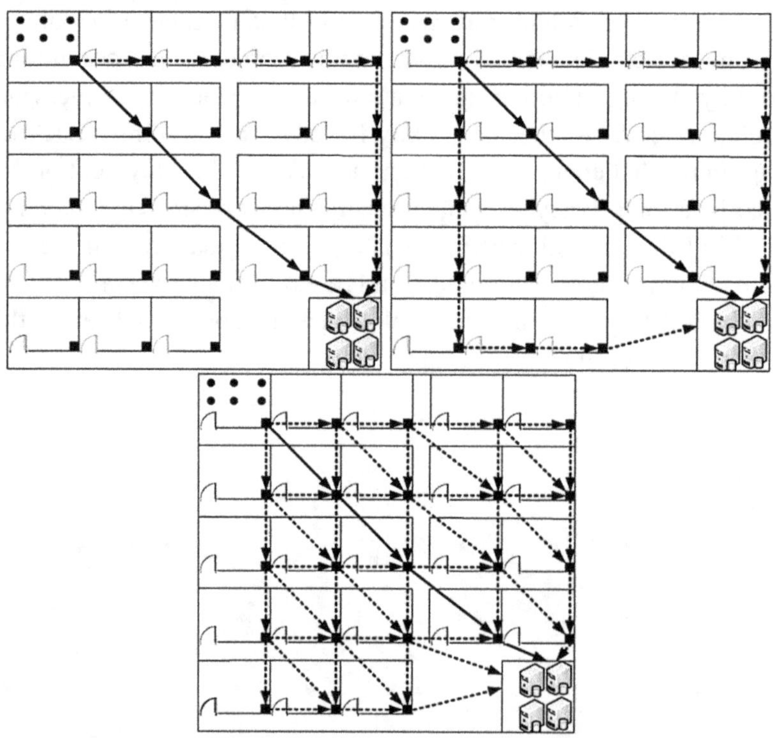

Algorithm 1. Eavesdropper strategy

```
eavesdropperLocation=destination;
while (eavesdropperLocation != sourceLocation) do
        overhear(node[eavesdropperLocation]);
        message= node[eavesdropperLocation].ReceivedMessage();
        if (message == isNewMessage()) then
                MoveTo(message.GetSenderNode();));
                eavesdropperLocation= message.GetSenderNode();
        end
end
```

When a packet reaches the database room, the eavesdropper is moving toward the direction of which the packet came from. With that back tracing technique, the eavesdropper will eventually reveal the exact location of the room that is sending the packets.

In order to enhance source location privacy in the above scenario, we are proposing the use of an opportunistic mesh networking scheme. Opportunistic routing is a multi-hop routing that changes the path between the source and the destination dynamically, according to network conditions in each time slot.

We are using four types of packets during the packet relaying process: Request To Send(RTS), Confirm To Send(CTS), DATA and ACK. RTS/CTS are used during the handshake process between neighbor nodes while ACKs are used for verification of DATA delivery.

When a node ***s*** has to transmit a packet, first it broadcasts a RTS packet, in which it includes its own address and the destination address, ***d***, and then node ***s*** keeps listening. All the surrounding nodes which are in the range of s are able to hear this request,conforming a set of candidate nodes E_s. There is a subset $V_s \leq E_s$ conformed by any node $i \in Es$ that satisfying the condition$c_{i,d} < c_{s,d}$so, $V_s = \{i \in E_s | c_{i,d} < c_{s,d}\}$

If a node is in V_s subset and is available for receiving a packet, and there aren't any packets in its buffer waiting to be send, it should send a CTS packet back to the sender node ***s***. In order to prioritize the nodes based on their distance from the destination,each node $i \in V_s$ initialize a timer, with timeout period T_i, which is inverse proportional to the difference $c_{s,d} - c_{i,d}$ and can be determined as follows:

$$T_i = \frac{C_0}{C_{s,d} - C_{i,d} + SIFS}, i \neq d$$

where C_0 is a constant and SIFS is the smallest time interval composed of the module processing time and the transceiver RX/TX switch time. In the next step, node i backs off for the period T_i. If the data channel is free after that period, node i sends a CTS to the sender node, otherwise it quits. After that procedure, the sender node s will receive the first CTS from the node which is closer to the destination, and this will be the next hop relay node and it will receive the DATA packet. When the next node receives the DATA packet it replies with an ACK to the sender and follows the same procedure until the DATA packet reaches the destination. In the case that the sender node receives more than one CTS packets simultaneously there are certain mechanisms in the sender node, such as cyclic redundancy check (CRC) that can detect this collision and differentiate the nodes. When a node i sends a CTS it waits for time T_w to receive the DATA packet from the sender node s, otherwise it goes back to its previous mode. T_w is the time needed for the sender to transmit the data to that node and can be defined as:

$$T_w = d(s,i) \cdot D_0 + SIFS$$

where $d(s, i)$ is the distance between the sender and the relay candidate and D_0 is a constant. In the same way, the sender node has to wait for T_c time to get a CTS packet before it broadcasts a RTS again. T_c is the time needed for a node which is located at the limit of the range R of the sender node, and can be defined as:

$$T_c = R \cdot C_0 + SIFS$$

where R is the range of the sensor. The time that a sender node will wait for an ACK before it retransmits the DATA can be also defined as:

$$T_A = d(s,i) \cdot A_0 + SIFS$$

where A_0 is a constant.

We illustrate the algorithm for the source node s and any relay candidate node in Algorithm 2 and Algorithm 3, respectively.

7. CONCLUSION

This research in the wireless medical environment introduces new ideas in conjunction to what is already available in the RFID technology and wireless networks.

With the reduction in cost of radio frequency identification (RFID) technology, it is expected the increased use of RFID technology in healthcare in monitoring patients and assisting in health care administration. An intelligent agent using fuzzy logic techniques is implemented for profile matching based on Mohammadian and Jentzsch, 2008. A second intelligent agent system is developed for data/user access control and classification to improve the application of regulatory data requirements for security and privacy of data exchange. Finally an approach for passing data packets in the hospital is proposed.

Algorithm 2. Source node

```
if (isNewMessage(msg)) then
        BroadcastRTS();
        interval=T_R;
        reason=Listen(interval);
        while (reason!=CTS) do
                BroadcastRTS();
                interval=T_R;
                reason=Listen(interval);
        end
        RelayNode= CTS.Sender.Node;
        SendMessage(msg.RelayNode);
        interval=T_A;
        reason=Listen(interval);
        while (reason!=ACK) do
                SendMessage(msg.RelayNode);
                interval=T_A;
                reason=Listen(interval);
        end
        GoToSleepMode();
end
```

Algorithm 3. Candidate node

```
if (isRTS(rts)) then
        Ti = CalculateBackoff();
        wait(Ti);
        Channel=ChannelSensing();
        if (Channel==IDLE) then
                SendCTS(rts.SenderNode);
                interval=Tw;
                reason=Listen(interval);
                if (reason==DATA) then
                        SendACK(DATA.SenderNode);
                 else
                        GoToSleepMode();
                 end
         else
                 GoToSleepMode();
         end
end
```

This new approach will provide more secure data transmission in the hospital. Part of the research study presented in this paper has already been published in journals and presented in conferences and book chapters over the past few years (Mohammadian & Jentzsch, 2007, 2008, 2012). The reference to these publications is provided in this manuscript when appropriate

Finally it is noted that data classification is a fundamental requirement for adequate information privacy and security. The consequences for not fully implementing a data classification scheme can be severe in financial sense and organization's reputation. However many organizations do not have a classification scheme for their data. Development of such data classification can be very expensive. The cost of data classification is to first to develop classification schema and then the training cost involved for the personnel to recognize data and classify data. In many case large amount of data in organization are unclassified. On the other hand some small to medium size organizations tend to classify all their data as confidential and encrypted all their data even though some of these data do not require encryption. We have proposed a method based on Mohammadian and Hatzinakos 2009 in this research study to provide a suitable data classification based on the meta-data values and the fuzzy logic system. Allocation of meta-data values for classification is not a simple or intuitive process. It requires a careful evaluation of data against a broad range of organization and regulatory policies which can make this process complex. But once data classification is accomplished the implementation of security and privacy can then successfully is performed. It should be noted that the proposed data classification method requires knowledgeable employees to recognize and classify data accordingly. A substantial effort is required to classify existing data and to continue classify new data and re-classify some existing data. Future work in this area will explore the notion of time related to classified data based on the proposed fuzzy data classification to allow for re-classification of data after certain time.

ACKNOWLEDGMENT

The authors would like to acknowledge the initial research work performed on this project at the University of Canberra by MIT students. The authors would like also to acknowledge the research work on Data Classification that was performed initially by A/Professor Masoud Mohammadian and Professor Dimitrios Hatzinakosat the University of Toronto and used in this research study. Part of the research study presented in this paper is published in journal and conferences in the past few years. The reference to these publications is provided in this manuscript when appropriate.

REFERENCES

Angeles, R. (2007). An empirical study of the anticipated consumer response to RFID product item tagging. *Industrial Management & Data Systems, 107*(4), 461–583. doi:10.1108/02635570710740643

Bhuptani, M., & Moradpour, S. (2005). *RFIDfield guide - deploying radio frequency identification systems*. New York, NY: Prentice Hall.

Bigus, J. P., & Bigus, J. (1998). *Constructingintelligent software agents with Java – A programmers guide to smarter applications*. San Francisco, CA: Wiley.

Cline, J. (2007). *Growing pressure for data classification*. Retrieved from http://www.computerworld.com/action/article.do?articleId=9014071&command=viewArticleBasic

Deng, J., Han, R., & Mishra, S. (2004). Intrusion tolerance and anti-traffic analysis strategies for wireless sensor networks. *Proceedings of the 2004 International Conference on Dependable Systems and Networks* (p. 637). Washington, DC: IEEE Computer Society. 10.1109/DSN.2004.1311934

Do, H., & Rahin, E. (2002). Coma: A system for flexible combination of schema matching approaches. *Proceedings of 28th Conference on Very Large Databases*, Morgan Kaufmann, USA (pp. 610-621). 10.1016/B978-155860869-6/50060-3

Doan, A-H., Lu, y., Lee, T., & Han, J. (2003). Profile-based object matching for information integration. *IEEE Intelligent Systems Magazine*, 54-59.

Ferraiolo, D. F., & Kuhn, D. R. (1992). *Role based access control. 15th National Computer Security Conference* (pp. 554-563). Baltimore, MD. USA.

Finkenzeller, K. (1999). *RFID handbook*. San Francisco, CA: John Wiley and Sons.

Glover, B., & Bhatt, H. (2006). *RFID Essentials*. California, USA: O'Reilly Media, Inc.

Mickey, K. (2004). RFID grew in 2002. *Traffic World, 5*, 20–21.

Mohammadian, M., & Hatzinakos, D. (2009). Data classification process for security and privacy based on a fuzzy logic classifier. *International Journal of Electronic Finance, 3*(4), 374–386. doi:10.1504/IJEF.2009.028977

Mohammadian, M., & Hatzinakos, D. (2013). An adaptive intelligent fuzzy logic classifier for data security and privacy in large databases. In I. Harvey, A. Gavoukian, G. Tomok, G. Borrett, D. Hatzinakos, & H. Kwan (Eds.), *Smart data privacy meets evolutionary robotics* (pp. 161–172). New York, NY: Springer.

Mohammadian, M., Hatzinakos, D., & Spachos, P. (2012). *Computational intelligence for user and data classification in hospital software development. Artificial Intelligence Applications and Innovations - AIAI* (pp. 145–154). Halkidiki, Greece: Springer.

Mohammadian, M., & Jentzsch, R. (2007). Intelligent agent framework for RFID. In L. Al-Hakim (Ed.), Web mobile-based applications for healthcare management (pp. 316-335). Hershey, PA: IGI Global.

Mohammadian, M., & Jentzsch, R. (2008). Intelligent agent Framework for secure patient-doctor profillingand profile matching. *International Journal of Healthcare Information Systems and Informatics, 1*, 1–10.

Mohammadian, M., & Jentzsch, R. (2012). User and data classification for a secure and practical approach for patient - doctor profiling using an RFID framework in hospital. In S. Samikar, D. Bennett, M. Gaynor, & M. Gaynor (Eds.), *Cases on healthcare information technology for patient care management* (pp. 224–253). Hershey, PA: IGI Global.

Pramatari, K. C., Doukidis, G. I., & Kourouthanassis, P. (2005). Towards 'smarter' supply and demand-chain collaboration practices enabled by RFID technology. In P. Vervest, E. Van Heck, K. Preiss, & L. F. Pau (Eds.), *Smart business networks* (pp. 197–210). New York, NY: Springer. doi:10.1007/3-540-26694-1_14

Schuster, E. W., Allen, S. J., & Brock, D. L. (2007). *Global RFID: The value of the EPCglobal network for supply chain management*. New York, NY: Springer.

Shaalana, K., El-Badryb, M., & Rafeac, A. (2004). A multiagent approach for diagnostic expert systems via the internet. *Expert Systems with Applications, 27*(1), 1–10. doi:10.1016/j.eswa.2003.12.018

Shao, M., Yang, Y., Zhu, S., & Cao, G. (2008). Towards statistically strong source anonymity for sensor networks. Proceedings of *INFOCOM2008 the 27th Conference on Computer Communications* (pp. 51-55). Phoenix, AZ.

Shepard, S. (2005). *RFID: Radio frequency identification*. New York, NY: McGraw-Hill.

Tindal, S. (2008). ZDNet Australia. *One in ten RFID projects tag humans*. Retrieved from http://www.zdnet.com.au/news/hardware/soa/RFID-people-tagging-benefits-health-sector-/0,130061702,339285273,00.htm

Weinstein, R. (2005). RFID: A technical overview and its application to the enterprise. *IT Professional Magazine, 7*(3), 27–33. doi:10.1109/MITP.2005.69

Whiting, R. (2004). MIT = RFID + Rx. *Information Week, 988*, 16.

Wooldridge, M., & Jennings, N. (1995). Intelligent software agents: Theory and Practice. *The Knowledge Engineering Review, 10*(2), 115–152. doi:10.1017/S0269888900008122

Zadeh, L. A. (1965). Fuzzy sets. *Information and Control, 8*(3), 338–352. doi:10.1016/S0019-9958(65)90241-X

KEY TERMS AND DEFINITIONS

Discretionary Access Controls: Discretionary Access Controls provide users with permits to access allow or disallow other users access to data items stored in a hospital. Such user or user groups with certain access permissions is allowed to access data items. It is possible that authorized users can then pass such data items to other users that may not have permission access to such a data. One way to restrict access to data items can be based an intelligent agent being capable to identity user/s data access level to which data can be provided. In this case each data item contains metadata information about its security and privacy. An intelligent agent can then control data access of users by checking metadata information attached to data.

Location Privacy: Location privacy is an important security issue. Lack of location privacy can lead to subsequent exposure of significant traffic information on the network and the physical world entities. For instance, cardiologic data packet coming out of a hospital in a mesh network enable an eavesdropper to analyze and find out at-risk heart patients, if the source location of those packets can be determined. Toward that goal, a number of source-location communication protocols have been proposed (Deng & Mishra, 2004; Shao, Yang, Zhu, & Cao, 2008), where the main idea is a mixture of valid and fake messages. Each node transmits either a valid or a fake message, consistently. The main disadvantage of this approach is that the broadcasting of fake messages consumes significant amount of the limited energy in each sensor node. Moreover, because each node has to transmit a packet in every time slot, the effect is increase in number of collisions, and decrease in the packet delivery ratio. Therefore, these approaches are not suitable especially for large scale wireless sensor networks.

Multi-Agent System (MAS): A Multi-agent System (MAS) consists of a set of intelligent agents software that are connected together to perform tasks in an environment to achieve a common goal. To achieve a common goal successfully the agents cooperate and collaborate with each other.

Panda-Hunter Model: Routing based protocols can provide source-location privacy. The Panda-Hunter model formalizes the problem in sensor networks and it proposes a phantom routing technique based on both flooding and single path routing. Phantom routing involves two phases: a random walk phase, and a subsequent flooding/single path routing.

Patient or Doctor Profiling: A patient or doctor profile is a collection of information that can be used in a decision analysis situation between the doctor, domain environment, and patient. A static profile is kept in pre-fixed data fields where the period between data field updates is long such as months or years. A dynamic profile is constantly updated as per evaluation of the situation in which the situation occurs. The updates may be automated or performed manually. The automated user profile building is especially important in real time decision-making systems. The profiling of patient doctor model is based on the patient/doctor information.

RFID: RFID or Radio Frequency Identification is a progressive technology that is easy to use and well suited for collaboration with intelligent software agents. Basically an RFID can be read-only, volatile read/write; or write once / read many times. RFID are non-contact; and non-line-of-sight operations. Being non-contact and non-line-of-sight will make RFIDs able to function under a variety of environmental conditions and while still providing a high level of data integrity (Finkenzeller, 1999; Glover & Bhatt, 2006).

This research was previously published in the Handbook of Research on Advancing Health Education through Technology edited by Victor C.X. Wang, pages 452-479, copyright year 2016 by Medical Information Science Reference (an imprint of IGI Global).

Chapter 15
The Development of a Secure Hospital Messaging and Communication Platform

Hoda Moghimi
RMIT University, Australia

Nilmini Wickramasinghe
Epworth HealthCare, Australia & Deakin University, Australia

ABSTRACT

Pagers and phone conversations have been the stalwarts of hospital communication. With good reason - they are simple, reliable and relatively inexpensive. However, with the increasing complexity of patient care, the need for greater speed and the general inexorable progress of health technology, hospital communication systems appear to be increasingly inefficient, non-secure, and inadequate. Thus, this study is proposed to answer the key research question: How can ICT (information communication technology) solutions ameliorate the current challenges regarding communication inefficiencies within healthcare? To answer this question, the study will design and develop a bespoke ICT solution for a specific context using three strong theories; communication theory, activity theory and agency theory to make a robust body of knowledge for the proposed solution. Further, it will serve to establish proof of concept, usability and feasibility of the proffered solution. The study participants will be selected from medical and nursing staff.

INTRODUCTION

The penetration rate of mobile devices such as smartphone and tablet computers has increased globally and will continue to increase in the future. The number of global smartphone subscribers is expected to reach 3.5 billion by 2019 ("Forrester Research," 2016). Due to the attractive features such as cost-effective sensors and wireless communication capabilities mobile devices have received great attention in the healthcare context. Kang et al. (2010) have notion that the ability to monitor patients' health remotely is making mobile devices popular in the health domain. Mobile devices such as smartphone can

DOI: 10.4018/978-1-5225-6198-9.ch015

measure heart rate, count the steps walked, and can tell calories consumed and utilized (Yu Rang, Yura, Guna, Jae Ho, & Soo-Yong, 2015). Mobile devices are being used in health care in different ways. For instance, patients are using mobile devices for consultation with doctors (Korzep, 2010). On the other hand, doctors are using mobile technology for real time monitoring/tele monitoring of patients (Slaper & Conkol, 2014; Zangbar et al., 2014). Therefore, mobile devices are changing the way healthcare is delivered and are offering mobility, flexibility, convenience and real time communication in healthcare.

The concept of using mobile devices in healthcare can be viewed as follows: any equipment (with different sensors) which can be worn as wrist band, implanted in the body or embedded with the living species, measuring different physical changes of the patients, monitoring physical activities, analyzing, alerting and communicating with healthcare professionals and patients from remote places (Dwivedi, Shareef, Simintiras, Lal, & Weerakkody, in press). The use of mobile devices in healthcare is benefiting the healthcare domain in many ways such as reducing cost and traveling time for patients and healthcare professionals. Hence, the concept of using mobile devices in healthcare is more beneficial than traditional healthcare services where patients have to wait in long queues and doctors remained overburdened.

The use of mobile devices in healthcare has enhanced the scope of health care services and in the near future will make health services more flexible (Agosti, Graziano, Artiaco, & Sorrentino, 2009; Ben-Zeev et al., 2013; I.-L. Wu, Li, & Fu, 2011). For example, Rana, Hume, Reilly, and Soar (2015) proposes an ensemble sensing network named w-health (wireless health) which can combine smart phone, smart watch and smart glass in one network and can be used in telehealth in the near future. Andersen, Lindgaard, Prgomet, Creswick, and Westbrook (2009) think nurses and clinician tasks performed during ward rounds, require highly mobile computing devices. With a clinical alert system, a message can be sent to doctors or nurses mobile device and they can talk while they walk. Further, use of mobile devices such as smartphones, pagers, tablets and Wi-Fi phones can speed up admit and discharge process, response quickly to patients, promote direct communication between clinicians and can communicate test results efficiently. However, the Academy of Australian Technology, Science and Engineering (ATSE) report says that assistive technologies such as mobile devices could play a vital role in reducing health care cost when they are socially accepted and widely adopted. Although, globally mobile devices are benefiting the healthcare system and widely used in healthcare however, their adoption is slow in the Australian healthcare system. Mobile devices such as tablets are used by doctors for only a small portion of clinical tasks such as sending reminders to the patients for next consultation. The Australian health department is in favor of using mobile devices (HCSC 2014). However, the majority of consultations are conducted face-to-face even after implementation of videoconferencing in primary care in Australia (Smith, Armfield, Croll &Gray 2012). An extensive systematic literature review reveals most of the mobile devices projects in the Australian healthcare system are implemented on a pilot basis. Hence, the uptake of mobile devices and their further use in the Australian healthcare is unknown.

Therefore, the objectives of this chapter are:

- To understand the perception of healthcare professionals about the use of the mobile devices in the Australian healthcare environment.
- To contribute to the knowledge of the adoption of mobile devices in the healthcare context.

The overall aim of this chapter is to develop a conceptual framework for the adoption of mobile devices in the Australian healthcare environment.

BACKGROUND

Literature review reveals factors influencing adoption of mobile devices can be studied both at hospital level and individual level (Andersen et al., 2009; Brown III, Yen, Rojas, & Schnall, 2013; Kay, 2011; Rana et al., 2015; Shareef, Kumar, & Kumar, 2014; West, 2012; I.-L. Wu et al., 2011). This chapter is mainly focused on the individual user's (healthcare professionals) level of adoption of mobile devices in the Australian healthcare context because healthcare professionals such as doctors and nurses are one of the key users to use technology in healthcare. From literature it appears that even after implementation of mobile devices in the Australian healthcare system they are not used up to their full potential. Therefore, understanding the factors influencing adoption of mobile devices at the healthcare professional's level may increase the use of mobile devices in healthcare. The increased use of mobile devices can ensure the long term profitability and reduction in healthcare cost (Boulos, Wheeler, Tavares, & Jones, 2011; Lim et al., 2011; Xue et al., 2012). Hence, the conceptual framework in this chapter is developed from the healthcare professional's perspective.

Use of mobile devices in healthcare is a new concept globally including Australia (Dwivedi et al., in press). Therefore, an extensive literature review is conducted for the adoption of various kinds of Health Information Technology in the healthcare domain. Literature review conducted for the purpose of this chapter resulted many factors influencing adoption of various technologies in the healthcare environment (Baig, 2010; Brewster, Mountain, Wessels, Kelly, & Hawley, 2014; Furukawa, Raghu, Spaulding, & Vinze, 2008; M. P. Gagnon et al., 2012). However, all factors given in reviewing the literature cannot be considered important to understand the adoption of mobile devices in the Australian healthcare environment because of the cultural differences (Michael J. Ackerman, Rosemarie Filart, Lawrence P. Burgess, Insup Lee, & Poropatich, 2010; Peddle, 2007; Tiong, Hafeez-Baig, Gururajan, & Soar, 2006). Further, some factors for the adoption of technology in healthcare are given from other perspectives such as organizational perspective and not from the healthcare professionals' perspectives. To understand factors in this study the authors have given importance to three dimension: technology (mobile device), context (healthcare) and individual (healthcare professionals) (N. K. Bradford, Young, Armfield, Herbert, & Smith, 2014).

First Component: Technology is considered important to design the conceptual framework in this chapter because the healthcare domain is concerned with the health of a person and technology used should have good supporting features. Ben-Zeev et al. (2013) developed a smartphone system for self-management of Schizophrenia named FOCUS and (during first testing of system) found that individuals were facing problems in using mobile devices in healthcare. The problems were: with abbreviations, large amount of text, difficult wording, font size, smaller touch space, small diameter of the buttons and smaller touch sensitivity. This means mobile devices used should have favoring features with the healthcare domain which healthcare professional can easily understand and operate. Further, health care professionals are responsible for the life of the patient and the technology used should be secure (Bønes, Hasvold, Henriksen, & Strandenæs, 2007). Therefore, technology context is considered important and the factor considered in technology component are: features of mobile devices and security issues.

The Second Component: Context (environment) is considered important to design the conceptual framework because uses of mobile devices have different influence on different domains. To choose this component mobile devices uses are linked with the healthcare context. In this component the main emphasis is given on how the use of mobile devices bring changes in clinical processes and healthcare professionals style of working. Healthcare professionals like the technology that can do clinical work efficiently and is compatible with the clinical work process. Mobile devices have potential to do clinical tasks efficiently. So the use of mobile devices in clinical context is considered important and factors considered in 'context' component are: advantages, complexity and compatibility.

The Third Component: Healthcare professionals are considered important to design the conceptual framework because they are the main users of the technology in the healthcare system. If heath care professionals consider that the use of mobile devices in healthcare is advantageous their intention will be strong to adopt mobile devices (I.-L. Wu et al., 2011). The factors considered in the individual component are: self-efficacy, individual readiness, social influences, intention, age, gender

Theory of planned behavior (TPB) and diffusion of innovation theory (DOI) are used as an underlying theory for developing the conceptual framework in this chapter. Several researchers have used various theories and models such as Technology Acceptance model (TAM), Theory of Reasoned Action (TRA), Theory of Planned Behavior (TPB), Diffusion of Innovation Theory (DOI), Unified Theory of Technology Acceptance and Utilization (UTAUT) to understand individual adoption of technology behavior (Daim, Basoglu, & Topacan, 2013; Deng, Mo, & Liu, 2013; Dwivedi et al., in press; Saad, Alias, & Ismail, 2013; Sanders et al., 2012). TAM is the most widely used model in literature to understand adoption of technology behavior. However, TAM is not suitable to understand adoption behavior in this research because TAM does not include the influence of social norm. The influence of social norms is important in this research context and can be explained using TRA. However, TRA does not include the basic predictors that is perceived usefulness and perceived ease of use from TAM. The basic predictors of TAM and TRA are included in the TPB. Therefore, TPB is the first underlying theory in this research. Further, Wu, Li and Fu (2011) explains psychological state is critical to determine individual behavior for the adoption of technology. TAM and DOI can easy explain the psychological state of the individual behaviour. However, TAM and DOI are extremely similar in some constructs and also supplements each other. While relative advantages is similar to perceived usefulness, complexity is similar to perceived ease of use (Wu, Wang & Lin 2007). Therefore, Both TAM and DOI can explain the psychological state of the individual behaviour without TAM. However, TAM cannot explain the compatibility factor which cannot be ignored in this research context because if technology used in healthcare is not compatible with the healthcare process then it is useless. For example, if a doctor is treating patients in emergency and technology used in that treatment is causing any inconvenience for it may divert doctor attention and may leads to wrong treatment. Therefore, compatibility of the technology with the healthcare process is important. Hence, DOI theory can explain the second component that is 'context' in this research (Ghodeswar & Vaidyanathan, 2007). Therefore, the second underlying theory in this research is DOI.

Note: The third component: technology in this research is a component which is derived from literature review on adoption of various kinds of technologies in the healthcare and not from any adoption theory or model.

MAIN FOCUS OF THE ARTICLE

Through the extensive literature review, it is found that there is limited research for the adoption of mobile devices (by healthcare professional) in the Australian healthcare context. From literature, it also appears that some healthcare facilities are already using mobile health services but their scope is limited to that environment. A major implication of such narrow adoption is lack of understanding of how mobile devices are adopted in the healthcare environment. Further, it appears from literature that even after implementation of technology in the Australian health organizations the adoption is not as much as expected. For example, although in July 2011, the Australian government has introduced funding for video-consultations in primary care yet, the adoption is not as much as expected (Smith et al. 2014).The majority of consultations in the Australia are conducted face to face compared with video consultation (Smith et al. 2012). Therefore, this study is focused on:

- Understanding factors influencing adoption of the mobile devices in the Australian healthcare environment.
- Developing a conceptual framework which can support adoption of mobile devices in the healthcare domain.

The various constructs used in the conceptual framework developed in this chapter and their operationalization is given in the next section.

DISCUSSION ON SOLUTIONS

The conceptual framework in this chapter is developed in two steps.

Step 1: Thorough Study of the Literature to Understand Factors Influencing Adoption of Various Kinds of Technology in the Healthcare Environment

In this phase various factors influencing adoption of technology are identified. This step provides a broad view of all the factors which can influence adoption of technology in the healthcare environment. Factors identified in this step are related to adoption of technologies such as electronic health records, wireless technology, health information technology, telehealth, home tele monitoring, eHealth mobile computing and mobile health (Daim et al., 2013; Deng et al., 2013; Heidarian & Mason, 2013; Singh, Lichter, Danzo, Taylor, & Rosenthal, 2012; Wang, Redington, Steinmetz, & Lindeman, 2010; Zinszer, Tamblyn, Bates, & Buckeridge, 2013). In step1 it is identified that some factors for the adoption of technology in the healthcare appears to be individual factors and some of them appears to be organizational factors. The list of the factors for the adoption of various technologies in the healthcare environment is given in the appendix 1.

Step 2: Selection of Appropriate Factor for Understanding Adoption of the Mobile Devices in the Australian Healthcare from the Healthcare Professionals' Perspective

To select the most appropriate factors the authors have given emphasis to understand technology, health environment and healthcare professionals' components. To understand the factors from technology aspect the authors have tried to understand the factors from the previous articles which were focused on mobile computing, mobile wireless devices, and wireless devices (M.-P. Gagnon, Ngangue, Payne-Gagnon, & Desmartis, 2016; Goswami & Chandra, 2013; I.-L. Wu et al., 2011; J.-H. Wu, Wang, & Lin, 2007; Yangil & Chen, 2007). To understand the Australian health context, the authors have tried to read and critically analyze the research articles based on use of mobile devices or mHealth from 2007-2016. A systematic literature review to understand the use of mobile devices in the Australian health domain is given in in Appendix 2. To understand the factors from the healthcare professionals' perspective the authors have emphasized on various theories/model and literature based on understanding individual behavior for the use of technology in healthcare.

It was a difficult decision to choose which factors can be selected and which can be discarded. Therefore, each construct is operationalized to understand the meaning of the construct. Construct operationalization is a process of achieving clear and practical meaning of the constructs to the chosen context. It includes an agreed definition of the constructs from literature, clear meaning of the constructs that is also suitable for the research and identifying key concept properties describing constructs (Ayers & Olander, 2013; Bhattacherjee, Hikmet, Menachemi, Kayhan, & Brooks, 2007). Operationalization of constructs is also helpful to understand constructs and apply them in a particular research context. Therefore, each construct considered for the development of conceptual framework in this research is operationalized below:

1. **Intention:** Intention in this research refers to the measure of the likelihood to use mobile devices in the healthcare process. User intention is a good indicator of how a system will be accepted. Intention to adopt mobile devices in healthcare may depend upon many factors. Therefore, intention to adopt mobile devices is a dependent variable and all the other factors explained below are independent variables.
2. **Readiness:** Readiness refers to the individual propensity to embrace and use mobile devices in health care. Individual readiness to adopt technology is influenced by the individual's positive and negative feelings. Positive feelings allow individuals to adopt technology while negative feelings distract from adopting technology. Parasuraman (2000) has considered positive feelings as 'optimism', one of the contributors of technology readiness. Optimism as 'a positive view of using technology and a belief that it offers people increased control, flexibility and efficiency in their lives. This clearly shows that individual readiness/ attitude to adopt technology is positively influenced by advantages provided by technology. On the other hand, negative feelings which create inconvenience make individuals less confident to adopt technology (Caison, Bulman, Pai, & Neville, 2008). In Australia, mobile devices use in healthcare is still in its infancy. Thus, healthcare professionals' readiness may be weak or strong depending on whether health professionals perceive disadvantages or advantages of mobile devices. Therefore, it can be concluded that induvial intention to adopt mobile devices in the healthcare may be influenced by readiness for the adoption of

Figure 1. Conceptual framework for adoption of mobile devices in the Australian healthcare context

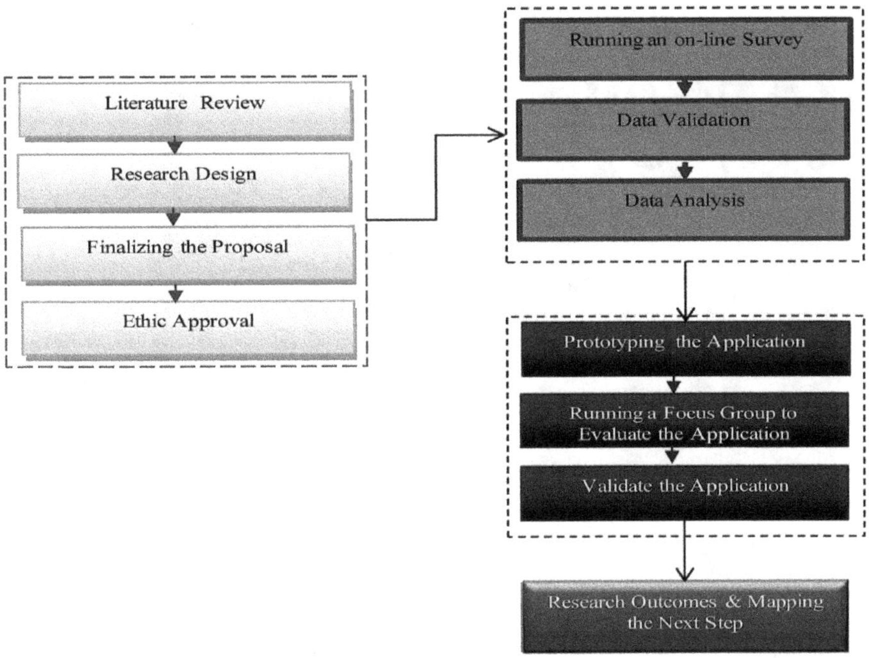

Figure 2.

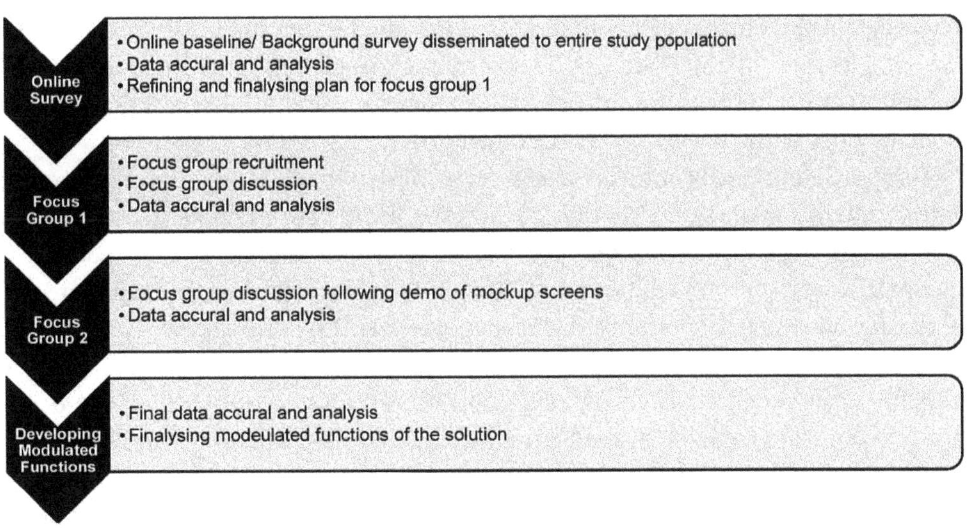

mobile devices which may further be influenced by the advantages and disadvantages of using the mobile devices in the healthcare domain.

3. **Advantages:** Advantages refer to the benefits of mobile devices in the healthcare processes. Advantages of technology can be measured using various parameters such as cost, features and compatibility. Benefits offered by technology positively impact individuals to adopt technology

Figure 3.

(Wang et al., 2010; I.-L. Wu et al., 2011). Previous research studies explain the benefits of using technology in healthcare and their positive influence on individual perception (Daim et al., 2013; M. P. Gagnon et al., 2012; D. Kim, 2009). The Australian healthcare domain is under tremendous pressure due to rising healthcare costs, shortage of qualified staff and a growing demand of services (AIHW, 2014; Croll, Norton, Gray, Bryett, & Smith, 2012). Mobile devices such as smart phone in the Australian health context can help in promoting and facilitating health services, quality of life and reducing healthcare cost (Gee, 2015). Therefore, the various benefits of using mobile devices in the Australian healthcare context may influence healthcare professionals' intention for the use of mobile devices in the healthcare environment.

4. **Safety Issues:** Safety issues in this research refer to individual understanding of patient's privacy and confidentiality while using mobile devices for healthcare purpose. In health domain where patients' data is transferred from one point to another its integrity, safety, security, privacy and confidentiality needs to be maintained. While using mobile devices in healthcare safety of data being transferred is very important (M.-P. Gagnon et al., 2016; Michael J. Ackerman et al., 2010; Peddle, 2007). For example, if a patient is in emergency unit and patient's data gets modified intentionally or unintentionally then patient will get wrong treatment which can decrease his/her chances of recovering from the disease quickly (Cinque, Coronato, & Testa, 2013). Further, use of mobile devices could also create another security risks such as malware, spear phishing and losing equipment risk (Testa, Cinque, Coronato, & De Pietro, 2015) and these risk cannot be avoided in any country's health context unless a strong security measures are not available. Hence, the use of mobile devices in healthcare is at the same time a great opportunity and a great risk (disadvantage) in the health domain. Therefore, it is important to study security factor while understanding adoption of mobile devices in healthcare.
5. **Features:** Features refer to the functional features such as battery life, screen size of mobile devices which may influence healthcare professionals for the adoption of mobile devices. In the healthcare the most important concern is a person's health and adopted technology should have good sup-

porting features with healthcare process (I.-L. Wu et al., 2011). Zhang et al. (2014) have idea that adoption of technology in healthcare depend on performance of technology and technology features are responsible for performance of technology. (Ghodeswar & Vaidyanathan, 2007) Therefore, whatever technology is used it should have good supporting features in healthcare (Massey, Khatri, & Montoya-Weiss, 2007). EMarketer 2012b cited in Ally and Gardiner (2012) support the notion that features of mobile devices such as touch screen, voice and user interface influence adoption of mobile devices in the market. Further, each healthcare domain is responsible for the life of people and if technology features create any kind of inconvenience with the health care process, then technology is useless in the health environment. Hence, it is important to consider the impact of functional features for the adoption of mobile devices in the Australian healthcare context.

6. **Self-Efficacy:** Bandura have introduced the concept of self-efficacy, which is defined as the expression of self-ability and confidence in successfully reaching goals in certain circumstances (Bandura 1997). Many researchers have used self-efficacy as a variable for information system adoption, and they have de-fined it as the degree of perceiving the need for a certain ability, technique, or piece of knowledge when an individual attempt to use a certain information technology system. Self-efficacy has been considered as an important factor for the adoption of mobile banking and smartphones (Luarn & Lin 2005; D. 1989). Many researchers have demonstrated self- efficacy as an important construct for the adoption of technology (Lim et al., 2011; J.-H. Wu et al., 2007; Yangil & Chen, 2007). Mobile devices are a new technology in the healthcare and in the Australian healthcare it is in its infancy stage. Therefore, some of the individuals may not be confident to use the new technology and hence their intention may be influenced by the self-efficacy to use mobile devices. However, if training is provided to the individual their confident level to use technology is increased.
7. **Training:** Training in this research refers to the hospital's support to update the healthcare professionals about the use of mobile devices in healthcare. Training is useful for the employees (Zolfo et al., 2010). Zolfo et al. (2010) in their research resulted that providing training to health worker is useful in scaling up the HIV/AIDS care in resource-limited settings. Providing training to use mobile devices in healthcare is also important because there is a difference between health professionals and the typical technology users. Health professional work is critical and time sensitive and if mistake is done it may cost the life of a patient. Further, in the Australian health system is facing shortage of staff. Therefore, proving training may reduce the complexity level of healthcare professionals to operate the mobile devices in the health domain and can bring productivity and efficiency even with the limited staff.
8. **Complexity:** Complexity refers to the degree to which healthcare professionals may experience difficulties in using mobile devices for health care services. Rogers (2003, p. 257) states complexity as the degree to which an innovation is perceived as relatively difficult to understand and use. In the context of adoption of mobile devices in healthcare, if an individual has to struggle with the use of technology in the health process, then the individual's intention for the adoption of mobile devices will be weak. However, complexity can be reducing by providing training to the health professionals. Many researchers proposed complexity as one of the major factors for adoption of technology in healthcare, with antonyms like simple/ easy to use/perceived easiness, perceived ease of use (Daim et al., 2013). Complexity can be an important factor for adoption of mobile devices in the healthcare environments because if HCPs find it difficult to operate mobile devices in healthcare process the intention for the adoption will be low. In the Australian health environment use

of mobile devices in on a pilot basis and the future of the pilot projects is unknown. Therefore, it is important to understand the complexity construct to expand the mobile health on large scale in Australia. Thus, complexity influences adoption of mobile devices and is included in the conceptual framework.

9. **Social Influences:** Social influences are defined as the influences of people (who are in the social circle such as colleagues of healthcare professionals) on healthcare professionals for the adoption of mobile devices in healthcare services. In the theory of reasoned action (TRA), TAM2, TPB, C-TAM-TPB the term social norm is used and has the same meaning as social influences. Social influences are the influences of peers on the individual (Morris, Venkatesh, & Ackerman, 2005). Social influences have a great influence on individual adoption of innovation (Rogers, 2003) and is therefore, considered one of the essential constructs in adoption of mobile devices in the Australian healthcare context.
10. **Compatibility:** Compatibility refers to the degree to which mobile devices are perceived as consistent with the existing healthcare process and needs of the healthcare professionals. Literature reveals HCPs show less interest for the adoption of new technology (Rogers, 2003; I.-L. Wu et al., 2011). However, if the technology is compatible with healthcare professional's work process, they like to adopt it (Xue et al., 2012). Additionally, in the health care environment new technology is adopted if it supports the existing system and adds newness to the system and mobile devices have that potential. Therefore, compatibility construct is added in the conceptual framework.
11. **Age:** Age in this research is defined as how old a health professional is who is working in the Australian healthcare. Age is also one of the important factors that affects the adoption of information systems. Several researchers found young people more actively use mobile telecommunications (Brown, Cajee, Davies, & Stroebel, 2003; Nikou, 2015). However, elderly people are resistant to technology change. Skiba (2015) says that '60- plus crowd' crowd is not technology averse. Further, most of the old people are dominated by various kinds of physical problems such as poor eyesight (Leventhal, 2008; Zhang et al., 2014). If a person has poor eyesight he/she will more likely to prefer mobile devices with the bigger screen size (Boulos et al., 2011). Caison et al. (2008) results shows that young male medical students have more curiosity for using technology. Therefore, researcher expect that age would be a significant factor in the adoption of mobile devices among the Australian healthcare professionals.
12. **Gender:** Gender in this research is defined as whether a healthcare professional working in the Australian health domain is male, female or other. Gender difference is also considered one of the factor for understanding the adoption of mobile devices in the Australian healthcare domain. Imhof, Vollmeyer, and Beierlein (2007) explains male computer users use computer more frequently for games, music and video players, or Internet shopping than their female counterparts. Lim et al. (2011) think women who do not have prior experience with technology may feel more uncomfortable in using technology. Further, previous literature indicates that men are task oriented and like to finish their office work efficiently and timely and if technology is compatible with their style of working and can help to finish their office work timely their intention to use technology is strong (Roberts & Bell, 2000). In the healthcare domain, healthcare professionals need to interact with various patients and their family members and friends to explain the disease condition of a patient. Therefore, gender difference could play a role for the use of mobile devices in the health domain.
13. **Other Factors:** Other factors refer to the factors which can be further added in the conceptual framework developed in this chapter. Other factors may be organizational factors, cultural differ-

ences and government policies and procedure to implement mobile devices in the healthcare system. These factors are not included yet in the conceptual framework because every country has its own health system. Others factors mentioned above may be influencing or may not be influencing in the Australian healthcare context. These factors may be added in the conceptual framework further by conducting exploratory study.

FUTURE RESEARCH DIRECTIONS

This research can provide avenues for future researchers for conducting further research. Future research can be conducted for the adoption of mobile devices in the Australian healthcare domain from other perspectives such as patients and management perspective. Also, the conceptual framework developed in this research can further be extended by conducting exploratory study to find out the other factors in the Australian health context.

CONCLUSION

This chapter concludes that there are many factors, which are influencing healthcare professionals for the adoption mobile devices in the Australian healthcare environment. The main factors which may influence healthcare professionals in the Australian healthcare context for the adoption of mobile devices proposed from literature are: intention, individual readiness, advantages, safety issues, features, self-efficacy, complexity, training, compatibility, social influences, age and gender. Other factors in the conceptual framework developed in this chapter can be added depending on the Australian health context.

ACKNOWLEDGMENT

The authors wish to express their thanks to Mr Paul Paddle who discussed with them his ideas for a smart paging platform. This study was in part motivated by his need to understand the background of this domain, its critical success factors and how one might successfully navigate the terrain.

REFERENCES

Agosti, V., Graziano, S., Artiaco, L., & Sorrentino, G. (2009). Biological mechanisms of stroke prevention by physical activity in type 2 diabetes. *Acta Neurologica Scandinavica*, *119*(4), 213–223. doi:10.1111/j.1600-0404.2008.01080.x PMID:18700881

Al-Qurishi, M., Al-Rakhami, M., Al-Qershi, F., Hassan, M. M., Alamri, A., Khan, H. U., & Xiang, Y. (2015). A framework for cloud-based healthcare services to monitor non communicable diseases patient. *International Journal of Distributed Sensor Networks*, *2015*, 1–11. doi:10.1155/2015/985629

Ally, M., & Gardiner, M. (2012). The moderating influence of device characteristics and usage on user acceptance of smart mobile devices. *Paper presented at the ACIS 2012*, Geelong, Australia.

Andersen, P., Lindgaard, A.-M., Prgomet, M., Creswick, N., & Westbrook, J. I. (2009). Mobile and fixed computer use by doctors and nurses on hospital wards: Multi-method study on the relationships between clinician role, clinical task, and device choice. *Journal of Medical Internet Research, 11*(3), 1–15. doi:10.2196/jmir.1221 PMID:19674959

Armfield, N. R., Gray, L. C., & Smith, A. C. (2012). Clinical use of Skype: A review of the evidence base. *Journal of Telemedicine and Telecare, 18*(3), 125–127. doi:10.1258/jtt.2012.SFT101 PMID:22362829

Ayers, S., & Olander, E. K. (2013). What are we measuring and why? Using theory to guide perinatal research and measurement. *Journal of Reproductive and Infant Psychology, 31*(5), 439–448. doi:10.1080/02646838.2013.834041

Baig, A. H. (2010). *Study to investigate the adoption of wireless technology in the Australian healthcare system [Unpublished doctoral dissertation].* Toowoomba, Australia, University of Southern Queensland.

Ben-Zeev, D., Kaiser, S. M., Brenner, C. J., Begale, M., Duffecy, J., & Mohr, D. C. (2013). Development and usability testing of FOCUS: A smartphone system for self-management of schizophrenia. *Psychiatric Rehabilitation Journal, 36*(4), 289–296. doi:10.1037/prj0000019 PMID:24015913

Bhattacherjee, A., Hikmet, N., Menachemi, N., Kayhan, V. O., & Brooks, R. G. (2007). The differential performance effects of healthcare information technology adoption. *Information Systems Management, 24*(1), 5–14. doi:10.1080/10580530601036778

Bønes, E., Hasvold, P., Henriksen, E., & Strandenæs, T. (2007). Risk analysis of information security in a mobile instant messaging and presence system for healthcare. *International Journal of Medical Informatics, 76*(9), 677–687. doi:10.1016/j.ijmedinf.2006.06.002 PMID:16931132

Bort-Roig, J., Gilson, N., Puig-Ribera, A., Contreras, R., & Trost, S. (2014). Measuring and influencing physical activity with smartphone technology: A systematic review. *Sports Medicine (Auckland, N.Z.), 44*(5), 671–686. doi:10.100740279-014-0142-5 PMID:24497157

Boulos, M. N., Wheeler, S., Tavares, C., & Jones, R. (2011). How smartphones are changing the face of mobile and participatory healthcare: An overview, with example from eCAALYX. *Biomedical Engineering Online, 10*(1), 1–14. doi:10.1186/1475-925X-10-24 PMID:21466669

Bradford, N., Armfield, N. R., Young, J., Ehmer, M., & Smith, A. C. (2012). Safety for home care: The use of Internet video calls to double-check interventions. *Journal of Telemedicine and Telecare, 18*(8), 434–437. doi:10.1258/jtt.2012.GTH102 PMID:23209268

Bradford, N. K., Young, J., Armfield, N. R., Herbert, A., & Smith, A. C. (2014). Home telehealth and paediatric palliative care: Clinician perceptions of what is stopping us? *BMC Palliative Care, 13*(1), 1–19. doi:10.1186/1472-684X-13-29 PMID:24963287

Bradford, S., & Rickwood, D. (2014). Electronic psychosocial assessment tool: Concept development and identification of barriers to successful implementation. *Journal of Technology in Human Services, 32*(4), 275–296. doi:10.1080/15228835.2014.967906

Brewster, L., Mountain, G., Wessels, B., Kelly, C., & Hawley, M. (2014). Factors affecting front line staff acceptance of telehealth technologies: A mixed-method systematic review. *Journal of Advanced Nursing*, *70*(1), 21–23. doi:10.1111/jan.12196 PMID:23786584

Brown, I., Cajee, Z., Davies, D., & Stroebel, S. (2003). Cell phone banking: Predictors of adoption in South Africa—an exploratory study. *International Journal of Information Management*, *23*(5), 381–394. doi:10.1016/S0268-4012(03)00065-3

Brown, W. III, Yen, P.-Y., Rojas, M., & Schnall, R. (2013). Assessment of the health IT usability evaluation model (health-ITUEM) for evaluating mobile health (mhealth) technology. *Journal of Biomedical Informatics*, *46*(6), 1080–1087. doi:10.1016/j.jbi.2013.08.001 PMID:23973872

Caison, A. L., Bulman, D., Pai, S., & Neville, D. (2008). Exploring the technology readiness of nursing and medical students at a Canadian university. *Journal of Interprofessional Care*, *22*(3), 283–294. doi:10.1080/13561820802061809 PMID:18569414

Castelnuovo, G., Manzoni, G. M., Pietrabissa, G., Corti, S., Giusti, E., Molinari, E., & Simpson, S. (2014). Obesity and outpatient rehabilitation using mobile technologies: The potential mHealth approach. *Frontiers in Psychology*, *5*, 1–11. doi:10.3389/fpsyg.2014.00559 PMID:24959157

Chetty, G., & Yamin, M. (2015). Intelligent human activity recognition scheme for ehealth applications. *Malaysian Journal of Computer Science*, *18*(1), 59–69.

Cimperman, M., Brenčič, M. M., Trkman, P., & Stanonik, M. L. (2013). Older adults' perceptions of home telehealth services. *Telemedicine Journal and e-Health*, *19*(10), 786–790. doi:10.1089/tmj.2012.0272 PMID:23931702

Cinque, M., Coronato, A., & Testa, A. (2013). *A failure modes and effects analysis of mobile health monitoring systems Innovations and Advances in Computer, Information, Systems Sciences, and Engineering* (pp. 569–582). Springer.

Clarke, J., Proudfoot, J., Birch, M.-R., Whitton, A., Parker, G., Manicavasagar, V., ... Hadzi-Pavlovic, D. (2014). Effects of mental health self-efficacy on outcomes of a mobile phone and web intervention for mild-to-moderate depression, anxiety and stress: Secondary analysis of a randomised controlled trial. *BMC Psychiatry*, *14*(1), 1–19. doi:10.118612888-014-0272-1 PMID:25252853

Clough, B. A., & Casey, L. M. (2015). The smart therapist: A look to the future of smartphones and mHealth technologies in psychotherapy. *Professional Psychology, Research and Practice*, *46*(3), 147–153. doi:10.1037/pro0000011

Daim, T. U., Basoglu, N., & Topacan, U. (2013). Adoption of health information technologies: The case of a wireless monitor for diabetes and obesity patients. *Technology Analysis and Strategic Management*, *25*(8), 923–938. doi:10.1080/09537325.2013.823150

Castro, D., Miller, B., & Nager, A. (2014, May). Unlocking the potential of physician-to-patient telehealth services.

Davis, F. D. (1985). *A technology acceptance model for empirically testing new end-user information systems: theory and results* [Doctor of philosophy]. Massachusetts Institute of Technology.

Deng, Z., Mo, X., & Liu, S. (2013). Comparison of the middle-aged and older users' adoption of mobile health services in China. *International Journal of Medical Informatics*, *83*(3), 210–224. doi:10.1016/j.ijmedinf.2013.12.002 PMID:24388129

Dwivedi, Y. K., Shareef, M. A., Simintiras, A. C., Lal, B., & Weerakkody, V. (in press). A generalised adoption model for services: A cross-country comparison of mobile health (m-health). *Government Information Quarterly*.

Fjeldsoe, B. S., Miller, Y. D., O'Brien, J. L., & Marshall, A. L. (2012). Iterative development of MobileMums: A physical activity intervention for women with young children. *The International Journal of Behavioral Nutrition and Physical Activity*, *9*(1), 151–161. doi:10.1186/1479-5868-9-151 PMID:23256730

Furukawa, M.F., Raghu, T.S., Spaulding, T.J., & Vinze, A. (2008). Adoption of health information technology for medication safety in U.S. Hospitals, 2006. *Health Affairs*, *27*(3), 865–875.

Gagnon, M.-P., Ngangue, P., Payne-Gagnon, J., & Desmartis, M. (2016). m-Health adoption by healthcare professionals: A systematic review. *Journal of the American Medical Informatics Association*, *23*(1), 212–220. doi:10.1093/jamia/ocv052 PMID:26078410

Gagnon, M. P., Desmartis, M., Labrecque, M., Car, J., Pagliari, C., Pluye, P., ... Legare, F. (2012). Systematic review of factors influencing the adoption of information and communication technologies by healthcare professionals. *Journal of Medical Systems*, *36*(1), 241–277. doi:10.100710916-010-9473-4 PMID:20703721

Ghodeswar, B. M., & Vaidyanathan, J. (2007). Organisational adoption of medical technology in healthcare sector. *Journal of Service Research*, *7*(2), 57–81.

Gold, J., Lim, M. S. C., Hellard, M. E., Hocking, J. S., & Keogh, L. (2010). What's in a message? Delivering sexual health promotion to young people in Australia via text messaging. *BMC Public Health*, *10*(1), 792–802. doi:10.1186/1471-2458-10-792 PMID:21190584

Goodall, K., Ward, P., & Newman, L. (2010). Use of information and communication technology to provide health information: what do older migrants know, and what do they need to know? Quality in Primary Care, 18(1), 27-32.

Goswami, S., & Chandra, B. (2013). Convergence dynamics of consume innovativeness vis-à-vis technology acceptance propensity: An empirical study on adoption of mobile devices. *The IUP Journal of Marketing Management*, *12*(3), 63–87.

Gururajan, R., & Hafeez-Baig, A. (2014). An empirical study to determine factors that motivate and limit the implementation Of ICT in healthcare environments. *BMC Medical Informatics and Decision Making*, *14*(1), 1–14. doi:10.1186/1472-6947-14-98 PMID:25540040

Haffey, F., Brady, R., & Maxwell, S. (2013). A comparison of the reliability of smartphone apps for opioid conversion. *Drug Safety*, *36*(2), 111–117. doi:10.100740264-013-0015-0 PMID:23322549

Hebden, L., Balestracci, K., McGeechan, K., Denney-Wilson, E., Harris, M., Bauman, A., & Allman-Farinelli, M. (2013). 'TXT2BFiT' a mobile phone-based healthy lifestyle program for preventing unhealthy weight gain in young adults: Study protocol for a randomized controlled trial. *Trials*, *14*(1), 1–9. doi:10.1186/1745-6215-14-75 PMID:23286245

Hebden, L., Cook, A., Ploeg, H. P., King, L., Bauman, A., & Allman-Farinelli, M. (2014). A mobile health intervention for weight management among young adults: A pilot randomised controlled trial. *Journal of Human Nutrition and Dietetics*, *27*(4), 322–332. doi:10.1111/jhn.12155 PMID:23992038

Heidarian, A., & Mason, D. (2013). Health information technology adoption in New Zealand optometric practices. *Clinical & Experimental Optometry*, *96*(6), 557–565. doi:10.1111/cxo.12059 PMID:24730034

Imhof, M., Vollmeyer, R., & Beierlein, C. (2007). Computer use and the gender gap: The issue of access, use, motivation, and performance. *Computers in Human Behavior*, *23*(6), 2823–2837. doi:10.1016/j.chb.2006.05.007

Kang, H. G., Mahoney, D. F., Hoenig, H., Hirth, V. A., Bonato, P., Hajjar, I., & Lipsitz, L. A. (2010). In situ monitoring of health in older adults: Technologies and issues. *Journal of the American Geriatrics Society*, *58*(8), 1579–1586. doi:10.1111/j.1532-5415.2010.02959.x PMID:20646105

Kay, M. (2011). mHealth: new horizons for health through mobile technologies. World Health Organization, pp. 66-71.

Kim, D. (2009). Adoption of personal information system: Innovation diffusion theory and task technology fit. *Academy of Information and Management Sciences*, *13*(2), 50–73.

Kim, S., & Garrison, G. (2008). Investigating mobile wireless technology adoption: An extension of the technology acceptance model. *Information Systems Frontiers*, *11*(3), 323–333. doi:10.100710796-008-9073-8

Kluge, E. H. (2011). Ethical and legal challenges for health telematics in a global world: Telehealth and the technological imperative. *International Journal of Medical Informatics*, *80*(2), 1–5. doi:10.1016/j.ijmedinf.2010.10.002 PMID:21067967

Korzep, K. (2010). The future of technology and the effect it may have on replacing human jobs. *Technology and Health Care*, *18*(4-5), 353–358. PMID:21209484

Lim, S., Xue, L., Yen, C. C., Chang, L., Chan, H. C., Tai, B. C., ... Choolani, M. (2011). A study on Singaporean women's acceptance of using mobile phones to seek health information. *International Journal of Medical Informatics*, *80*(12), 189–202. doi:10.1016/j.ijmedinf.2011.08.007 PMID:21956003

Lu, B.-Y., & Wu, H.-D. (2015). Auscultation using modern mobile communication. *Acoustics Australia / Australian Acoustical Society*, *43*(3), 303–309. doi:10.100740857-015-0024-5

Lupton, D. (2013). Quantifying the body: Monitoring and measuring health in the age of mHealth technologies. *Critical Public Health*, *23*(4), 393–403. doi:10.1080/09581596.2013.794931

Massey, A. P., Khatri, V., & Montoya-Weiss, M. M. (2007). Usability of online services: The role of technology readiness and context. *Decision Sciences*, *38*(2), 277–308. doi:10.1111/j.1540-5915.2007.00159.x

Michael, J. (2010). Developing next-generation telehealth tools and technologies: Patients, systems, and data perspectives. *Telemedicine Journal and e-Health*, *16*(1), 93–95. doi:10.1089/tmj.2009.0153 PMID:20043711

Morris, M. G., Venkatesh, V., & Ackerman, P. L. (2005). Gender and age differences in employee decisions about new technology: An extension to the theory of planned behavior. *IEEE Transactions on Engineering Management*, *52*(1), 69–84. doi:10.1109/TEM.2004.839967

Nikou, S. (2015). Mobile technology and forgotten consumers: The young-elderly. *International Journal of Consumer Studies*, *39*(4), 294–304. doi:10.1111/ijcs.12187

Palmier-Claus, J. E., Rogers, A., Ainsworth, J., Machin, M., Barrowclough, C., Laverty, L., ... Lewis, S. W. (2013). Integrating mobile-phone based assessment for psychosis into people's everyday lives and clinical care: A qualitative study. *BMC Psychiatry*, *13*(1), 1–12. doi:10.1186/1471-244X-13-34 PMID:23343329

Parasuraman, A. (2000). Technology Readiness Index (TRI) a multiple-item scale to measure readiness to embrace new technologies. *Journal of Service Research*, *2*(4), 307–320. doi:10.1177/109467050024001

Peddle, K. (2007). Telehealth in context: Socio-technical barriers to telehealth use in labrador, Canada. [CSCW]. *Computer Supported Cooperative Work*, *16*(6), 595–614. doi:10.100710606-006-9030-3

Rana, R., Hume, M., Reilly, J., & Soar, J. (2015). Whealth-transforming telehealth services. *arXiv preprint arXiv:1506.05543*.

Roberts, J. E., & Bell, M. A. (2000). Sex differences on a mental rotation task: Variations in electroencephalogram hemispheric. *Developmental Neuropsychology*, *17*(2), 199–223. doi:10.1207/S15326942DN1702_04 PMID:10955203

Rogers, E. M. (2003). *Diffusion of innovations* (5th ed.). New York: Free Press.

Saad, N. M., Alias, R. A., & Ismail, Z. (2013). Initial framework on identifying factors influencing individuals' usage of telehealth. *Paper presented at the 2013 International Conference on Research and Innovation in Information Systems (ICRIIS)*, Kuala Lumpur, Malaysia. 10.1109/ICRIIS.2013.6716704

Sanders, C., Rogers, A., Bowen, R., Bower, P., Hirani, S., Cartwright, M., ... Hendy, J. (2012). Exploring barriers to participation and adoption of telehealth and telecare within the whole system demonstrator trial: A qualitative study. *BMC Health Services Research*, *12*(1), 220. doi:10.1186/1472-6963-12-220 PMID:22834978

Shand, F. L., Ridanir, R., Tighe, J., & Christensen, H. (2013). The effectiveness of a suicide prevention app for indigenous Australian youths: Study protocol for a randomized controlled trial. *Trials*, *14*(1), 1–15. doi:10.1186/1745-6215-14-396 PMID:24257410

Shareef, M. A., Kumar, V., & Kumar, U. (2014). Predicting mobile health adoption behaviour: A demand side perspective. *Journal of Customer Behaviour*, *13*(3), 187–205. doi:10.1362/147539214X14103453768697

Singh, R., Lichter, M. I., Danzo, A., Taylor, J., & Rosenthal, T. (2012). The adoption and use of health information technology in rural areas: Results of a national survey. *The Journal of Rural Health*, *28*(1), 16–27. doi:10.1111/j.1748-0361.2011.00370.x PMID:22236311

Skiba, D. J. (2015). Connected Health 2015: The Year of Virtual Patient Visits. *Nursing Education Perspectives*, *36*(2), 131–133. doi:10.5480/1536-5026-36.2.131 PMID:24783732

Slaper, M. R., & Conkol, K. (2014). mHealth tools for the pediatric patient-centered medical home. *Pediatric Annals*, *43*(2), 39–43. doi:10.3928/00904481-20140127-09 PMID:24512160

Steele, R., & Lo, A. (2013). Telehealth and ubiquitous computing for bandwidth-constrained rural and remote areas. *Personal and Ubiquitous Computing*, *17*(3), 533–543. doi:10.100700779-012-0506-5

Testa, A., Cinque, M., Coronato, A., & De Pietro, G. (2015). *A Formal Approach for a Dependability Assessment of a mHealth Monitoring System Mobile Health* (pp. 171–194). Springer.

Tiong, I., Hafeez-Baig, A., Gururajan, R., & Soar, J. (2006, August). Preliminary investigation to explore perceptions of security issues associated with wireless technology in healthcare in Australia. *Paper presented at the HIC '06.*

Wang, A., Redington, L., Steinmetz, V., & Lindeman, D. (2010). The ADOPT model: Accelerating diffusion of proven technologies for older adults. *Ageing International*, *36*(1), 29–45. doi:10.100712126-010-9072-1

West, D. (2012). How mobile devices are transforming healthcare. *Issues in Technology Innovation*, *18*(1), 1–14.

Willcox, J. C., Campbell, K. J., McCarthy, E. A., Wilkinson, S. A., Lappas, M., Ball, K., ... Crawford, D. A. (2015). Testing the feasibility of a mobile technology intervention promoting healthy gestational weight gain in pregnant women (txt4two) - study protocol for a randomised controlled trial. *Trials*, *16*(1), 1–8. doi:10.118613063-015-0730-1 PMID:25947578

Worringham, C., Rojek, A., & Stewart, I. (2011). Development and feasibility of a smartphone, ECG and GPS based system for remotely monitoring exercise in cardiac rehabilitation. *PLoS ONE*, *6*(2), 1–6. doi:10.1371/journal.pone.0014669 PMID:21347403

Wu, I.-L., Li, J.-Y., & Fu, C.-Y. (2011). The adoption of mobile healthcare by hospital's professionals: An integrative perspective. *Decision Support Systems*, *51*(3), 587–596. doi:10.1016/j.dss.2011.03.003

Wu, J.-H., Wang, S.-C., & Lin, L.-M. (2007). Mobile computing acceptance factors in the healthcare industry: A structural equation model. *International Journal of Medical Informatics*, *76*(1), 66–77. doi:10.1016/j.ijmedinf.2006.06.006 PMID:16901749

Xue, L., Yen, C. C., Chang, L., Chan, H. C., Tai, B. C., Tan, S. B., ... Choolani, M. (2012). An exploratory study of ageing women's perception on access to health informatics via a mobile phone-based intervention. *International Journal of Medical Informatics*, *81*(9), 637–648. doi:10.1016/j.ijmedinf.2012.04.008 PMID:22658778

Yangil, P., & Chen, J. V. (2007). Acceptance and adoption of the innovative use of smartphone. *Industrial Management & Data Systems*, *107*(9), 1349–1365. doi:10.1108/02635570710834009

Yu Rang, P., Yura, L., Guna, L., Jae Ho, L., & Soo-Yong, S. (2015). Smartphone Applications with Sensors Used in a Tertiary Hospital--Current Status and Future Challenges. *Sensors, 15*(5), 9854-9869.

Zangbar, B., Pandit, V., Rhee, P., Aziz, H., Hashmi, A., Friese, R. S., ... Joseph, B. (2014). Smartphone surgery: How technology can transform practice. *Telemedicine Journal and e-Health, 20*(6), 590–592. doi:10.1089/tmj.2013.0234 PMID:24693938

Zhang, S., McClean, S. I., Nugent, C. D., Donnelly, M. P., Galway, L., Scotney, B. W., & Cleland, I. (2014). A predictive model for assistive technology adoption for people with dementia. *IEEE Journal of Biomedical and Health Informatics, 18*(1), 375–383. doi:10.1109/JBHI.2013.2267549 PMID:24403437

Zinszer, K., Tamblyn, R., Bates, D. W., & Buckeridge, D. L. (2013). A qualitative study of health information technology in the Canadian public health system. *BMC Public Health, 13*(1), 509. doi:10.1186/1471-2458-13-509 PMID:23705692

Zolfo, M., Iglesias, D., Kiyan, C., Echevarria, J., Fucay, L., Llacsahuanga, E., & Lynen, L. (2010). Mobile learning for HIV/AIDS healthcare worker training in resource-limited settings. *AIDS Research and Therapy, 7*(1), 35–40. doi:10.1186/1742-6405-7-35 PMID:20825677

This research was previously published in the Handbook of Research on Healthcare Administration and Management edited by Nilmini Wickramasinghe, pages 243-267, copyright year 2017 by Medical Information Science Reference (an imprint of IGI Global).

APPENDIX 1: LIST OF FACTORS

1. (J.-H. Wu et al., 2007)
   1.1. Compatibility
   1.2. Perceived usefulness
   1.3. Perceived ease of
   1.4. MHS self-efficacy
2. (S. Kim & Garrison, 2008)
3. Cognitive Influence Process
   3.1. Job relevance
   3.2. Perceived ease of use
   3.3. Perceived usefulness
4. Technological Influence Process
   4.1. Ubiquity
   4.2. Reachability
5. (Furukawa et al., 2008)
   5.1. Technology
   5.2. Geographic location
   5.3. Patient safety
6. (Michael J. Ackerman et al., 2010)
   6.1. Lack of technology integration
   6.2. Interoperability,
   6.3. Standardization
   6.4. Limited financing
   6.5. Lack of data standards
   6.6. Cultural barriers
   6.7. Usability
   6.8. Ease of use
   6.9. Security
   6.10. Privacy
   6.11. Trust
7. (Wang et al., 2010)
   7.1. Cognitive, physical limitation
   7.2. Health status
   7.3. Disease conditions
   7.4. Technology literacy
   7.5. Perceived Usefulness of technology
8. (I.-L. Wu et al., 2011)
   8.1. Personal innovativeness
   8.2. Availability

9. (Kluge, 2011)
   - 9.1. Liability
   - 9.2. Interoperability
   - 9.3. Legacy
10. (Singh et al., 2012)
    - 10.1. Financial constraints
    - 10.2. Return on investment issues
    - 10.3. Initial data entry labour intensive
    - 10.4. Initial loss of productivity
    - 10.5. Training burden for physicians
11. (Davis, 1985; M. P. Gagnon et al., 2012)
    - 11.1. Perceived usefulness
    - 11.2. Perceived ease of use
12. (Goswami & Chandra, 2013)
    - 12.1. User interface clarity
    - 12.2. Usage friendliness
    - 12.3. Social influence
    - 12.4. Support of product and service providers
    - 12.5. Learning predispositions
13. (Heidarian & Mason, 2013)
    - 13.1. Technology updates
    - 13.2. Cost
    - 13.3. Lack of time
    - 13.4. Equipment integration
14. (Zinszer et al., 2013)
    - 14.1. Lack of national vision
    - 14.2. Lack of leadership
    - 14.3. Insufficient investment
    - 14.4. Poor conceptualization of the priority areas
15. (Daim et al., 2013)
    - 15.1. Usefulness
    - 15.2. Quality of Services
    - 15.3. Compatibility with User's Life Style
    - 15.4. Quality of Support
    - 15.5. Quality of Information
    - 15.6. Image of Technology
    - 15.7. Usages Time
    - 15.8. Accessibility
    - 15.9. Easy to Use
16. (Goswami & Chandra, 2013)
    - 16.1. Interface clarity
    - 16.2. User friendliness
    - 16.3. Social influences

16.4. Support of product and services provided
16.5. Learning predisposition
17. (Deng et al., 2013)
17.1. Physical condition
17.2. Resistance to change
17.3. Technology anxiety
17.4. Self-actualization needs
18. (Cimperman, Brenčič, Trkman, & Stanonik, 2013)
18.1. Security
18.2. Usability
19. (Brewster et al., 2014)
19.1. Staff-Patient interaction
19.2. Credibility
19.3. Negative impact of service change
19.4. Autonomy
19.5. Technical issues
20. (Daniel Castro, Ben miller, & Nager, 2014)
20.1. Standard of care
20.2. State licensing policies
20.3. Compatibility
20.4. Interoperability
21. (N. K. Bradford et al., 2014)
21.1. Technical (setting up equipment, username and password)
21.2. Individual (culture, linguistic and social variations)
21.3. Service factors (Lack of staff)
22. (M.-P. Gagnon et al., 2016)
22.1. Perceived usefulness
22.2. Ease of use
22.3. design and technical concerns
22.4. Cost
22.5. Time
22.6. Privacy and security issues
22.7. Familiarity with the technology
22.8. Risk-benefit assessment
22.9. Interaction with others (colleagues, patients, and management).

APPENDIX 2

Table 1. A systematic literature review to understand the use of mobile devices in the Australian health context

Title and Authorship	Research Design	Results
Intelligent human activity recognition scheme for eHealth applications (Chetty & Yamin, 2015)	A Samsung Galaxy S2 smartphone was used for quantitative data collection, which contains an accelerometer and a gyroscope for measuring 3-axial linear acceleration and angular velocity respectively at a constant rate of 50Hz, which is sufficient for capturing human body motion.	For single model experiments, we found optimal attribute selection Techniques based on information theory based ranking leads to better performance. For multimodal experiments, a combination of supervised and unsupervised yields best performance.
'TXT2BFiT' a mobile phone-based healthy lifestyle program for preventing unhealthy weight gain in young adults: study protocol for a randomized controlled trial (Lana Hebden et al., 2013)	Nine month two-arm parallel-group randomized controlled trial in Sydney, New South Wales, Australia is conducted. Participants are recruited from primary medical care clinic using random sampling.	The results of this study are not given in the paper. The results may be found in the further publications
Integrating mobile-phone based assessment for psychosis into people's everyday lives and clinical care: a qualitative study (Palmier-Claus et al., 2013)	Qualitative interviews were conducted in order to explore participants' perceptions and experiences of the devices. 24 community based individuals with non-affective psychosis completed a randomised repeated-measure cross-over design study, where they filled in self-report questions about their symptoms via text-messages on their own phone, or via a purpose designed software application for Android smartphones, for six days.	Three themes emerged from the data are: i) the appeal of usability and familiarity, ii) acceptability, validity and integration into domestic routines, and iii) perceived impact on clinical care. The mobile-phone assessment procedure itself may needto be tailored to the individual, potentially through machine learning, in order to maximise compliance
A Comparison of the Reliability of Smartphone Apps for Opioid Conversion (Haffey, Brady, & Maxwell, 2013)	Between 28 January and 5 February 2012, the major online app stores (Android 'Google Play' store, iOS 'App Store', BlackBerry 'Blackberry App World', Windows Phone 'Marketplace', Symbian 'Ovi (Nokia)' store and Bada 'Samsung Apps' store) are searched using relevant terms (opioid, morphine, medical calculator, conversion) to identify apps that had opioid dose/formulation conversion capabilities.	We have demonstrated that opioid calculator conversion outputs are highly variable between apps. Evidence of medical involvement in app creation or referencing of primary sources is lacking in many cases, and a statistically significant difference was found in conversion outputs of hydromorphone in those apps with and without stated medical involvement Despite large developments in Internet technology and growth in the knowledge of how to develop more effective web-based interventions, overall website quality was low and the majority of freely available physical activity websites lack the components associated with behavior change.
The effectiveness of a suicide prevention app for indigenous Australian youths: study protocol for a randomized controlled trial (Shand, Ridanir, Tighe, & Christensen, 2013)	A Pilot study will be conducted in this research. Use of the app (that is, login and logout times, activities completed, time spent in each section, answers to self-Assessment questions) will be tracked within the device, and downloaded to secure servers whenever internet connectivity is established. The app will require a security pin for access and any data stored on the device will be encrypted and hidden within the app coding. Answers to self-assessment questions will be coded so that they are unintelligible to anyone not connected with the study.	The primary outcome measure is a reduction in frequency and intensity of suicidal thoughts. Secondary outcome measures are the reduction of depression, anxiety and impulsivity. .

continued on following page

Table 1. Continued

Title and Authorship	Research Design	Results
Telehealth and ubiquitous computing for bandwidth-constrained rural and remote areas (Steele & Lo, 2013)	review article	In this article, we review the benefits of ubiquitous computing for rural and remote telehealth including social media based preventative, peer support and public health communication, mobile phone platforms for the detection and notification of emergencies, wearable and ambient biosensors, the utilization of personal health records including in conjunction with mobile and sensor platforms, chronic condition care and management information systems, and mobile device–enabled video consultation.
Obesity and outpatient rehabilitation using mobile technologies: the potential mHealth approach (Castelnuovo et al., 2014)	review article	To conclude, further studies should investigate both possible advantages and applications of Internet and mHealth technologies in the treatment of obesity. In spite of promising preliminary reports, the evidence-base for the effectiveness of mHealth applications is meagre and it remains too early to be able to recommend it for use in clinical settings
Safety for home care: the use of Internet video calls to double-check interventions (N. Bradford, Armfield, Young, Ehmer, & Smith, 2012) -RCH in Brisbane	This research is a trial research. The devices were taken on scheduled home visits to patient homes and video calls with a second clinician were conducted to double-check various items associated with the clinical care of the patient. Over a 14-month period, 88 video calls were conducted during which a total of 600 checks were completed. The items checked included medication names, doses, segmentations on syringes and details of ventilator settings.	The quality of the video call was acceptable on 97% of occasions. On three occasions (3%) it was not possible to establish a connection and the double check was not achieved. The use of Internet video calls is feasible for double-checking and has the potential to improve patient safety and reduce costs
Clinical use of Skype: a review of the evidence Base (Armfield, Gray, & Smith, 2012)	This articles is a literature review article. Five electronic databases were searched to identify relevant studies (CochraneLibrary, DARE,Medline,EMBASE,CINAHL). The search terms were: Internet; video; remote consultation; Skype; teleconsultation; telehealth; telemedicine.	One small (n=7) controlled clinical trial had assessed the effect of nursing communication using Skype on elderly patients with dementia and their careers. While there were many case reports and small studies, no firm evidence either in favour of, or against the use of Skype for clinical telehealth was found.
The smart therapist: A look to the future of smartphones and mHealth technologies in psychotherapy (Clough & Casey, 2015)	Review article.	Research indicates that consumers want to engage in mHealth interventions, although only a limited number of empirically supported options currently exist. Weakness Much of the research to date is lacking in methodological rigor, and reliance on standard research designs often used in other areas of psychological research may constrain development of interventions. Strength: The main areas of research identified included the use of SMS technologies, ecological momentary assessment, online and software application based interventions and the recent integration of sensor and data mining technologies.

continued on following page

Table 1. Continued

Title and Authorship	Research Design	Results
Quantifying the body: monitoring and measuring health in the age of mHealth technologies (Lupton, 2013)	Review article.	Given that mobile digital technologies are so novel, research directed at how people actually use them for health purposes – how they 'domesticate' them and incorporate them into their everyday lives – has yet to be published. What types of people self-track? How do the devices they use come to acquire meaning in the context of everyday use? What are the social lives of these commodities? Aspects of how and to what extent these devices are incorporated in concepts of selfhood and embodiment also remain to be fully explored.
Testing the feasibility of a mobile technology intervention promoting healthy gestational weight gain in pregnant women (txt4two) - study protocol for a randomised controlled trial (Willcox et al., 2015)	txt4two is a parallel randomised control trial	Findings will inform the development of larger-scale Digitally based programmes to improve the delivery of healthy pregnancy nutrition, physical activity and healthy GWG. The findings of this trial will contribute to the literature on promotion of healthy lifestyles in pregnant women.
Iterative development of MobileMums: a physical activity intervention for women with young children (Fjeldsoe, Miller, O¿Brien, & Marshall, 2012)	Focus groups with women with young children to explore their perceptions of and needs for physical activity interventions were conducted. Women were recruited via posters and flyers displayed in general practice clinics, Playgroup locations, the public hospital antenatal clinic, and the public library of a community located in Queensland, Australia. To be eligible to participate, women had to have at least one child aged less than 5 years, and be able to speak and understand English. In pre-testing To be eligible to participate, women had to have at least one child aged less than 5 years, and be able to speak and understand English.	Key findings identified throughout the development process that shaped the MobileMums program were the need for: behaviour change techniques to be grounded in Social Cognitive Theory; tailored SMS content; two-way SMS interaction; rapport between SMS sender and recipient; an automated software platform to generate and send SMS; and, flexibility in location of a face-to-face delivered component.
A mobile health intervention for weight management among young adults: a pilot randomised controlled trial (L. Hebden et al., 2014)	This was a prospective randomised controlled trial, conducted from July to the end of December, 2011. In total, 192 gender-specific SMS text messages were developed (24 SMS for each of four behaviours and two genders). Text messages were tailored to the processes of change identified in the Transtheoretil Model (Prochaska et al., 1992), and moved from addressing cognitive to behavioural processes to facilitate movement through stages of change. Four smartphone applications were developed by the investigators (one per behaviour). Briefly, each application enabled users to record their behaviour (e.g. daily minutes of physical activities performed, daily servings of fruit and vegetables, or weekly frequency, and energy and fat content, of take-away meals) and to then receive instantaneous tailored motivational advice, as well as feedback in reference to population health guidelines	All analyses were conducted using SAS, version 9.2 Data were analysed according to the intention-to-treat principle with baseline values imputed for missing follow-up data. Descriptive statistics were used to summarise baseline differences in participant characteristics between intervention and control groups. Changes within groups pre- to post-intervention were tested using paired sample t-tests (continuous normal data), Wilcoxon signed ranks Z-tests (ordinal data; i.e. fruit and vegetable servings) or Pearson chi-square tests (categorical data). The effect of the mHealth intervention on primary and secondary outcomes was analysed using analysis-of-covariance models for quantitative outcomes and logistic regression models for dichotomous outcomes

continued on following page

Table 1. Continued

Title and Authorship	Research Design	Results
Effects of mental health self-efficacy on outcomes of a mobile phone and web intervention for mild-to-moderate depression, anxiety and stress: Secondary analysis of a randomised controlled trial (Clarke et al., 2014)	Randomized controlled trial is conducted in this research.	Mental health self-efficacy influences symptom outcomes of a self-guided mobile phone and web-based psychotherapeutic intervention and may itself be a worthwhile target to increase the effectiveness and efficiency of online treatment programs.
A framework for cloud-based healthcare services to monitor non communicable diseases patient (Al-Qurishi et al., 2015)	A framework is proposed in this research.	Analysis is conducted using AHP techniques.
Electronic psychosocial assessment tool: Concept development and identification of barriers to successful implementation (S. Bradford & Rickwood, 2014)	Interview is conducted in this research. At the beginning of each interview, participants are informed of the purpose of the study.	By engaging the stakeholders in this early process researchers were able to identify a range of software functionalities that are likely to improve the applicability for users, as well as identify possible barriers to implementation.
An empirical study to determine factors that motivate and limit the implementation of ICT in healthcare environments (Gururajan & Hafeez-Baig, 2014)	Mix methodology is used in this research. 80 staff are interviewed. Once the initial motivators were identified, a focus group is conducted.	This research paper provides some initial findings of the factors that motivate and limit ICT implementation in healthcare organisations. A list of themes which can influence the ICT implementation in a healthcare environment was identified. Some of the themes were already identified in the literature review; however this research also identified new themes, in the form of internal and external factors, which contribute to the general research domain.
Adoption phenomena for wireless handheld devices in the healthcare environment (Gururajan & Hafeez-Baig, 2014)	Survey	All the three constructs, CP, SD and Comp for the adoption of wireless handheld devices represent good measurements for the intention-to-use these devices in the healthcare setting. Whereas SD is strongly influenced by CP and Comp as determinants for their intention, in spite of the direct effects of CP and Comp on the intention.
Use of information and communication technology to provide health information: what do older migrants know, and what do they need to know? (Goodall, Ward, & Newman, 2010)	Qualitative study (Focus group discussion) method is used in this research	
What's in a message? Delivering sexual health promotion to young people in Australia via text messaging (Gold, Lim, Hellard, Hocking, & Keogh, 2010)	Eight gender-segregated focus groups were held with 21 males and 22 females in August 2008.	Text messages were viewed as an acceptable and 'personal' means of health promotion, with participants particularly valuing the informal language. There was a preference for messages that were positive, relevant and short and for messages to cover a variety of topics. Participants were more likely to remember and share messages that were funny, rhymed and/or tied into particular annual events. The message broadcasting, generally fortnightly on Friday afternoons, was viewed as appropriate. Participants said the messages provided new information, a reminder of existing information and reduced apprehension about testing for sexually transmitted infections.

Chapter 16
Hadoop Map Only Job for Enciphering Patient-Generated Health Data

Arushi Jain
Ambedkar Institute of Advanced Communication Technologies and Research, India

Vishal Bhatnagar
Ambedkar Institute of Advanced Communication Technologies and Research, India

ABSTRACT

Today, Big Data is being leveraged in many industries from criminal justice to health care to real estate with powerful outcomes. Organizations are using Big Data to predict the future in turn making them smarter and efficient. All the health care data such as discharge and transfer patient data maintained in Computer based Patient Records (CPR), Personal Health Information (PHI), and Electronic Health Records (EHR). The use of Big Data analytics is becoming increasingly popular at health care centres, in clinical research, and consumer based medical product development. The biggest challenge with implementation of big data is that the nature of information of public health sector is of very sensitive nature and needs to be protected from unauthorized access and release of contents. Therefore, to provide solution to the deidentifying personal health big data in this paper we author make use of only mapper job framework for data encryption.

INTRODUCTION

In recent years, there is a proliferation in the amount of data generated. Voluminous complex data is generated from various sources that cannot be processed by traditional tools. Big Data is a current and hot topic of IT industry. Harnessing such huge amount of data is a tough job and thus it requires business intelligence (BI) or analytics. Business Intelligence is required to explore new knowledge, relationships and patterns among different data elements for example the banks have to make perpetual customer centric decisions and make sure they are well informed about all the risks and stakes. One of the best ways to reach out to the customers is to constantly communicate with them. This is obviously difficult as the

DOI: 10.4018/978-1-5225-6198-9.ch016

banks have thousands of branches and more than a million customers in countries which are populous and have a huge number of customers. Therefore, the use of Big Data is really important here. By asking the customers for feedback about products and services, banks can come to know about the customer's needs and preferences. In this paper, we focused on analyzing all the complaints from the customers that can help banks make well informed decisions in optimizing the services and business process along with region or time of the year specific schemes and offers which may attract more customers.

Big Data is a collection of Data sets so large and complex that traditional database management systems cannot handle efficiently. Big Data in an umbrella term used for the humongous amount of data generated from myriad of sources such the Web, mobile devices, sensors, enterprise applications and digital repositories. The data can be structured as well as unstructured. The data ranges from terabytes to exabytes of data. The relational database management systems (RDBMS) have proven to inefficient to handle such huge volumes of data. Another important factor which renders the conventional database systems unsuitable is that the majority of data being generated is unstructured; the RDBMS systems are only adept to handle structured data. Hence new tools and schemes for data analysis and management were in order. Big Data can be characterized by the 5 V's:

1. **Volume:** This refers to the amount of data. The sheer volume of data generated these days by real time applications and other data sources such as twitter feeds, photos, videos on social media, click streams of web, sensor-enabled equipment is so mammoth it runs to petabytes and exabytes of data. Big Data technology enables us to store this amount of data on distributed systems;
2. **Velocity:** It refers to the speed at which the data is generated. For example, the social media portals generate process able data at a very high rate, the amount of credit card transactions that happen in a second. All these require data to be analyzed at a very high rate;
3. **Variety:** Refers to the myriad of sources from which data is accumulated. Data is in different formats structured, semi-structured, unstructured. A full 90% world data has been generated over the past two years and majority of it is unstructured;
4. **Veracity:** Refers to the trustworthiness of data. There are uncertainty surroundings the data being generated these days with data being incomplete and inconsistent. The Big Data analytics empowers us to work with such kind of data;
5. **Value:** Refers to techniques of deriving value from data. There is an intrinsic value that the data may possess and must discover for analysis. This makes 'value' the most important 'V' of Big Data. Modern technologies have made it possible to find the value from data.

Today, Big Data is being leveraged in many industries from criminal justice to health care to real estate with powerful outcomes. Organizations are using Big Data to predict the future in turn making them smarter and efficient. Applications from Big Data are innumerable, from retail industry where Big Data helps retailers gain insights into the customer to needs and habits, to Banking, HealthCare & Hospitality. Government agencies are increasingly incorporating Big Data analytics to curb crime and maintain law and order through social media traffic analysis and other means. Therefore, to get actionable data and perform analytics requires specialized tools. There are thousands of Big Data tools available in the market. There are open source tools like Hadoop, a term which has become synonymous to Big Data, Cloudera.

RELATED WORK AND MOTIVATION

To manage the growing demands, there is a need to increase the capacity and performance of tools and methods employed for analysis of data. Chen et al. (2014), in their work "Big data: A survey" focused on big data and reviewed related technologies and examined the application of big data in various fields. Al-Jarrah et al. (2015), in their work "Efficient Machine Learning for Big Data: A Review" reviewed the data modeling in large scale data intensive field relating to model efficiency and new algorithm approaches. Hoffmann and Birnbrich (2012) to protect their customer from third party fraud proposed a conceptual link between retail bank activities in "The impact of fraud prevention on bank-customer relationships: An empirical investigation in retail banking". Srivastava and Gopalkrishnan (2015) revealed some of the best techniques which are used by the banks across the globe and can be used by the Indian banks to enhance their services offerings to the customers in "Impact of Big Data Analytics on Banking Sector: Learning for Indian Banks". Azar and Hassanien (2014) for dimensionality reduction presented a linguistic hedges neuro-fuzzy classifier with selected features (LHNFCSF). In this paper author compared the new classifier with the other classifiers for various classification problems in "Dimensionality reduction of medical big data using neural-fuzzy classifier". Hassanien et al. (2015) focused on application, challenges and opportunities of big data in "Big Data in Complex Systems: Challenges and Opportunities." Wahi et al. (2014) proposed a social media and its implication on customer relationship management in "Social Media: The core of enterprise 2.0.". Shabeera and Madhu Kumar (2015), in their work "Optimizing virtual machine allocation in MapReduce cloud for improved data locality" focused on improving data locality by allocating virtual machines for executing map reduce jobs. Aloui and Touzi (2015) proposed a methodology for designing ontology on a new platform called "FO-FQ Tab plug-in" and then querying them smartly based on conceptual clustering and fuzzy logic in "A Fuzzy Ontology-Based Platform for Flexible Querying". Ghallab et al. (2014), in their work "Strictness petroleum prediction system based on fussy model" predicted the status of crude oil and then compared it with other petroleum values. Huang et al. (2015) summarized the latest application of big data in health science. The authors also reviewed the latest technologies of big data and discussed the future perspective of health sciences in "Promises and challenges of big data computing in health science". Jagadish (2015) in "Big Data and Science: Myths and Reality" explored myths about big data and exposed the underlying truth. Jin et al. (2015) introduced the concept of big data and described the challenges as well as solution to these challenges in "Significance and challenges of big data research". Ryan and Lee (2015) presented a Multi-tier resource allocation as resource management technique for distributed systems in "Multi-tier resource allocation for data-intensive computing". Tiwari and Joshi (2015) in "Data security for software as a service" discussed security vulnerabilities of software as a service (SaaS) and its solution. Wahi et al. (2015) focused on whether the organization could able to address challenges posed by big data successfully or not. It also focused on the reasons why it is necessary to transit from the enterprise 1.0 stage to enterprise 2.0 stage in "Big Data: Enabler or Challenge for Enterprise 2.0." Deepak and John (2016) illustrated that information system is one of the most significant problem in fuzzy domain. Authors illustrated a case where hesitant membership value arrived from attribute value whose membership values are a family of set and also discusses the homomorphism between hesitant information systems in "Information Systems on Hesitant Fuzzy Sets". Bhanu and Tripathy (2016) in "Rough Set Based Similarity Measures for Data Analytics in Spatial Epidemiology" carried out epidemiological studies to understand a pattern and transmission of disease instances.

HADOOP ECOSYSTEM

The huge amount of data, besides being useful comes with some drawbacks also. Firstly, storage of such a huge amount of data is an issue. Special hardware is to be installed with lots of memory which in turn will cost a fortune and cannot be afforded by everyone. Secondly, processing this huge amount of data is also of a major concern. A normal home computer, on an average, would require 4 minutes to process a data of 3 GB. If a similar hardware is put into place to process data amounting to 3 Petabytes, which is equivalent to 3145728 GB of data, it will take nearly 8 years to process the data. There are other issues also with Big Data which need to be handled and clearly highly dedicated hardware is required for this purpose which is not feasible all the time.

Hadoop is one of the most popular cluster systems (see Figure 1), which includes the high-performance calculation model Map Reduce and Spark. Hadoop is an excellent and robust analytics platform for Big Data which can process huge data sets at a really quick speed by providing scalability. It can manage all aspects of Big Data such as volume, velocity and variety by storing and processing the data over a cluster of nodes whose numbers can run into thousands. The best thing about it is that the nodes in the cluster need not possess complex and costly hardware but simply commodity hardware which neither costly nor difficult to procure. Hadoop has a proper architecture, which consists of two components, MapReduce and the Hadoop Distributed File System.

Hadoop Distributed File System

The Hadoop distributed file system (HDFS) is a highly scalable and distributed file system for Hadoop. With the introduction of YARN (Yet Another Resource Negotiator) in later releases, Hadoop was integrated with a number of wonderful projects which can be used for storing, processing and analyzing data

Figure 1. Hadoop ecosystem (Taie, 2015)

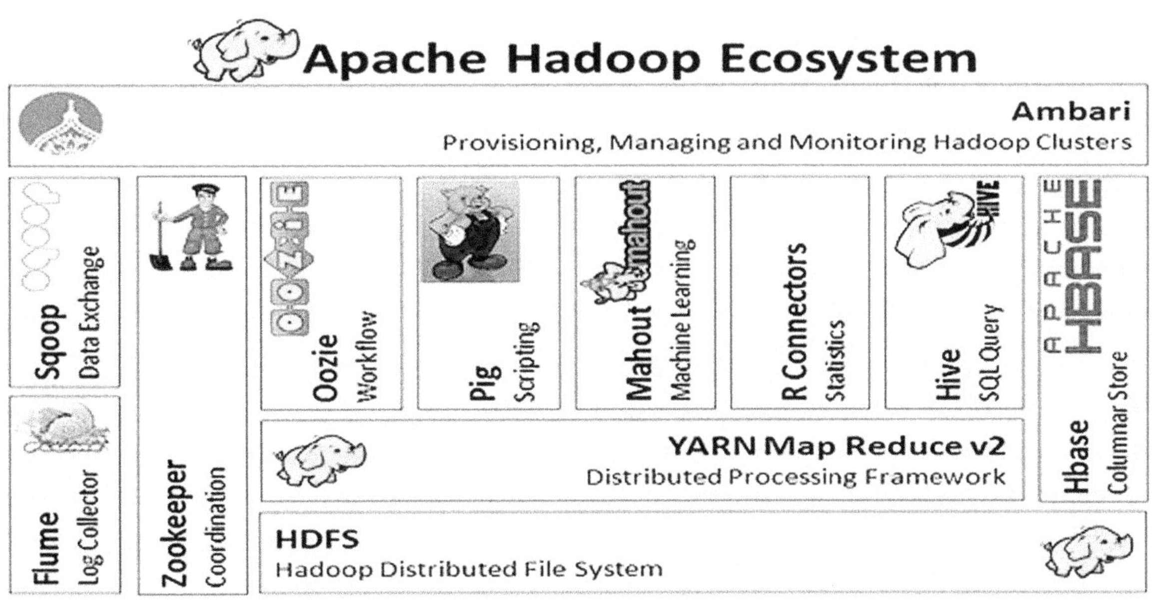

a lot more efficiently, thus aiding in exploration of data for undiscovered facts at a smooth pace. Hadoop is an open source project written in Java and is currently under the Apache Software Foundation and is free to use. The initial design idea is mainly used to deploy on the cheap hardware. The framework implements a distributed file system, referred to as "HDFS," which has high fault tolerance and high speed and scalability to store large amounts of data and implements a computation model of parallel processing large data sets; this model is of high speed computing in big data processing field. In addition to these advantages, the distributed system framework is open source software. With the advent of big data, it has become a good and reliable solution, which can store and process big data with ease. While in this day and year it has become a famous big data processing framework in the field of the big data analytics, at the same time, Hadoop ecosystem, with continuous improvement and optimization, has become better. With the advent of second generation, a lot more projects have been introduced into the Hadoop ecosystem.

Map Reduce

The term MapReduce is used to allude to two separate things: the model of programming and the implementation of Hadoop framework. The design of MapReduce is for the following features: Simplifying the development of applications, carry out the distribution of data for applications and also for a fault-tolerant data processing by the applications. The core of MapReduce is the term Map and Reduce separately. The software developers who are involved in programming write the programs rather jobs, which consist of two parts: Map function and Reduce function. The work of the framework is to take care of the minute details of work parallelization. The framework is also supposed to be considerate about scheduling jobs and their parts on the nodes, monitor them and recover from failure too. The developers are blessed with not focusing on 'how to implement complex codes'. The developers aim at algorithms and logic behind the business. The user-written code is called by the framework and not another way around. This is analogous to Java application servers wherein servlets are activated whenever an HTTP request is received: the container is in charge of setting up, tearing down and also providing a runtime environment for the code supplied by the user. Just like servlet writers are not required to implement the low-level details of socket I/O, event handling loops and the complex task of thread synchronisation, MapReduce developers do the programming in a clear-cut, simple interface and the 'container' does the hard task. MapReduce was proposed in a paper written by two Google engineers in the year 2004, which was titled as 'MapReduce: Simplified Data Processing on Large Clusters'. This paper dealt with the two models: programming and framework. Hadoop MapReduce is an open source application of the technology proposed in this paper. The implementation of Hadoop MapReduce is very closely related to this paper.

The features which MapReduce showcases are as follows:

- **Simplicity:** MapReduce is very simple for programmers. In MapReduce, there is no need of socket programming, no threads or any complex logic. There is also no need for any special technique to deal with the humongous amount of data. Programmers use the functional programming rules to construct applications that are used to process data. These applications work on one record at a time. The Map function works on these records to give key-value pairs as its output. Whereas, the Reduce part operates on the intermediate key-value pairs, working upon all the values that have the same key and then giving out the output. These primitives are the used to deploy functionalities like projection, aggregation, etc.;

- **Scaling:** In the MapReduce the tasks do not need communicating with each other. Neither do they share their states with each other. This property can be very helpful. Due to this property the tasks can be serialized and parallelized on distinct machines. More machines can be added to as well as removed from the cluster. The applications running can take immediate advantage of the new machines;
- **Parallelization and Distribution:** Programmers focus on the Map and Reduce functions for doing the task. The framework is in charge of dividing a MapReduce job into tasks and then executing each task on nodes;
- **Fault Tolerant:** Failures happen all the time, one cannot eliminate failure but instead one needs to work upon dealing with the failures. MapReduce supports executing the failed tasks on healthy nodes again. The task at hand either gets completed or fails, there is no provision of partial completion.

Operation of the MapReduce can be explained in simple terms. In MapReduce, the user submits a job at hand. This job contains the job, a map fuction and a reduce function. The MapReduce framework first breaks the job into tasks. These tasks are then supplied to the healthy nodes. The health of the nodes are monitored by the Job Tracker. The Task Tracker of the node receives the task. Task Tracker deploys the map and reduce functions and give out the output. A job works upon an input dataset and usually outputs one as well.

Stages of MapReduce

A job of MapReduce is made up of distinct stages:

1. User job submission;
2. Map function execution;
3. Shuffling and sorting.

The MapReduce framework is loaded with many APIs for submitting jobs. There are APIs to interact with the cluster as well. Jobtracker, a master process, is assigned with the task of accepting the job submissions. These job submissions take place over the network. Due to this the client, may or may not run on the cluster node. The framework then has the splitting task in hand. The task of splitting the dataset is onto the framework now. The input splits of the datasets can be processed in parallel. In Hadoop the component which is responsible for splitting is called Input Format, there are libraries for this.The number of splits determines the number of map tasks. This means that there will be as much map tasks as there are splits. A map task runs the map function provided by the user on records in the splits. These map tasks are parallely executed. Each input split is processed at the same time by the map functions on distinct machines. Input Format turns each record into key-value pair for map function. The map function's input is a key-value pair and the output produced is 0 or more key-value pairs. This output is called the intermediate key-value pairs. The map phase is where one does the filtering process. The next phase is the shuffling and sorting phase. This phase is responsible for ensuring that the reducer gets the key valued pairs in sorted manner. This is an intermediate phase. The shuffling and sorting is a part of reduce tasks. Before running the reduce tasks the reduce tasks run the shuffling and sorting. When started, each reducer is given one of the part on which it is supposed to work. First the intermediate key-value pairs are copied from each worker for their assigned partition. This copying takes a fair amount of time as this is done on the network.

Finally, we want to run the reducer function on the intermediate data produced by the map function. Some guarantees are required for this:

- Reducer will see all the values for each key. For example, if key is 'A' then reducer will look for all the keys with 'A';
- One key will be processed by individual reducer and not more than one;
- Reducer will need the keys in sorted order.

These guarantees are the work of shuffling and sorting phase.

After this comes the reducer stage where in the sorted data is processed by the user written reducer function. Each reducer is bound to produce separate output file. This is done so that the reducer do not have to coordinate access to file sharing. Thus the complexity is reduced to a very large extent and speed is also enhanced. The output format is responsible for determining the format of the output file.

A specific implementation of MapReduce programming model proposed in the paper written by the engineers of Google is Hadoop MapReduce. The computation component of MapReduce programming model is implemented by Apache Hadoop. By combining HDFS and MapReduce a very powerful tool is implemented. HDFS being similar to GFS of Google and Hadoop MapReduce an equivalent of MapReduce. Hadoop MapReduce has specific techniques to improve the machine overhead. The technique is that Hadoop MapReduce is aware of HDFS and is well aware of namenode. Thus, it can place the job at the best possible machine on the cluster. This technique makes sure that workers need not be worried about placing the data by themselves thus saving their overhead. Just like computers traditional distributed systems, Hadoop MapReduce is also equipped with a framework and job (see Figure 2). The master node is responsible for coordination while the worker nodes do their tasks.

Figure 2. Map reduce

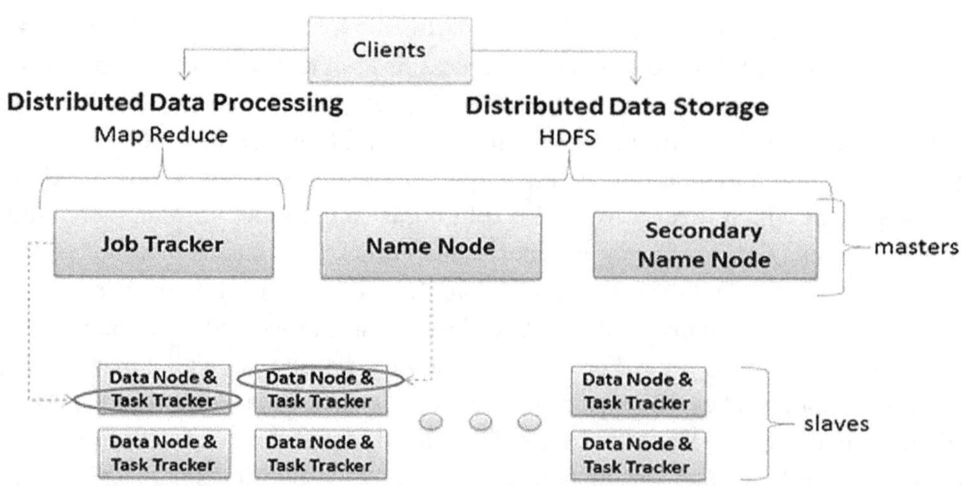

One of the most prominent techniques of Hadoop MapReduce is data locality. Here in the job is done at the place where the data is located. That means all the computation of data is performed on the machine where the data resides. Many powerful data performance computing systems have similar master slave models but the processing of data and data storage is never at the same site. In traditional data processing techniques the data is usually stored at one large repository and the processing is done by bringing in the data to the processing machine from the repository. This leads to many problems like time in transfer of data over network is large. When many nodes want to fetch the data from the repository then there can be a chaos. The solution to these problems is Hadoop MapReduce programming.

There are two types of daemons in Hadoop MapReduce:

1. **Job Tracker:** It is the master process which is responsible for accepting the jobs from the user. This is also assigned the task of distributing jobs to various worker nodes. It acts like the administrator. One MapReduce cluster is assigned only one job tracker. This job tracker runs on a machine which is reliable as failure of the master will result on loss of all the jobs;
2. **Task Tracker:** This daemon accepts the job and performs the MapReduce programming. It is answerable to job tracker. It is the slave daemon. Task tracker also has to report its health and job status to the job tracker.

Analogous to the relationship of datanode and namenode in HDFS, in MapReduce we have job tracker and task tracker. Task trackers send heartbeats to job tracker giving the status that they are active and ready to take the job. Upon receiving the job, the job tracker first checks for the task trackers active by listening to their heartbeats. The task trackers alive are assigned with the job. During the job, the task tracker sends the completion status at regular interval. It also sends the heartbeat to the job tracker. Upon completion, the job tracker is sent with the job information and then the job tracker responds to the user. Job tracker doesn't take up the job neither the results. It only assigns the job and acts like a medium between user and task tracker.

Hive

This tool is the SQl dialect of Hadoop. Apache Hive is a tool of data warehousing constructed on the Apache Hadoop. Its purpose is to provide data summaries, queries and analysis of data. Apache Hive was developed by Facebook. It is now maintained by Apache foundation. Various companies use Hive such as Netflix. Amazon has a forked version of Hive for its Amazon Elastic MapReduce which is deployed on Amazon Web Services.Apache Hive is capable of analyzing large data sets which are stored in Hadoop's HDFS and other file systems like Amazon S3. Hive has a SQL equivalent called HiveQL. HiveQL is capable of converting schemas and queries into queries to MapReduce, Apache Tez and Spark Jobs. These three execution engines can run on Hadoop YARN. It is also blessed with techniques of indexing which accelerate the queries. Hive can work with different storage types like plain text files, RCF, HBsae, etc. Hive has features which significantly reduce time taken to perform semantic checks at the time of query execution, which further facilitates metadata storage in RDBMS. Hive has capabilities of operating on compressed data which is stored in the Hadoop ecosystem with the help of algorithms like snappy, BWT, etc. Hive also supports user defined functions. These fuctions are capable o manipulating data.

Sqoop

This tool is used for the transfer of large data sets between Hadoop and the RDBMS. Sqoop uses command line interface for this kind of transfer. It has escalating loads of a single table or at times free form SQL queries as well as jobs which are saved, which are capable of running many times to import updates which are made to a database after the checkpoint. Checkpoint is the point of last import. Tables and Hbase and Hive are populated using these imports. On the other hand exports are usually employed to put data into Hadoop from some realtional database. Name of Sqoop comes from SQl + hadOOP. Pentaho has methods for connecting which are based on Sqoop. These methods include- Sqoop Import, Sqoop Export.Sqoop's connectors are used by various companies like Microsoft. Microsoft uses a connector to connect SQL Server Database to Hadoop. Transfers can be made using these connectors. Couchbase Inc. also deploys a similar kind of connector for its product Couchbase Server.

Pig

This is a data flow tool which works over Map Reduce. It's scripting language is PIG Latin which is a high-level language which works over Map Reduce. PIG is a platform of Hadoop for making data-flows which involve ETL-extract, transform and load. Its high-level language, PIG Latin, works over Map Reduce. It is fairly easy to write in PIG Latin for those who are not versed with JAVA and Map Reduce. Programs of Pig Latin are transformed into various functions of Map Reduce which in turn carry out the tasks. PIG Latin also comprises of the functionalities of SQL like Join and Filter. Pig Latin's program is converted into Java code for a MapReduce program behind the scene. Thus, it is the tool for programmers not versed with Java, as it requires writing programs in simple language similar to SQL for RDBMS. The functionality of PIG can be extended by writing programs in various languages like Java, Ruby, Python, etc. Technique of PIG was developed in 2006 by researchers at Yahoo. This was a way for their programmers not knowing Java to write MapReduce programs. Pig is managed from Apache foundation since 2007. If PIG is compared with SQL, then we come various across conclusions. PIG uses lazy evaluation as compared to SQL. PIG uses the principle of data mining tool, that is ETL, extract, transform and load, while SQL doesn't use such principles. PIG is also capable of storing data at any point of pipeline. PIG is also said to declare its execution plans beforehand. PIG is procedural as compared to declarative SQL.

HBase

It is a database of Hadoop for storing the big data. HBase is an open source database made in Java. It is non-relational in nature. HBase is modeled after Big Table which is Google's product. It is developed by Apache foundation and has compatibility with Apache Hadoop. HBase runs on top HDFS which is the file system of Hadoop. It provides capabilities similar to that of Big Table for the Hadoop. HBase is known for providing fault-tolerant data storage. This data is sparse in nature that is the data include small amount of information which is found between large collection of data. Example can be a task to find 100 largest items in a group of 5 billion records. HBase has features like compression, operations that are in-memory. HBase has Bloom filters which can be applied on each column basis similar to Big-Table. Input and output for MapReduce jobs in Hadoop can be acted by the HBase tables. HBase tables are versatile and can also be accessed via means of various APIs in Java like REST, Avro, gateway, etc.

HBase is a store which is column oriented key valued. HBase is famous for its connectivity with Hadoop and HDFS. HBase Has high throughput and lower latency. This is because it runs on top of HDFS and is accustomed for read and write operations which are faster even on large datasets.

HBase cannot be considered to be a replacement of any SQL database. HBase and SQL databases have their own individual importance. A SQL layer for HBase is provide by the Phoenix which is also a product of Apache foundation. Phoenix is also responsible for providing HBase with Java connector JDBC. This driver can be integrated into various applications and analytical tools. HBase is widely used by many of the companies and websites. Its prominent examples would be messaging platform deployed by Facebook. HBase does not support SQL scripting like the traditional databases do. Though, it has similar scripts written in Java for applications.

Zookeeper

Zookeeper is a software project which is maintained by Apache foundation. It is also one of the open source projects like the other projects of Apache. It is regarded as software for distributed configuration services. It is aimed at also providing synchronous service and naming registry for big distributed systems. It was earlier one of the sub-projects under Hadoop but now it is an independent project. The architecture of Zookeeper is supportive of high availability through services which are redundant. Thus the clients can ask other Zookeeper leader if the first one fails to respond. Nodes of Zookeeper are known to store their data in name spaces which are hierarchical, just like the tress data structure. Clients are required to read and write data onto these nodes and thus accomplish a configuration service which is shared in nature. Service of Zookeeper is used by companies like Rackspace and Yahoo. Even eBay is a part of users of Zookeeper services.

RESEARCH METHODOLOGY

The data we are going to analyse is health sector data. It contains confidential information such patient contact details, Id and disease procured (see Figure 3). As per protocol the health sector data is very sensitive as it contains personal information such as personal health records, demographic data and needs

Figure 3. Research methodology

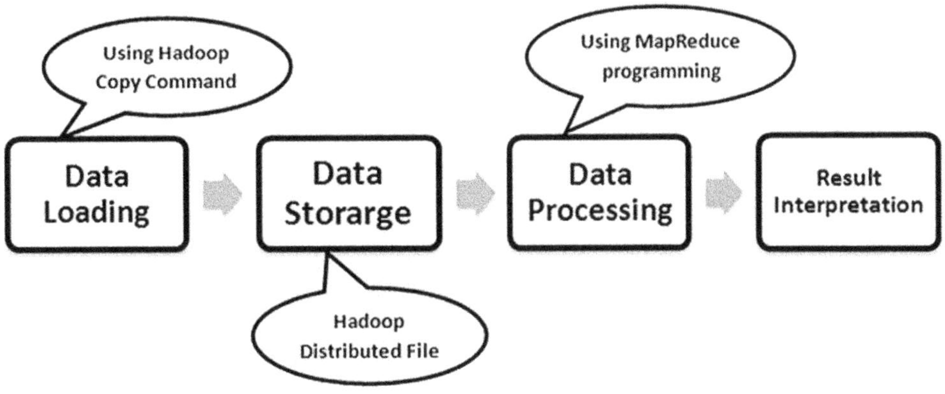

to be secured, therefore we encrypt the data using Hadoop framework. All the health care data such as discharge and transfer patient data maintained in Computer based Patient Records(CPR), Personal Health Information (PHI), and Electronic Health Records (EHR). The use of Big Data analytics is becoming increasingly popular at health care centres, in clinical research, and consumer based medical product development. The biggest challenge with implementation of big data is that the nature of information of public health sector is of very sensitive nature and needs to be protected from unauthorized access and release of contents. Due to the sheer volume of data, traditional systems have proven to be in effective in handling and encryption of such huge amounts of data effectively. Therefore, to provide solution to the deidentifying personal health big data we make use of the Hadoop framework for data encryption.

The input file which is .csv file will be the dataset acting as an input. A dataset is a collection of related sets of information that is composed of separate elements but can be manipulated as a unit by a computer (see Table 1). The dataset used in the project comprises of medical records of patients. It is a real-time healthcare database based upon records of over 45K patients per day. You'll be able to see information like disease trends over time and by patient, what diseases are being diagnosed, and patient records like their email address, date of birth, patient id, etc.

Table 1 is an example of data in the dataset. It contains patient id, patient name, date of birth, phone number, email address, patient sex, disease, etc.

ALGORITHM USED TO ENCIPHERING PATIENT GENERATED HEALTH DATA

The Advanced Encryption Standard (AES) is a modern symmetric block cipher technique that is an advancement over traditional DES approach. It was developed by cryptographers- Joan Daemon and Vincent Rijmen of Belgium. Owing to its benefits over DES and enhanced security the Algorithm was adopted by US government for its use and is now being used worldwide. The plaintext in AES is input as a 128-bit or 16-byte block. The Algorithm is referred to as AES-128, AES-192, AES-256 depending upon the key length being 128,192,256 bits respectively. The AES is based on the traditional principles of Symmetric key cryptography that is- Substitution & Permutation, using a combination of both. Feistel ciphers are not used in AES unlike DES where it formed the backbone of the encryption/decryption process. A single 128-bit block is represented as a 4x4 square matrix. The block is copied into a state array which in turn is changed at stage of the encryption or decryption process. The key also is represented as an array, the number of rounds in AES cipher depend upon the key length for example a 16-byte key requires 10 rounds. There is an initial AddRoundKey transformation before the first round. There are certain distinct functions that are carried out in the first N-1 rounds- SubBytes, ShiftRows, MixColumns, AddRoundKey. The Nth round is different from the previous round wherein there are only three transformations. The input to each transformation is a 4x4 matrix which produces a similar 4x4 output matrix. The final round gives the ciphertext as the output. The key expansion function generates, one

Table 1. Sample data set

11111	bbb1	12-10-1950	1234567890	bbb1@xxx.com	1111111111	M	Diabetes	78
11112	bbb2	12-10-1984	1234567890	bbb2@xxx.com	1111111111	F	PCOS	67

more the total number of round, number of keys. The algorithm works in different phases. Algorithm takes the data as the input file in the format of .csv. The output of the algorithm is in encrypted format:

Phase 1 - Map Phase: The first phase is map phase. The initial task of the phase is to convert the strings into tokens. Tokens are easy to process and give better results. The next step of the phase is to start a loop till all the tokens are read and in this loop, convert each token into encrypted key.

Phase 2 - Encryption Phase: This encryption is the second phase of the algorithm. In this phase, any of the encryption various keys can be used. The keys used here are AES, ECB, PKCS5 Padding. After selecting the key the token is encrypted using the AES algorithm.

Phase 3 - Output Phase: The tokens after being encrypted are sent to the third and the final phase where in the display of the tokens in encrypted format occurs.

This algorithm does not use the reducer phase as the job here is to encrypt the data and then display the encrypted file, thus no reducer is used here. This algorithm also demonstrates that Hadoop can also work without a reducer phase.

Code used

```
public static class Map extends Mapper<Object, Text, NullWritable, Text> {
public void map(Object key, Text value, Context context) throws IOException,
InterruptedException
 {
        StringTokenizer itr = new StringTokenizer(value.toString(),",");
        List<Integer> list=new ArrayList<Integer>();
        Collections.addAll(list, encryptCol);
        System.out.println("Mapper:: one:"+value);
        String  newStr="";
        int counter=1;
        while (itr.hasMoreTokens())
 {
        String token=itr.nextToken();
        System.out.println("token"+token);
        System.out.println("i="+counter);
        if(list.contains(counter))
{
        if(newStr.length()>0)
        newStr+=",";
        newStr+=encrypt(token, key1);

}
else
{
```

```
1.          if(newStr.length()>0)
2.          newStr+=",";
3.          newStr+=token;
}
8. Counter=counter+1;
}
9. context.write(NullWritable.get(), new Text(newStr.toString()));

}
}
public static String encrypt(String strToEncrypt, byte[] key)
{
try
{
1.          Cipher cipher = Cipher.getInstance("AES/ECB/PKCS5Padding");
2.          SecretKeySpec secretKey = new SecretKeySpec(key, "AES");
3.          cipher.init(Cipher.ENCRYPT_MODE, secretKey);
4.          String encryptedString = Base64.encodeBase64String(cipher.
doFinal(strToEncrypt.getBytes()));
5.          String(cipher.doFinal(Base64.decodeBase64(strToEncrypt)));
6.          return encryptedString.trim();
}
}
```

Output

From Figures 4 and 5, it is clearly visible that patient's personal healthcare information is successfully deidentified using Hadoop map job only.

Figure 4. Plaint text file

11111	bbb1	12/10/1950	1234567890	bbb1@xxx.com	1111111111	M	Diabetes	78
11112	bbb2	12/10/1984	1234567890	bbb2@xxx.com	1111111111	F	PCOS	67
11113	bbb3	712/11/1940	1234567890	bbb3@xxx.com	1111111111	M	Fever	90
11114	bbb4	12/12/1950	1234567890	bbb4@xxx.com	1111111111	F	Cold	88
11115	bbb5	12/13/1960	1234567890	bbb5@xxx.com	1111111111	M	Blood Pressure	76
11116	bbb6	12/14/1970	1234567890	bbb6@xxx.com	1111111111	F	Malaria	84
11117	bbb7	12/15/1980	1234567890	bbb7@xxx.com	1111111111	M	Swine Flu	64
11118	bbb8	12/16/1990	1234567890	bbb8@xxx.com	1111111111	F	Fever	33
11119	bbb9	12/17/2000	1234567890	bbb9@xxx.com	1111111111	F	Fever	29
11120	bbb1	12/10/1950	1234567890	bbb1@xxx.com	1111111111	M	Diabetes	78
11121	bbb2	12/10/1984	1234567890	bbb2@xxx.com	1111111111	F	PCOS	67
11122	bbb3	712/11/1940	1234567890	bbb3@xxx.com	1111111111	M	Fever	90
11123	bbb4	12/12/1950	1234567890	bbb4@xxx.com	1111111111	F	Cold	88
11124	bbb5	12/13/1960	1234567890	bbb5@xxx.com	1111111111	M	Blood Pressure	76
11125	bbb6	12/14/1970	1234567890	bbb6@xxx.com	1111111111	F	Malaria	84
11126	bbb7	12/15/1980	1234567890	bbb7@xxx.com	1111111111	M	Swine Flu	64
11127	bbb8	12/16/1990	1234567890	bbb8@xxx.com	1111111111	F	Fever	33
11128	bbb9	12/17/2000	1234567890	bbb9@xxx.com	1111111111	F	Fever	29

Figure 5. Encrypted file of plain text file

CONCLUSION

All the health care data such as discharge and transfer patient data maintained in Computer based Patient Records (CPR), Personal Health Information (PHI), and Electronic Health Records (EHR).The use of Big Data analytics is becoming increasingly popular at health care centres, in clinical research, and consumer based medical product development. The biggest challenge with implementation of big data is that the nature of information of public health sector is of very sensitive nature and needs to be protected from unauthorized access and release of contents. Due to the sheer volume of data traditional systems have proven to be ineffective in handling and encryption of such huge amounts of data effectively. Therefore, to provide solution to the deidentifying personal health big data in this paper we author made use of the only mapper job framework for data encryption.

ACKNOWLEDGMENT

The authors would like to thank Ambedkar Institute of Advanced Communication Technologies and Research, New Delhi to successfully accomplish the research work.

REFERENCES

Al-Jarrah, O. Y., Yoo, P. D., Muhaidat, S., Karagiannidis, G. K., & Taha, K. (2015). Efficient Machine Learning for Big Data: A Review. *Big Data Research*, *2*(3), 87–93. doi:10.1016/j.bdr.2015.04.001

Aloui, A., & Touzi, A. G. (2015). A Fuzzy Ontology-Based Platform for Flexible Querying. *International Journal of Service Science, Management, Engineering, and Technology*, *6*(3), 12–26. doi:10.4018/IJSSMET.2015070102

Aye, K. N., & Thein, T. (2015). A platform for big data analytics on distributed scale-out storage system. *International Journal of Big Data Intelligence.*, *2*(2), 127–141. doi:10.1504/IJBDI.2015.069088

Azar, A.T., Hassanien, A.E. (2014). Dimensionality reduction of medical big data using neural-fuzzy classifier. *Soft computing, 19*(4), 1115-1127. DOI:.10.1007/s00500-014-1327-4

Barker, K. J., Amato, J., & Sheridon, J. (2008). Credit card fraud: Awareness and prevention. *Journal of Financial Crime*, *15*(4), 398–410. doi:10.1108/13590790810907236

Bhanu, S. K., & Tripathy, B. K. (2016). Rough Set Based Similarity Measures for Data Analytics in Spatial Epidemiology. *International Journal of Rough Sets and Data Analysis*, *3*(1), 114–123. doi:10.4018/IJRSDA.2016010107

Chen, M., Mao, S., & Liu, Y. (2009). Big Data: A Survey. *Springer-. Mobile Networks and Applications*, *19*(2), 171–209. doi:10.100711036-013-0489-0

Deepak, D., & John, S. J. (2016). Information Systems on Hesitant Fuzzy Sets. *International Journal of Rough Sets and Data Analysis*, *3*(1), 55–70.

Fedoryszak, M., Tkaczyk, D., & Bolikowski, L. (2013). Large Scale Citation Matching Using Apache Hadoop. Springer- Research and Advanced Technology for Digital Libraries, LNCS (Vol. 8092, pp. 362-365).

Ghallab, S. A., Badr, N. L., Salem, A. B., & Tolba, M. F. (2014). Strictness petroleum prediction system based on fussy model. *International Journal of Service Science, Management, Engineering, and Technology*, *5*(4), 44–65. doi:10.4018/ijssmet.2014100104

Hashem, I. A. T., Yaqoob, I., Anuar, N. B., Mokhtar, S., Gani, A., & Khan, S. U. (2015). The rise of big data on cloud computing: Review and open research issues. *Information Systems*, *47*, 98–115. doi:10.1016/j.is.2014.07.006

Hassanien, A. E., Azar, A. T., Snasel, V., Kacprzyk, J., & Abawajy, J. H. (2015). *Big Data in Complex Systems: Challenges and Opportunities. In Studies in Big Data* (Vol. 9). Springer.

Hoffmann, A. O. I., & Birnbrich, C. (2012). The impact of fraud prevention on bank-customer relationships. *International Journal of Bank Marketing*, *30*(5), 390–407. doi:10.1108/02652321211247435

Huang, T., Lan, L., Fang, X., An, P., Min, J., & Wang, F. (2015). Promises and challenges of big data computing in health science. *Big Data Research.*, *2*(1), 2–11. doi:10.1016/j.bdr.2015.02.002

Ibrahim, S., Jin, H., Lu, L., Qi, L., Wu, S., & Shi, X. (2009). Evaluating MapReduce on Virtual Machines: The Hadoop Case. In Cloud Computing, LNCS (Vol. 5931, pp. 519-528).

Jacobs, A. (2009). The pathologies of big data. *Communications of the ACM - A Blind Person's Interaction with Technology, 52*(8), 36-44.

Jagadish, H. V. (2015). Big Data and Science: Myths and Reality. *Big Data Research*, *2*(2), 49–52. doi:10.1016/j.bdr.2015.01.005

Jin, X., Wah, B. W., Cheng, X., & Wang, Y. (2015). Significance and challenges of big data research. *Big Data Research*, *2*(2), 59–64. doi:10.1016/j.bdr.2015.01.006

Kolomvatsos, K., Anagnostopoulos, C., & Hadjiefthymiades, S. (2015). An Efficient Time Optimized Scheme for Progressive Analytics in Big Data. *Big Data Research*, *2*(4), 155–165. doi:10.1016/j.bdr.2015.02.001

Labrinidis, A., & Jagadish, H. V. (2012). Challenges and opportunities with big data. *ACM- Proceedings of the VLDB Endowment*, *5*(12), 2032-2033.

Lee, Y., & Lee, Y. (2013). Toward scalable internet traffic measurement and analysis with Hadoop. *ACM SIGCOMM Computer Communication*, *43*(1), 5–13. doi:10.1145/2427036.2427038

Madhavi, D., & Ramana, B. V. (2015). De-Identified Personal health care system using Hadoop. *Iranian Journal of Electrical and Computer Engineering*, *5*(6), 1492–1499.

Ryan, T., & Lee, Y. C. (2015). Multi-tier resource allocation for data-intensive computing. *Big Data Research*, *2*(3), 110–116. doi:10.1016/j.bdr.2015.03.001

Shabeera, T. P., & Madhu Kumar, S. D. (2015). Optimizing virtual machine allocation in MapReduce cloud for improved data locality. *International Journal of Big Data Intelligence.*, *2*(1), 2–8. doi:10.1504/IJBDI.2015.067563

Srivastava, U., & Gopalkrishnan, S. (2015). Impact of Big Data Analytics on Banking Sector: Learning for Indian Bank. *Big Data. Cloud and Computing Challenges*, *50*, 643–652.

Taie, M. (2015). Hadoop Ecosystem: an Integrated Environment for Big Data. Retrieved from http://blog.agroknow.com/?p=3810

Tiwari, P. K., & Joshi, S. (2015). Data security for software as a service. *International Journal of Service Science, Management, Engineering, and Technology*, *6*(3), 47–63. doi:10.4018/IJSSMET.2015070104

Wahi, A. K., Medury, Y., & Misra, R. K. (2014). Social Media: The core of enterprise 2.0. *International Journal of Service Science, Management, Engineering, and Technology*, *5*(3), 1–15. doi:10.4018/ijssmet.2014070101

Wahi, A. K., Medury, Y., & Misra, R. K. (2015). Big Data: Enabler or Challenge for Enterprise 2.0. *International Journal of Service Science, Management, Engineering, and Technology*, *6*(2), 1–17. doi:10.4018/ijssmet.2015040101

This research was previously published in the International Journal of Information Retrieval Research (IJIRR), 7(4); edited by Zhongyu (Joan) Lu, pages 72-86, copyright year 2017 by IGI Publishing (an imprint of IGI Global).

Section 4
Surveys and Case Studies

Chapter 17
Users' Satisfaction With the Electronic Health Record (EHR) in the Kingdom of Bahrain

Aysha Ebrahim Abdulla
University of Bahrain, Bahrain

Shurooq Yousif Ahmed
University of Bahrain, Bahrain

Maryam Abdulrahman Alnoaimi
University of Bahrain, Bahrain

Hayat Ali
University of Bahrain, Bahrain

ABSTRACT

Today, many hospitals seek to adopt the latest and most sophisticated technologies in order to raise the service quality and users' satisfaction. The Electronic Health Record (EHR) had a substantial impact on the health sector and has enhanced the efficiency and effectiveness of healthcare providers. The purpose of this research is to examine the factors that affect users' satisfaction with the current Health Record System in the Kingdom of Bahrain. A research model was built based on three popular models of users' satisfaction toward information systems. Toward achieving the research objective, a quantitative approach was followed to collect data from an online survey. Accordingly, 152 responses were collected from the users of EHR in public hospitals and health centres in Bahrain. The results of the survey were analyzed using SPSS and SmartPLS 3.0. It was concluded that the most effective factors in the users' satisfaction with EHR were directly service quality and technical support, with system and information quality indirectly through trust.

DOI: 10.4018/978-1-5225-6198-9.ch017

INTRODUCTION

Government and health institutes in different countries seek to improve the quality of their healthcare using the latest technologies, and systems that can serve all the medical staff in hospitals. One of these systems is the Electronic Health Record (EHR), which is an integrated system used in hospitals by doctors and medical staff; it collects, controls and manages patients' complete health information (Paul and Lansky, 2005). Using these systems can reduce time and cost, increase the efficiency of performance, improve the quality of healthcare services provided for patients, and ensure access to medical information and exchange of experiences (Nassiliou, 2009). The Kingdom of Bahrain followed this trend by applying this system in their hospitals. The most important ICT project at present is the National Health Information System (I-Seha) that provides high quality and efficient health services, making health information available and accessible to healthcare providers' and easing their workload (MOH, 2015).

To ensure that the system's services are appropriate to the different users, it is important to examine their level of satisfaction and the factors that affect their satisfaction; this not only measures the success of the system but also indicates whether any additional technical support or financial incentives are required. This is the topic of this research.

ELECTRONIC HEALTH RECORD

The Healthcare Information and Management Systems Society (HIMSS) (2012) defines EHR as longitudinal electronic record of patient health information produced by encounters in one or more care settings. Included in this information are patient demographics, progress notes, problems, medications, vital signs, past medical history, immunizations, laboratory data, and radiology reports. The EHR automates and streamlines the clinician's workflow. The EHR has the ability to independently generate a complete record of a clinical patient encounter, as well as supporting other care-related activities such as decision support, quality management, and clinical reporting.

The Healthcare Financial Management Association (HFMA) (2006), on the other hand, focuses more widely on the functions of an Electronic Health Record, stating for example that EHR enables the process of order-entry management, such as medication ordering. It can effectively give the medical staff access to all patients' information. It can store the patients' health records in databases, facilitate administrative processes such as billing, scheduling and resource management, provide effective electronic exchange of patient data and develop decision-support systems. EHR itself provides many benefits, supporting decision making, reducing expenditure and costs, minimizing the time spent on many processes, and providing comprehensive information for doctors to facilitate medical diagnosis (Lorenzi, Kouroubali, Detmer and Bloomrosen, 2009). It also enables healthcare providers to deal more easily with patient's records, sharing the information with other hospitals, healthcare providers, nursing homes and specialists from outside the hospital whenever it is required (HealthIT, 2015). In addition, any changes or modifications that occur to patients' recorded information are updated automatically and made available to all other healthcare providers. Tracing the progress of patients' health is important in order to provide them with appropriate treatment. It also speeds up the large number of tasks that physicians and nurses are responsible for; at the same time, the patients can coordinate and manage much of their own health information, as well as contacting their healthcare professionals (Computing Reviews, 2015). Thus it will increase the efficiency of medical staff's workflow, free management time, and improve business strategic planning and operational processes (Thakkar & Davis, 2006). However, as with any newly deployed technology, many problems can emerge. Thus, with EHR, concerns about security have been raised; Ajami and Arab-Chadegani (2013), for example, stress that the system must provide high security, confidentiality and privacy, as healthcare services deal with critical and sensitive personal data. Also, EHR is more complicated than paper-based records (Lorenzi, Kouroubali, Detmer, and Bloomrosen, 2009), affecting the medical staff's ability to use it (Ajami and Arab-Chadegani, 2013). For example, the process of installing EHR in health institutions is causing difficulties because of the complexity of this kind of software package, given the amount of coding involved in such a system (Ash, Berg & Coiera, 2004). A person without appropriate computer knowledge and familiarity with the system's user interface, or simply without adequate typing skills to enter patient information, prescriptions and notes into the system, might be afraid of losing their position, see the system as a very complex application. The medical staff therefore need training to improve their skills in using the system (Ajami & Bagheri-Tadi, 2013). Furthermore, the system is very expensive in financial terms, because the cost includes coordination costs, purchase price, negotiating costs, monitoring costs, upgrade costs, governance costs, and the cost of recovery from errors in the software, computer crashes or programming errors. If overloaded, the system becomes sluggish. Finally, the interaction between doctors and patients when using EHR could lead to many problems, such as using the computer in front of the patient appearing to be rude, with absence of eye contact with the patient (Linder, Schnipper, Tsurikova & Melnikas, 2006).

OVERVIEW OF USER STATISFACTION

Definitions of User Satisfaction

The term satisfaction was originally derived from the Latin, denoting "release from uncertainty" (Oxford Library of Words and Phrases, 1993). English Collins Dictionary (2000) defines satisfaction as "the act of satisfying or state of being satisfied; the fulfilment of a desire; the pleasure obtained from such fulfilment; and a source of fulfilment". Gelderman (1998) presented a definition from the organizational viewpoint; he stated that it is a "reflection of the context to which the information needs of a manager have been met". Kotler (2009) describes it as "a person's feeling of pleasure or disappointment resulting from comparing a product's perceived performance or outcome in relation to his/her expectations". In the context of information systems, user satisfaction has been extensively discussed by researchers as a measuring tool for system success (Ives, Olson and Baroudi, 1983) and system effectiveness (Amoroso and Cheney, 1992). Yuthas and Eining (1995) argue that system effectiveness can be assessed by its usage, performance and user satisfaction. In contrast, Abugabah and Sanzogni (2009) stress that considering only these elements to determine the success of the system may be inapplicable, due to the absence of other factors in different environments. According to Chin and Lee (2000), end-user satisfaction is "the overall affective evaluation an end-user has regarding his or her experience related with the information system". Finally, Au, Ngai and Cheng (2002) indicate that user satisfaction is "the overall affective and cognitive evaluation of the pleasurable level of consumption-related fulfilment experienced with the IS".

User Satisfaction in the Healthcare Sector

In the healthcare sector, user satisfaction is used in the process of system evaluation, beside other factors such as system quality (Ammenwerth et al., 2004). Rahimi and Vimarlund (2007) reviewed research into electronic systems published between 2003 and 2005, and found that evaluation research was strongly emphasized in user satisfaction. However, Ammenwerth et al. (2003) reported that, as a result of their reviews of previous research, health information systems evaluation was associated with three problems, due to complexity in three fields: the evaluation object, the evaluation project, and the motivation for evaluation. To solve these problems, researchers argue for the employment of user satisfaction tools. According to Stricklin, Bieber and Struk (2003) in their research on homecare nurses, a strongly satisfied user, in this case the nurse, positively increases and facilitates the success and quality of healthcare. Schubart and Einbinder (2000) discuss the impact of personal skills and working environment, which enhance satisfaction with health information systems. The deployment of computerized patient records has influenced user satisfaction positively. It helps in facilitating the improvement of the healthcare work (Delpierre et al., 2004). Furthermore, Rahimi and Vimarlund's (2007) review conclude that the introduction of a new computerized patient record and health decision-support system led to high user satisfaction. Perry (2012) points to the importance of considering user satisfaction in the evaluation of a clinical decision-support system (CDSS) In addition, Ward, Stevens, Brentnall and Briddon (2008) stress the importance of a suitable interface design for the health decision-support system, because of its impact on user satisfaction. Kushniruk and Kannry (2011) define the user interface as "the component of a man-machine system taken as a whole, responsible for communication with the user of the system".

THE MODEL OF SYSTEM USERS' SATISFACTION

The literature on models of user satisfaction with information systems in general is first reviewed, followed by their satisfaction with health information systems. In both contexts, the focus is on the models applied and validated.

Gallagher (1974) published the first study which discussed user satisfaction, in which he examined the output reports generated from information systems by 75 managers. In the 1980s, studies on the concept of user satisfaction increased, and a number of instruments were introduced by different researchers to build generalized tools to measure information system success and, consequently, user satisfaction (Bakke and Stensrud, 2008). As a result of the large number of studies and measures, DeLone and McLean (1992) note the difficulty of designing a definitive model. Researchers have built instruments for measuring the satisfaction of end users, based on the user's environment in interacting with the computer. One of the earliest of these studies was by Bailey and Pearson (1983), who created a measurement of Computer User Satisfaction (CUS) by developing a tool consisting of 39 evaluation elements. They tested it, through interviews, on 32 managers working in eight different institutes and found 625 factors, later reduced to 39 The test involved indirect or intermediate users. Mahmood, Burn, Gemoets and Jacquez (2000) note that after reviewing 45 field studies of user satisfaction from 1982 to 2000, Bailey and Pearson's measurements proved to be the most popular tool. However, Gable, Sedera and Chan (2003) believe that these dimensions are more suitable for assessing system success than user satisfaction. Doll and Torkzadeh (1988) studied and validated End-User Computing Satisfaction

(EUCS) and they argued that Bailey and Pearson's model omitted one factor, ease of use. Other than that, content, accuracy, format and timeliness were the building blocks. Torkzadeh and Doll (1991) define EUCS as "the affective attitude towards a specific computer application by someone who interacts with the application directly". This definition does not specify the type of application the user is evaluating. Later, Torkzadeh and Doll (1993) refined this definition, clarifying the type of system user, whether management, professional or operation personnel.

Considering user satisfaction as a dependent variable, Delone and McLean (1992) reviewed 180 studies in the field of information system success and effectiveness, and concluded that user satisfaction should be one of six additional variables used to assess effectiveness, together with system quality, information quality, system usage, individual impact, and organizational impact (Mashhour and Zaatreh, 2008). The relations between dimensions were described as follows: first, system quality leading to system use and user satisfaction were interdependent; and second, use and satisfaction determined the individual impact, that led to organizational impact. Some researchers agreed to add the service quality category to Delone and McLean'ssuccess model (Watson, Pitt and Kavan, 2001). Consequently, Delone and McLean proposed an updated version of their model in 2003. Raija (2011) points to the flexibility of the model, which can be altered to fit different types of information system and users. Aggelidis and Chatzoglou (2012) integrated these models, providing additional valid and reliable factors to Doll and Torkzadeh's 1988 version for hospital information systems: system speed, interface, user documentation, user training, information department support, and maintenance support. Aggelidis and Chatzoglou (2012) grouped the criteria from DeLone and McLean's (1992) model in two major constructs: ?? and system quality. These two factors cover the level of user satisfaction. Furthermore, Bliemel and Hassanein (2006) developed a model to investigate the factors that influence overall consumer satisfaction with online Health Information Retrieval (HIR). Their model comprises eight factors (specific content, content quality, technical adequacy, appearance, satisfaction with system quality, satisfaction with information quality, trust) that directly or indirectly affect overall satisfaction.

KINGDOM OF BAHRAIN'S EXPERIENCE

The Kingdom of Bahrain is taking social, political and economic initiatives to improve the quality of citizens' lives. The priority in developing the health sector is to ensure high-quality healthcare services for all citizens; to use all available resources in achieving the healthcare mission through the use of IT projects, improving services in all hospitals and health centres for all citizens. The most important project at present is the National Health Information System (I-Seha) that is aiming to provide efficient, high-quality health services, making health information available and accessible, supporting healthcare providers' work. Bahrain launched the first version of I-Seha at one health centre, after which it was distributed among 25 health centres, two main hospitals (King Hamad University Hospital and Salmanyia Medical Complex), and four peripheral hospitals in Bahrain. Previously, healthcare management was based on manual and paper-based systems which led to inefficiency in the administrative work, lost requests and patient files, and slowed down the quality of healthcare for patients. A major goal of the new system is to improve the quality of health services at the lowest cost. For example, it is now possible to reduce the cost of buying radiology films, which is more than a quarter of a million dinars a year. Medical staff are optimistic about I-Seha and believe it will raise the level of health and assist providers of healthcare services. Nevertheless, the system needs continuous improvement, especially considering its geographical extent, the number of patients visiting hospitals and clinics every day, and training medical staff to use the system as efficiently as possible (MOH, 2015).

Meanwhile, the Bahrain Defence Force hospital implemented its own Electronic Health Record, AL-Care; in which continues to evolve to meet the needs of all staff (Rehman, 2015). AL-Care was developed and fully implemented in 1998. It integrates various modules divided into three categories, clinical, non-clinical and administrative, which include the Clinic Appointment Module, Patient Registration, Laboratory Results, Blood Bank System, Pharmacy System, Bed Status Information, Appointment SMS Module, Inpatient Billing Module, Online Leave System and many more (Alhashimi, 2009).

RESEARCH MODEL AND HYPOTHESES

The Research Model

The proposed model for this research has been constructed from a number of factors derived from three of the previously discussed models: those of Delone and McLean (2003), Bliemel and Hassanein (2006), and Aggelidis and Chatzoglou (2012). DeLone and McLean's model has been used by many researchers to measure the success of different information systems. Three factors, information quality, system quality and service quality, are generally considered the most important measures for IS success. Researchers also focus on the EUCS model of Aggelidis and Chatzoglou, which includes two major variables related to information quality: content, given the huge amount of information the system manages; and accuracy, which is the most important factor in measuring user satisfaction (Bailey, 1983). Bliemel and Hassanein's strong model produced impressive results, so technical adequacy and appearance are adopted from them to measure system quality. Researchers also recommend the direct relationships between trust and system quality with overall satisfaction.

The extensive review and understanding of previous literature revealed many studies that point to the existence of approved relations between the constructs, and these have been adopted in this research model. Table 1 represents these relationships and their sources.

Accordingly, this research model consists of 11 hypotheses, as depicted in Figure 1.

The Research Hypotheses

In this study, eleven factors are proposed which may play a significant role in measuring users' satisfaction with EHR in Bahrain.

Content

Content is defined as specifications of the main concepts that comprise the significant data components in EHR, such as patient information, medication administration, admission nursing note, nursing care plan, physical assessment, data input, test results, daily charts, complaints, problems, documentation, reports, prescriptions, insurance, demographic data, medical history, etc. (Yoon, Nah, & Chin, 2013). Information content is a measure of the importance of the information offered to the users and the correctness and extent of the information in the system (Gorla, Somers and Wong, 2010). Many authors

Table 1. The relationship between the research model's features

Positive Relationship	Reference
Technical Adequacy → System Quality	(Bliemel and Hassanein, 2006); (Aladwani and Palvia, 2002); (Abd Aziz, Sulaiman and Musa, 2008).
Appearance → System Quality	(Aladwani and Palvia, 2002); (Bliemel and Hassanein, 2006); (Abd Aziz, Sulaiman and Musa, 2008).
System Quality → Service Quality	(Nunes and Javier, 2014). (Akter, D'Ambra and Ray, 2010)
System Quality →Trust	(Kassim, Jailani, Hairuddin, and Zamzuri, 2012) (Chuang and Fan, 2011)
Content → Information Quality	(Bliemel & Hassanein, 2006) (Aggelidis & Chatzoglou, 2012) (Gorla, Somers, & Wo, 2010)
Accuracy → Information Quality	(Aggelidis & Chatzoglou, 2012) (Ballou & Pazer, 1985) (Caby, Pautke, & Redman, 1995) (DeLone & McLean, 1992) (Doernberg. & Ziegler, 1980,) (Fox, Levitin, & Redman, 1994) (Goodhue, 1995) (Hilton, 1979) (Miller, 1996) (Wang & Strong, 1996) (Zmud, 1978)
Information Quality → Service Quality	(Bhattacharya, Gulla, & Gupta, 2012) (Webb & Webb, 2004) (Chakraborty, Srivastava, & Warren, 2005) (Ho & Lee, 2007) (Park, & Gretzel, 2007) (Kaisara & Pather, 2011) (Barnes, & Vidgen, 2006) (Cao, Zhang, & Seydel, 2005)
System Quality → Trust	(Kassim, Jailani, Hairuddin, and Zamzuri, 2012) (Ayyash, Ahmad and Singh, 2012) (Chuang and Fan, 2011)
Trust → User Satisfaction	(Kassim et al., 2012) (Alsajjan, 2014) (Bliemel and Hassanein, 2007)
Technical Support → User Satisfaction	(Sinha and Kurian, 2014) (Nour Eldin, 2007) (Aggelidis & Chatzoglou, 2012)
Service quality → User Satisfaction	(DeLone and McLean, 2003) (Teo, Srivastava and Jiang, 2008)

Figure 1. Research model

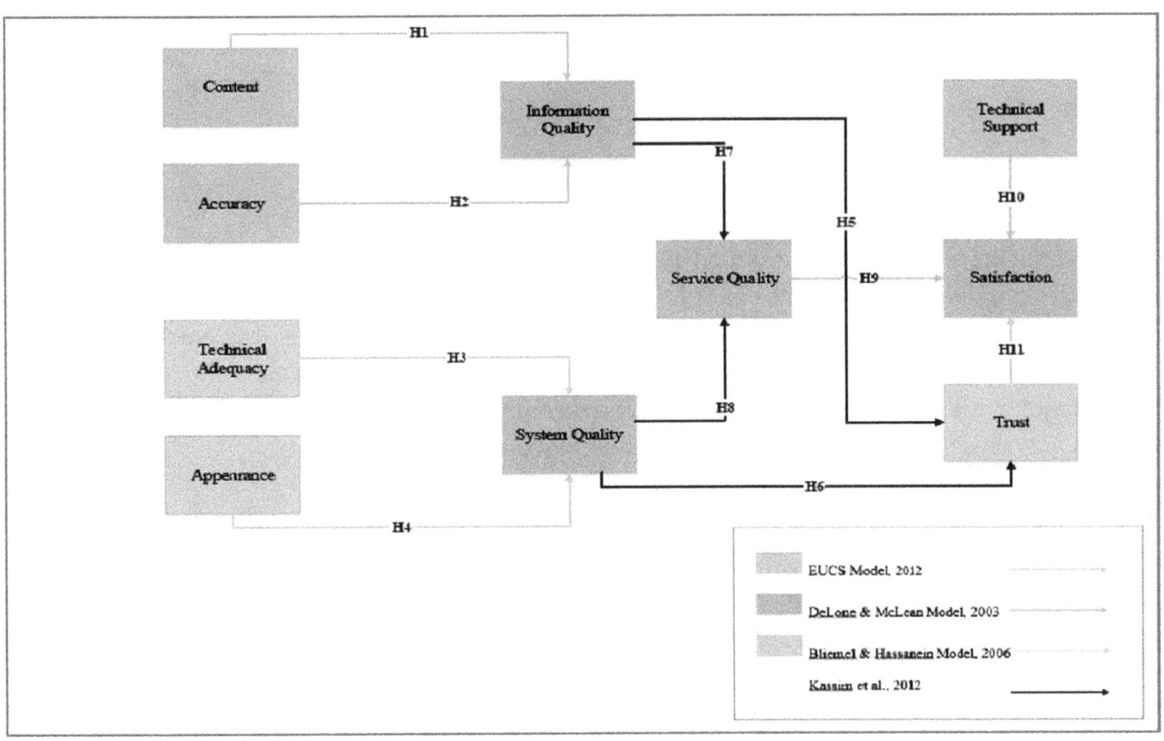

consider that content is an important factor that measures information quality (Bliemel & Hassanein, 2006; Aggelidis & Chatzoglou, 2012; Gorla, Somers & Wo, 2010). Therefore, the following hypothesis is proposed:

H1: Content has a positive effect on information quality.

Accuracy

Accuracy is defined as the degree of conformity of the measured value quantity to the true value quantity (Yoon, Nah & Chin, 2013). Many authors consider accuracy as a significant factor in measuring information quality (Bailey & Pearson, 1983; Zmud, 1978; Hilton, 1979; Doernberg & Ziegler, 1980; Ballou & Pazer, 1985; Delone & McLean, 1992; Fox, Levitin, and Redman, 1994; Goodhue, 1995; Caby, Pautke, and Redman, 1995; Miller, 1996; and Wang & Strong, 1996). Accordingly, the following hypothesis is proposed:

H2: Accuracy has a positive effect on information quality.

Technical Adequacy

Some researchers define the technical adequacy as the extent to which pages can load fast, easy access to site, capabilities of searching as well as customizations and personalization features (Aladwani and Palvia, 2002). In addition, this term is used to define the level of reliability and validity that a particular evaluation requires (Rabinowitz, Sato, Berkes and WestEd, 2001). Some studies discovered a relationship between this factor and the quality of the system (Bliemel & Hassanein, 2006; Aladwani and Palvia, 2002). Based on this, the following hypothesis is proposed:

H3: Technical adequacy has a positive effect on system quality.

Appearance

Appearance is the second factor or dimension that is used to assess the quality of a system in terms of interface design, such as appropriate use of font and colour, organization, and attractiveness of the system features (Aladwani and Palvia, 2002). All of these system visual characteristics contribute in shaping satisfaction with system quality. Some researchers consider that appearance can have a positive impact on satisfaction with system quality (Bliemel & Hassanein, 2006; Aladwani and Palvia, 2002; Abd Aziz, Sulaiman and Musa, 2008). Therefore, the following hypothesis is proposed:

H4: Appearance has a positive effect on system quality.

Some authors have found a relationship between information quality and trust (Kassim, Jailani, Hairuddin and Zamzuri, 2012; Ayyash, Ahmad and Singh, 2012; Chuang and Fan, 2011). Others (Bhattacharya, Gulla & Gupta, 2012; Webb & Webb, 2004; Chakraborty, Srivastava & Warren, 2005; Ho & Lee, 2007; Park & Gretzel, 2007; Kaisara & Pather, 2011; Barnes & Vidgen, 2006; Cao, Zhang & Seydel,

2005) have found a relationship between information quality and service quality. For this model, the researchers have chosen two factors from Doll and Torkzadeh's model, content and accuracy, to measure information quality. Therefore, the following hypotheses are proposed:

H5: Information quality has a positive effect on trust.
H6: Information quality has a positive effect on service quality.

System Quality

System quality is focused on measuring the technical success of any information system in terms of flexibility of the system, its integration, error recovery, convenience and language. It can contribute to specifying the level of user satisfaction and the intention to use the system (Ramayah and Lee, 2012). This includes some factors that help in the evaluation of system quality: reliability, response time, ease of terminal use, data accuracy, and ease of use (Negash, Ryan and Igbaria, 2003), the quality and the maintainability of the program code, and documentation, and whether the system contains bugs or not (Nunes and Javier, 2014). Other researchers compiled their own indexes, some focusing on the technical side and design features of the system: technical adequacy and appearance to measure satisfaction with system quality. In this research model, system quality has the following dimensions for the measurement of the construct: technical adequacy and appearance.

Chuang and Fan (2011) believe that system quality acts as the main determinate and has a positive impact on trust. System quality should have a relationship with service quality (Akter, D'Ambra and Ray, 2010) and trust. Thus the following hypotheses are proposed:

H7: System quality has a positive effect on trust.
H8: System quality has a positive effect on service quality.

Service Quality

DeLone and McLean (2003) stated that service quality is a critical element in the assessment of IS service quality and effectiveness. They define service quality as "the overall support delivered by the service provider, [which] applies regardless of whether this support is delivered by the IS department, a new organizational unit, or outsourced". Many researchers agreed to use the SERVQUAL tool (Parasuraman, Zeithaml and Berry, 1988) to measure IS service with a slight modification to suit the IS field (Gorla, 2012). Therefore, the following hypothesis is proposed:

H9: Service quality positively affects user satisfaction with IS

Technical Support

Dillinger (2010) states that supporting and training end users in the use of EHR are important. Therefore, the following hypothesis is proposed:

H10: Technical support influences end-user satisfaction.

Trust

In general, trust is defined as "the extent to which one is willing to ascribe good intentions and to have confidence in the words and actions of other people" (Cook & Wall, 1980). According to Hunter (2007), a trustworthy information system is reflected in its integrity, reliability of outputs, credibility, confidence and accountability of information and records. In addition, McKnight (2005) states that user trust relies upon the information system. Therefore, trust is considered as an important element which influences the level of user satisfaction (Kassim, Jailani, Hairuddin and Zamzuri, 2012). Thus, the following hypothesis is postulated:

H11: Trust has a positive effect on user satisfaction.

User Satisfaction

User satisfaction reflects the extent to which users believe that the information system meets their informational requirements (Ives, Olson and Baroudi, 1983). It has a large impact on users' intention to utilize the actual application of the system, as well as to gain positive results for the organization as a whole (DeLone and McLean, 2003).

RESEARCH METHODOLOGY

This research is primarily a quantitative study based on the research objectives; this methodology collects numerical data in order to define and describe phenomena. Analysis of the collected data is represented by mathematically based methods such as statistics, tables and graphics (ACAPS, 2012). This methodology requires a large sample size to be effective, distributed according to the targets of either a specific or random sample size (Dudwick, Kuehnast, Jones and Woolcock, 2006).

Instrument

A questionnaire was used to gather valid data about user satisfaction using the EHR system in Bahrain. Based on the research model, the questionnaire starts with a section on personal information, followed by groups of questions in sections corresponding to the model's features: content, accuracy, information quality, appearance, technical adequacy, system quality, service quality, technical support, trust, and overall satisfaction). There are two types of question: multiple choice for the personal information, and rating on a 1 to 5 Likert scale for their level of satisfaction with each feature. As shown in Figure 1, the research model was derived from models in previous studies, with variables adopted to suit this research.

Validity

To measure the validity of the variables, convergent validity and discriminative validity were assessed.

Convergent Validity

Convergent validity is defined as the level of conformity between two or more measures of the same latent variable (Carmines & Zeller, 1979); items are pointers to a latent variable which ought to share a high extent of variance (Hair el al., 2006). The convergent validity of the items was assessed using two criteria. First, the outer loadings should be 0.70 or higher (Hulland, 1999). Secondly, the average variance extracted (AVE) for every latent variable should be above 0.50 (Fornell and Larcker, 1981).

Average Variance Extracted (AVE)

AVE was calculated using SmPLS 3. The values should be greater than 0.50 (Bagozzi and Yi, 1988), although according to Fornell and Larcker (1981), 0.4 is acceptable if the composite reliability is more than 0.6. (Huang et al., 2013). Using Table 2, each latent variable was evaluated by the AVE method, resulting in values between 0.553 and 0.749, above the satisfactory standard of 0.50, the convergent validity test had satisfactory results.

Factor Loadings

Outer loading was suitable for the reflective measurement model to show the loading for each item. The outer loading level. shows the item level of contribution to the factor, acceptable at 0.50 or above (Afthanorhan, 2013). The results showed that every item's outer loading was significantly higher than 0.50. These results indicated strong convergent validity for the constructs in our model.

Discriminative Validity

The discriminative validity of the items was measured using two criteria, cross-loading and correlations among variables (Fornell and Larcker, 1981).

Table 2. Average variance extracted (AVE)

AVE	
Accuracy	0.567
Appearance	0.602
Content	0.553
Information quality	0.591
Satisfaction	0.749
Service quality	0.685
System Quality	0.612
Technical Adequacy	0.575
Technical support	0.648
Trust	0.628

Correlation Matrix (Fornell and Larcker)

The AVE for each latent variable was higher than the correlation values in the column. It was also higher than those in the rows, with two exceptions. The values of the correlation between the latent variables system quality and technical adequacy (0.817) was slightly greater (0.035 and 0.059) than the square roots of the AVEs of these same latent variables (0.782 and 0.758). Generally, the values of the correlation between system quality and technical adequacy had little distinction, so the researchers left the model as it is with no adjustment made. Thus, the discriminative validity was settled

Reliability

A reliability test was performed in order to clarify the accuracy of the selected items. According to Miles and Hubberman (1994), reliability is the extent to which the instrument yields the same results on repeated trials by different people. Furthermore, Cronbach's alpha was used to calculate and to assess the internal consistency reliability because it is the most suitable when using Likert-type scales (Gliem and Gliem, 2003). Cronbach's alpha is acceptable in the range between 0.70 and 0.95 (Tavakol and Dennick, 2011). Composite reliability or factor reliability is another internal consistency reliability test, and this is acceptable at 0.7 or higher, although values of 0.60 to 0.70 are acceptable in exploratory research (Kwong and Wong, 2013). Both composite reliability and Cronbach's alpha are important in checking the reliability of the model (Afthanorhan, 2013). Table 3 indicateds that the model constructs were considered to be reliable.

Sample Size

This research adopted a non-random snowball sampling method. To acquire the desired samples from the public health sector, the initial respondents were family and friends who were asked to continue forwarding the same online survey they had received. According to the Regional Health Systems Observatory (EMRO) (2007) Bahrain provides public healthcare services through the Bahrain Defence Force Hospital (BDF), Salmaniya Medical Complex, Psychiatric Hospital, Geriatric Hospital, four maternity

Table 3. Composite reliability and Cronbach's alpha

	Composite Reliability	Cronbach's Alpha
Accuracy	0.839	0.745
Appearance	0.883	0.834
Content	0.861	0.800
Information Quality	0.853	0.770
Satisfaction	0.923	0.888
Service quality	0.867	0.769
System Quality	0.904	0.873
Technical Adequacy	0.890	0.853
Technical support	0.880	0.821
Trust	0.870	0.801

hospitals, 21 healthcare centres and two dental hospitals and, recently, King Hamad University Hospital. The characteristics of the target population were users of EHR, which includes doctors, nurses, laboratory workers, clerks and others; the end users' nationality and gender were disregarded during distribution of the survey.

Data Collection and Analysis

The research survey was established using Google forms placed online to be accessed through a unique link. It was available to all current end users of EHR in Bahrain's public hospitals and health centres. Based on the adopted sampling method, word of mouth and social media applications were employed to distribute the survey link for the collection of data. Descriptive analysis was performed with the Statistical Package for the Social Sciences (SPSS). As already explained, reliability and validity tests were conducted using SmartPLS 3 to model the relation and regression between variables. It facilitated the task of data analysis, providing results in HTML, Excel and LaTeX formats, and a parameterized path model (Temme, Kreis and Hildebrandt, 2006).

RESEARCH RESULTS

Demographic Analysis

Table 4 shows the demographic data from the research sample. Almost a third of respondents were in the age range 25-29 a cohort considered to be the most familiar with technology. 65.1% were female, and in terms of occupation 34.2% were doctors and 32.2% were nurses.

The Evaluation of the Structural Model

A widely used technique in structural models is to capture the linear regression of different path models with latent variables or constructs (Hox and Bechger, 2000). The structural model was designed, and validated by SmartPLS 3, using the following criteria:

Table 4. Demographic information

Variable	Category	Frequency	Percentage
Age	20-24 25-29 30-34 35 and above	32 50 28 42	21.1% 32.9% 18.4% 27.6%
Gender	Male Female	53 99	34.9% 65.1%
Occupation	Doctor Nurse Lab Specialist Clerk Other	52 49 19 21 11	34.2% 32.2% 12.5% 13.8% 7.2%

Co-Linearity Test (VIF)

Co-linearity or (multi-collinearity) creates problems when there is a high correlation between two or more predictor variables (Allison, 2012), meaning that there is redundant information in the responses or that they measure the same thing; this may affect the structure of the model. In order to observe or assess co-linearity; the variance inflation factor (VIF) is used to calculate the correlation between predictors, where VIF > 5 means that there is a high correlation (Roman and Dănuleţiu, 2013). As shown in Table 5, the values of VIF predictors are all less than 3, which is the acceptable point (no high correlation).

Significance of Path Coefficient

After recognizing multi-collinearity, a bootstrapping procedure was established to test the statistical significance of each path coefficient (Chin, 1998). The number of bootstrap samples should be 5,000 and the number of bootstrap cases should be the same as the number of valid observations (Hair et al., 2011). The researchers selected 5,000 samples and 400 cases to estimate the path coefficient. Path coefficient analysis is an important statistical tool used to describe links between variables in the structural model, and the influence of each variable on others (Joshi, 2005). P value was used to measure the significance of the relationship, where $P \leq 0.05$ (Padi, 2003). Regarding the t-statistic value, all paths in the structural model should be greater than 2 (Rossiter, 2002). As shown in Table 6, all the relationships but two were significant at $P \leq 0.05$.

For system quality the result showed a significant and positive relationship with technical adequacy, where ($T = 11.173$, $P < 0.05$) was accepted, while the relationship between system quality and appearance was rejected ($T = 1.733$, $P > 0.05$). The relationship of content ($T = 2.956$, $P < 0.05$) and accuracy ($T = 3.678$, $P < 0.05$) was significant and positive and accepted. As for the importance of service quality in determining satisfaction, there was a significant and positive relationship with system quality and service

Table 5. The Values of VIF

	Accuracy	Appearance	Content	Information Quality	Satisfaction	Service Quality	System Quality	Technical Adequacy	Technical support	Trust
Accuracy				1.456						
Appearance							2.104			
Content										
Information Quality				1.456		1.724				1.724
Satisfaction										
Service Quality										
System Quality					1.897	1.724				1.724
Technical Adequacy							2.104			
Technical support					1.997					
Trust					2.052					

Table 6. Significance of Path Coefficient

	Standard Error (STERR)	T Statistics (|O/STERR|)	P Values	Accepted/ Rejected
Accuracy -> Information Quality	0.111	3.678	0.000	Accepted
Appearance -> System Quality	0.079	1.733	0.083	Rejected
Content -> Information Quality	0.104	2.956	0.003	Accepted
Information Quality -> Service Quality	0.085	0.379	0.705	Rejected
Information Quality -> Trust	0.079	4.637	0.000	Accepted
Service quality -> Satisfaction	0.090	3.210	0.001	Accepted
System Quality -> Service quality	0.077	10.187	0.000	Accepted
System Quality -> Trust	0.094	5.127	0.000	Accepted
Technical Adequacy -> System Quality	0.064	11.173	0.000	Accepted
Technical support -> Satisfaction	0.096	2.026	0.043	Accepted
Trust -> Satisfaction	0.085	5.434	0.000	Accepted

quality (T= 10.187, P < 0.05), so it was accepted; however, there was no significant relationship with? (T= 0.379, P > 0.05), and so it was rejected. Likewise, the result for trust showed a significant and positive relationship with system quality (T= 5.127, P ≤ 0.05) and was accepted. The significant and strong positive relationship between? and trust (T= 4.637, P > 0.05) was also accepted. For satisfaction there was a significant and positive relationship with trust (T= 5.434, P ≤ 0.05) and with service quality and satisfaction (T= 3.210, P < 0.05) so both were accepted. However, the relationship between technical support and satisfaction (T= 2.026, P > 0.05) was rejected.

Total Effects

Some researchers evaluate both the construct's direct and indirect effect on another, and the total effect is the sum of the two (Hair, Jr, n.d.). As can be seen from Table 7, all hypothesized paths were significant except for the direct path from appearance to system quality (B=0.137), and to service quality (B=0.032). However, as shown in Table 7, there were indirect effects between technical adequacy and trust via system quality, as well as user satisfaction.

Predictive Accuracy

Predictive accuracy is used to measure the significance of the structural model, explaining variation by observing R square, the rate of variance between independent and dependent variables (Rossiter, 2002; Schemper, 2003). The results obtained from PLS analysis show that the R square value for the dependent variable was 0.401, which means that the content accuracy was explained by 40.1% of the variance in information quality. Also, for the dependent variable user satisfaction, it was 0.687 which means that service quality, technical support and trust are interpreted by 68.7% of the variance in user satisfaction. The R square value for the dependent variable service quality it was 0.649, which means that both? and system quality are explained by 64.9% of the variance in service quality. For the dependent variable system quality, it was 0.676, which means that together, technical adequacy and appearance are explained by

Table 7. Total effect of constructs

	Accuracy	Appearance	Content	Information Quality	Satisfaction	Service Quality	System Quality	Technical Adequacy	Technical support	Trust
Accuracy	1.00			0.409	0.073	0.013				0.150
Appearance		1.00			0.061	0.107	0.137			0.066
Content			1.00	0.306	0.055	0.010				0.112
Information Quality				1.00	0.179	0.032				0.368
Satisfaction					1.00					
Service Quality					0.289	1.00				
System Quality					0.449	0.784	1.00			0.483
Technical Adequacy					0.322	0.563	0.718	1.00		0.347
Technical support					0.194				1.00	
Trust					0.461					1.00

67.6% of the variance in system quality. Furthermore, for the dependent variable trust it was 0.598, which means that the pair of? and system quality were explained by 59.8% of the variance in trust. All the R square values in Table 8 indicate a greater influence between both dependent and independent variables.

The structural model in Figure 2 was created using SmartPLS 3, and it shows different latent variables with their items, as well as the path coefficient values for relationships in the model.

DISCUSSION

According to the results outlined in the previous section, content and accuracy have a positive and significant relationship with information quality. Many researchers' findings (Bliemel & Hassanein, 2006; Aggelidis & Chatzoglou, 2012; Gorla, Somers & Wo, 2010) support the relationship between content and information quality, while authors such as Aggelidis and Chatzoglou (2012), Ballou and Pazer (1985), Caby, Pautke and Redman (1995) and DeLone and McLean (1992) support the relationship between accuracy and information quality. Thus, the more accurate, reliable, useful, concise and complete is the

Table 8. Predictive accuracy (R Square)

R Square	
Information Quality	0.401
Satisfaction	0.687
Service quality	0.649
System Quality	0.676
Trust	0.598

Figure 2. Structural model

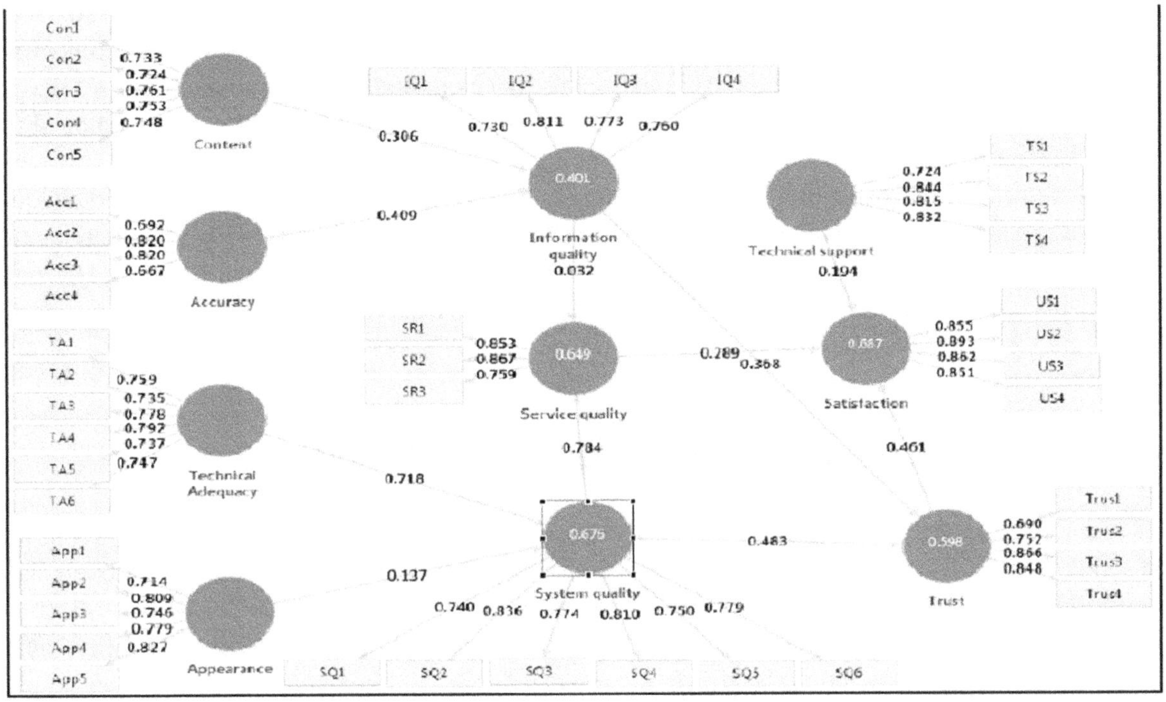

information, the more qualified is it, which in turn will affect user satisfaction toward with EHR. In addition, it is suggested that system quality is affected positively by appearance and technical adequacy, and this is consistent with the results of Bliemel and Hassanein (2006), Aladwani and Palvia (2002) and Abd Aziz, Sulaiman and Musa (2008). Thus, better technical adequacy in the speed of the system and ease of access will lead to high quality of the system. However, there is no significant relationship between appearance and system quality, possibly because the users of EHR believe that system quality focuses on technical success measured in terms of ease of use and error recovery as well as response time, while appearance mostly assesses design features that are not important to them in assuring system quality.

Furthermore, the researchers propose that service qualityis affected positively by information quality and system quality, which is consistent with the findings of Akter, D'Ambra and Ray (2010). Thus, a system's high quality (both hardware and software) results in high quality of service. For example, using modern technology to provide fast access to information will enhance service quality, which in turn will lead to greater user satisfaction. However, there is no significant relationship between information quality and service quality, possibly because dependence on EHR is relatively new; users may be unaware of the optimal use of the system, failing to take advantage of the system's services in an efficient and effective way. Additionally, the researchers suggest that trust is affected positively by? and system quality. There is a positive significant relationship between information quality and trust, and between system quality and trust. Some authors support the relationship between information quality and trust, including Kassim, Jailani, Hairuddin and Zamzuri (2012), and Ayyash, Ahmad and Singh (2012), while others support the relationship between system quality and trust: Kassim, Jailani, Hairuddin and Zamzuri (2012) and Chuang and Fan (2011). Thus, good system quality (such as using modern and sophisticated technology in which the system become more efficient and reliable and responds quickly to user requests) as well as having relevant, complete, and high-level quality of information, will increase the confidence of users in the system, which becomes highly trusted. Furthermore, the researcher proposed that user satisfaction is affected positively by technical support, as well as by trust and service quality, and the results proved this. Several authors support the relationship between service quality and user satisfaction (DeLone and McLean, 2003; Teo, Srivastava and Jiang, 2008). Others support the relationship between technical support and user satisfaction (Nour El Din, 2007; Aggelidis & Chatzoglou, 2012). This means that a high level of technical support (for example, having proper training before using the system, and adequate instructions to help users become more confident with the system), will result in high user satisfaction. The relationship between trust and user satisfaction is strongly supported by Kassim, Jailani, Hairuddin and Zamzuri (2012) and Alsajjan (2014). Thus, the more users trust the system content, the more satisfied will they be with the system.

CONCLUSION, RECOMMENDATIONS AND FUTURE WORK

This research investigated factors affecting user satisfaction with EHR Bahrain, by capturing their opinions using a survey based on a model built by the researcher, adapted from well-known models in the literature. From the results, it is concluded that users' trust in EHR has an influence on their satisfaction with it, as well as service quality and technical support. Information quality and system quality have an indirect effect on user satisfaction through trust.

According to the research results, the following recommendations are proposed. The Ministry of Health should focus on:

- Providing users with complete, relevant, high quality, timely information, given their significant impact on satisfaction.
- Increasing users' trust by providing them with reliable output and information content to make EHR trustworthy.
- Providing users with good-quality of services through EHR, to meet their requirements of service quality.
- Increasing system quality by making EHR easy to use, reliable, efficient, fast to access to information, and responsive to users' requests.
- Enhancing the technical support by making EHR easy to access, always available, with valid links (hyperlinks), adequate search facilities, fast to accomplish multiple tasks, and personalized or customized in a way that meet users' requirements in accomplishing their tasks.

Considering these recommendations and the factors surrounding EHR users in public hospitals and health centres, this research provides the Ministry of Health with an understanding of how these factors are responsible for user satisfaction. It also provides an explanation of the aspects that need to be supported to enhance overall satisfaction and work performance of EHR users, which in turn will be reflected in the quality of health care delivery.

On the other hand, this research has limitations and constraints. First, there was a lack of access to EHR users which meant that the sample size was too small. In addition, the snowball sampling method applied might not result in a generalized picture of satisfaction with EHR in the public health sector. Therefore, as future work, the sample size should be increased, using a different sampling method to include more users in measuring their satisfaction. Furthermore, a system for monitoring user satisfaction can be developed for hospitals, so they can monitor their own system over different periods of time and include factors other than those included in this model, to customize EHR for specific healthcare contexts and environments.

REFERENCES

Abd Aziz, N., Sulaiman, A., & Musa, G. (2008). Predicting the Influence of Travel Web Site Effectiveness Characteristics on Technology Acceptance and Its Marketing Implications. *Journal Pengurusan*, *27*, 105–128.

Abugabah, A., & Sanzogni, L. (2009). Enterprise Resource Planning Systems (ERP) and User Performance: A Literature Review. *Proceedings of ACIS '09*.

ACAPS. (2012). Qualitative and Quantitative Research Techniques for Humanitarian Needs Assessment.

Afthanorhan, W. (2013). A Comparison of Partial Least Square Structural Equation Modeling (PLS-SEM) and Covariance Based Structural Equation Modeling (CB-SEM) for Confirmatory Factor Analysis. *International Journal of Engineering Science and Innovative Technology*, *2*(5), 198–202.

Aggelidis, V., & Chatzoglou, P. (2012). Hospital information systems: Measuring end user computing satisfaction (EUCS). *Journal of Biomedical Informatics*, *45*(3), 566–579. doi:10.1016/j.jbi.2012.02.009 PMID:22426283

Ajami, S., & Bagheri-Tadi, T. (2013). Barriers for Adopting Electronic Health Records (EHR) by Physicians. *Acta Inform Med*, *21*(2), 129–134. doi:10.5455/aim.2013.21.129-134 PMID:24058254

Akter, S., D'Ambra, J., & Ray, P. (2010). User perceived service quality of m-Health services in developing countries. *Proceedings of the 18th European Conference on Information Systems* (pp. 1-12). Pretoria, South Africa: University of Pretoria.

Aladwani, A., & Palvia, P. (2002). Developing and Validating an Instrument for Measuring User-Perceived Web Quality. *Information & Management*, *39*(6), 467–476. doi:10.1016/S0378-7206(01)00113-6

Alhashimi, M. (2009). *Brief profile*. Retrieved from http://drmukhtaralhashimi.com/yahoo_site_admin/assets/docs/My_Resume_for_Web_site.12911341.pdf

Allison, P. (2012). When Can You Safely Ignore Multicollinearity? *Statistical horizons*. Retrieved from http://statisticalhorizons.com/multicollinearity

Alsajjan, B. (2014). Satisfaction - Trust Model: Developing Customer Satisfaction and Trust Indices for Mobile Service Providers in the UK. *International Review of Management and Business Research*, *3*(2), 1092–1110.

Ammenwerth, E., Brender, J., Nykänen, P., Prokosch, H., Rigby, M., & Talmon, J. (2004). Visions and strategies to improve evaluation of health information systems. *International Journal of Medical Informatics*, *73*(6), 479–491. doi:10.1016/j.ijmedinf.2004.04.004 PMID:15171977

Amoroso, D., & Cheney, P. (1992). Quality end user-developed applications. *ACM SIGMIS Database*, *23*(1), 1–11. doi:10.1145/134347.134350

Ash, J., & Berg, M. (2004). Some Unintended Consequences of Information Technology in Health Care: The Nature of Patient Care Information System-related Errors. *Journal of the American Medical Informatics Association*, *11*(2), 104–112. doi:10.1197/jamia.M1471 PMID:14633936

Au, N., Ngai, E. W. T., & Cheng, T. C. E. (2002). A critical review of end-user information system satisfaction research and a new research framework. *International Journal of Management Sciences*, *30*, 451–478.

Ayyash, M., Ahmad, K., & Singh, D. (2012). Investigating the Effect of Information Systems Factors on Trust in E-Government Initiative Adoption in Palestinian Public Sector. *Research Journal of Applied Sciences. Engineering and Technology*, *5*(15), 3865–3872.

Ayyash, M., Ahmad, K., & Singh, D. (2012). Investigating the Effect of Information Systems Factors on Trust in E-Government Initiative Adoption in Palestinian Public Sector. Research *Journal of Applied Sciences. Engineering and Technology*, *5*(15), 3865–3872.

Bagozzi, R. P., & Yi, Y. (1988). On the evaluation of structural equation models. *Journal of the Academy of Marketing Science*, *16*(1), 74–94. doi:10.1007/BF02723327

Bailey, J. E., & Pearson, S. W. (1983). Development of a Tool for Measuring and Analyzing Computer User Satisfaction. *Management Science*, *29*(5), 530–545. doi:10.1287/mnsc.29.5.530

Bakke, A., & Stensrud, E. (2008). *Measurement of User Satisfaction with Enterprise Portals: An Empirical Study.* Retrieved from http://www.nik.no/

Ballou, D. P., & Pazer, H. L. (1985). Modeling data and process quality in multi-input, multioutput information systems. *Management Science*, *31*(2), 150–162. doi:10.1287/mnsc.31.2.150

Barnes, S. J., & Vidgen, R. T. (2006). Data Triangulation and Web Quality Metrics: A Case Study in E-Government. *Information & Management*, *43*(6), 767–777. doi:10.1016/j.im.2006.06.001

Bhattacharya, D., Gulla, U., & Gupta, M. P. (2012). E-service quality model for Indian government portals: Citizens' perspective. *Journal of Enterprise Information Management*, *25*(3), 246–271. doi:10.1108/17410391211224408

Bliemel, M., & Hassanein, K. (2006, December 9). An Empirical Study of Consumer Satisfaction with Online Health Information Retrieval. *Proceedings of the Fifth Annual Workshop on HCI Research in MIS* (pp. 70-74(. Retrieved from http://sighci.org/uploads/published_papers/ICIS2006/SIGHCI_2006_Proceedings_paper_11.pdf

Bliemel, M., & Hassanein, K. (2007). Consumer Satisfaction with Online Health Information Retrieval: A Model and Empirical Study. *e-Service Journal*, *5*(2), 53–84. doi:10.2979/ESJ.2007.5.2.53

Bliemel, M., & Hassanein, K. (2007). Consumer Satisfaction with Online Health Information Retrieval: A Model and Empirical Study. *e-Service Journal*, *5*(2), 53–84. doi:10.2979/ESJ.2007.5.2.53

Caby, B. C., Pautke, R. W., & Redman, T. C. (1995). Strategies for improving data quality. *Data Quality, 1*(1), 4–12.

Cao, M., Zhang, Q., & Seydel, J. (2005). B2C E-Commerce Web Site Quality: An Empirical Examination. *Industrial Management & Data Systems, 105*(5), 645–661. doi:10.1108/02635570510600000

Carmines, E., & Zeller, R. (1979). *Reliability and Validity Assessment*. USA: SAGE Publications. doi:10.4135/9781412985642

Chakraborty, G., Srivastava, P., & Warren, D. (2005). Understanding corporate B2B Web sites' effectiveness from North American and European perspectives. *Industrial Marketing Management, 34*(4), 420–429. doi:10.1016/j.indmarman.2004.09.008

Chin, W. (1998). W. The partial least squares approach for structural equation modeling. In G. A. Marcoulides (Ed.), *Modern methods for business research* (pp. 295–236). London: Lawrence Erlbaum Associates.

Chin, W. W., & Lee, M. K. O. (2000) A proposed model and measurement instrument for the formation of IS satisfaction: the case of end-user computing satisfaction. *Proceedings of the 21st International Conference on Information Systems*, Brisbane, Queensland, Australia (pp. 553–563).

Chuang, H., & Fan, C. (2011). The mediating role of trust in the relationship between e-retailer quality and customer intention of online shopping. *African Journal of Business Management, 5*(22), 9522–9529.

Chuang, H., & Fan, C. (2011). The mediating role of trust in the relationship between e-retailer quality and customer intention of online shopping. *American Journal of Business, 5*(22), 9522–9529.

Computing reviews. (2015). Computing Reviews, the leading online review service for computing literature. Retrieved from http://www.computingreviews.com/hottopic/hottopic_essay_11.cfm

Cook, J., & Wall, T. (1980). New work attitude measures of trust, organizational commitment and personal need non-fulfillment. *Journal of Occupational Psychology, 53*(1), 39–52. doi:10.1111/j.2044-8325.1980.tb00005.x

Delone, W. H., & McLean, E. R. (1992). Information systems success: The quest for the dependent variable. *Information Systems Research, 3*(1), 60–95. doi:10.1287/isre.3.1.60

DeLone, W. H., & McLean, E. R. (2003). The DeLone and McLean Model of Information Systems Success: A Ten-Year Update. *Journal of Management Information Systems, 19*(4), 9–30.

Delpierre, C., Cuzin, L., Fillaux, J., Alvarez, M., Massip, P., & Lang, T. (2004, October 01). A systematic review of computer-based patient record systems and quality of care: More randomized clinical trials or a broader approach? *International Journal for Quality in Health Care, 16*(5), 407–416. doi:10.1093/intqhc/mzh064 PMID:15375102

Dillinger, T. (2010). *Defining and Building an Implementation Plan for a Small Office to go to an Electronic Medical Record Environment*. Retrieved from http://garyfarrar.com/td.pdf

Doernberg, D. L., & Ziegler, D. H. (1980). Due process versus data processing: An analysis of computerized criminal history information systems. Retrieved from http://digitalcommons.pace.edu/lawfaculty/54/

Doll, W. J., & Torkzadeh, G. (1988). The measurement of end-user computing satisfaction. *Management Information Systems Quarterly*, *12*(2), 259–274. doi:10.2307/248851

Dudwick, N., Kuehnast, K., Jones, V., & Woolcock, M. (2006). Analyzing Social Capital in context: A Guide to Using Qualitative Methods and Data. *World Bank*. Retrieved from http://siteresources.worldbank.org/WBI/Resources/Analyzing_Social_Capital_in_Context-FINAL.pdf

English Collins Dictionary. (2000). satisfaction definition. Retrieved from http://dictionary.reverso.net/english-definition/satisfaction

Fornell, C., & Larcker, D. F. (1981). Evaluating Structural Equation Models with Unobservable Variables and Measurement Error. *JMR, Journal of Marketing Research*, *18*(1), 39–50. doi:10.2307/3151312

Fox, C., Levitin, A., & Redman, T. (1994). The notion of data and its quality dimensions. *Information Processing & Management*, *30*(1), 9–19. doi:10.1016/0306-4573(94)90020-5

Gable, G., Sedera, D., & Chan, T. (2003) Enterprise systems success: a measurement model. In S.T. March, A. Massey, & J.I. DeGross (Eds.), Proceedings Twenty-Fourth International Conference on Information Systems, Seattle, USA (pp. 576-591).

Gallagher, C. (1974). Perceptions of the Value of a Management Information System. *Academy of Management Journal*, *17*(1), 46–55. doi:10.2307/254770

Gelderman, M. (1998). The relation between user satisfaction, usage of information systems and performance. *Information & Management*, *34*(1), 11–18. doi:10.1016/S0378-7206(98)00044-5

Gliem, J. and Gliem, R. (2003). Calculating, Interpreting, and Reporting Cronbach's Alpha Reliability Coefficient for Likert-Type Scales. *Proceedings of the Midwest Research to Practice Conference in Adult, Continuing, and Community Education* (pp. 82-87).

Goodhue, D. L. (1995). Understanding user evaluations of information systems. *Management Science*, *41*(12), 1827–1844. doi:10.1287/mnsc.41.12.1827

Gorla, N. (2011). An assessment of information systems service quality using SERVQUAL+. *ACM SIGMIS Database*, *42*(3), 46–70. doi:10.1145/2038056.2038060

Gorla, N. (2011). An assessment of information systems service quality using SERVQUAL+. *ACM SIGMIS Database*, *42*(3), 46. doi:10.1145/2038056.2038060

Hair, J. F. Jr, Black, W. C., Babin, B. J., Anderson, R. E., & Tatham, R. L. (2006). *Multivariate data analysis* (6th ed.). Upper Saddle River, NJ: Pearson-Prentice Hall.

Health, I. T. (2015). *Definition and Benefits of Electronic Medical Records (EMR)*. Retrieved from http://www.healthit.gov/providers-professionals/electronic-medical-records-emr

Healthcare Financial Management Association. (2006) Overcoming Barriers to Electronic Health Record Adoption: Results of survey and roundtable discussion conducted by the Healthcare. Retrieved from http://www.providersedge.com/ehdocs/ehr_articles/Overcoming_Barriers_to_EHR_Adoption.pdf]

Hilton, R. W. (1979). The determinants of information system value: An illustrative analysis. *Journal of Accounting Research*, *17*(2), 411–435. doi:10.2307/2490511

HIMSS Electronic Health Record Association. (2012). HIMSS EHR Association Definitional Model and Application Process. Retrieved from http://www.himssehra.org/docs/EHRVA_application.pdf

Ho, C. I., & Lee, Y. L. (2007). The Development of an E-Travel Service Quality Scale. *Tourism Management*, *28*(6), 1434–1449. doi:10.1016/j.tourman.2006.12.002

Hox, J. and Bechger, T. (2000). An Introduction to Structural Equation Modeling. *Family Science Review*, 11, 354-373.

Hulland, J. (1999). Use of partial least squares (PLS) in strategic management research: A review of four recent studies. *Strategic Management Journal*, *20*(2), 195–204. doi:10.1002/(SICI)1097-0266(199902)20:2<195::AID-SMJ133.0.CO;2-7

Hunter, G. (2007). *Impact of Information Technology on Local Government Archival Records* [online]. Long Island: The Closest to Home Project. [Accessed 23 Apr. 2015], Retrieved from http://www.statearchivists.org/lga/

Ives, B., Olson, M. and Baroudi, J. (1983). The measurement of user information satisfaction. *ACM*, 26(10), 785-793.

Joe, F. (n.d.). [*Using the SmartPLS Software Structural Model Assessment.* SAGE.]. *Hair, Jr.*

Joshi, B. (2005). Correlation, regression and path coefficient analyses for some yield components in common and Tartary buckwheat in Nepal. *Fagopyrum*, [online] 22, pp.77-82.

Kaisara, G., & Pather, S. (2011). The E-Government Evaluation Challenge: A South African Batho Pele-Aligned Service Quality Approach. *Government Information Quarterly*, *28*(2), 211–221. doi:10.1016/j.giq.2010.07.008

Kassim, E., Jailani, S., Hairuddin, H., & Zamzuri, N. (2012). Information system acceptance and user satisfaction: The mediating role of trust. *Social and Behavioral Sciences*, *57*, 412–418.

Kotler, P. (2000). *Marketing Management*. Englewood Cliffs, NJ: Prentice-Hall, Inc.

Kushniruk, A., & Kannry, J. (2011). *Human Aspects of Health Care Information Systems. Encyclopedia of life support system* (Vol. 2). Medical Science.

Kwong, K. and Wong, K. (2013). Partial Least Squares Structural Equation Modeling (PLS-SEM) Techniques Using SmartPLS. *Marketing Bulletin*, 24(1).

Linder, J., Schnipper, J., Tsurikova, R., & Melnikas, A. (2006). *Barriers to electronic health record use during patient visits* (pp. 499–503). AMIA Annul Symposuium Proceedings.

Lorenzi, N., Kouroubali, A., Detmer, D., & Bloomrosen, M. (2009). How to successfully select and implement electronic health records (EHR) in small ambulatory practice settings. *BMC Medical Informatics and Decision Making*, *9*(1), 15. doi:10.1186/1472-6947-9-15 PMID:19236705

Mahmood, M., Burn, J., Gemoets, L., & Jacquez, C. (2000). Variables affecting information technology end-user satisfaction: A meta-analysis of the empirical literature. *International Journal of Human-Computer Studies*, *52*(4), 751–771. doi:10.1006/ijhc.1999.0353

Mashhour, A., & Zaatreh, Z. (2008). A Framework for Evaluating the Effectiveness of Information Systems at Jordan Banks: An Empirical Study. *Journal of Internet Banking and Commerce*, *13*(1), 2–4.

McKnight, D. (2005). Trust in Information Technology. In G. Davis (Ed.), The Blackwell Encyclopedia of Management (1st ed., pp. 329-331). Malden, MA: Blackwell: Management Information Systems.

Miles, M. B., & Huberman, A. M. (1994). *Qualitative data analysis: An expanded sourcebook* (2nd ed.). London: Sage.

Ministry of Health-Kingdom of Bahrain. (2015). Moh.gov.bh. Retrieved from http://www.moh.gov.bh/ar/HealthEstablishment/Establishments/Maternity.aspx?print=true

Nassiliou, A. (2009). *E-health for a better quality of healthcare. The European files: e-health in Europe*. Retrieved from http://ec.europa.eu/health-eu/doc/ehealth.pdf

Negash, S., Ryan, T., & Igbaria, M. (2003). Quality and effectiveness in Web-based customer support systems. *Information & Management*, *40*(8), 757–768. doi:10.1016/S0378-7206(02)00101-5

Nour El Din, M. M. (2007). Physicians' use of and attitudes toward electronic medical record system implemented at a teaching hospital in Saudi Arabia. *The Journal of the Egyptian Public Health Association*, *82*(5-6), 347–364. PMID:18706293

Nunes, G., & Javier, M. (2014). Hospital Information System satisfaction in Brazil: Background and Moderating Effects. *JRFHHA*, *2*(1), 1–9. doi:10.5005/jp-journals-10035-1007

Padi, F.Francis Kwame Padi. (2003). Correlation and Path Coefficient Analyses of Yield and Yield Components in Pigeonpea. *Pakistan Journal of Biological Sciences*, *6*(19), 1689–1694. doi:10.3923/pjbs.2003.1689.1694

Parasuraman, A., Zheitmal, V. A., & Berry, L. L. (1988). SERVQUAL: A multiple item scale for measuring consumer perceptions of service quality. *Journal of Retailing*, *64*(1), 12–40.

Park, A., & Gretzel, U. (2007). —Success Factors for Destination Marketing Web Sites: A Qualitative Meta-Analysis. *Journal of Travel Research*, *46*(1), 46–63. doi:10.1177/0047287507302381

Perry, A. (2012). *Exploring User Interfaces for Search and Content Based Clinical Decision Support in Electronic Health Records [Master thesis]*. Norwegian University of Science and Technology. Retrieved from http://www.diva-portal.org/smash/get/diva2:658751/FULLTEXT01.pdf

Rabinowitz, S., Sato, E., & Berkes, E. (2001). *Choosing Assessments for Measuring Students Growth.* Retrieved from https://www.azed.gov/highly-qualified-professionals/files/2012/02/technical-adequacy-of-assessments-for-student-growth-082411.pdf

Rahimi, B., & Vimarlund, V. (2007). Methods to Evaluate Health Information Systems in Healthcare Settings: A Literature Review. *Journal of Medical Systems*, *31*(5), 397–432. doi:10.100710916-007-9082-z PMID:17918694

Raija, H. (2011). *Reflecting with the Delone and Mclean Model*. Helsinki: International Workshop on Practice Research. Retrieved from http://www.vits.org/uploads/PracticeResearch2011/HalonenReflectingSuccessModel.pdf

Ramayah, T., & Lee, J. (2012). System characteristics, satisfaction and e-learning usage: A structural equation model. *The Turkish Online Journal of Educational Technology, 11*(2), 196–206.

Regional Health Systems Observatory- EMRO. (2007). Health Systems Profile- Bahrain. Retrieved from http://apps.who.int/medicinedocs/documents/s17291e/s17291e.pdf

Rehman, A.M. (2015). *Health Information Technology at the Royal Medical Services*. Bahrain Defense Force Royal Medical Services. Retrieved from https://www.bdfmedical.org/en/index.php/organization-of-bdfrms/divisions-of-royal-medical-services/information-technology

Roman, A. and D ănuleţiu, A. (2013). An Empirical Analysis of the determinates of Bank profitability in Romania. *Annales Universitatis Apulensis Series Oeconomica, 15*(2), 580–593.

Rossiter, J. (2002). The COARSE procedure for scale development in marketing. *International Journal of Research in Marketing, 19*(1), 305–335. doi:10.1016/S0167-8116(02)00097-6

Rossiter, J. (2002). The COARSE procedure for scale development in marketing. *International Journal of Research in Marketing, 19*(1), 305–335. doi:10.1016/S0167-8116(02)00097-6

Schemper, M. (2003). Predictive accuracy and explained variation. *Statistics in Medicine, 22*(14), 299–308. doi:10.1002im.1486 PMID:12854094

Schemper, M. (2003). Predictive accuracy and explained variation. *Statistics in Medicine, 22*(14), 299–308. doi:10.1002im.1486 PMID:12854094

Schubart, J., & Einbinder, J. (2000). Evaluation of a data warehouse in an academic health sciences center. *International Journal of Medical Informatics, 60*(3), 319–333. doi:10.1016/S1386-5056(00)00126-X PMID:11137474

Sinha, R. and Kurian, S. (2014). Assessment of end user satisfaction of hospital information system. *Management in health*, 18(3), 26-33.

Stricklin, M., Bierer, S., & Struk, C. (2003). Home Care Nurses' Attitudes Toward Computers. *Computers, Informatics, Nursing, 21*(2), 103–111. doi:10.1097/00024665-200303000-00012 PMID:12802951

Tang, P. C., & Lansky, D. (2005, September 01). The Missing Link: Bridging The Patient-Provider Health Information Gap. *Health Affairs, 24*(5), 1290–1295. doi:10.1377/hlthaff.24.5.1290 PMID:16162575

Tavakol, M., & Dennick, R. (2011). Making sense of Cronbach's alpha. *International Journal of Medical Education, 2*(53-55), 2042–6372.

Temme, D., Kreis, H., & Hildebrandt, L. (2006). *PLS Path Modeling – A Software Review*. SFB 649 Discussion Paper 2006-084. Berlin, Germany: Institute of Marketing, Humboldt-Universität.

Teo, T., Srivastava, S., & Jiang, L. (2008). Trust and Electronic Government Success: An Empirical Study. *Journal of Management Information Systems, 25*(3), 99–132. doi:10.2753/MIS0742-1222250303

Thakkar, M., & Davis, D. C. (2006). Risks, Barriers, and Benefits of EHR Systems: A Comparative Study Based on Size of Hospital. *Perspectives in Health Information Management, 3*, 1–10. PMID:18066363

Torkzadeh, G., & Doll, W. (1991). Test-Retest Reliability of the End-User Computing Satisfaction Instrument. *Decision Sciences*, *22*(1), 26–37. doi:10.1111/j.1540-5915.1991.tb01259.x

Torkzadeh, G., & Doll, W. J. (1993). The Place and Value of Documentation in End-User Computing. *Information & Management, 24*, 147–158.

Torkzadeh, G., & Doll, W. J. (1993). The Place and Value of Documentation in End-User Computing. *Information & Management*, *24*(3), 147–158. doi:10.1016/0378-7206(93)90063-Y

Wang, R., & Strong, D. (1996). Beyond Accuracy: What Data Quality Means to Data Consumers. *Journal of Management Information Systems*, *12*(4), 5–33. doi:10.1080/07421222.1996.11518099

Ward, R., Stevens, C., Brentnall, P., & Briddon, J. (2008). The attitudes of health care staff to information technology: A comprehensive review of the research literature. *Health Information and Libraries Journal*, *25*(2), 81–97. doi:10.1111/j.1471-1842.2008.00777.x PMID:18494643

Watson, R., Pitt, L., & Kavan, C. (2001). Measuring Information Systems Service Quality: Lessons from Two Longitudinal Case Studies. *Management Information Systems Quarterly*, *22*(1), 61–79. doi:10.2307/249678

Webb, H. W., & Webb, L. A. (2004). Sitequal: An integrated measure of web site quality. *Journal of Enterprise Information Management*, *17*(6), 430–440. doi:10.1108/17410390410566724

Yoon, H., Nah, J., & Chin, W. (2013). Measuring End User Satisfaction with Hospital's Mobile Health System in Korea. *International Journal of u- and e- Service. Science and Technology*, *6*(6), 21–30.

Yuthas, K., & Eining, M. (1995). An experimental evaluation of measurements of information system effectiveness. *Journal of Information Systems*, *9*(2), 69–8.

Zmud, R. (1978). Concepts, theories, and techniques: An empirical investigation of the dimensionality of the concept of information. *Decision Sciences*, *9*(2), 187–195. doi:10.1111/j.1540-5915.1978.tb01378.x

This research was previously published in the International Journal of E-Health and Medical Communications (IJEHMC), 7(3); edited by Joel J.P.C. Rodrigues, pages 29-51, copyright year 2016 by IGI Publishing (an imprint of IGI Global).

Chapter 18
Web Healthcare Applications in Poland:
Trends, Standards, Barriers and Possibilities of Implementation and Usage of E-Health Systems

Anna Sołtysik-Piorunkiewicz
University of Economics in Katowice, Poland

Małgorzata Furmankiewicz
University of Economics in Katowice, Poland

Piotr Ziuziański
University of Economics in Katowice, Poland

ABSTRACT

This publication consists three main areas of interest: management of patient information in Polish health care system, novel ideas and recent trends on healthcare Web-based applications in Poland and healthcare information behavior of users of self-diagnosis and self-treatment systems in Poland. The methodology adopted includes a literature review for the utilization of Web-based healthcare applications in Poland as well as the trends of medical information systems and healthcare system in Poland. Furthermore the results of a survey research for the management of patient information in Poland are provided. Respondents have been asked about their interested and experiences on the new Polish information electronic health record system or others information systems dedicated to the management of the healthcare processes in Poland. Also another survey researches are presented. Respondents have been asked which internet tools they use for self-treatment and self-diagnosis and are also asked to rate their credibility.

DOI: 10.4018/978-1-5225-6198-9.ch018

SUMMARY

The chapter presents the development and analysis of the use of ICT tools in e-health in the light of the results of studies conducted in the years 2013 - 2014. The authors described the trends and standards of medical information systems and healthcare system. In this chapter authors characterized the different kind of information system in health care, i.e. Electronic Verification of Beneficiary Entitlements (Polish original name of system is "Elektroniczna Weryfikacja Uprawnień Świadczeniobiorców", abbr. e-WUŚ), Integrated Patient System (Polish original name of system is "Zintegrowany Informator Pacjenta", abbr. ZIP), Health insurance card, Electronic European Health Insurance Card, and the system of electronic prescription e-Prescription. Based on literature review authors proposed the classification of ICT tools used in e-health. The results of a survey research for the management of patient information in Poland were also described in this chapter. Additionally the analysis of the factors determining the behavior of Internet users in the field of medical knowledge sharing via the Internet, communication tools used in e-health and medical diagnostics over the Internet were conducted and described. In the last part of this chapter authors proposed SWOT analyses "Web healthcare applications: Strengths, Weaknesses, Opportunities, and Threats" of determinants influencing the Internet user behavior in health care, taking into account the popularity and credibility of ICT tools in self-diagnosing and self-treatment. In the summary of this chapter authors concluded the e-health readiness in Polish information society.

INTRODUCTION

There is a Virgil quote which says: "The greatest wealth is health", indicating the importance of health to most if not all people. As a result most of us since the dawn of time look for information about healthy lifestyle, preventive healthcare and especially pieces of information connected with diseases and treatment. Nowadays, the majority of those interested in seeking information for healthcare is taking advantage of up-to-date technologies and hence are using the Internet. Although the available healthcare information has been radically increased over the years, the abundance of information impose additional difficulties when seeking trusty and reliable web-based resources as well as the information found in the internet should be evaluated since it could be dangerous. On the other hand, medical service providers see potential opportunities and try to apply modern technologies to enhance the healthcare services and reduce the associated costs. E-health for example provides interesting opportunities relating the modern information technologies with doctors' knowledge. E-health applications can include specialized medical portals, internet expert systems for disease diagnose based on symptoms, doctors' Internet advice, online (video) consultation with doctors, etc. (Furmankiewicz, Sołtysik-Piorunkiewicz, & Ziuziański, 2014; Ziuziański, Furmankiewicz, & Sołtysik-Piorunkiewicz, 2014). Multimedia and Web technologies can help in the specific healthcare applications to grow up the impact of their usability (Sołtysik-Piorunkiewicz, 2009, 2014a, 2014b, 2014c, 2015a, 2015b) and can change the favorable circumstances of user behavioral intention. The favorable circumstance is one of the dimensions of the Venkatesh unified user acceptance of information technology model (Venkatesh, Morris, Davis G., & Davis F.D, 2003). This model is presented in Figure 1. Unified theory of acceptance and use of technology (UTAUT) was proposed by American professor K. Viswanath (Venkatesh, Morris, Davis G., & Davis F.D, 2003) and his collaborators at University of Maryland. UTAUT is based on occurrence four factors influencing

Figure 1. UTAUT model
Source: own study based on: V. Venkatesh, M.G. Morris, G.B. Davis and F.D. Davis, User acceptance of information technology: Toward a unified view, „MIS Quarterly" 2003, nr 27, pp. 425–478.

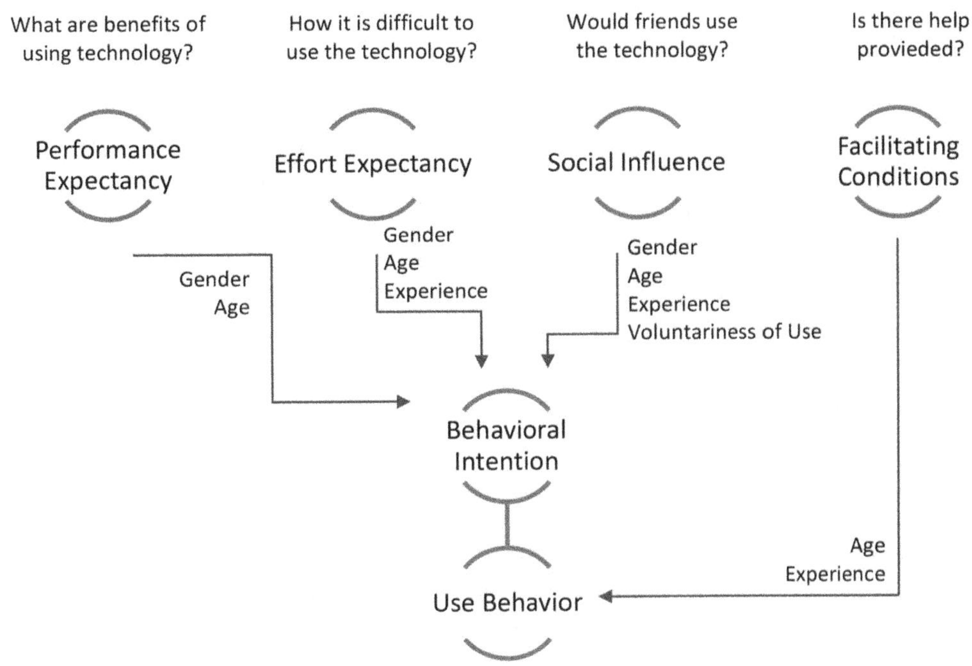

on intention using given technology: facilitating conditions, social influence, effort expectancy, and performance expectancy.

Nowadays web healthcare applications are enriched by multimedia like texts, sounds, still images, animations, videos. Multimedia plays an increasingly important role in supporting medical diagnosing, treatment and prevention (Przelaskowski, 2011). Authors of blogs, specialized medical portals / websites about health and members of online forums / discussion groups use many kinds of multimedia forms to transfer knowledge. They use forms like photographs videos, text to describe illnesses, symptoms, methods of treatment, share experience.

It is worth mentioning that there are medical encyclopaedias, i.e. MedlinePlus (National Library of Medicine, n.d.), where Internet users, especially medical students and doctors besides definitions can find surgery videos, anatomy videos and interactive tutorials. It could be very helpful for them owing to multi-stream transmission of information. Multimedia play significant role in e-learning. Multimedia forms raise curiosity, interest and it has enormous potential (Opoka, 2008). On the other side multimedia could be used in diagnosing area. One of specific example of multimedia usage in this area is study of medical diagnostics imaging named echocardiography. The transmission of information in this form is a deeply interactive (Przelaskowski, 2011).

Usage of multimedia make possible to online transfer information between patient and doctors during online medical visits. Patients can send i.e. their results, photos of symptoms, radiographs, etc.

Usage of multimedia is clearly noticeable in e-commerce. Online pharmacies use online advertising through i.e. text, logos, animations, videos, photographs, sounds. Drug price comparisons websites consists, i.e. photos and descriptions of products.

BACKGROUND

IT systems, which are used to streamline, improve the functioning of health care, are the area of interest as well in Poland (Kawiorska, 2004; Pankowska, 2004), as in many countries across the world and Europe, e.g. Germany (Deutsche Krankenhausgesellschaft, 2002), Switzerland (Thatcher, 2013), Netherlands (Shekelle, Morton, & Keeler, 2006). The transformation, due to implementation of IT systems, into a uniform, integrated and flexible structure, is undoubtedly beneficial for the safety and comfort of patients and streamlining activities in a health care system (Kaplan & Harris-Salamone, 2009). Benefits from implementing IT systems fulfill the expectations of patients and health services providers, as well as improving the whole process of decision making by participants of the health care system. One of the most important e-health systems is electronic health record (EHR). The research of United States academic medical centers reported generally positive attitudes towards using the electronic health record (EHR) as a structured, distributed documentation systems that differ from paper charts (Han & Lopp, 2013) in the ambulatory setting and significant concerns about the potential impact of the EHR on their ability to conduct the doctor-patient encounter (Rouf, Chumley, & Dobbie, 2008).

However the clinical documentation, an essential process within electronic health records (EHRs), takes a significant amount of clinician time. There is luck of methods to optimize documentation ways to deliver effective health care (Pollard et al., 2013). The results of a large-scale statewide analysis in 2006 in United States showed the adoption of health information technologies in physician's offices (Menachemi & Brooks, 2006; Menachemi, Perkins, van Durme, & Brooks, 2006). They discovered that the use of quality enhancing technologies such as PDAs, use of e-mail with patients and EHR was less common. But the state of adoption of health information technologies in physician's offices is still changing. The American Medical Information Association (AMIA) recommendations are based on research and publication, best practices, advocacy, education, certification, databases and knowledge integration. With the United States joining other countries from Europe, e.g. The Great Britain, in national efforts to reap the many benefits that use of health information technology can bring for health care quality and savings, to minimize complexity and difficulties of implementing even smaller-scale systems (Kaplan & Harris-Salamone, 2009).

The last research about factors affecting physician professional satisfaction and their implications for patient care, health systems, and health policy, based on cooperation between RAND Health, a division of the RAND Corporation and AMA (American Medical Association), was published in 2013 (Friedberg et al, 2013). A major factor limiting efficiency and quality gains from clinical information technologies is the lack of full use by the clinicians (Kralewski et al., 2008).

Projects for implementing IT systems in health care for management of patient information are complex and complicated undertakings. These processes should be carried out in stages. To achieve best effects, individual stages should be implemented efficiently and consistently. Moreover, appropriate legislative changes are required and – importantly – persuading decision makers about necessity of such solutions. Any IT systems implemented in health care must comply with legal standards in force and be adaptable to changes made to the national law. Due to the mass scale of health care, they must also be efficient both for patient and health protection system (Sołtysik-Piorunkiewicz, 2012).

There are many benefits of health information technology that can be bring more health care quality and savings, sobering reports recall the complexity and difficulties of implementing even smaller-scale systems (Kaplan & Harris-Salamone, 2009). The ability of EHRs to improve the quality of care in ambulatory care settings was demonstrated improvements in provider performance when clinical information management and decision support tools were made available within an EHR system (Shekelle, Morton, & Keeler, 2006). The developers, implementers and certifiers of EHRs should focus on increasing the adoption of robust EHR systems and increasing the use of specific features rather than simply aiming to deploy an EHR regardless of functionality to maximize health care quality (Poon et al., 2010). Over the past several years, Sittig and others (2011) have carried out an extensive qualitative research program focusing on the barriers and facilitators to successful adoption and use of advanced, state-of-the-art clinical information systems based on commercially available EHR vendors and the internally developed EHRs. They concluded that if the well-designed commercially-available systems are coupled with the other key socio-technical concepts required for safe and effective EHR implementation and use, and organizations have access to implementable clinical knowledge, the transformation of the healthcare enterprise that so many have predicted, is achievable using commercially-available, state-of-the-art EHRs (Sittig et al., 2011).

Accelerating the adoption of IT health care systems will require greater public-private partnerships, new policies to address the misalignment of financial incentives, and a more robust evidence base regarding IT implementation (Goldzweig, C.L., Towfigh, A., Maglione, M., & Shekelle P.G., 2009). But it seems more and more important the development of Web-based health care system for common usage of patients, doctors and others users in health care system.

One of the more important cases in area of healthcare is National Health Service (NHS) in England which was started in 1948. The NHS has grown to become the world's largest publicly funded health service. In 1998 was launched the "NHS Direct". This service was the one of the largest single e-health services in the world, handling more than half a million calls each month. The NHS Direct service closed in 2014. Now instead, a new non-emergency number 111 was introduced to make it easier for people to access local NHS healthcare services in England (NHS.uk, n.d.).

Another examples of the tools used in e-health care expert systems, namely computer programs whose task is to solve the intellectual-based problems (Mulawka, 1996). Expert systems can also be used as a support for decision making, e.g. by prompting the user potential solutions (Cieciura, 2006). As an illustration of an expert system related to health, the MYCIN system can be used (Buchanann & Shortlife, 1984). Its aim was to diagnose contagious blood diseases, and to recommend the medicaments, as well as to suggest doses, taking into account the following parameters: sex, weight, or blood group (Karnowka, Shafer, & Sularz, n.d.). Also the others expert systems were used in England: INTERNIST (Pople at al., 1975), FUNAGES (Dimitroula, Bassiliades, Vlahavas, & Dimitrakos, 2001), CASNET (Weiss, Kulikowski, & Safir, 1978), PUFF (Kunz at al., 1978).

TRENDS AND STANDARDS OF MEDICAL INFORMATION SYSTEMS AND HEALTHCARE SYSTEM IN POLAND

The directions of implementing IT systems in Polish health care are now concentrated mainly on putting into effect the assumptions of the European Commission concerning e-Health. The main issues presented in the „e-Health Poland" plan which require implementation by 2015 include the following (Directions, 2009):

1. Ensuring citizens easier access to health care information.
2. Improving effectiveness of the health care system with regard to electronic flow of information.

3. Creating procedures and guidelines, gathering and giving access to good practices to improve management of a health care centre thanks to implementing information and communication systems.
4. Modernizing the system of medical information to analyze the demand for provided health services.
5. Practical realization of the development of IT solutions in protection of health in line with the guidelines the European Commission which will allow the Republic of Poland to be included into the area of interpretational Electronic Health Record.

Currently, due to accepting for implementation Programmes financed from structural measures, the following activities are of key importance: implementation of the Programme of Health Protection Informatization and creating the conditions for the development of health protection e-services – especially telemedical systems (teleconsultations, telemonitoring, online patients' registration) e-prescriptions and electronic health records, which will be linked with a new identity card (Sołtysik-Piorunkiewicz, 2012).

There are some lacks of information health care systems in Poland:

- Certain data redundancy – similar data is stored in different registers, kept by two different entities, e.g. Central List of Insured Persons and Central Register of Insured Persons (The National Health;
- Federation and Health Insurance Company;
- Incoherence of data between different registers – the same data in different registers may vary (e.g. change of address is not automatically updated in all registers in which the address appeared);
- Lack of cooperation with reference registers – generally, registers do not refer to base reference registers;
- Lack of cooperation between registers in health protection – registers in health protection do not use source information which already exists in other health protection registers;
- Lack of a uniform data model – registers and databases in health protection do not use a uniform data model;
- Lack of structure and relationships between registers in health protection.

There are systems for pharmacists' shops being implemented, i.e. EuroMedica (Apteka Euromedica, 2012), based on modern Internet technologies. Clinics implement InfoMedica or OSOZ systems. However, it seems necessary to implement such IT systems in health protection for management of patient information in Polish health care system e.g. a Health Patient Information System (System Zdrowotny Informator Pacjenta, 2013) which will ensure (Directions, 2011) creation of information conditions that will allow taking long-term optimal decisions in health policy, irrespective of the adopted organizational model of health protection and principles of financing it; creation of a stable information system in health protection, characterized by a flexible approach to the organization of the system of health protection resources, including a model of financing services from public funds, and resilience to disturbance in data gathering and archiving, caused by system changes in health care; decreasing the information gap in the health protection sector that makes it impossible to build an optimal model of health protection; organizing an existing system for information gathering, processing and analyzing (e.g. in Business Intelligence systems); building a system ensuring electronic communication and possibility of exchanging documents and reports between health protection entities and the proprietary body.

Nowadays two projects are being developed in Poland in accordance with the law on publicly funded health care benefits: Electronic Verification of Beneficiary Entitlements (Polish original name of system is “Elektroniczna Weryfikacja Uprawnień Świadczeniobiorców”, abbr. e-WUŚ) and Integrated Patient System (Polish original name of system is “Zintegrowany Informator Pacjenta”, abbr. ZIP).

The system Electronic Verification of Beneficiary Entitlements was tested from the 15th of October to the end of 2012 and its aim is to streamline the system for confirming the entitlement to health care. Thanks to the system, doctors will no longer be responsible for checking whether the patient is entitled to health care. The project was the result of the cooperation of the Ministry of Administration and Digitalization with the Ministry of Health, National Health Fund (ZUS), Social Insurance Institution and Agricultural Social Insurance Fund (KRUS), and was consulted with the Chief Inspectorate of Personal Data Protection. From the 1st of January 2013, patients visiting doctors should present their PESEL number and identity card, driving licence or passport in order to confirm their entitlement to publicly funded health care (Korczak, 2014; Ministerstwo Admnistracji i Cyfryzacji, 2012).

Thanks to the information exchange between the Central List of the Insured of the National Health Fund and ZUS and KRUS registers, it will be possible to check online in a hospital or clinic reception whether the patient is insured. There will be no need for the RMUA slip (although it still will be valid, as will be the pensioner identity card and other documents entitling to benefits) (Sidorko, 2012).

If the system doesn’t confirm the entitlement to health care (e.g. when the employer has not registered the employee for insurance), the citizen’s declaration that he/she is insured will suffice (art. 7 and 8 of the relevant law). The introduction of the e-WUŚ system made the procedure of issuing the European Health Insurance Card (Polish EKUZ) more efficient. The citizen doesn’t have to produce the RMUA slip in a National Health Fund institution; instead he/she can show a document proving his/her identity.

Integrated Patient System is a Polish nationwide service providing registered users with historical data about their treatment and financing of treatment, collected since 2008 by the National Health Fund. The Integrated Patient System was implemented at the 1st of July 2013 and is successfully developed, and now has some important functionality, e.g. quick access to information about right to health care for the citizen, knowledge of treatment and benefits granted and the medicines prescribed, and information about the amounts that have been transferred to finance treatment of patient. To have access to information about health care, medical records and financing of the services provided, the citizen first should to read the rules of use of EHR, and then go to the registration page and fill out the included electronic form. After completing the application the citizen should take the ID card to a branch of the National Health Fund in order to access data (user ID and temporary password). After receiving the access, the citizen can start to use the service. Both the registration and use of the site are free for the citizen. Medical information is available in the latest issue and is so-called “sensitive data”. So according to the principles of data protection and guidance of the Inspector General, the data are under special protection. Therefore the personal visit of each citizen to the headquarters of the Fund is required (Zintegrowany Informator Pacjenta, n.d.).

Health Insurance Card

Electronic health insurance card was implemented in the Silesia voivodeship about 10 years ago as part of works connected with informatization of Silesia voivodeship department of National Health Fund. Electronic health insurance card is used to verify insurance status of an authorized card holder in the system of Silesia voivodeship department of National Health Fund. It also provides personal data and is used for authorization of the services provided as part of performing a contract with Silesia voivodeship department of National Health Fund (Narodowy Fundusz Zdrowia, 2013). This card may be issued to an insured person with a Universal Electronic System for Registration of the Population national identification number who can prove his/her residence on the territory of the Silesia voivodeship.

Since 2004, it has been planned in Poland to implement country-wide Health Insurance Card modelled on the Silesian card. To continue the works ordered by Health Minister on February 8, 2005, a Group was set up to draw up a strategy for the development of a medical information system in health protection and prepare a conception of implementing European Health Insurance Card and Health Insurance Card. The tasks of this group included drawing up a strategy for the development of a medical information system in health protection and preparing a conception of implementing European Health Insurance Card and Health Insurance Card. It was decided that representatives of Health Ministry would be involved in the works of a Group for developing a conception and design of electronic Health Insurance Card (e-HIC), electronic Medical Services Register (e-MSR) and programmes of their implementation, set up by order of the President of the National Health Fund No 40/2004 of 25 November 2004 for the purpose of preparing a functional conception, scope of application and design of e-HIC, linked with European Health Insurance Card, and a conception and design of a system for electronic registration and medical services monitoring (e-MSR) and programmes (strategy, plan, schedule) for implementing e-HIC and e-MSR (Reply of the Secretary, 2013). Up to now, the implementation has not been completed.

Another proposal of implementing electronic medical report was using the function of a health insurance card in an electronic identity card as part of the implementation of the programme MSR II. This project was described in a publication on the connection between Health Insurance Card and pl.ID. Its basic aim is to ensure the verification of the parties, place and sequence of a medical transaction (patient and professionals) by means of a cryptographic secret carrier. This project was based on the use of a crypto processor card (Health Insurance Card, Professional's Card) in an environment that was safe for creation of electronic signature. It was also assumed that pl.ID would constitute an electronic document which might be used for verifying a person (including limited identification), creating personal and qualified signatures, and entitles to cross the borders of the countries united by the Schengen Agreement, as well as serves the function of HIC. As a result of a decision by National Council of Ministers of December 2009, the project was linked with pl.ID, in which there was a separate space for HIC. The National Health Fund was developing and handing to the Ministry of the Interior and Administration the document „Technical and functional requirements for Health Insurance Card, HIC application and their environment.

Electronic European Health Insurance Card

One of the elements of a Polish health care system in managing patient information is electronic European Health Insurance Card. It was created as a response to the necessity of verification of a patient in the European Union and in the countries of the Schengen area.

Currently the project of Health Insurance Card for the patient developed in connection with the pl.ID system is suspended in Poland. However, the work on the system of European Health Insurance Card is still underway due to the development of a new conception of the system of patient identification in the countries of European Free Trade Association (Departament Współpracy Międzynarodowej Centrali NFZ, 2012). The most important part of implementation of pl.ID in polish National Registers System is now in the stage of consultation and analysis of future functionalities (Ministerstwo Spraw Wewnętrznych, 2013).

This card may be issued to an insured person with a national identification number Universal Electronic System for Registration of the Population who can prove his/her residence on the territory of the Silesia province. To continue the works ordered by the Health Minister on February 8, 2005, a Group was set up to draw up a strategy for the development of a medical information system in health care, and prepare a conception of implementing European Health Insurance Card and Health Insurance Card. The tasks of this group included drawing up a strategy for the development of a medical information system in health protection and preparing a conception of implementing European Health Insurance Card and Health Insurance Card. The main aim of the implementation of this solution was to (Ujejski, 2011):

- Facilitate the process of registration and confirming the right to health services (effectiveness);
- Improve the reliability of accounting data sent to the National Health Fund (integrity and indisputability);
- Secure access to data (confidentiality);
- Increase the chance of a rescue in emergencies;
- Satisfaction of the entitled, possibility of getting remote access to own data;
- Increase the effectiveness of medical centers (automation of activities);
- Decrease the number of frauds, mainly in the area of accounting for services that have not been provided;
- Improve the quality of data.

Ultimately, health insurance card should be an element of the information system in the health protection system in Poland. However, using such a card entails access to sensitive data, thus one of basic requirements is ensuring security of the information system used in medical services. It should also be able to integrate with the medical services register and medication register. It may also be an instrument for checking insurance as part of processing European Health Insurance Card instead of the National Health Fund slip of the monthly report for the insured person. Electronic medical report should be integrated with the information system for pharmacies and health care centers as well as the electronic system for prescriptions and electronic system for medical appointment referrals (Sołtysik-Piorunkiewicz, 2014c).

The Electronic Verification of Beneficiary Entitlements

At the same time, the Electronic Verification of Beneficiary Entitlements project is being developed in Poland, i.e. Electronic Verification of Beneficiary Entitlements, in compliance with the law on the healthcare services financed from the public funds. From January 1, 2013 patients visiting doctors should present Universal Electronic System for Registration of the Population number and identity card, driving licence or passport, in order to confirm their entitlement to receiving healthcare from public funds [Do lekarza, 2012]. The tests of the system were conducted from October 15 to the end of 2012, and its aim is to introduce orderliness into the system of confirming the entitlement to receive healthcare services. It also takes from doctors the responsibility for checking whether the patient is entitled. The project was created from the cooperation of the Ministry of Administration and Digitalization with the Ministry of Health, National Health Fund (Polish ZUS), Social Insurance Institution and Agricultural Social Insurance Fund, and was consulted with, among other things, the Chief Inspectorate of Personal Data Protection. Thanks to information exchange among the Central List of the Insured of the National Health Fund and Agricultural Social Insurance Fund registers, it will be possible to check online in a hospital or clinic reception whether the patient is insured. There will be no need for the National Health Fund slip of the monthly report for the insured person (although it still will be valid, just like the pensioner identity card and other documents entitling for receiving services).

In the case when the system did not confirm the entitlement to receiving the services (e.g. when the employer did not register the employee for insurance), the statement of the insured person will suffice (art. 7 and 8 of the relevant law).

The introduction of the Electronic Verification of Beneficiary Entitlements system made the procedure of issuing the National Health Fund slip of the monthly report for the insured person more efficient. The citizen doesn't have to produce the monthly report for the insured person slip in the National Health Fund institution, instead he/she can show identity document.

The System of Electronic Prescription E-Prescription

The e-Prescription system is one of the first projects of building a modern IT system in health protection. It is being implemented under the project „Electronic Platform for Gathering, Analysing and Sharing digital resources about medical events", which is part of the country-wide Programme for Health Protection Informatization. The entity responsible for implementing the electronic prescription system is the Centre for Health Information Systems, set up by the Health Minister [Poznaj system e-recepta, 2012]. All the information gathered in the e-Prescription system is protected in accordance with security standards in force. The information in the e-Prescription system is exchanged using encrypted SSL protocol (Secure Sockets Layer), which is a widely used data transmission standard. This allows for the communication in the e-Prescription system to take place only between entitled participants of the prototype and an unauthorized access to data is prevented. Moreover, cryptographic techniques are used to verify and ensure information consistency. The technique of encryption is based on process of transforming plain text or data into cipher data that cannot be read by anyone other than the sender and the intended receiver (Laudon & Laudon, 2011). Usage of encryption is necessary to protect digital information that are stored, transferred, or sent over the Internet. The capability to generate secure session is built into Internet client browser software and servers. SSL is designed to establish a secure connection between both two computers of the client and the server.

The e-Prescription system allows electronic prescription to function parallel with its formal counterpart in the form of a paper prescription. The use of electronic prescription along with a paper prescription does not disturb the existing model of prescriptions circulation which is based only on paper prescriptions. Currently, due to the lack of appropriate regulations regarding the functioning of prescriptions only in an electronic form, an electronic document does not constitute a prescription in the light of law. Writing a prescription in an electronic form takes place by means of a doctor's medical computer program, and in the case when a doctor does not have internal software in his/her surgery, communication with the e-Prescription system occurs through an on-line application.

Pharmacists work with the e-Prescription system directly through their chemist's program, and can also dispense electronic prescriptions through a dedicated on-line application. Patients get access to information about the history of their pharmacotherapy through the e-Prescription Internet Account.

The prototype of the e-Prescription system was launched in mid-March 2011, but the actual processing of prescriptions using this system started on 18 April 2011. The delays were connected with the need to install and launch the system in the entities that signed a cooperation agreement with the Centre for

Health Information Systems, and necessity of providing training to medical personnel and pharmacists on how to operate the system.

The e-Prescription system supported 20 facilities, i.e. 2 clinics, 2 doctor's surgeries and 16 chemist's shops. This implementation can be regarded as a pilot implementation across the country in 2012 (Sołtysik-Piorunkiewicz, 2012). However, for the system to be successfully implemented across the whole country, electronic prescription should be regulated by new legal regulations, and in the whole country there should be access to the public database of medicines, patient's insurance card and a tool for doctors authentication.

THE WEB-BASED HEALTHCARE APPLICATIONS IN POLAND FOR SELF-DIAGNOSIS AND SELF-TREATMENT

There is a variety of Internet systems, web pages and portals and other applications connected with healthcare and e-health. This tools, especially used in self-diagnosis and self-treatment has been classified into nine categories:

1. Blogs,
2. Doctors' Internet advices,
3. Drug price comparisons websites,
4. Internet encyclopedias (i.e. Wikipedia),
5. Internet systems indicate disease based on symptoms,
6. Online forums / discussion groups,
7. Online pharmacies,
8. Online video consultation with doctors,
9. Specialized medical portals/ websites about health.

Online pharmacies constitute the next type of the tools used in e-health. The Polish example of an online pharmacy is www.doz.pl, which apart from enabling to order a product, provides a range of information from the "Encyclopedia" concerning a given preparation, e.g. composition, form, effect, dosing, and possible side effects, which accounts for the most of the information placed on the drug information leaflet. The above-mentioned pharmacy also allows asking a "Question to a Pharmacist". Furthermore, it provides a "Related Products" mechanism, which presents a list of products similar to the one already chosen by the user (Doz.pl (n.d.).

The e-health application also includes drug search engines, whose task is to find the equivalent of the refunded drug. An example of a search engine is www.leku.pl, which allows finding a cheaper equivalent of a drug using a name of a preparation or the name of the active substance being an ingredient of the medicament (Leku.pl, n.d.).

Expert systems can also be used as a support for decision making, e.g. INTERNIST in internal medicine (Pople at al., 1975), and others: MYCIN (Buchanann & Shortlife, 1984), FUNAGES (Dimitroula, Bassiliades, Vlahavas, & Dimitrakos, 2001), CASNET (Weiss, Kulikowski, & Safir, 1978), PUFF (Kunz at al., 1978).

Nowadays, the www.dooktor.pl portal operates and provides a free information system. Basing on primary symptoms, the www.dooktor.pl portal allows identification of a particular disease. Therefore, it has the attributes of an expert system (Dooktor.pl (n.d.); Furmankiewicz, Sołtysik-Piorunkiewicz, & Ziuziański, 2014).

The Internet network allows for a round-the-clock contact with a doctor. This is enabled by www.qzdrowiu.pl portal, which provides video-consultations. The user can choose a doctor and make an appointment for a specific date of a video-consultation (Qzdrowiu.pl, n.d.).

The user can also use tools such as comparison shopping engine for comparing the drug prices, Internet forums, and encyclopedias, e.g. www.wikipedia.pl. Comparison shopping engines enable people to find the cheapest product. Additionally, one can share health-related information when using online forums, ask questions, and find alternative treatments. Moreover, the user can write about the progress in a fight against the illness or addiction, and to exchange opinions about doctors or drugs.

Modern technologies, including the Internet, have been widely used in the field of public health, which is reflected in the concept of e-health. To be more specific, the e-health notion can be defined as activities in the spheres such as diagnosis, treatment, control, prevention, and leading a healthy lifestyle with the use of information and communication technologies (Rozpędowska-Matraszek, 2012; Zdrowie publiczne, n.d.).

The Internet plays a key role in the field of health care and treatment, as evidenced by the results of a survey conducted by the ARC Rynek i Opinia company. According to the company's research conducted in 2012, over 60% of the Internet users declare searching the Internet for health information. On the other hand, more than half (52%) of respondents, who search for information on diseases, admitted that they happen to heal themselves or their relatives using methods that have been found via Internet, at the same time resigning from the consultation with a doctor or pharmacist. As many as 91% of respondents claimed that the method found via the Internet roved to be effective (ARC Rynek i Opinia, 2012a). It should also be noted that over 90% of the respondents expressed their willingness to contact a specialist online at the health care unit's official website (e.g. hospital, clinic), if there was such a possibility (ARC Rynek i Opinia, 2012b).

THE RESULTS OF A SURVEY RESEARCH FOR THE MANAGEMENT OF PATIENT INFORMATION IN POLAND

Nowadays usage of artificial intelligent systems such as expert systems, agent systems and others systems for decision support diagnostics support is very interesting topic. This chapter presents results of two kinds of researches: first connected with systems dedicated to the management of the healthcare processes and second on connected with self-diagnosis and self-treatment tool usage.

The main study of usage trends of electronic health care systems was provided in academic year 2013-14 in University of Economics in Katowice. About 200 second year students of master degree were invited to complete a questionnaire after explaining the topic about e-health during one of the course. The authors elicited themes for the questionnaire by asking a focus group of second year students how they use and know the implementation of electronic health care systems for citizens in Poland and in the world and what is the most important functionality of electronic health care system for them as the patients. Three themes emerged: using of EKUZ during vacations and going abroad, access to online resources about medical health care record and personal medical history, and implementation of other IT systems in e-health. The authors added a two theme: using of e-WUŚ for the publicly funded health care and the Integrated Patient System (EHR) for finding the historic data about amount of treatment of patient. The author created a 5-item questionnaire, based on these themes. Each answer had the three possibility to choose in questionnaire: “Yes”, “No”, and “I don’t know”. The survey was prepared on web platform Net-ankieta.pl. The authors emailed an electronic survey link to consenting students.

The Polish Information Electronic Health Record System or Others Information Systems Dedicated to the Management of the Healthcare Processes in Poland

In April 2014, the research team conducted a study on the management of information concerning the patient. The study was conducted using an online questionnaire (Computer-Assisted Web Interview method, abbr. CAWI). The questionnaire was completed by 229 respondents. Men accounted for 36% of the survey sample, whereas women constituted 64% of the group.

The respondents gave answers to the question whether they use health services in the Polish health care system based on health insurance in accordance with the data in the e-WUŚ system, a system of electronic verification of patients’ rights (Elektroniczna Weryfikacja Uprawnień Świadczeniobiorców, abbr. e-WUŚ). The results are presented in Figure 2.

The e-WUŚ system was implemented January 1, 2013, and is used by health care units cooperating with the National Health Fund (Narodowy Fundusz Zdrowia, abbr. NFZ). Around 57% of respondents using health care benefits acknowledge their right for a free execution of these services via e-WUŚ system. When using health care benefits, 27% of respondents do not know whether they use the e-WUŚ system or not, whereas 16% of the surveyed claimed they do not use e-WUŚ system.

Subsequently, the respondents were asked about the use of health services in the European health care system based on the European Health Insurance Card (abbr. EHIC). The results are presented in Figure 3.

As many as 22% of respondents using health services in the European health care system confirm their rights resulting from EHIC.

Then, respondents were asked whether they are registered in ZIP system, the integrated patient information (Zintegrowany Informator Pacjenta, abbr. ZIP), which, like e-WUŚ was implemented in Poland in 2013. The respondents’ answers are presented in Figure 4.

Figure 2. The use of the e-WUŚ by respondents

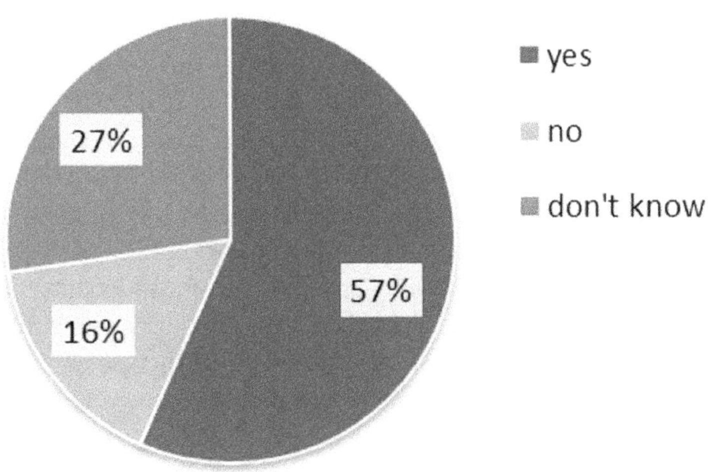

Figure 3. The use of the EHIC by respondents

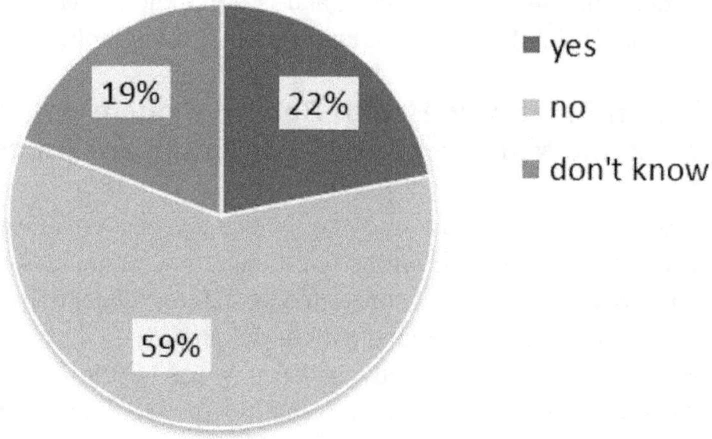

Figure 4. The registration in the ZIP system

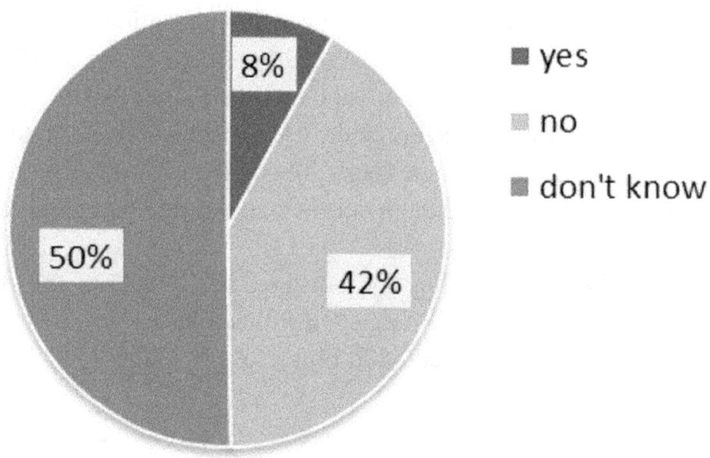

Being registered in the ZIP system was declared by 8% of the people participating in the survey. Around 42% of respondents declared that they are not registered in the ZIP system. However, it is disturbing that a half of the respondents do not know whether they are registered in the system or not. This may result from ignorance concerning the implementation and purpose of the system.

Respondents also answered a question concerning the consistency of the data gathered in the ZIP system with the actual course of the respondent's medical history. The results are presented in Figure 5.

As many as 8% of the respondents, the identical percentage of respondents that declared to be registered in the ZIP, confirmed the consistency of the data stored in the system with the actual state.

The last question to be answered by the surveyed concerned the use of other systems designed to manage the patient information. The respondents' answers are presented in Figure 6.

Figure 5. Consistency of data in the ZIP system

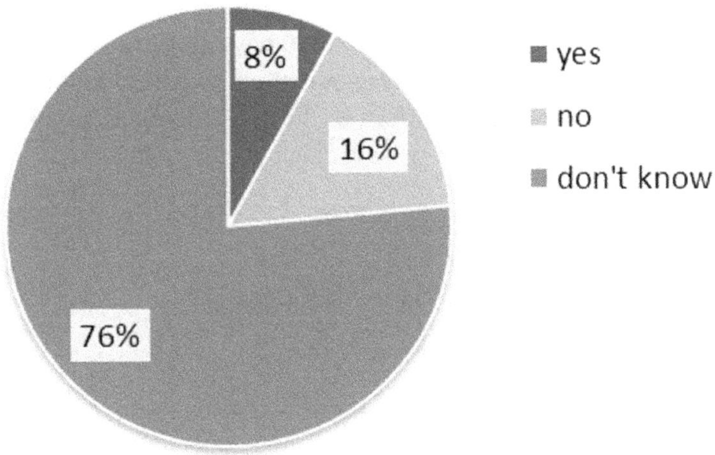

Figure 6. The use of other systems for patient information management

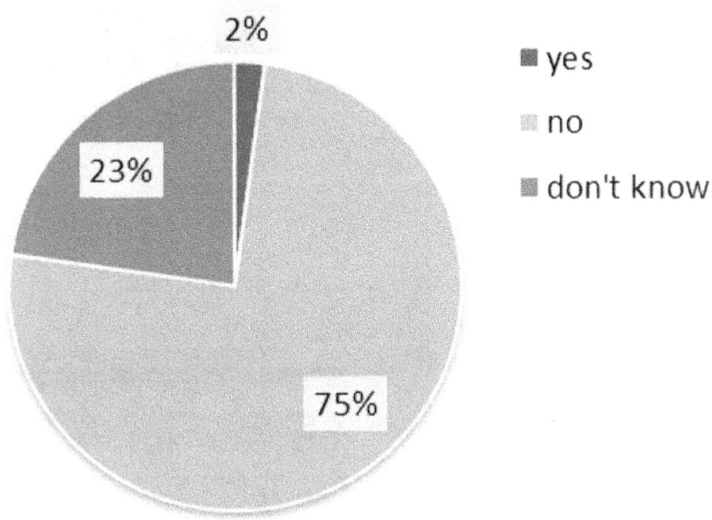

Only 2% of the respondents claim that they use other systems for patient information management. A huge majority, i.e.75% of the surveyed state that they do not use other systems, and 23% do not know if they use other information technology tools or not.

Self-Diagnosis and Self-Treatment Tool Usage Researches

In 2013 research team conducted a study on the evaluation of the use ICT (especially Internet technologies) in the self-treatment and self-diagnosis process. One year later team conducted researches once again. Researches main aim is checking popularity and reliability of these tools and Internet evaluation in the context of traditional sources of information about health and its role that it assumes in modern human life (Furmankiewicz & Ziuziański, 2013). The researches were carried out with the use of an Internet questionnaire (CAWI method). The survey was conducted in late April and early May 2014 and lasted less than a month. Previous study lasted for 10 days in March 2013. The survey in 2013 had a pilot character. It was conducted among 212 respondents. In 2014 research sample was significantly increased to 526 respondents. Gender distributions of these two samples are presented in the Figure 7.

Respondents were asked about their use of self-diagnosis and self-treatment tool. The results are presented in Figure 8.

Figure 7. Structure of respondents by gender and sample numbers

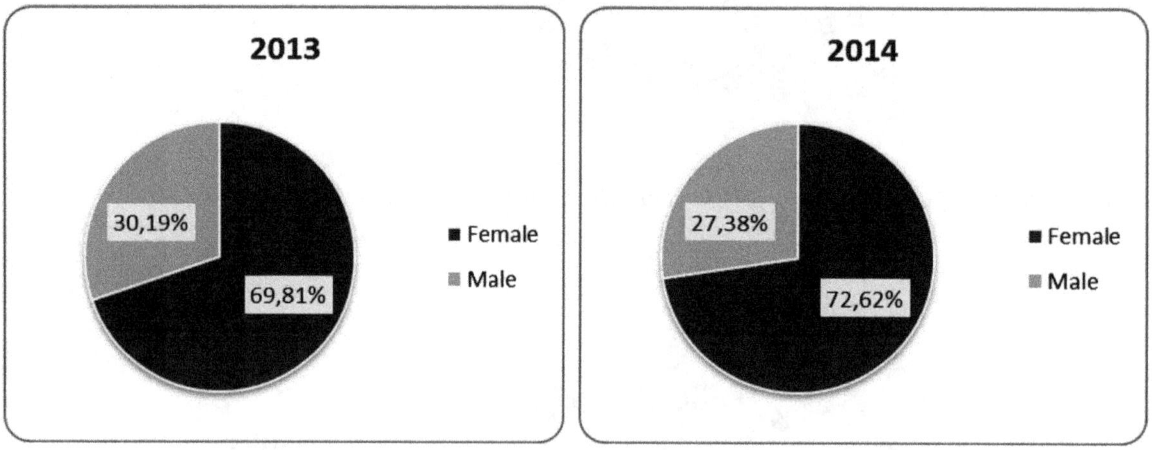

Figure 8. Self-diagnosis and self-treatment tools using by respondents

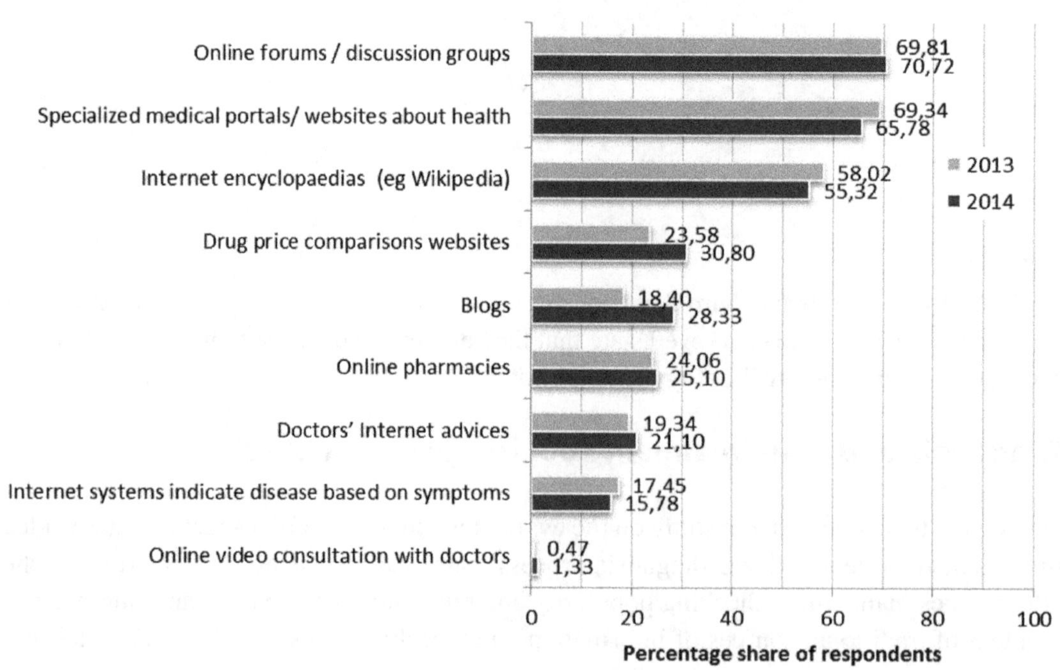

According to the respondents the most commonly used tools include: online forums / discussion groups, specialized medical portals/ websites about health and internet encyclopedias (i.e. Wikipedia). Least likely respondents use online video consultation with doctors, Internet systems indicate disease based on symptoms and doctors' Internet advices. The results compared to a year earlier survey do not differ too much, which confirms that the most popular tools, the use of which fluctuates around 70% are online forums / discussion groups, specialized medical portals/ websites about health.

Noteworthy respondents evaluation of the reliability of self-diagnosis and self-healing tools. Their task was to evaluate the different tools on a scale from 1 to 5 then calculated the average scores of all respondents. The results are presented in the Figure 9.

The most reliable of these tools became online video consultation with doctors, specialized medical portals/ websites about health and Doctors' Internet advices. According to respondents least credible are blogs, internet systems indicate disease based on symptoms and online forums / discussion groups.

Comparison of test results from the previous year to this year's research suggests only conclude that the assessment of the respondents in the two studies are similar.

In addition to evaluation of the credibility of the individual tools used for self-diagnosis and self-medication, respondents were asked to evaluate the credibility of various sources of information about health on a scale from 1 to 5. Evaluations have calculated the average. The lowest was rated television, then print media and the internet. The above were evaluated medical books (including encyclopedias, guides) and the highest doctor. The results are presented in the Figure 10.

The results from last year are almost identical to current research findings. It appears that the Internet is perceived better than television and on the same level as the print media in the context of credibility.

Invariably, the most reliable source of health information is a doctor.

A very interesting question in questionnaire was question about the effectiveness of tips obtained on the Internet. As many as 44% of respondents said that the tips are usually effective, 27% of respondents said that it is not looking for advice on the Internet. The results are presented in graph X. Another

Figure 9. Respondents ratings about the reliability of self-diagnosis and self-healing tools

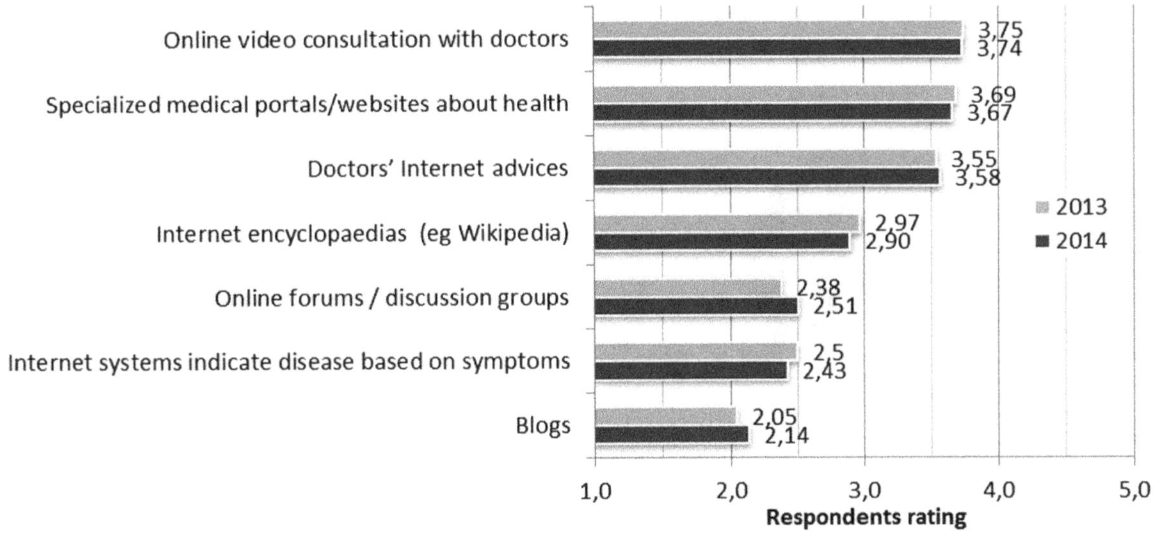

Figure 10. Evaluation of the reliability of source knowledge about health according to respondents

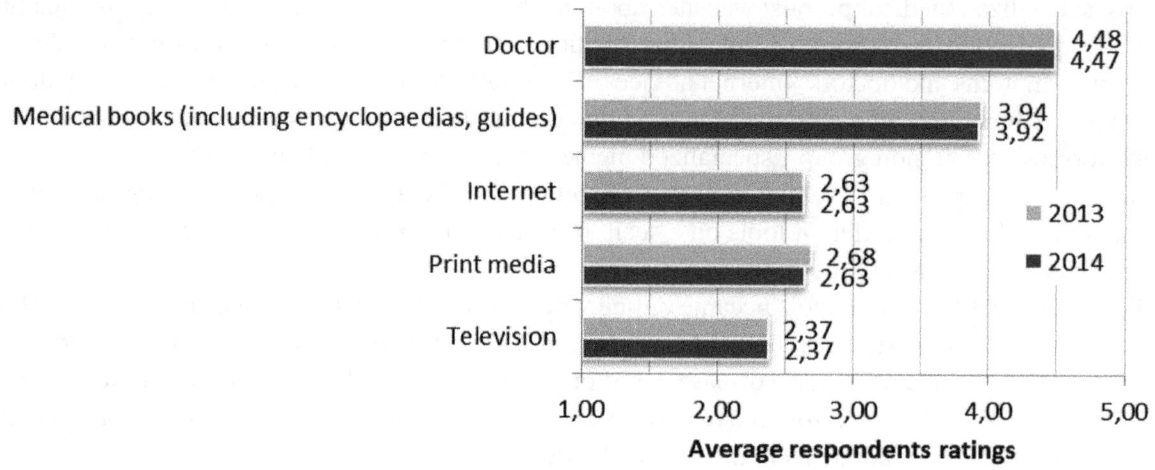

question about self-diagnosing was a question about failing visiting doctor because of advice obtained on the Internet. Of the 18% of respondents answered that happened to them (previous year: 20.75%). Respondents' answers are presented in the Figure 11.

Failure of medical visits can be interpreted in terms of both positive and negative. Because the Internet has replaced to some extent, e.g. GP or medical specialist queue waiting times can be reduced. You can also avoid a situation in which a patient visits a doctor with trivial reason that he does not endanger the lives and health. On the other hand, disregard the symptoms of disorders of homeostasis or misinterpretation of symptoms can result in serious consequences and lead in extreme cases, even death. It is also

Figure 11. Tips obtained on the Internet: effectiveness and risks according to respondents

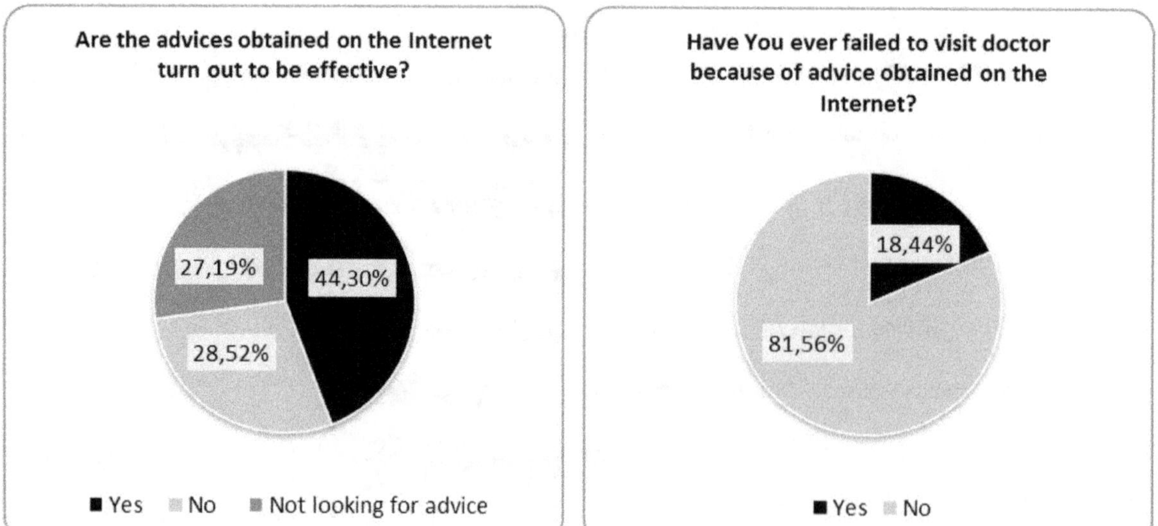

alarming that only 13% of respondents using health-related websites verify competencies' advice. One of the questions in questionnaire was dedicated to respondents' level of agreement or disagreement on a Likert scale regarding to statement: *The anonymity on the Internet allows freer discussion of shameful topics about the health.* 84,03% of respondents strongly agreed or agreed with this statement.

WEB HEALTHCARE APPLICATIONS: STRENGTHS, WEAKNESSES, OPPORTUNITIES, THREATS

The authors of this publication as a form of summary previous considerations chosen SWOT analysis.

Strengths of Web-Based Application

The strengths of Web-based application for e-health are easy way of communication between: patient and health care unit (i.e. arrange visit, obtain medical results), National Health Fund and healthcare providers (i.e. checking if patient is insured), and Government and other stakeholders in Polish healthcare system (i.e. information portals, e-mails). Source of knowledge for patients, doctors, pharmacists and other people are involved into health care. Also power of healthy lifestyle promoting tools is helping to monitor sport activity and food consumed.

Weaknesses of Web-Based Application

There are a few weaknesses of Web-based application. First it is hard to verify truthfulness and original source of knowledge, and second patients without access to Internet cannot take advantage of the benefits of the Web-based application.

Opportunities of Web-Based Application

There are a lot of opportunities of Web-based application: chance to solve not very important Internauts' health problems without medical consultation, and connecting people with the same health problem from all over the world in order to mentally support each other and get knowledge about methods of treatment. And one of the most important during self-diagnosis and self-treatment is the possibility of contact with doctor without need of leaving home

Threats of Web-Based Application

We can describe also some threats of Web-based application, e.g. doctor's authority reduction. The Web-based application also could be dangerous because of failing to visit doctor thanks to advice obtained on the Internet, or impersonating doctors, fake articles and other issues generating false knowledge, and it could be the way of selling bad quality or fake and dangerous medicines.

The results of analysis of Web-based application for health care is presented in the Figure 12.

Figure 12. Results of SWOT analysis of Web-based application in health care

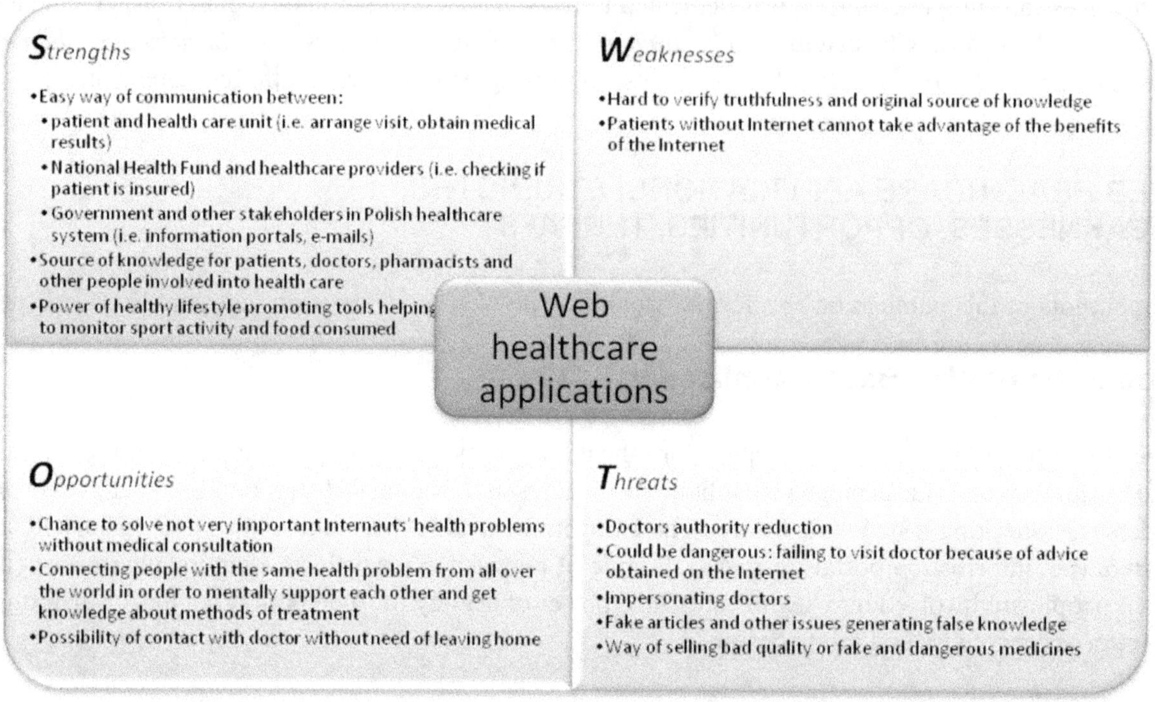

FUTURE RESEARCH DIRECTIONS

EHR implementation is essential to improving patient safety but is still highly heterogeneous across health care systems and providers, and this heterogeneity leads to equally variable implications for patient safety (Sittig & Singh, 2012). As a conclusion of the study about trends of using electronic health care systems by patient, students of University of Economics in Katowice reported some needs of usage of the European Health Insurance Card, but they mostly reported the lack of knowledge about implementation and functionality of information systems in Polish health care. Only part of them was interested in finding the information about health care in EHRs data bases of Integrated Patient System, and only a few of them had the own id and tried to use it on the online platform for e-health services.

The future research should be conducted to discover the reason of this state of knowledge about Polish electronic health care system.

CONCLUSION

Healthcare Information Behavior of Users of Self-Diagnosis and Self-Treatment Systems in Poland

The most important reason of future changes of healthcare information behavior could be to learn the citizens about functionality of implemented electronic systems in health care (usage of courses at the universities to make a progress in that field), facilitate the process of registration and confirmation of the right to health services (effectiveness), improve the reliability of accounting data sent to the National Health Fund (integrity and incontrovertibility), ensure a secure access to data (confidentiality), increase the satisfaction of those entitled to health care and enable a remote access to own data (increase of patient's interest), increase the effectiveness of medical centers (automation of activities), decrease the number of frauds, mainly in the area of accounting for services that have not been provided improve the quality of data.

Internet is a common used by patients or people seeking information about health. It allows freedom especially in the discussion of embarrassing health topics. It should be emphasized that the respondents indicated that the Internet is a more reliable source of information than television and it is at similar level in relation to print media. Conducted researches shows that Polish Internet users mostly use specialized medical portals/websites about health and online forums/discussion groups, while the least likely they use online video consultation with doctors. It is worth noting that video consultation is considered as the most reliable tool, while online forums/discussion groups, internet systems indicate disease based on symptoms and blogs are the least reliable.

The Web-based application in e-health could be a valuable source of knowledge, but it should be rather complement of the knowledge, not a substitute for a medical visit. In order to avoid the dangers arising from the use of the Internet in the area of e-health, patients should use it with caution: trust credible sources, respectively filter pieces of information and consult with specialists. Due to the fact that the respondents showed great confidence in online video consultation with doctors, it can be assumed that maybe there will be a some breakthrough in healthcare and traditional stationary clinic could be replaced by a virtual clinic in a future.

Presented researches in e-health area in Poland should be deepened and extended. Usage of web tools for self-diagnosis and self-treatment and government e-health systems usage by Poles is a very interesting area.

E-Health Readiness in Polish Information Society

The computerisation of the Polish health care are still concentrated mainly on meeting the objectives set by the European Commission concerning e-Health. The main issues presented in the „e-Health Poland" plan which must be implemented by 2015, generally:

1. Creating procedures and guidelines, collecting and sharing good practices to improve the management of a health care centre through the implementation of information and communication Web-based systems.
2. Modernizing the medical information system to ensure the analysis of the demand for health services
3. Practical realization of the development of IT solutions in health protection following the guidelines of the European Commission to allow the Republic of Poland to be included into the area of interoperational Electronic Health Record.

Additionally the Web-based application could ensure citizens an easier access to health care information, improve the effectiveness of the health care system with regard to an electronic flow of information.

REFERENCES

Apteka Euromedica. (2014). Aptekaeuromedica.pl. Retrieved from http://www.aptekaeuromedica.pl

Buchanann, B., & Shortlife, E. (1984). *Rule-Based Expert Systems. The Mycin Experiments of the Stanford Heuristic Project.* Reading: Addison-Wesley; Retrieved from http://aitopics.org/sites/default/files/classic/Buchanan/Buchanan00.pdf

Cieciura, M. (2006). *Podstawy technologii informacyjnych z przykładami zastosowań.* Warszawa: Wyd. Opolgraf SA.

Departament Współpracy Międzynarodowej Centrali, N. F. Z. (2012). Departament Współpracy Międzynarodowej Centrali NFZ. Retrieved from https://www.ekuz.nfz.gov.pl/

Handbuch zur Kalkulation von Fallkosten. Version 2.0 (2002). Deutsche Krankenhausgesellschaft. Retrieved from http://www.aokgesundheitspartner.de/imperia/md/gpp/bund/krankenhaus/drg_system/kalkulation/kalkulationshandbuch_2_0.pdf

Dimitroula, V., Bassiliades, N., Vlahavas, I., & Dimitrakos, S. (2001). FUNAGES: An expert system for fundus fluorescein angiography. *Health Informatics Journal, 7*(3-4), 214–221. http://lpis.csd.auth.gr/publications/shimr2001.pdf doi:10.1177/146045820100700317

Direction *Directions of development of e-services in health protection in West Pomeranian Voivodeship for the years 2011-2020.* (2011). Szczecin. Retrieved from www.wz.wzp.pl/download/index/biblioteka/7620

Directions of informatisation "e-Health Poland" for the years 2011-2015. (2009). Retrieved from www.csioz.gov.pl/publikacja.php

Dooktor.pl. (n.d.). *Dooktor.pl.* Retrieved from http://www.dooktor.pl/diagnoza.html

Doz.pl. (n.d.). *Dbam o zdrowie.* Retrieved from http://www.doz.pl

Friedberg, M. W., Chen, P. G., Van Busum, K. R., Aunon, F. M., Pham, Ch., Caloyeras, J. P., et al. (2013). *Factors Affecting Physician Professional Satisfaction and Their Implications for Patient Care, Health Systems, and Health Policy*, The RAND Corporation. Retrieved from http://www.rand.org/content/dam/rand/pubs/research_reports/RR400/RR439/RAND_RR439.pdf

Furmankiewicz, M., Sołtysik-Piorunkiewicz, A., & Ziuziański, P. (2014). Artificial Intelligence Systems for Knowledge Management in e-Health: The Study of Intelligent Software Agents. In *Proceedings of the 18th International Conference on Systems* (part of CSCC '14), Santorini, pp. 551-556.

Furmankiewicz, M., & Ziuziański, P. (2013). Ocena wykorzystania technologii teleinformatycznych w procesie autodiagnozy i samoleczenia w świetle badania opinii internautów. H. Sroki & T. Porębska-Miąc (Eds.) Systemy Wspomagania Organizacji SWO 2013, Katowice: Wydawnictwo Uniwersytetu Ekonomicznego w Katowicach.

Goldzweig, C. L., Towfigh, A., Maglione, M., & Shekelle, P. G. (2009). Costs and benefits of health information technology: new trends from the literature. Retrieved from http://www.ncbi.nlm.nih.gov/pubmed/19174390

Han, H., & Lopp, L. (2013). Writing and reading in the electronic health record: an entirely new world. *Medical Education Online*. Retrieved from http://www.ncbi.nlm.nih.gov/pmc/ articles /PMC3566375/

Kaplan, B., & Harris-Salamone, K. (2009). Health IT Success and Failure: Recommendations from Literature and an AMIA Workshop. Retrieved from http://www.ncbi.nlm.nih.gov/pmc/articles/PMC2732244/

Karnowka, A., Shafer, K., & Sularz, B. (n.d.). *System ekspertowy Mycin*. Retrieved from zsi.tech.us.edu.pl/~anowak/index.php?s=file_download&id=21

Kawiorska, D. (2004). *Narodowe Rachunki Zdrowia*. Kraków: Kantor Wydawniczy ZAKAMYCZE.

Korczak, K. (2014). *Internetowe narzędzia wspomagające opiekę zdrowotną*. Warszawa: Wolters Kluwer.

Kralewski, J. E., Dowd, B. E., Cole-Adeniyi, T., Gans, D., Malakar, L., & Elson, B. (2008). Factors influencing physician use of clinical electronic information technologies after adoption by their medical group practices. *Health Care Manage Rev*. Retrieved from http://www.ncbi.nlm.nih.gov/pubmed/18815501

Kunz, J. C., Fallat, R. J., McClung, D. H., Osborne, J. J., Votteri, R. A., & Nii, H. P. … Feigenbaum, E.A. (1978). A Physiological Rule-based System for Interpreting Pulmonary Function Test Results, Stanford University.

Laudon, K., & Laudon, J. (2011). Essentials of management information systems [9th Edition]. New Jearey: Pearson.

Leku.pl. (n. d.). Retrieved from http://www.leku.pl

Menachemi, N., & Brooks, R. G. (2006). EHR and other IT adoption among physicians: results of a largescale statewide analysis. *J Healthc Inf Manag*. Retrieved from http://www.ncbi.nlm.nih.gov/pubmed/16903665

Menachemi, N., Perkins, R. M., van Durme, D. J., & Brooks, R. G. (2006). Examining the adoption of electronic health records and personal digital assistants by family physicians in Florida. *Inform Prim Care*. Retrieved from http://www.ncbi.nlm.nih.gov/pubmed/16848961

Ministerstwo Administracji i Cyfryzacji. (2012). *Do lekarza z PESELem, czyli e-WUŚ*. Retrieved from http://mac.gov.pl/dzialania/e-administracja/do-lekarza-z-peselem/

Ministerstwo Spraw Wewnętrznych. (2013). *Nowa odsłona projektu pl.ID*. Retrieved from https://msw.gov.pl/pl/aktualnosci/11600,Nowaodslona-projektu-plID.html

Mulawka, J. J. (1996). *Systemy ekspertowe*. Warszawa: Wydawnictwo Naukowo-Techniczne.

Narodowy Fundusz Zdrowia. (2013). Karta Ubezpieczenia Zdrowotnego. Retrieved from http://www.nfzkatowice.

National Library of Medicine. (n.d.). Medline Plus. Retrieved from http://www.nlm.nih.gov/medlineplus

NHS.uk. (n.d.). The NHS in the 1990s. Retrieved from http://www.nhs.uk/NHSEngland/thenhs/nhshistory/Pages/NHShistory1990s.aspx

Opoka, J. (2008). Multimedia w edukacji. *Studia i Materiały Centrum Edukacji Przyrodniczo-Leśnej*, 10, 1[17], p. 120.

Pankowska, M. (2004). Value-driven management in e-healthcare. *Studies in Health Technology and Informatics*, *105*, 3–11. PMID:15718588

Pollard, S. E., Neri, P. M., Wilcox, A. R., Volk, L. A., Williams, D. H., & Schiff, G. D. … Bates, D.W. (2013). How physicians document outpatient visit notes in an electronic health record. Retrieved from http://www.ncbi.nlm.nih.gov/pubmed/22542717

Poon, E. G., Wright, A., Simon, S. R., Jenter, C. A., Kaushal, R., & Volk, L. A. … Bates, D.W. (2010). Relationship between use of electronic health record features and health care quality: results of a statewide survey. *Med Care.* Retrieved from http://www.ncbi.nlm.nih.gov/pubmed/20125047

Pople, H. E. Jr., Myers, R.D., Miller, R.A. (1975). DIALOG: A Model of Diagnostic Logic for Internal Medicine. *International Joint Conferences on Artificial Intelligence* 4.

Poznaj system e-Recepta (2012). *Poznaj system e-recepta.* Retrieved from http://e-recepta.gov.pl/web/sim/poznaj-system-e-recepta

Przelaskowski, A. (2011). *Techniki multimedialne.* Warszawa: PW OKNO (pp. iii, 3).

Qzdrowiu.pl. (n. d.). Retrieved from http://www.qzdrowiu.pl

Reply of the Secretary. (2013). Reply of the Secretary of State Zbigniew Podraza in Health Ministry – on behalf of the Minister – to the question no 9353 concerning resigning from introducing a medical services register 02.03.2005. Retrieved from http://orka2.sejm.gov.pl/IZ4.nsf/main/7719F1C7

Rouf, E., Chumley, H. S., & Dobbie, A. E. (2008). Electronic health records in outpatient clinics: perspectives of third year medical students, *BMC Med Educ.* Retrieved from http://www.ncbi.nlm.nih.gov/pubmed/18373880

Rozpędowska-Matraszek, D. (2012). e-Zdrowie w kształtowaniu zatrudnienia w opiece zdrowotnej. In Zieliński Z. (Ed.), Rola informatyki w naukach ekonomicznych i społecznych. Innowacje i implikacje interdyscyplinarne, Tom 1/2012, Kielce: Wydawnictwo Wyższej Szkoły Handlowej w Kielcach.

Rynek i Opinia, A. R. C. (2012a). *Internet leczy skutecznie.* Retrieved from http://www.arc.com.pl/internet_leczy_skutecznie-40999450-pl.html

Rynek i Opinia, A. R. C. (2012b). *Lekarz online poszukiwany.* Retrieved from http://www.arc.com.pl/lekarz_online_poszukiwany-41999453-pl.html

Shekelle, P. G., Morton, S. C., & Keeler, E. B. (2006). Costs and benefits of health information technology. [Full Rep]. *Evid Rep Technol Assess*, (132): 1–71. Retrieved from http://www.ncbi.nlm.nih.gov/pubmed/17627328 PMID:17627328

Sidorko, A. (2012). *Druk RMUA co miesiąc tylko na wyraźne żądanie pracownika.* Retrieved from http://serwisy.gazetaprawna.pl/praca-i-kariera/artykuly/613096,druk_rmua_co_miesiac_tylko_na_wyrazne_zadanie_pracownika.html

Sittig, D. F., & Singh, H. (2012). *Electronic Health Records and National Patient-Safety Goals.* Retrieved from http://www.ncbi.nlm.nih.gov/pmc/articles/PMC3690003/#!po=50.0000

Sittig, D. F., Wright, A., Meltzer, S., Simonaitis, L., Evans, R. S., & Nichol, W. P. … Middleton B. (2011). Comparison of clinical knowledge management capabilities of commercially-available and leading internally-developed electronic health records. *BMC Med Inform Decis Mak*. Retrieved from http://www.ncbi.nlm.nih.gov/pubmed/21329520

Sołtysik-Piorunkiewicz, A. (2009). Kontroling w organizacji i zarządzaniu. Koncepcja informatyzacji. Sosnowiec: Oficyna Wydawnicza „HUMANITAS", 2009, p. 177.

Sołtysik-Piorunkiewicz, A. (2012). Implementation of IT system in health protection. In R. Knosala (Ed.), *Innovations in management and production engineering* (pp. 568–576). Opole: Oficyna Wydawnicza Polskiego Towarzystwa Zarządzania Produkcją.

Sołtysik-Piorunkiewicz, A. (2014a, June 24). The Telecom Business Strategies: a Comparative Study of Corporate Blogs. *Proceedings of the 2014 Multimedia, Interaction, Design and Innovation International Conference on Multimedia, Interaction, Design and Innovation* (pp. 1-11). ACM. 10.1145/2643572.2643592

Sołtysik-Piorunkiewicz, A. (2014b). Technologie mobilne w zarządzaniu organizacją opartą na wiedzy. In R. Knosala (Ed.), *Komputerowo Zintegrowane Zarządzanie. Część VIII. Zarządzanie wiedzą, transfer wiedzy, systemy wspomagania podejmowania decyzji* (pp. 263–274). Opole: Oficyna Wydawnicza Polskiego Towarzystwa Zarządzania Produkcją.

Sołtysik-Piorunkiewicz, A. (2014c). The management of patient information in Polish health care system. International Journal Information Theories and Applications, 21(2).

Sołtysik-Piorunkiewicz, A. (2015a). Knowledge management impact of information technology Web 2.0/3.0. The case study of agent software technology usability in knowledge management system. *AIP Conference Proceedings*, *1644*(1), 219–227. doi:10.1063/1.4907840

Sołtysik-Piorunkiewicz, A. (2015b): The evaluation method of Web 2.0/3.0 usability in e-health knowledge management system. *Online Journal of Applied Knowledge Management*, 3(2), 168-180.

System Zdrowotny Informator Pacjenta. (2013). System: Zdrowotny Informator Pacjenta. Retrieved from https://zip.nfz.poznan.pl/ap-zip-user/

Thatcher, M. (2013). IT Governance in Acute Healthcare: A Critical Review of Current Literature. In C. George, D. Whitehouse, P. Duquenoy (Eds.), *eHealth: Legal, Ethical and Governance Challenges* (pp. 349-370). Retrieved from http://link.springer.com/chapter/10.1007/978-3-642-22474-4_15

Ujejski, M. (2011). Funkcje karty ubezpieczenia zdrowotnego realizowane w elektronicznym dowodzie osobistym. Proceedings of *Materiały konferencyjne "IT w służbie zdrowia GigaCon"*, Warszawa. Retrieved from http://sdcenter.pl/system/lectures/attachment1s/1729/original/Funkcje_KUZ_w_pl.ID_23.03.2011_final_pdf?1301043192

Venkatesh, V., Morris, M. G., Davis, G. B., & Davis, F. D. (2003). User acceptance of information technology: Toward a unified view. *Management Information Systems Quarterly*, *27*, 425–478.

Weiss, S. M., Kulikowski, C. A., & Safir, A. (1978). A Model-Based Consultation System the Long-term Management Glaucoma. *International Joint Conferences on Artificial Intelligence*, 5.

Zdrowie publiczne (n. d.). Retrieved from http://www.ec.europa.eu/health-eu/care_for_me/e-health/index_pl.htm

Zintegrowany Informator Pacjenta. (n.d.). *Zintegrowany Informator Pacjenta.* Retrieved from https://zip.nfz.gov.pl

Ziuziański, P., Furmankiewicz, M., & Sołtysik-Piorunkiewicz, A. (2014). E-health artificial intelligence system implementation: Case study of knowledge management dashboard of epidemiological data in Poland. *International Journal of Biology and Biomedical Engineering*, *8*, 164–171.

KEY TERMS AND DEFINITIONS

Decision Support Systems (DSS): Information system supporting activities connected with decision-making process.

E-Health: An emerging area in the intersection of public health, medical informatics and business.

Electronic Health Record (EHR): Also named electronic medical record (EMR). A systematic set of electronic health information about population or individual patient.

Expert System: A computer program which perform tasks in many categories using knowledge base.

Information Society: A society where information is a significant activity equal or even more valuable than material goods.

Management of Patient Information System: It is the information management system dedicated for patient and health protection, generally brings satisfaction and their implications for patient care, health systems, and health policy. IT systems in health care is implemented for management of patient information and gives many benefits; the using of information technology can bring for health care more quality and savings, implementing of even smaller-scale systems can minimize complexity and difficulties of health care processes.

Medical Portal: Specially-designed Web page connected with medicine and/or health which is aimed to distribute pieces of information from many sources in homogeneous way.

Online Consultation with Doctors: The form of consultation when patient and doctor are using computer and Internet to communicate using text-messaging, sound and/or view.

Self-Diagnosis or Self-Treatment Tool: A webpage, forum, computer/mobile application (like expert system), etc. using by patients to auto diagnosing or auto heal themselves.

This research was previously published in Experimental Multimedia Systems for Interactivity and Strategic Innovation edited by Ioannis Deliyannis, Petros Kostagiolas, and Christina Banou, pages 258-283, copyright year 2016 by Information Science Reference (an imprint of IGI Global).

Chapter 19
A Survey on Prediction Using Big Data Analytics

M. Supriya
Anna University, India

A.J. Deepa
Ponjesly College of Engineering, India

ABSTRACT

This article describes how nowadays, the growth of big data in bio-medical and healthcare community services is increasing rapidly. The early detection of diseases and patient care are analyzed with the help of accurate analysis of medical data includes diagnosed patients' details. The analysis of accuracy rate is considerably reduced when the quality of medical data is unclear since every part of the body has unique characteristics of certain regional diseases that may suppress the prediction of diseases. This article reviews the detailed survey of different prediction methods developed for analyzing the accuracy rate of disease affected patients in 2015-2016 mainly focuses on choosing the efficient predictions based on the quality of medical data not only provides the overall view of prediction methods but also gives the idea of big data analytics in medical data further discusses the methods, techniques used and the pros and cons of prediction methods.

1. INTRODUCTION

Nowadays 50% of Americans have one or more chronic diseases, and 80% of American medical care fee is spent on chronic disease treatment. With the improvement of living standards, the incidence of chronic disease is increasing. The United States has spent an average of 2.7 trillion USD annually on chronic disease treatment. This amount comprises 18% of the entire annual GDP of the United States. The healthcare problem of chronic diseases is also very important in many other countries

In China, chronic diseases are the main cause of death, according to a Chinese report on nutrition and chronic diseases in 2015, 86.6% of deaths are caused by chronic diseases. Therefore, it is essential to perform risk assessments for chronic diseases. Electronic health records (EHRs) are digital versions of a patient's medical history, maintained over time by health care providers that contain information

DOI: 10.4018/978-1-5225-6198-9.ch019

relevant to a patient's care, including to demographics, diagnoses, medical procedures, medications, vital signs, immunizations, laboratory results, and radiology images. With the growth in medical data collecting electronic health records is more convenient. One of the applications is to identify high-risk patients which can be utilized to reduce medical cost since high-risk patients often require expensive healthcare (Chen Hao, Hwang, & Wang, 2016).

Convolutional Neural Network-based Multimodal Disease Risk Prediction (CNN-MDRP) algorithm is applicable for structured and unstructured data. The disease risk model is obtained by the combination of structured and unstructured features and the accuracy is analysed.

2. RELATED WORKS

2.1. Big Data Analytics

The massive set of data is generated from different organizations throughout the world this huge and heterogeneous data is called Big Data. Big data plays an important role in achieving predictive analysis in the healthcare domain. The transformation of using sophisticated technologies by healthcare provides to gain insights from clinical datasets and make informed decisions had changed by big data analytics with the help of Hadoop framework. Effective healthcare management can be achieved by providing effective data driven services to people by predicting their needs (Wang & Hajli, 2016). Big data analytics in healthcare is defined as the ability to acquire, store, process and analyze large volume of health data in various forms, and deliver meaningful information to users, which allow them to discover business values and insights in a timely manner. The theoretical foundation of Big data analytics-enabled business value (BDA-BV) model comprises of two elements resource based theory (RBT) and capability building view. RBT has been the principal theoretical foundation for explaining how resources can be transformed into a sustained competitive capability building view has been utilized to complement the pitfalls of RBT.

The real-world implementation of big data analytics in healthcare big data analytics capabilities are derived from the various tools and functionalities are mainly triggered by a data processing component. Data aggregation: The tools use to transform different types of healthcare data into a data format that can be read by the data analysis platform. Data processing: The tool use to process all kinds of data and perform appropriate analysis to harvest insight and decision. Data visualization: The tool used to produce reports about daily healthcare services to aid managers decisions and actions.

2.1.1. Big Data and Smart Healthcare Systems

Big data and smart healthcare systems (Pramanik, Lau, & Demirkan, 2005) are independently attracting extensive attention from both academic and industry the combination of both big data and smart systems can expedite the prospects of the healthcare industry. A big data enabled smart healthcare system framework (BSHSF) that offers theoretical representations of an intra and inter organized business model in the healthcare context. The technology and infrastructure of smart cities the reconstruct thinking behind existing healthcare systems (eg: mhealth, ehealth) and telemedicine to create a new and comfortable ubiquitous concept which is called as smart health. Also, smart health integrates idea from ubiquitous computing and ambient intelligence applied to predictive personalized preventive and participatory healthcare systems. There are three technical (3T) areas-Intelligent Agents, Machine Learning and Text Mining-all of which are contributing to the promotion of healthcare technologies (Ahmed, Yaqoob, Khan, Imran, & Vasilakos, 2016). The management of big data in a continuously expanding network gives rise to non-trivial concerns regarding data collection efficiency, data processing, analytics and security. To address these concerns researchers have examined the challenges associated with the successful deployment of IOT (internet of things). IOT is one of the biggest sources of big data which are rendered useless without analytics power. IOT interacts with big data when voluminous amounts of data are needed to be processed transformed and analysed in high frequency. The big data processing and analytics solutions for IOT identify the numerous requirements for big data and analytics in IOT applications. The three main types of analytics are descriptive analytics, predictive analytics and perspective analytics. One of the IOT applications is smart healthcare which is used to predict disease, help insurance companies make better policies and pick up the warning signs of any serious illnesses during their early stages.

3. EMOTION AWARE IN HEALTHCARE SERVICES

In system design, for Big Data Application in Emotion-Aware Healthcare (BDAEH), it improves the logic reasoning and the emotional computing for resource utilization. To improve the network performance the SDN (Software Defined Networks) and 5G technologies are used. The BDAHE functions such as: healthcare data collection, healthcare data transmission, healthcare data storage, healthcare data analysis and human machine interaction. The BDAEH system returns the analysis result to the users for further treatment. The emotion factor is considered here to improve the performance of healthcare services (see Figure 1).

Figure 1. Functions of the BDAEH system

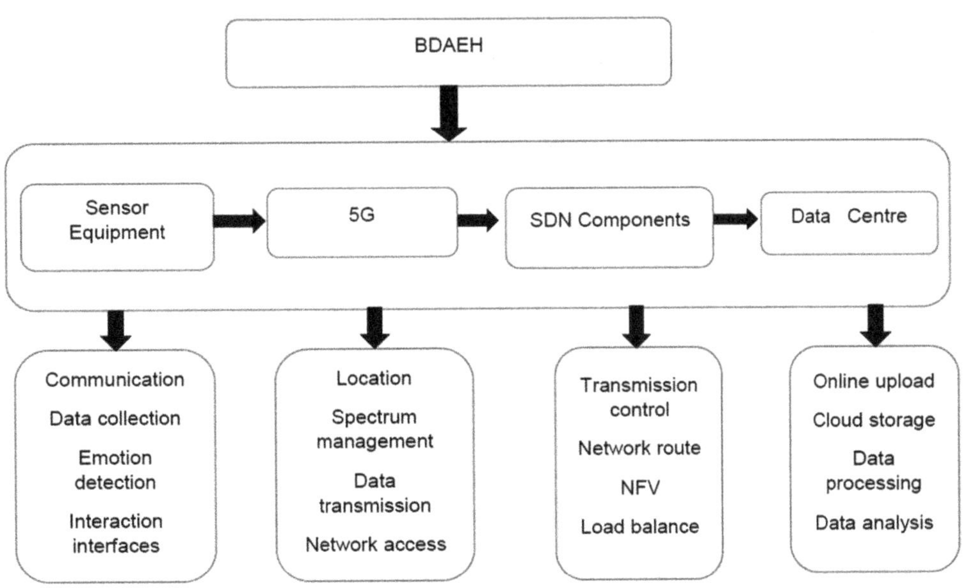

The function of BDAEH system processing the flow of healthcare data is integrating emotion computing and emerging communication technologies into the processing procedure of healthcare data, which improves the efficiency of healthcare data collecting, transmitting, storage. The data centre provides the storage, integration and analysis functions of healthcare data (Lin, Xia, Wang, Tian, & Song, 2016). Through analysing and mining the massive data, the system not only gets the users' emotional state and physical condition real-timely but also explores all kinds of relations between them hidden in the healthcare data. To improve the performance of healthcare data transmission, the SDN (Software Defined Networks) is utilized in the BDAEH system. Meanwhile, the SDN provides transmission control, network route, load balance of the system can effectively access network anytime and anywhere and transmit the healthcare data received from the lowest layer (sensing equipment). Utilizing the powerful data processing ability and valid emotion aware, the BDAEH system helps users to know their own physical conditions, provides personalized improvement suggestion for them to keep a stable physical condition, and sends the physical information to the families Translations and content mining are permit to doctors. Doctor are another type user of the system, can timely acquire the patient's condition and gain convenience for formatting reasonable treatment scheme so that patients can enjoy more effective treatments. Big healthcare data are produced when the number of users is huge and the system operation time is long. These data can be validly utilized to predict the state of human physical health even across generations, which have an active impact on medicine area. The BDAEH system is also expected to increase patient participation and improve individual health quality.

4. LARGE SCALE ELECTRONIC HEALTH RECORDS DATABASE

By an overview of two hybrid intelligent techniques that might be utilized us a new generation of predictive analytics in big data and decision making processes in social informatics and e-health systems (Wu et al., 2016). E-health is the use of information technology such as digital media the internet, robotics and virtual reality for the purpose of preventing, promoting, treatment, maintenance and improving health care services. Inpatients diseases and procedures from large hospitalization records database building in neural embedding language models (Stojanovic et al., 2016). The inpatient mortality for certain procedures and medical conditions such as length of stay and total charges of an inpatient stay and can be considered as important metrics for evaluating quality of care by using regression methods goodness-of-fit metric R^2 is defined.

$$R^2 = 1 - \frac{\Sigma_i (y_i - \widehat{y_i})^2}{\Sigma_i (y_i - \mu)^2} \tag{1}$$

Where y_i . and $yi\hat{} are true predicted values \; for record\, ri$. $\mu is the mean value for all records in the set R$.

4.1. Accuracy Measure

$$accuracy = \frac{t_p + t_n}{t_p + f_p + t_n + f_n} \tag{2}$$

t_p and t_n denotes true positives and true negatives
f_p and f_n denotes false positive and false negative

Electronic Health Records (EHR)provide a rich source of temporal data that present a unique opportunity to characterize disease patterns and risk of imminent disease. While many data mining tools have been adopted for EHR-based disease early detection, Linear Discriminant Analysis (LDA)is one of the most commonly used statistical methods, but it is difficult to train an accurate LDA model for early disease diagnosis (Xiong, Zhang, Huang, Leach, & Barnes, 2017). EHRs are digital versions of a patient's medical history, maintained over time by health care providers, that contain information relevant to patient care, including to demographics, diagnoses, medical procedures, medications, vital signs, immunizations, laboratory results, and radiology images. To train accurate predictive model many machine learning methods such as Support Vector Machine (SVM)random forest, Bayesian network, Gaussian process the prediction accuracy is 7.5%-43.5%.

$$Accuracy = \frac{TP + TN}{TP + TN + FP + FN} \tag{3}$$

$$FI - Score = \frac{2 * TP}{2 * TP + FP + FN} \tag{4}$$

Due to the advance in wearable and wireless sensor technology it is possible to monitor multiple vital signs (eg: heart rate, blood pressure) of a patient anytime, anywhere, vital signs (Forkan, Khalil, & Atiquzzaman, 2017) are an essential part of daily monitoring and disease prevention. When multiple Vital Sign data from many patients are accumulated for a long period they evolve into Big Data the visiBiD (Vital Sign Big Data) can accurately identify dangerous clinical events of a home monitoring patient the best accuracy (95.85%) was obtained through a random forest (Jayanthi, VijayaBabu, & Sambasiva, 2017). EHR (Electronic Health Records) data management system is to provide insights and predict outcomes from past patient data. An EHR data management system is used to process massive amounts of healthcare data, patient data can be uploaded to hive from a variety of sources like flat files, web pages real time applications and databases. The data warehouse is used for transaction processing and analytics hive is used only for analytics. And it generates graphs and charts which are useful for doctors and research to build on top of hadoop data platform and is similar to a distributed file system. The data in hive are queried using hiveQL is submitted via the command line interface hive stores data in a distributed system easy to retrieve with simple SQL queries the high-risk patients are identified and improve the patient lifestyle to avoid risk.

4.2. Protecting Patient Data

To obtain the wealth of health information (Kanti, 2017) integrating sharing and availing data are the essential tasks that ultimately demand the concept of distributed system. The privacy and security of data needs to be considered when data accessed from various location in the distributed system. All electronic medical records HER or EMR should be guarded through owner-ship controlled encryption enabling secure storage transmission. Access to the creation and maintenance of HER should preserve not only content authenticity but also data integrity and customizable patient privacy throughout HER integration process. The provision of high-quality but low cost medical services for the patients through interaction among the practitioners across the country and wide spread use of big data analytics across the healthcare organization and the healthcare industry search engines. Also, the social networks help to gather people reactions and monitor the conditions of epidemic diseases. Big data analytics in healthcare is evolving into a promising field for providing insight from very large data set and improving outcomes while reducing costs. More importantly, the big concern in context of big data is privacy and security (Cano, Tenyi, Vela, Miralles, & Roca, n.d.). Re-use of various sources of health information for purposes in addition to the direct clinical care of specific patients or the direct investigation of specific biomedical research hypotheses. Current application includes epidemiological and pharmaco-vigilance studies facilitating recruitment to randomized controlled trials, carrying out audits and benchmarking studies, financial and service planning. The health apps should support evidence-based health services and that are not purely designed in the interest of market entities, and by establishing reference policies and platforms to support communication between health apps and EHRs.

5. PREDICT THE FUTURE HEALTH CONDITION OF THE PATIENTS

A cloud enabled big data analytics platform is the best way to analyze the structures and unstructured data generated from healthcare management systems (Sahoo, Mohapatra, & Lin Wu, 2016). A probabilistic data collection mechanism is designed and correlation analysis of these collected data is performed. A stochastic prediction model is designed to foresee the future health condition of the most correlated patients based on their current health status with the accuracy of 98% and maintains 90% of CPU and bandwidth utilization to reduce the analysis time. The correlation factor in the analysis is given by,

$$\Gamma ek_{ij}\left(t\right) = \begin{cases} +1 & if\, positive\, coorelation \\ 0 & if\, no\, coorelation \end{cases}$$

5.1. Predicting Septic Shock in Any ICU Patient

Sepsis and septic shock (Ho, Cheng, & Ghosh, 2014) are common and potentially fatal conditions that often occur in intensive care unit (ICU) patients. Sepsis is a systemic response to infection that is a common and life threatening in hospital complication causing more deaths than prostate cancer breast cancer and HIV/AIDS. Sepsis is one of the leading causes of mortality in intensive care unit (ICU) patients. To build a predictive model for septic shock onset generalizable to larger groups of ICU patients make use of the MIMIC-II database (Saeed et al. 2011). Septic shock is sepsis-induced hypotension persisting despite adequate fluid resuscitation along with the presence of hypo perfusion abnormalities or organ dysfunction (Bone at al.1992). A predictive model using a recursive partitioning and regression tree (RPART) to identify early predictions from clinical data 1,864 hospitalized non-ICU septic patients to predict septic shock from noisy gathered clinical data (Jayanthi et al., 2017). Hybrid Prediction modelling uses text mining for unstructured data. It uses historical and present data to predict future regarding activity behaviour and trends, predictive analytics does not expect anything about data but it allows the data led the way. It uses statics, machine learning, neural computing, robotics, computational mathematics and artificial intelligence to explore all data and find meaningful relationships and patterns. Taxonomy of predictive analytics contains two types: supervised learning and unsupervised learning. Supervised learning is a process of creating predictive models using a set of historical data and produce predictive results. Unsupervised learning does not use the previously known result to train its models, it uses descriptive statics (see Table 1).

5.2. Multi-Outcomes Prediction in Healthcare

Health analysis (Saha Gupta Phung & Venkateshs, 2016) often involves prediction of multiple outcomes of mixed-types, the framework is formulated as iterative optimization problems and solved -using an efficient Block Coordinate Descent method (BCD). This method is based on modelling multivariate outcomes consisting of a mixture of categorical and continuous outcomes for predicting multiple outcomes framework can simultaneously model several mixed-types outcomes eg., future emergency presentations(binary), emergency admission(count), emergency length-of-stay-days(non-negative) and emergency time-to-next-admission-day(non-negative).One consideration is here that the EHR data is static where patient examples have similar distributions and outcome entries are also complete.

5.3. Bone Fracture Prediction

The EHR data used to develop an effective disease risk management model that not only models the progression of the disease but also predicts the risk of the disease for early disease control or prevention (Li, Li, Ramanathan, & Zhang, 2013). EHR data use integrated features to effectively predict osteoporosis and bone fractures a multi-tasking framework for osteoporosis the not only extracts the integrated features for progressive bone loss and bone fracture prediction but also selects the individual informative Risk Factors(RFs) that are valuable for both patients and medical researchers these integrated features constructed from original RFs will become the most effective features for bone disease prediction. In addition, Bone disease memory (BDM) memorizes the characteristics of those individuals who suffer from bone diseases. The goal of BDM is to monitor those RFS which cause people to get osteoporosis.

5.4. Predicting Asthma

Asthma is one of the most prevalent and costly chronic conditions (Ram, Zhang, & Williams, & Pengetnze, 2015) in the United States which cannot be cured. Current national asthma disease surveillance system can have data availability lags up to two weeks rapid progress has been made in gathering non-traditional, digital information to perform disease surveillance. Multiple data sources are used to predict the number of asthma related Emergency Department(ED) visits in a specific area. Twitter data, Google

Table 1. Predictive analytics in health care

Model	Data size	Sources	Quality
Old	Limited data	Claims data, patient data	Poor, Unstructured data cannot be accessed
Modern	Large data	EMR+claims+Inpatient+outpatient+ED	Excellent, Unstructured data can be accessed

search interests and environmental sensor data were collected to predict the number of asthma ED visits based on near real time environmental and social media data with approximately 70% precision. "Twitter/Environment data model" predicts the risk for a daily number of ED visits being High, Low or Medium called as population-level risk model. This model is used for better planning and proactive treatment in specific geo-locations at specific time periods.

6. PREVENTIVE AND MAINTENANCE MECHANISMS

A manufacturing big data solution is used for active preventive maintenance in manufacturing environments (Wan, 2016). The analyse method used for collection of manufacturing big data according to the data characteristics and perform data processing in the cloud, including the cloud layer architecture the real time active maintenance mechanism and the off-line prediction. In general manufacturing big data can be divided into three types: device data, product data and command data. Device data can be collected in real-time including device alarm, device logs and devices status, in order to evaluate the health condition of manufacturing equipment. The alarm signals and key state information are encapsulated as the items requiring maintenance the mechanism of off-line analysis and prediction based on historical data stored in the distributed file system which enhance the prediction activities. To estimate MOAP (Mechanism of Offline-line Analysis and Prediction) the remaining effective working time T^k is given by

$$T^k = \left(1 - \eta^k\right).(\ni .T^{-k} + \left(1 - \ni\right).\frac{1}{m}\sum_{i=1}^{m} T^K{}_i e \quad (5)$$

7. DATA ANALYTICS IN MACHINE LEARNING

Prediabetes is a major epidemic and is associated with adverse cardio-cerebrovascular outcomes. The prediction using several machine learning(ML) methods (Hu et al., 2016) of rapid progression of carotid intima-media thickness in impaired glucose(IGT) participants(382) with IGT underwent carotid intima-media thickness(CIMT) ultrasound evaluation at baseline and at 15-18 months, and were divided into rapid progressors (RP,n=39,58+-17.5m change) and non-rapid progressors (NRP,n=343,5.8+-20m change p<0.01 versus RP). ML methods were applied to a clinical trial dataset and showed good performance in identifying predicting impaired glucose tolerance participants to rapid carotid plaque progression. Naïve Bayes method showed superior performance over multilayer perceptron, random forest methods and feature selection improved predictive performance. The utility of ML methods in data analytics for clinical applications will improve the performance in prediction.

7.1. Spark Based Machine Learning

With the availability of large health care datasets, (Nair, Shetty, & Shetty, 2017) the real time remote health status prediction system built around open source Big Data processing engine, the Apache spark deployed in the cloud which focus on applying machine learning model on streaming big data. The user tweets his health attributes and the application receives the same in real time, extracts the attributes and applies machine learning model to predict users health status by using decision tree model on the data and sends a direct message to the user regarding his/her health status. Decision tree is an easy to interpret and popular machine learning method for classification to perform the prediction.

8. BIOMEDICAL BIG DATA FOR PRECISION MEDICINE

Rapid advances of high throughput technologies and wide adoption of electronic health records have led to fast aggregation of –omic and EHR data. Omic data contain a comprehensive catalog of molecular profiles (e.g. genomic, transcriptomic, epigenomic, proteomic and metabolomics) in biological samples that provide basis for precision medicine (Wu et al., 2016). The personalized, predictive, preventive and participatory medicine (a.k.a.P4 medicine) model that aims to transform current reactive care to future proactive medicine and ultimately to reduce health care expenditure band improve patients health outcomes. The new precision medicine model precisely classify patients into subgroups sharing a common biological basis of diseases for more effective treatment. Omic and EHR big data analytics is challenging due to data frequency, quality and dimensionality. And the heterogeneities of integrative analysis of multi-omic data helps in understanding of cancer mechanisms.

Users generate a high volume of biomedical data during health monitoring which can be used by the mHealth serves for training predictive models for diseases diagnosis and treatment (Gong Fang & Guo, 2015). The biomedical sensing data raise serious privacy concerns, because they reveal sensitive information such as health status and lifestyles of the sensed subjects. A scheme that keeps the training samples private while enabling accurate construction of predictive models the logistic regression models which are widely used for predicting dichotomous outcomes in healthcare and decompose the logistic regression problem into small sub problems over two types of distributed sensing data (i.e.) horizontally partitioned data and vertically partitioned data. The subproblems are solved using individuals' private data and thus mhealth users can keep their private data locally and only upload (encrypted) intermediate results to the mHealth server for model training. The logistic regression, a classic machine learning technique which is appropriate for predicting dichotomous outcomes and thus widely used for making decisions in medical diagnosis and prognosis. A privacy preserving collaborative learning scheme that utilizes continuous sensing data from multiple patients forward training logistic regression models in mHealth. The two different cases of distributed sensing data; horizontally partitioned: All patients have a database of sensing data that are sensed by the same set of sensors. Vertically partitioned data: Each patient only owns a few sensors and has a database sensed with partial sensors. It is a private scheme for learning a logistic regression model based on distributed biomedical sensing data and enables mHealth users to control their raw data and only share necessary intermediate results during the training process and protecting the private information of intermediate results during the aggregation process.

8.1. Predict and Control the Diseases

Predicting the survival of heart transplant patients is an important but challenging problem. Data mining models (Dag, 2016) can effectively analyse and extract information from large complex transplantation datasets. There is a study that used to predict 1-, 5- and 9- years patients graft survival following a heart transplant surgery via the deployment of analytical models that are based on four powerful classification algorithms(i.e.: decision trees, artificial neural networks, support vector machine and logistic regression). The logistic regression combined with synthetic minority over sampling are employed achieves the best classification for the 1-5 and 9 years outcomes prediction, with Area Under the Curve(AUC) values of 0.24, 0.676 and 0.838 respectively. By applying sensitivity analysis to the data analytical models, the most important predictors and their associated contribution for the 1-, 5-, and 9-year graft survival of heart transplant patients are identified in Table 2.

9. PREDICT AND CONTROL THE CHRONIC DISEASES

Chronic diseases (Baechle Agarwal, & Xingquanzhu, 2017) are the imperative reason for death in the world. Chronic obstructive pulmonary disease (COPD) is a chronic lung disease that affects airflow to the lungs. Discovering the co-occurrence of COPD with other diseases symptoms, and medication is invaluable to medical staff. A sample of highest scoring diseases is shown in Table 3. The baseline method shows diseases which have a high population gain (such as diabetes), to occur higher in the baseline method than COED.

The highest scoring diseases using Co-occouring Evidence Discovery(COED)are lung and respiratory diseases. The Apache Hadoop ecosystem is leveraged for COED. Hadoop Distributed File System (HDFS) is used for the storage and distribution of de identified patient discharge summaries. Apache Spark is utilized for MapReduce operations and pyspark python interface is used for programming (Table 4).

The accuracy rate of predictive models using effective mechanisms such as Naïve Bayes, Decision Tree, Support Vector machine (SVM) and Artificial Neural Network (ANN) are given in Table 5.

Table 2. Observations of survivals, failures, and censored

Time point	Survivals	Failures	Censored
1-year	12,103	1,617	1,860
5-years	5,519	3,081	6,980
9-years	1,663	3,837	10,080

Table 3. Top ten results for diseases and disorders

Rank	Baseline	COED
1	Hypertension	Respiratory failure
2	Diabetes mellitus	Hypertension
3	Coronary disease	Pneumonia
4	Heart fibrillation	Congestive heart failure
5	Arteriopathic disease	Diabetes mellitus
6	Congestive heart failure	Chronic respiratory failure
7	Pneumonia	Acute respiratory distress
8	Respiratory failure	Acute chronic respiratory failure
9	Anemia	Chronic respiratory insufficiency
10	Kidney disease	Heart fibrillation

Table 4. Platforms for Big data Analytics

Platform	Advantages	Limitations
Apache Hadoop (Map Reduce)	Horizontally scalable; fault-tolerant; free and open-source	Most effective for batch-mode processing; not appropriate for real- time, online analytics
IBM InfoSphere Platform	Includes purpose-built tools to handle streaming information; integrates with open-source tools such as Hadoop	Commercial licensing
Apache Spark Streaming	Integrates with the Hadoop stack; allows one code base for both batch-mode and online analysis	Depends on more expensive hardware with large amounts of RAM to work efficiently
Tableau, Qlik View and other visual analytics tools	Visualization of large and complex data sets	Generally incomplete solutions, requiring other tools to effectively handle data storage

Table 5. Accuracy table

SI. No	Classifiers	Accuracy%
1	Naïve Bayes	65.07
2	Support Vector Machine	87.32
3	Decision Tree	87.46
4	Artificial Neural Network	76.02

10. CONCLUSION

Recent big data platforms are supported by several techniques and analytical tools, these platforms are highly helpful to extract knowledge and information from the complex environment. In this paper, prediction methods related to medical data are deeply discussed and also explained in detail in to that the advantages and limitation of big analytics prediction methods were reviewed in detail also several methods related to big data analytics can also be reviewed and analysed in future.

REFERENCES

Ahmed, E., Yaqoob, I., Hashem, I. A. T., Khan, I., Ahmed, A. I. A., Imran, M., & Vasilakos, A. V. (2016). The Role of Big Data Analytics in Internet of Things.

Baechle, C., & Agarwal, A. (2017). Big data driven co-occurring evidence discovery in chronic obstructive pulmonary disease patients. Journal of Big Data, 4(1).

Cano, I., Teny, A., Vela, E., Miralles, F., & Roca, J. (2017). Perspectives on Big Data applications of health information.

Chen, M., Hao, Y., Hwang, K., Wang, L., & Wang, L. (2016). *Disease Prediction by Machine Learning Over Big Data from Healthcare Communities*.

Dag, A., Oztekin, A., Yucel, A., Bulur, S., & Megahed, F. M. (2017). Predicting Heart Transplantation Outcomes Through Data Analytics.

Forkan, A. R. M., Khalil, I., & Atiquzzaman, M. (2017). *ViSiBiD: A Learning Model for Early Discovery and Real-Time Prediction of Severe clinical Events Using Vital Signs as Big Data.*

Gong, Y., Fang, Y., & Guo, Y. (2015). *Private Data Analytics on Biomedical Sensing Data Via Distributed Computation.*

Ho, J. C., Lee, C. H., & Ghosh, J. (2014). Septic shock prediction for patients with missing data. *ACM Transactions on Management Information Systems*, *5*(1).

Hu, X., Reaven, P. D., Saremi, A., Liu, N., Abbasi, M. A., Liu, H., ... & ACT NOW Study Investigators. (2016). Machine Learning to Predict Rapid Progression of carotid Atherosclerosis in patients with Impaired Glucose Tolerance. EURASIP Journal, (1).

Jayanthi, N., Babu, B. V., & Rao, N. S. (2017). Survey on clinical prediction models for diabetes prediction. Journal of Big Data, 4(1).

Li, H., Li, X., Ramanathan, M., & Zhang, A. (2013). Prediction and information Risk Factor Selection of Bone Diseases.

Lin, K., Xia, F., Wang, W., Tian, D., & Song, J. (2016). *System Design for Big Data Application in Emotion-Aware Healthcare.*

Nair, L. R., Shetty, S. D., & Shetty, S. D. (2017). Applying spark based machine learning model on streaming big data for health status prediction. In Computers & Electrical Engineering.

Pramanik, M. I., Lau, R. Y., Demirkan, H., & Azad, M. A. K. (2005). Smart Health: Big Data Health Paradigm within Smart Cities.

Ram, S., Zhang, W., Williams, M., & Pengetnze, Y. (2015). Predicting Asthma-Related Emergency Department Visits Using. PMID:25706935

Saha, B., Gupta, S., Phung, D., & Venkatesh, S. (2017). A Framework for Mixed-type Multi-outcome Prediction with Applications in Healthcare.

Sahoo, P. K., Mohapatra, S. K., & Lin Wu, S. (2016). *Analyzing Healthcare Big Data with Prediction for Future Health Condition.*

Sarkar, B. K. (2017). Big data for secure healthcare system: A conceptual design. *Complex & Intelligent Systems*, *3*(2), 133–151.

Song, H., Basanta-Val, P., Steed, A., & Jo, M. (2017). Next-Generation Big Data Analytics: State of the Art, Challenges, and Future Research Topics.

Stojanovic, J., Gligorijevic, D., Radosavljevic, V., Djuric, N., Grbovic, M., & Obradovic, Z. (2016). Modeling Healthcare quality Via Compact Representations of Electronic health Records.

Wan, J., Tang, S., Li, D., Wang, S., Liu, C., Abbas, H., & Vasilakos, A. V. (2017). A manufacturing big data solution for active preventive maintenance.

Wang, Y., & Hajli, N. (2017). Exploring the path to big data analytics success in healthcare. *Journal of Business Research*, *70*, 287–299.

Wang, Y., Kung, L., Wang, W. Y. C., & Cegielski, C. G. (in press). *An integrated big data analytics-enabled transformation model: Application to health care*.

Wu, P. Y., Cheng, C. W., Kaddi, C. D., Venugopalan, J., Hoffman, R., & Wang, M. D. (2017). Omic and Electronic Health Record Big Data Analytics for Precision Medicine.

Xiong, H., Zhang, J., Huang, Y., Leach, K., & Barnes, L. E. (2017). Daehr: A Discriminant Analysis Framework for Electronic Health Record Data and an Application to Early Detection of Mental Health Disorders. *ACM Transactions on Intelligent Systems and Technology*, *8*(3), 47.

This research was previously published in the International Journal of Big Data and Analytics in Healthcare (IJBDAH), 2(1); edited by Mu-Yen Chen, pages 1-15, copyright year 2017 by IGI Publishing (an imprint of IGI Global).

Chapter 20
Translating Technology in Professional Practices to Optimize Infection Prevention and Control:
A Case Study Based on the TRIP-ANT Framework

Randa Attieh
CHU de Québec, Université Laval, Canada

Marie-Pierre Gagnon
Université Laval, Canada

Geneviève Roch
Université Laval, Canada

Sarah L. Krein
University of Michigan Medical School, USA

ABSTRACT

The aim of this study was to explain how the Polymerase Chain Reaction (PCR) technology was translated into professional practices to prevent and control vancomycin-resistant enterococci outbreaks via an actor-network, based on the integrated framework TRIP-ANT. A single case study was conducted in three purposefully selected sites implementing the PCR-VRE assay. The complete dataset comprised semi-structured interviews with 28 participants and a review of hospital and external documents. A content analysis was conducted. The authors' findings indicate the emergence of four main themes, including illustration of who was involved in the adoption process, attribution of roles and responsibilities, interaction/communication/ collaboration mechanisms, and changes in professional practices. Their findings also address five challenges that arose from each theme. The translation of PCR technology into professional practices relies on the enrolment of an organisational, clinical, managerial and financial support network, and on the evolution of practices, communications, and roles and responsibilities.

DOI: 10.4018/978-1-5225-6198-9.ch020

INTRODUCTION

The development of healthcare technology plays a key role in the achievement of evidence-based outcomes in patient care (Melnyk and Fineout-Overholt, 2011). The attention paid by managers to healthcare quality and efficiency highlights the need to master the adoption of new technologies, especially in infection prevention and control (IPC) in order to improve the quality of care and productivity of the healthcare system (Kyratsis, Ahmad and Holmes, 2012).

Technology adoption relates to a hospital's decision to acquire a technology and make it available to healthcare professionals for supporting or enhancing their task performance (Ghodeswar and Vaidyanathan, 2008). In recent years, several steps have been taken to support the prevention and control of nosocomial infections (NI), such as the use of rapid screening tests based on the Polymerase Chain Reaction (PCR) molecular technology to enhance vancomycin-resistant enterococci (VRE) infection prevention interventions (Versalovic and Lupski, 2002). Rapidly obtaining results compensates for the resources allocated and reduces the time involved in adopting IPC practices, making it possible to optimize the handling of VRE cases (Diekema et al., 2004). Adoption of technologies has been associated with improved organisational performance (Antoniou and Ansoff, 2004) because it affects the quality, care, costs, and competitive position of the organisation (Ghodeswar and Vaidyanathan, 2008). However, implementing a new technology requires organisational change, particularly surrounding the new IPC procedures, organisation of hospital services and even attribution of new roles within a network of actors, at the macro, meso and micro levels (Attieh, Gagnon and Krein, 2014). As with other innovation in healthcare organisations (Greenhalgh et al., 2005), understanding how IPC technology adoption can transform healthcare practices and outcomes for patients requires a theoretical basis. In previous work (Attieh, Gagnon and Krein, 2014), we discussed the introduction of a conceptual framework that could be applied to understand the dynamics involved in the adoption process of IPC technology. In an effort to identify the outcomes of such a process, we developed an integrated framework based on the Translating Infection Prevention into practice (TRIP) model, developed by Krein et al (Krein et al., 2006) that is based on Rogers' diffusion of innovation (DOI) model (Rogers, 1995), and the Actor-Network Theory (ANT) (Akrich, Callon and Latour, 2006; Callon and Latour, 1986). The TRIP-ANT framework (Figure 1) shifts the focus to the inter-relatedness of the technical and social factors in the adoption process of technologies in healthcare (Attieh, Gagnon and Krein, 2014). Thus, identifying how new technology integration can translate into different responses to change and into the practices of a network of actors involved directly and/or indirectly in the new intra- and inter-organisational processes surrounding the handling of IPC practices (Attieh, Gagnon and Krein, 2014).

The TRIP framework (Krein et al., 2006) was fine-tuned and adjusted to better understand IPC technology adoption and implementation, given the organisational and environmental context. Adapting the framework to the context of infection prevention technology and practice adoption, two phases in the process stand out: the decision-making phase and the implementation phase. Unlike the TRIP framework, which emphasises the intrinsic qualities of the technology adoption and its capacity to spread by contagion (Rogers, 1995), the ANT model (Akrich, Callon and Latour, 1988b; Callon, 1999) focuses on the capacity to unite several allies who will then depend on interactions, negotiations and adjustments of a socio-technical network hoping to advance the technology. This constitutes the concept of translation. Translation involves four processes that are intertwined and interact with each other, "problematization", "interessment", "enrolment" and "mobilization" (Callon and Latour, 1986). Looking at the first step of the translation process, problematisation calls on characteristics of the external envi-

Figure 1. Integrated TRIP-ANT model

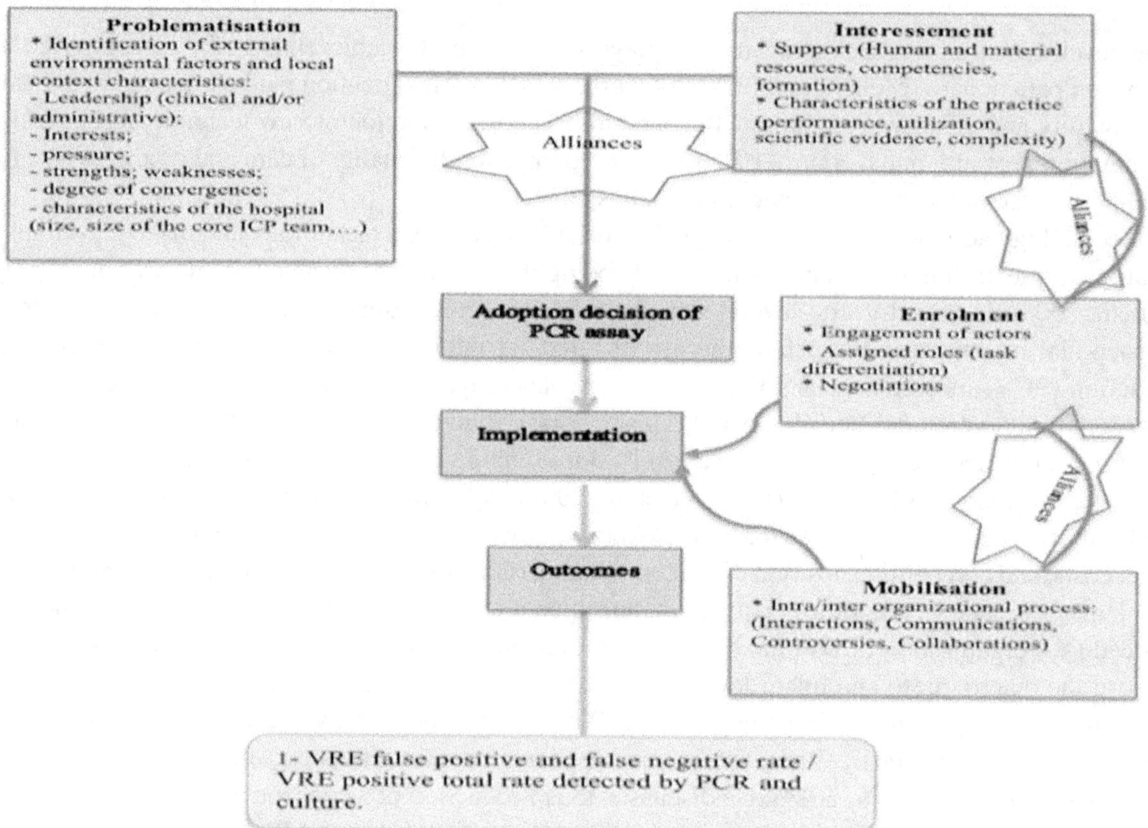

ronment and the local context into which the innovation will be translated (Attieh, Gagnon and Krein, 2014). Interessment is the phase in which the translator – who is selected based on his or her credibility and legitimacy – defines strategies to attract other actors to rally around the network in support of the shared objective (Akrich, Callon and Latour, 2002a; Akrich, Callon and Latour, 2002b). The interessment strategies draw upon both organisational characteristics and the characteristics of the practice to be adopted (Attieh, Gagnon and Krein, 2014). Enrolment is the creation of alliance networks, the aim of which is to build up agreement among the stakeholders concerning their interests. In this phase, once the decision to adopt the new technology is made, leaders or spokespersons start negotiations relating to the adjustment mechanisms for roles and responsibilities to consolidate the network around technology adoption (Akrich, Callon and Latour, 2002a; Akrich, Callon and Latour, 2002b). Mobilisation consists of garnering cooperation from interdisciplinary teams, resolving controversies and setting up interaction mechanisms to achieve effective implementation of the innovation (Akrich, Callon and Latour, 2002a; Akrich, Callon and Latour, 2002b).

Between the initial decision to adopt PCR technology to manage VRE outbreaks and achieving better performance of healthcare services, the aim of this study was to explain how the technology was translated into professional practices via an actor-network to improve the implementation of infection prevention and control interventions, based on the integrated TRIP-ANT framework. Our reserach question was:

How was the adopted PCR-VRE technology translated into IPC professional practices via a network of actors? In this regard, one hypothesis was formulated: The TRIP-ANT framework allows us to better understand the translation of technology and its impact on the professional IPC practices.

METHODS

Study Design

We used a qualitative methods design, consisting of a single case study in three purposefully selected sites implementing PCR-VRE. Yin (2009) explains that case study is needed to understand complex phenomena in real-life context. In fact, case studies are the preferred strategy when "how or why" questions are being asked. The complete dataset comprised semi-structured interviews with 28 participants and a review of hospital and external documents. A content analysis was conducted of interview transcripts and documents. The COREQ 32-item checklist (Tong, Sainsbury and Craig, 2007) for qualitative studies was followed to report the methodological standards used in this study.

Study Setting

The study was conducted in three facilities that are part of one large academic healthcare organisation in Quebec (Canada). The facilities have a bed capacity of 258 in #1, 265 in #2 and 370 in #3 providing general and specialized services for the population of Quebec City and surrounding area. They are geographically separated and each has their own nursing staff and care units. However, they do share a common administrative direction and the same IPC department affiliated to nursing directorate. The IPC team consists of nine nurses supervised by two executive assistants. In each facility, three nurses (one senior and two consultants) represent the IPC team.

Participants

The project lead (RA) and a senior nurse in IPC purposefully sampled key informants including managers (senior and middle), IPC team (senior and consultant nurses), microbiologists, laboratory technologists and nursing staff (chief and registered nurses). The selection of participants was guided by the following two criteria: 1) being involved in some aspects of the adoption decision process of the PCR technology; and 2) being witness of professional practices changes in infection prevention since the adoption of the technology. The exclusion criteria were: 1) not being involved in any aspect of the adoption decision process of the PCR technology or professional practices changes in infection prevention since the adoption of the technology; 2) not being present in one of the three facilities studied at the time of the adoption decision and/or technology implementation. Participants were initially recruited for interviews with the help of the IPC executive assistant and/or the senior nurse in IPC during site visits. During interviews, individual participants were invited to recommend further interviewees, actively seeking a range of different viewpoints. This strategy resulted in 28 participants (Figure 2). Of these, a total of 26 agreed to participate in our study. The remaining two – one microbiologist and one registered nurse (RN) – declined to participate due to lack of time. During the interviews, participants suggested two other relevant respondents i.e. a microbiologist and a RN in the emergency unit, who met the selection criteria.

Figure 2. Flow diagram

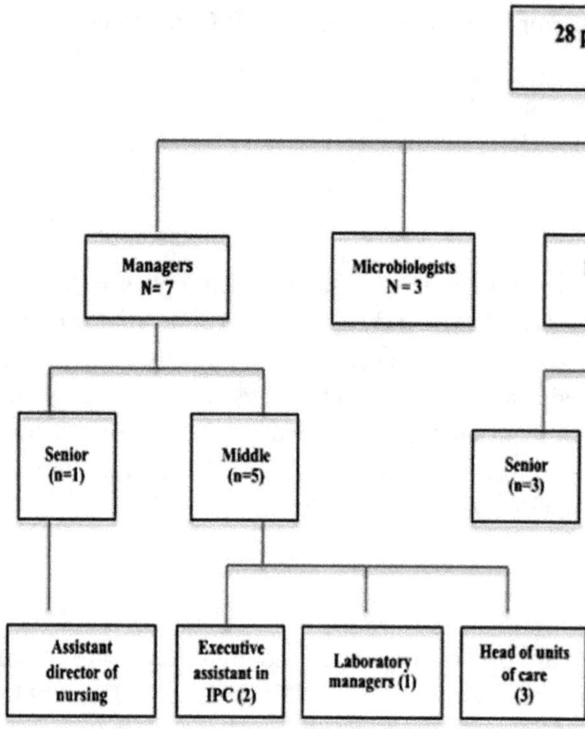

These individuals were added to the final sample of 28 participants. The site visits included informal observation and interactions with the IPC team and nursing staff in order to present the research study and to become familiar with the milieu. A written informed consent was obtained from the participants. Ethics approval was granted by the hospital center ethics committee.

Data Collection

Qualitative data were collected at the three selected sites through semi-structured individual interviews and a review of documentary sources. Individual interviews were conducted with all study participants by one trained interviewer (RA) using a semi-structured interview guide between July and November 2013. The interviewees were asked to identify who was involved in the adoption decision, how the adoption process occurred, and the changes resulting from the technology adoption. All but one interview were face-to-face. The remaining interview was conducted by telephone due to a lack of time. Interviews continued with participants until theoretical saturation was reached. The interviews lasted between 30 and 78 minutes (mean 48 min). All interviews were audio-recorded and transcribed verbatim. Verbatim quotes cited in the context of this article were translated from French to English language. For confidentiality reasons, all parties participating in the project have been anonymised.

The document analysis included reviewing hospital documents dating from 2010 to 2013, which were recommended by informants and resource people as relevant to the study. Documents available on the webpages of the hospital center being studied and the facilities websites were also consulted. In addition to these documents, other external documentary sources were consulted throughout the data collection period. Respondents were asked during interviews to suggest any other documents relevant to the study. A total of 33 documents/websites including healthcare facilities policies and protocols specific to prevention and control of VRE outbreaks, laboratory policies and manuals, meeting minutes that helped to contextualize the adoption decision-making process, provincial government reports and best practice guidelines related to VRE outbreaks were collected.

Data Analysis

The transcribed interviews and documents were coded, with the help of NVivo 10 software, using a pre-established list of codes based on the integrated TRIP-ANT framework but also allowing for the emergence of new codes during analysis. RA and a research associate independently coded all data then met to review discrepancies – overlapped and/or similar codes – through discussion or the involvement of a third reviewer (MPG) who served as an arbitrator, enhancing interrater reliability (Mays and Pope, 2000; Sofaer, 1999). The documentary review helped provide a better understanding of the technology adoption decision process. A content analysis was conducted. A total of 30 codes were identified. Data were sorted by the codes according to the types of actors involved, the characteristics of the local context, external environmental factors, organisational context, characteristics of the technology, change in practices, roles and mechanisms of interaction / communication / collaboration. The final classification aimed to link these typologies with the four phases of the translation process.

Study Validity

To ensure the validity of the study, we have followed the criteria posposed by Yin (2009), which include the following: data triangulation, use of a framework (TRIP-ANT) to guide data collection and analysis, and inter/intra coder agreement. These approaches aimed to strengthen the study's internal validity. External validity, including capacity to generalize results, was ensured by including a detailed description of the setting, the representativeness of all groups of actors, and by data saturation. This allowed us to identify the conditions required to generalize the results to other similar contexts and to maximize potential applicability and transferability of our findings to other innovation adoption in the context of IPC.

FINDINGS

Our findings highlight the emergence of four main themes, including 1) timeline of the adoption process illustrating who was involved and how the problem and its solution were defined, 2) attribution of roles and responsibilities, 3) interaction/ communication/ collaboration mechanisms, and 4) changes in professional practices. Our findings also address challenges that arose in relation to each theme. This includes the need to align and engage a wider actor-network in the adoption decision, the need to consider other forms of interessment strategies, visibility of roles and responsibilities, the improvement of interaction/ communication/ collaboration networks and the need for preparation to change at an early stage. Emerging themes, related challenges and strategies to address each challenge are outlined in Table 1.

Table 1. Emerging themes and related challenges from our study

Themes		Challenges and Strategies to Address Each Challenge	
1. Timeline of the adoption process	Quick Middle-Up decision emanating from middle managers	1.1 Need to align and engage a wider actor-network in the adoption decision	Enrolling actors from the meso (e.g. head of units and IPC senior and consultant nurses) and micro (e.g. technologists and nursing staff) levels in the adoption decision phase
	Adoption decision process relied on ten actors belonging to the macro level		Expansion of the network to enroll more human and non-human actors (new infrastructure) in the technology network
	A credible and legitimate translator (clinical middle manager) became indispensable	1.2 Need to consider other forms of interessment strategies in the technological adoption process	Allocation of adequate resources Continuous formal training Competencies Evaluation (audit /survey / statistical studies)
	Middle-Up decision modulated by political, economic and professional interessment strategies		
2. Attribution of roles and responsibilities	Attribution of multiple roles at multiple time points for human and non-human actors involved in the adoption process (See Table 2 for more details)	2.1 Visibility of roles and responsibilities	Reduce hierarchies between actors
			Define and coordinate roles and responsibilities of allies early in the translation process
			Enroll a sufficient body of allies in the adoption process and interrelate their attributed roles
3. Interaction/ communication/ collaboration mechanisms	Collaboration and communication difficulties between facilities in the standardization of the management process of VRE outbreaks	3.1 Improvement of interaction/ communication/ collaboration networks	Formal training, continuous interaction and communication between healthcare units
	Tensions between actors at the micro level		Engagement and involvement of link nurses in IPC
	Collaboration difficulties between nursing staff in the emergency units and technologists		Change of information communicators
	Interactions / communications within each facility with collaborative work and open discussions between actors involved at the implementation phase		Collaboration of the IPC team with the laboratory at an early stage
4. Significant changes in professional practices	Workloads in user work practices at the meso and micro levels	4.1 Need for preparation to change at an early stage	Customization of the technology before adoption
	Difficulties to adapt to the language of the new technology		Explanation to the staff on why and how to use the technology at an early stage
	Frustration and resistance to change at the micro level		Formal knowledge transfer for actors at the micro level
	Workarounds often unanticipated by management		
	Changes at both administrative and operational levels in IPC practices		

Timeline of PCR-VRE's Adoption Process: Quick Middle-Up Decision

The adoption decision took place over a period of one year from December 2010 until December 2011 and involved several groups of actors, relying on dozens of people belonging to the macro level: directors, middle managers, microbiologists, IPC team and external organisations (Accreditation Canada, the Committee of Nosocomial Infections of Quebec (CNIQ) and the National Public Health Institute of Quebec (NPHIQ)) (Doc01, Doc02). Throughout the adoption decision phase, it was a sequence of many meetings that served to familiarize a total of ten enrolled actors with the characteristics of the project and create alliances between actors. The potential technology users who were not part of the decision-making were informally informed of the technology adoption during 10-15 minutes meetings. Their opinions, suggestions and /or comments were sought after technology implementation:

Sure technologists have heard about PCR arrival, but from the moment they must implement a technology in the laboratory, from day zero, technologists will be notified. (M002-02)

Indeed, the adoption of PCR-VRE technology was a Middle-Up decision emanating from middle managers. Our findings provide important insight relating to the mobilizing effect of a legitimate and credible translator, represented by the executive assistant in IPC in our context, who helped to initiate, announce and transmit the change in the translation process. However, prior to the involvement of actors in the network surrounding the PCR-VRE technology, our findings also outlined the use of limited devices by the project leaders that helped to interest and convince an increasing number of allies to accept the proposed solution. In addition to the large scale savings for hospitals by reducing the isolation time of infected and/or colonized VRE patients, the external and internal pressures, the professional interests of the IPC team and the self-funding of the project appeared to modulate the speed of the adoption decision. During the interview with a laboratories manager, he mentioned that:

In their money-saving amalgamation approach, to say yes we (laboratory managers) will use a better screening test with a faster turnaround which will cost more, but will be self-financing from what will be saved on the wards. (MM001-03)

As a result of the close focus on the first theme from our study, two challenges arose as important to overcome in translating technologies in professional practices. This includes the need to align and engage a wider actor-network in the adoption decision and the need to consider other forms of interessment strategies.

Need to Align and Engage a Wider Actor-Network in the Adoption Decision

From an imposed relationship that involved PCR-VRE as a non-human actor and human actors in the social network of the hospital, a heterogeneous network evolved and the aligned interest was to track PCR-VRE so as to improve efficiencies in day-to-day operations. From a macro perspective, the enrolment of actors from the meso (e.g. head of units and IPC senior and consultant nurses), and the micro (e.g. technologists and nursing staff) levels in the PCR-VRE adoption decision phase was considered of little interest. An IPC manager explained that:

The decision should be made by the executive committee or by leaders. That's a far better level, involving the IPC, infectious disease specialists, microbiologists and managers. In the implementation, it takes people on the ground but not in decision-making because they don't have a clear grasp of the subject and they don't have all the necessary information to make an informed decision related to this technology. (MM003-04)

However, from a micro-level perspective, actors such as technologists, IPC senior and consultant nurses and nursing staff highlighted their need to be enrolled in the adoption decision phase. The two senior and the consultant IPC nurses justified this need with respect to understanding the new PCR language and being able to assist with diffusion, including providing leadership for other staff nurses. A senior nurse in IPC said:

I realize that if at any given time a new technology like that is introduced, I'll ask to be involved from the outset with the lab technologists, with managers, with technologists who can help bring us up to speed, yes with the technique itself, but also to understand how it is done. So then we can share the information and we can understand what is happening. (SN013-03)

As for technologists, they expressed interest in being involved in the whole process of PCR-VRE adoption as an indication of respect for their work and being prepared for practice changes:

Sometimes we (technologists) need to be involved in decisions even if we don't care what they'll do about it, sometimes just asking our opinion, it shows respect for our work. You know, the simple fact of asking us whether it will integrate well into our normal work practice, it's just a sign of respect and helps us be prepared for the change. (Tech005-01)

PCR-VRE technology has been viewed as a new component added to an established network consisting of healthcare staff and existing non-human actors e.g. paper, other technologies, and an information system. The integration of the PCR-VRE required the development of new interactions around it and ANT helped to gain a deeper insight into the processes involved.

Need to Consider Other Forms of Interessment Strategies

Factors at the macro context such as political and economic factors clearly had an impact on the enrolment of allies into the actor-network. However, organisational factors were also important, such as allocation of adequate human and financial resources, a continuous formal training and competencies, to consider as interessment strategies in shaping the network of allies involved in the technology adoption process in healthcare organisations. Our findings also point to the importance of other issues in helping to integrate PCR-VRE and thus improving the implementation of infection prevention interventions. This includes the need to consider evaluation as an important interessment device to use in the technology adoption process. Multiple forms of assessment have been identified as helping to facilitate successful technology translation into professional IPC practices for managing VRE outbreaks at the hospital center being studied. This includes audit / survey / statistical studies on IPC knowledge application as translated, a retrospective analysis of the saved time, cost-effectiveness study, reports on financial gaps, impact of preventive screenings on VRE outbreaks, reports on gains or savings and a study on staff resistance to change.

Attribution of Roles and Responsibilities

Once the decision to adopt the new technology was made, our findings indicate an adjustment of the roles and responsibilities assigned to each actor involved was necessary to consolidate the network around the technology adoption. Actors considered in the network include humans, technology, documents, or other entities that play a role in the negotiations that gave shape to the network in question. Our findings highlight that different actors played multiple roles at multiple time points. Table 2 presents a summary of human and non-human actors enrolled and their attributed roles and responsibilities during the adoption decision and implementation phases.

However, in the process of negotiation that ensued from getting actors interested and enrolled in the network surrounding the technology, our findings highlighted the visibility of roles and responsibilities. A presentation of this challenge is outlined below.

Visibility of Roles and Responsibilities

Actors highlighted that the implemented PCR-VRE technology tended to change professional roles and responsibilities, often making hierarchies more visible between actors. For instance, there was enrolment of only some individuals at the decision phase, despite belonging to the same level either meso or macro. During interview, a consultant nurse in IPC expressed her discontent from being missed in these enrolment activities and explained that:

Basically, we (consultant nurses) knew about the existence of this technology, but we have never been consulted. Maybe our senior nurses were, but not us. (Cons004-03)

In addition, the changes in roles and responsibilities were seen as particularly problematic by technologists who expressed concern that the resulting placement of extra responsibilities on to their shoulders was detracting from their more pressing laboratory work. A technologist said:

I'm responsible for two techniques and in my case PCR technology has given me more responsibility since it's up to us to know how it works. Basically we collaborate a lot with the IPC nurses because they know we are short staffed and if they want to do more screening they prepare patient screening schedules for us. (Tech005-01)

Technology users highlighted that roles and responsibilities of allies need to be defined and coordinated early in the translation process so they can realize what they can be in the network. In addition, they expressed the importance of enrolling a sufficient body of allies and interelating the various roles allocated to them so that they act in particular ways to maintain the network.

Table 2. A summary of roles attributed and responsibilities for actors involved

Actors Enrolled	Roles and Responsibilities: Adoption Decision Phase	Roles and Responsibilities: Implementation Phase
1- Middle managers		
Executive assistant in IPC	Translator of the problem and the solution proposed	
	Nursing leadership that allied clinical and bridges with medical management	
	Enabler enrolling others into the network.	
Medical-administrative co-managers of laboratories	Spokesperson presenting the financial consideration of the implementation plans for the PCR-VRE technology.	
2- Microbiologists		
	Spokesperson: Ally other actors	Medical implementer of new algorithms and protocols
3- IPC team		
One senior IPCnurse	Representative of nursing perspective: scientifically prove the usefulness of PCR in controlling VRE outbreaks	
Senior (n=3) and consultant nurses (n=3)		Leadership in IPC: Identify, prevent, and control VRE outbreaks in the three hospitals
		Spokespersons at the level of clinical experts and staff nursing: Provide information about innovation and its implication through 10-15 minutes training modules
		Monitors: Application of appropriate isolation measure and daily supports for staff nursing
4 -Laboratory coordinators		
		Spokespersons at the level of technologists: Management of workbenches and staff availability, training for assigned technologists on how to use PCR, preparation of operating methods
5- Nursing staff		
		Frontline defense for applying daily practices to prevent infections and transmission of VRE to other patients: adherence and compliance with the new developed algorithms and protocols
6- Technology supplier		
	Negotiator at the level of labs manager	Trainer for technologists on how to use the PCR technology
7- PCR		
	Mediator actor in a series of actions: enrolement of actors with different interests around common objective, shape interactions between political, strategic, operational and functional levels	Rapid and specific screening assay
	Vehicle of control of VRE outbreaks: optimization of infection prevention interventions, compensation for the resources allocated, reduction of the time involved in adopting IPC practices	
8- Technology's attributes: reagents, software encoding test, double screening, new working procedures		
		Infrastructures

Interactions/Communications/Collaborations Between Actor-Network Involved

At the mobilization phase of the translation process, actors were involved in a concrete action and collaborated to fulfill their role in modifying their previous practices. Since the adoption decision of PCR-VRE technology in 2010 until its implementation in 2011, actors highlighted a series of changes in IPC professional practices. However, the success of these changes required inter/intra-disciplinary communication and collaboration within and between facilities. Following the implementation of the technology, good inter/intra-disciplinary interactions/ communications were found within each facility with collaborative work and open discussions between laboratories, microbiologists, care units, and the IPC team. As stated by a technologist:

We (technologists) had already set up a communication system, we call the microbiologists, the IPC nurses and we call the wards. With VRE outbreaks, we also developed a teamwork system because we realized that we need each other. (Tech005-01)

Our findings emphasize a team approach between actors involved at the implementation phase including IPC team, microbiologists and technologists within and between facilities in order to achieve better results. During our interview, a senior nurse expressed the need to engage everybody in IPC and that IPC is everyone's job:

In the PCR-VRE adoption process, we had an interdisciplinary practice with the microbiologists, technologists and IPC nurses. As for me (senior nurse), working with microbiologists was not problematic. But for the technologists it involved adapting with all the … On the contrary, I think the new technology has made us communicate more with each other. We had no choice but to talk to each other. (SN025-02)

As previously mentioned, only certain actors were involved in the adoption decision phase, technologists, nurses, consultants and two of three senior IPC nurses, and heads of care units were not consulted prior to the decision-making. The IPC team accentuated the difficulty of the collaboration between facilities, particularly in the standardization of the management process of VRE outbreaks due to difference in context. A senior nurse explained that:

Well, we (senior nurses) were working… we were working with technologists, or microbiologists, depending on the area, we were the ones managing it...with the laboratory. But we didn't set up groups of experts to tell everyone what to do and communicate between facilities. You know, you can't manage things the same way everywhere. (SN025-02)

Also, some collaboration difficulties were found between nursing staff in the emergency units and technologists. Our results outlined tensions between those two groups of actors. A chief nurse said:

But it's difficult. It's hard to reach everyone. It's hard to have an impact on every department. Yes, we (staff nurses) do cooperate with other wards. With the laboratory, for several years now we have had difficulties with the laboratory. You get the feeling that technologists working on different shifts don't follow the same procedures. There's a lot of friction when we talk on the phone. (CN021-01)

The focus on the third theme from our study posed a significant challenge to overcome in translating technologies in professional practices. This includes the need to improve interaction/ communication/ collaboration networks.

Improvement of Interaction/ Communication/ Collaboration Networks

On the basis of our interviews, participants expressed their need to improve interaction/ communication/ collaboration networks for the implementation of appropriate IPC interventions. They highlighted the importance of formal and continuous interaction / communication between healthcare units about outbreaks to overcome this challenge. Furthermore, our findings outlined the importance of engaging and involving link nurses in IPC, change of information communicators, and collaboration of the IPC team with the laboratory at an early stage of the technology adoption. A chief nurse explained that:

I think the communication of information (on IPC) should be done by someone other than the usual IPC personnel. They (IPC team) should involve other staff nurses, so I think that for communication, it's really crucial we change the people who usually pass on information. In addition, maybe they should explain the impact that an outbreak may have on a ward...give a reminder of the impact it has, because we do not fully realize the extent (since the VRE outbreak does not affect us in the emergency unit). (CN021-01)

Also, similar statments about challenges of multidisciplinary formal and continuous communication were revealed by senior and consultant nurses in IPC, they mentioned that:

There's no doubt the communication of IPC information is always improved if it is approached formally. (Cons004-03)

I dream of the day when we'll have people to speak responsible for IPC, who represent us on each care unit. You know. (SN013-03)

Changes in Professional IPC Practices

Our results suggest difficulties with translation of the PCR-VRE technology into professional IPC practices throughout a full year, starting from December 2011 until December 2012. This, in turn, was perceived to impact implementation, as users (i.e. technologists, staff nursing, microbiologists and IPC team) found it difficult to adapt to the language of the new technology, which changed the way they were used to working. Over time, users mentioned that they became more proficient with the technology and to some extent found ways of accommodating it within their existing practices. Indeed, the choice of the technology came from managers. A consultant nurse, during her interview, stated that:

At any rate, I think that as soon as something new is implemented, there's always some resistance before people fully understand the difference between PCR-VRE and culture techniques. So I'd say it was, that's what was the most difficult at first, but now the new technology is well integrated. It was incomprehensible for them, like a sort of laboratory analysis, but just... For them, it did not make any sense at first, but with explanations and the goal and all that, I think people don't understand but then they finally bought into it. (Cons007-02)

From the adoption decision until the implementation of PCR-VRE, a series of changes took place at both administrative and operational levels in IPC in the hospital center being studied. A new management process for IPC practices was established. This included the management of infected and/or colonized VRE patients from their arrival at the hospital until their discharge and the management of samples from collection until communication of laboratory results. The translation of PCR-VRE technology into professional IPC practices also generated implementation of several changes a priori related to budget management, infrastructure, equipment and personnel. An IPC manager mentioned that:

I think the whole organisation is affected. The laboratory is the operational side, and once you change things, the result impacts the whole organisation. There's really the mobilization of whole a network of actors behind the results of a screening test. If IPC measures are properly applied in a care unit, this can result in less work for laboratory personnel. Errors have an impact and vice versa, a false positive result in the lab has an impact on units of care. (MM003-04)

Additionally, our results highlight that staff nurses showed, besides adjustment and adaptation, some frustration and resistance to change, particularly with regard to the number of patients to isolate and where to isolate due to limitations in isolation units/ or cohorts/ or single rooms. Indeed, they had to adjust according to patients' needs. As for technologists, they felt disempowered by the new technology, as it destabilized their relationships as well as made them less efficient. Apart from resistance to change, technologists expressed their discontent with respect to the use of the PCR-VRE technology due to many challenges, including limited use of the technology (i.e. 5 days/7 and on the day shift), limited capacity of the PCR at 14 samples/run with two runs per day, shortage of human resources, workload, new interpretation of results in terms of false negative and/or positive PCR-VRE results, and changes to their working hours during VRE outbreaks. During interviews, technologists explained that:

This implementation has caused a bit of resistance at work for technologists due to overload and adjustments to new tasks, new roles and even new organisation of work. We were glad to have this rapid technology but not the extra work especially during an outbreak. We never had extra personnel to do that, even the girl on the evening shift was not trained to do PCR. (Tech005-01)

We only do PCR on the day shift, from 8 to 4, then we only do it from Monday to Friday. Requests for PCR on the weekend, on Friday and Saturday, are automatically transferred to culture technique. (Tech019-03)

It was also noted that the adoption of PCR technology had an impact not only on the practices of human actors but also non-human actors. Several documents, such as algorithms, protocols and procedures were uniformly distributed in the hospital center being studied but in some cases adapted to the context of each hospital. In addition to these documents, data entry and results in the sites information system were modified.

By investigating at close the resulting changes that arose from translating this technology into professional practices, our findings indicate an important challenge to overcome, including the need for preparation to change at an early stage. This challenge is presented below.

Need for Preparation to Change at an Early Stage

Interviewees stated that the technology needed customization to the context of each site before being adopted in order to fit the perceptions of the healthcare professionals. Specifically, our findings highlighted that healthcare professionals had to be satisfied that the technology would enhance their workflow and empower them with the ability to provide better care, with efficiency. Also, as stated in our findings, providing an explanation to the staff on why and how to use the technology at an early stage of the adoption process could help to increase the adherence of actors to the common objective, thus minimizing resistance to change. An IPC manager explained that:

To carry out a change like that, I will inform (the care units and staff nursing), because the way I manage is going to be different. If I want them to understand me and to follow my way of doing things, most likely I'll put them in the picture by saying from now on we have a new technology and they need to know that if they receive a report on VRE van B they need to understand it. (MM003-04)

Our findings reveal that knowledge transfer between IPC team and nursing staff about the new technology arrival and related changes occurred often through informal training and through formal meetings on occasion. Our results also show that technologists were informally trained on how to interpret results through learning-by-doing. However, better training, particularly through formal interactive courses and more information to support the required changes may help to overcome this challenge. As mentioned by a chief nurse:

Training is really super interesting, but maybe in small modules of information. It helps us to understand so after that we can say: 'Ah, OK so that's why I did that. And that's why this can make that, but not that.' But we need to be interested. Personally I am (chief nurse), but most of the staff aren't. (CN021-01)

In sum, inherent to the TRIP-ANT approach, the adoption of PCR-VRE technology has shaped interactions of actor-network involved in the adoption. Technology translation in IPC professional practices to improve interventions in managing VRE outbreaks occurred with relevant involvement of actors and consideration of implementation at the point of the adoption decision, coupled with well structured and managed implementation plans. The success of VRE prevention and control relied on collaborative action and repeated interactions within and between hospitals to foster cooperation.

DISCUSSION

This study examined how a specific technology was translated into professional practices via an actor-network to improve the implementation of IPC interventions based on the integrated framework TRIP-ANT. Our findings highlight four main themes and five challenges that emerged from the data collected. This leads us to four main observations.

First, our findings suggest that technology translation focuses on understanding how actor-networks are created, strengthened and weakened (Tatnall and Gilding, 1999). The potential enrolment of the organisational, clinical, managerial and financial support network is perceived as a key enabler for the effective translation of technologies in professional practices. Each actor enrolled in the translation influences the technology – to shape it into the ultimate form that is adopted (Unnithan and Tatnall, 2104). Our results indicate that improving IPC is predicated on an understanding of the interactions within healthcare settings between actor-networks, including changes in the practices and procedures they perform, the technology, roles and responsibilities, the work environment, the organisation itself, communication and collaboration.

Second, our results shed light on the consequences and impact of PCR-VRE technology on IPC professional practices across a range of actors. The middle-up decision to procure PCR-VRE, modulated by political, economic and professional interessment strategies, resulted in significant changes to user work practices. Also, our results pointed out that the introduction of PCR-VRE in professional practices resulted in more visible hierarchies of roles and responsibilities combined sometimes contradictory. During early adoption of a particular technology, certain problems may be short-lived and attenuate with increased use (Creswell, Worth and Sheikh, 2010). Indeed, we found this to be the case in the context of our study. Users adapted to the new working organisation coupled with the implementation of PCR-VRE. New routines were developed and existing work practices changed to some extent. Similarly, Puffer et al (Puffer et al., 2007) and Cresswell et al (Creswell, Worth and Sheikh, 2010; Creswell, Worth and Sheikh, 2012) found in their respective studies that along with the adoption of a new technology in a healthcare organisation, users need to change their work processes to fit in with the technology. If users could not resist use, they have employed workarounds which were often unanticipated by management. These strategies had several effects on other factors within the hospital, including hierarchical structures and patterns of communication within the multi-disciplinary healthcare actors, the time and quality of interactions and inter/intra-disciplinary collaboration within and between facilities. It is commonly acknowledged that increased customization can facilitate the introduction of technology within actor work practices (Unnithan and Tatnall, 2104).

Third, the main actors involved in the adoption process of a given technology should recognize their interdependence, coordinate, interact and collaborate as an actor-network, within and between healthcare delivery services, even though their interests may diverge (Centers for Disease Control and Prevention (CDC), 1995; Damschroder et al., 2009). Effective communication throughout all adoption decisions between various actors at different levels was frequently mentioned in the literature as a facilitator for enforcing IPC efforts (Kyratsis, Ahmad and Harris, 2010; Storr, Wigglesworth and Kilpatrick, 2013; Uchida et al., 2011). However, poor communication and confusion of responsibilities delayed the time in which they could react to outbreaks (Waterson, 2010). Indeed, many actors interviewed in our study reported that infection control meetings increased the opportunity for interdisciplinary collaboration. Some also shared how such communication opportunities could be enhanced through formal training. IPC team can and should take an active role in helping to continually inform technology users about outbreaks, the rationale for technology selection (including results of any evaluations), and the implementation strategy in order to facilitate staff enrolement (Kyratsis, Ahmad and Harris, 2010). However, some participants in our study expressed frustration at not being included during the adoption decision. The organisational environment not only encompassed the need for material resources but also included support for human resources, skills and competencies. Many actors expressed frustration over increased workload and resistance to new work processes related to the use of PCR-VRE technology. Increasing inter/intra-disciplinary collaboration within and between facilities and effective communication through formal training may facilitate information sharing and give impetus to further improvements in IPC in general, and in VRE infection prevention in particular (Damschroder et al., 2009).

Fourth, the focus on micro contexts proved useful in order to explain how the translating technology impacted IPC professional practices. This perspective helped to shed light on the role of technology users in the technology translation and thereby helped uncover wider social processes (McLean and Hassard, 2004; Williams and Edge, 1996). Similar to what has been outlined by Creswell et al. (Creswell, Worth and Sheikh, 2010; Creswell, Worth and Sheikh, 2012), our findings show the need to expand the actor-network to enroll actors from the micro level, e.g. technologists and staff nursing, early in the adoption decision. This will help to provide an understanding of technological change, ensure staff concurrence with the technology as well as adequate staff, knowledge, and resources, reinforcing organisational efforts to make prevention and control of VRE outbreaks everybody's responsibility and ownership (Ahmad, Kyratsis and Holmes, 2012; Collins, 2008). As IPC requires a strategic and general management approach, this also needs to be reflected in how adoption decisions are made. Involvement and implications for technology adoption need to be considered early on (Holmes, 2007).

Reflection About the Combined Framework TRIP-ANT

We applied an integrated TRIP-ANT framework as a means to overcome some of the shorcomings of the adoption/diffusion approach that is widely applied in the context of healthcare innovation. However, combining frameworks that are considered opposite (Tatnall and Lepa, 2003) is also challenging, which led us to reflect on the strengths and limitations of our approach. On the positive side, the TRIP-ANT was able to address the individual weaknesses of each of the two frameworks used. The contribution of ANT lays in its dynamic approach that led to a better understanding of how PCR-VRE was translated into professional practices. However, certain limitations in the theory can be noted (Aubry and Potvin, 2012). One of the challenges concerns the lack of methodology or instructions on how to make the theory operational (Callon, 1986). The response of a number of researchers has been to combine ANT with other theories, in order to generate greater explanatory power (Bell, 2010; Greenhalgh and Stones, 2010). For instance, Rogers' DOI theory has been used to investigate the adoption of web-based distance-learning technologies (Samarawickrema and Stacey, 2007). We recognize that the DOI and its terminology have made their way into the minds of hospital managers and others. This may be due to one clear advantage of the DOI, which is its simplicity of description and apparent organisational actionability, reflected in its widespread use and application (Cranfield et al., 2015; Zanaboni and Wootton, 2012) including in the area of technology implementation in the context of IPC in hospitals. However, this reveals a shortcoming given the limited lens of the DOI by focusing more on innovation attributes and factors instead of the adoption process and by viewing the adoption as merely a communication process (Sarosa, 2012).

The integrated framework TRIP-ANT refers to the intentional combining of domains of the two perspectives to prompt alternative insights not generated by one perspective alone. Yet we see a similar potential for the ANT perspective. The concepts of interessment incentives, contextualization of the problem–solution and the context that facilitates or hinder the translation of the innovation have parallel meanings to those of TRIP perspective, including characteristics and resources of the adopting organisation and the degree to which actors believe that the innovation responds to immediate and significant pressures in their environment. Critically, it is not the similarity or dissimilarity between domains from different perspectives that is key, but the complementarity of these domains that is sought. This may result in novel lens to best understand the true complexity of innovation adoption in the context of IPC in hospitals. Albeit innovation diffusion and innovation translation are based on quite different and often opposing philosophies, we believe they are complementary. One accentuates the properties of the innovation and the change agent, and the other emphasizes on negotiations and network formation (Akrich, Callon and Latour, 1988b).

In summary, the integrated TRIP-ANT framework is an attempt that has, we contend, much to offer. It could be valuable in advancing the field of innovation translation in its true complexity in healthcare organisation. Also, it could help in highlighting links between different theories and exploring ways of applying more than one understanding of innovation translation in the context of IPC at the same time.

Strengths and Limitations

This study is the first in understanding how an involved network of actors may optimize the implementation of IPC measures. Our study contributes to knowledge in a context lacking a theoretical basic understanding of how the adoption of IPC technology can transform healthcare practices and outcomes for patients. We approached this work from an integrative theoretical perspective, combining ANT with the TRIP framework in order to explain how the technology is adopted at the micro level, as well as its influence, such as social change, at the macro level. Understanding how technology adoption decisions from a macro and micro perspective can modulate the optimization of IPC professional practices provided useful explanations of the process of change, which can be essential for implementing effective interventions in a variety of contexts. The TRIP-ANT framework, along with the triangulation of multiple data sources, proved useful in guiding our analysis of the impact of technology translation on the practices of a network of actors involved directly or indirectly in the new intra- and inter-organisational processes surrounding the handling of IPC practices.

Despite the advances that this study brings to the field of technology adoption in IPC, it has some limitations. First, the limitation related to interviews is addressed. During interviews, the participants were encouraged to express their perceptions through a climate of openness, thus providing a true representation of the study topic. To ensure a high quality study, a study must include the triangulation of multiple sources of data to shed light on the themes (Yin, 2009). Individual face-to-face interviews were the primary data source for this study. Document analysis provided additional support for the interview findings. To improve rigour in coding of themes, a senior investigator looked at all transcripts. No changes were made to the themes as a result of these processes.

Second, our analysis of the technology adoption decision mainly focused on the case of a specific technology (PCR-VRE) in the context of IPC as a wider technology. It is obvious that this deliberate choice raises questions concerning the scope of our results. For instance, how our study could be useful for other IPC technology adoption as well as for a wider context in healthcare? However, since our goal is to understand a phenomenon, this can allow the generalization of findings across similar contexts. Future work could focus on monitoring both process and outcomes, and assessing their relationships to establish that improved IPC processes lead to better healthcare outcomes (Collins, 2008). This will enable evaluating the direct impact of the initial problem identified, i.e. recurrent VRE outbreaks, and the proposed solution, adoption of PCR-VRE technology.

CONCLUSION

The findings of this study contribute to strengthening knowledge about the complex phenomenon of innovation adoption in the context of IPC in hospitals. We have outlined how the translation of PCR-

VRE technology can be used to better understand how to improve IPC practices through the formation of a network in which all actors ally to each other through their acceptance of a common objective, thus improving the quality of patient care. Also, this study highlights the importance of taking into account the broader context during the adoption decision of technologies in IPC, in addition to the evolution in practices, interactions, communications, roles and responsibilities. This study uncovers how the new technology affected the everyday activities of many actors and the visibility of hierarchies of roles and responsibilities between actors. In summary, technology adoption in IPC is the enrolment and mobilization of an interdisciplinary network of actors e.g. microbiologists, IPC team, nursing staff, technologists and managers. Behind a rapid PCR result, a whole process to get from signal (i.e. presence of past VRE) to action (i.e. order patient isolation) requires continued human monitoring and effective communication between actors. The TRIP-ANT framework played an important role in formalizing the understanding of how new technology implementation in the context of IPC can shape professional practices of a network of actors and the consequences of change on the organisation and on patients.

ACKNOWLEDGMENT

We thank the participating hospitals for supporting this work, and we thankfully acknowledge all those who participated in our study, shared their knowledge and thoughts with us and kindly gave their time. We are also grateful to Mrs. Danielle Goulet and Mrs. Nathalie Boissonneault who generously gave their time and collaborate infinitely. We gratefully acknowledge the substantial contribution of Marie Dermatis to the coding of data. Also a special thank to Mr. Bruno Hubert for his help with verbatim transcription.

REFERENCES

Ahmad, A., Kyratsis, A., & Holmes, A. (2012). When the user is not the chooser: Learning from stakeholder involvement in technology adoption decisions in infection control. *The Journal of Hospital Infection, 81*(3), 163–168. doi:10.1016/j.jhin.2012.04.014 PMID:22633278

Akrich, M., Callon, M., & Latour, B. (1988b). À quoi tient le succès des innovations. *Gérer et comprendre, 97*(12), 14-29.

Akrich, M., Callon, M., & Latour, B. (2006). *Sociologie de la traduction-Textes fondateurs*. Paris: Presses de l'École des mines.

Akrich, M., Callon, M., Latour, B., & Monaghan, A. (2002a). The key to success in innovation part I. The art of interessement. *International Journal of Innovation Management, 6*(2), 187–206. doi:10.1142/S1363919602000550

Akrich, M., Callon, M., Latour, B., & Monaghan, A. (2002b). The key to success in innovation part II. The art of choosing good spokespersons. *International Journal of Innovation Management, 6*(2), 207–225. doi:10.1142/S1363919602000562

Antoniou, P. H., & Ansoff, H. I. (2004). Strategic management of technology. *Technology Analysis and Strategic Management, 16*(2), 275–291. doi:10.1080/09537320410001682928

Attieh, R., Gagnon, M., & Krein, S. L. (2014). How can implementing an infection prevention and control (IPC) technology transform healthcare practices and outcomes for patients? *Healthcare Infection, 19*(3), 81–88. doi:10.1071/HI14003

Aubry, F., & Potvin, L. (2012). *Construire l'espace sociosanitaire: expériences et pratiques de recherche dans la production locale de la santé*. Montréal: Les Presses de l'Université de Montréal.

Bell, F. (2010, May 3-4). Network theories for technology-enabled learning and social change: Connectivism and actor network theory. *Proceedings of the Seventh International Conference on Networked Learning*. Aalborg, Denmark.

Callon, M. (1986). Éléments pour une sociologie de la traduction – La domestication des coquilles Saint-Jacques et des marins-pêcheurs en baies de Saint-Brieuc. *L'Annee Sociologique, 36*, 169–208.

Callon, M. (1999). Actor-network theory, the market test. In J. Law, & J. Hassard (Ed.), Actor network theory and after (pp. 181-195). Oxford: Blackweel Publishers.

Callon, M., & Latour, B. (1986). Les paradoxes de la modernité: Comment concevoir les innovations? *Prospective et Santé, 36*, 13–25.

Centers for Disease Control and Prevention (CDC). (1995). Recommendations of the Hospital Infection Control Practices Advisory Committee (HICPAC). Recommendations for Preventing the Spread of Vancomycin Resistance.

Collins, A. (2008). Preventing Health Care-Associated Infections. In *H. RGs. Preventing Health Care-Associated Infections Rockville (MD)*. US: Agency for Healthcare Research and Quality.

Cranfield, S., Hendy, J., Reeves, B., Hutchings, A., Collin, S., & Fulop, N. (2015). Investigating healthcare IT innovations: A conceptual blending approach. *Journal of Health Organization and Management, 29*(7), 1131–1148. doi:10.1108/JHOM-08-2015-0121 PMID:26556172

Creswell, K., Worth, A., & Sheikh, A. (2010). Actor-Network Theory and its role in understanding the implementation of information technology developments in healthcare. *BMC Medical Informatics and Decision Making, 10*(67), 1–11. PMID:20067612

Creswell, K., Worth, A., & Sheikh, A. (2012). Integration of a nationally procured electronic health record system into user work practices. *BMC Medical Informatics and Decision Making, 12*(15), 1–12. PMID:22217121

Damschroder, L., Banaszak-Holl, J., Kowalski, C., Forman, S., Saint, S., & Krein, S. L. (2009). The role of the champion in infection prevention: Results from a multisite qualitative study. *Quality & Safety in Health Care, 18*(6), 434–440. doi:10.1136/qshc.2009.034199 PMID:19955453

Diekema, D., Dodgson, K., Sigurdardottir, B., & Pfaller, M. A. (2004). sigurdardottir, B, and Pfaller, M. Rapid Detection of Antimicrobial-Resistant Organism Carriage: An Unmet Clinical Need. *Journal of Clinical Microbiology, 42*(7), 2879–2883. doi:10.1128/JCM.42.7.2879-2883.2004 PMID:15243032

Ghodeswar, B. M., & Vaidyanathan, J. (2008). Organizational adoption of medical technology in healthcare sector. *Journal of Service Research, 7*(2), 57–82.

Greenhalgh, T., Robert, G., Bate, S., MacFarlane, F., & Kyriakidou, O. (2005). *Diffusion of Innovations in Health Service Organizations*. Oxford: Blackwells. doi:10.1002/9780470987407

Greenhalgh, T., & Stones, R. (2010). Theorising big IT programmes in healthcare: Strong structuration theory meets actor-network theory. *Social Science & Medicine, 70*(9), 1285–1294. doi:10.1016/j.socscimed.2009.12.034 PMID:20185218

Holmes, A. H. (2007). Can organisational change reduce hospital acquired infections? *The Journal of Hospital Infection, 65*(Suppl. 2), 191–192. doi:10.1016/S0195-6701(07)60042-5 PMID:17540269

Krein, S. L., Olmsted, R. N., Hofer, T. P., Kowalski, C., Forman, J., Banaszak-Holl, J., & Saint, S. (2006). Translating infection prevention evidence into practice using quantitative and qualitative research. *American Journal of Infection Control, 34*(8), 507–512. doi:10.1016/j.ajic.2005.05.017 PMID:17015156

Kyratsis, Y., Ahmad, R., & Harris, A. D. Understanding the process of Innovation Adoption in 12 NHS trusts-technology selection, Procurement and Implementation to help reduce HCAIs. Retrieved from http://www.wp.dh.gov.uk/hcai/files/2011/07/Technology-Adoption-Study*DH-Approved-final-report_Feb-2011_including-Acknowledgments.pdf*

Kyratsis, Y., Ahmad, R., & Holmes, A.A. (2012). Technology adoption and implementation in organisations: Comparative case studies of 12 English NHS Trusts. *BMJ Open, 2*(2), 1–10. doi:10.1136/bmjopen-2012-000872 PMID:22492183

Mays, N., & Pope, C. (2000). Qualitative research in health care. Assessing quality in qualitative research. *British Medical Journal, 320*(7226), 50–52. doi:10.1136/bmj.320.7226.50 PMID:10617534

McLean, C., & Hassard, J. (2004). Symmetrical absence/symmetrical absurdity: Critical notes on the production of actor network theory. *Journal of Management Studies, 41*(3), 493–519. doi:10.1111/j.1467-6486.2004.00442.x

Melnyk, B., & Fineout-Overholt, E. (2011). *Evidence-Based Practice in Nursing and Healthcare: A Guide to Best Practice*. Lippincott Williams & Wilkins.

Puffer, G., Ferguson, J., Wright, B., Osborn, J., Anshus, A., & ... (2007). Partnering with clinical providers to enhance the efficiency of an EMR. *Journal of Healthcare Information Management, 21*(1), 24–32. PMID:17299922

Rogers, E. (1995). *Diffusion of Innovations*. New York: The Free Press.

Samarawickrema, G., & Stacey, E. (2007). Adopting web-based learning and teaching: A case study in higher education. *Distance Education, 28*(3), 313–333. doi:10.1080/01587910701611344

Sarosa, S. (2012). Adoption of social media networks by Indonesian SME: A case study. *Procedia Economics and Finance, 4*, 244–254. doi:10.1016/S2212-5671(12)00339-5

Sofaer, S. (1999). Qualitative methods: What are they and why use them? *Health Services Research, 34*(17). PMID:10591275

Storr, J., Wigglesworth, N., & Kilpatrick, C. (2013). Integrating human factors with infection prevention and control. *Proceedings of the 27th EGOS Colloquium*. Retrieved from http://lup.lub.lu.se/record/2062995

Tatnall, A., & Gilding, A. (1999). Actor-Network Theory and Information Systems Research. *Proc. 10th Australasian Conference on Information Systems*, Austria.

Tatnall, A., & Lepa, J. (2003). Tatnall, A and Lepa, J. The Internet, E-Commerce, and Older People: An Actor Network Approach to Researching Reasons for Adoption and Use. *Logistics Information Management, 16*(1), 56–63. doi:10.1108/09576050310453741

Tong, A., Sainsbury, P., & Craig, J. (2007). Consolidated criteria for reporting qualitative research (COREQ): A 32-item checklist for interviews and focus groups. *International Journal for Quality in Health Care, 19*(6), 349–357. doi:10.1093/intqhc/mzm042 PMID:17872937

Uchida, M., Stone, P., Conway, L., Pogorzelska, M., Larson, E., & Raveis, V. (2011). Exploring infection prevention-Policy implications from a qualitative study. *Policy, Politics & Nursing Practice, 12*(2), 82–89. doi:10.1177/1527154411417721 PMID:22042614

Unnithan, C., & Tatnall, A. (2014). Actor-Network Theory (ANT) Based Visualisation of Socio-Technical Facets of RFID Technology Translation: An Australian Hospital Scenario. *International journal of actor network theory and technological innovation, 6*(1), 31-53.

Versalovic, J., & Lupski, J. (2002). Molecular detection and genotyping of pathogens: More accurate and rapid answers. *Trends in Microbiology, 10*(Suppl. 10), S15–S21. doi:10.1016/S0966-842X(02)02438-1 PMID:12377563

Waterson, P. (2010). Infection outbreaks in acute hospitals: A systems approach. *Journal of Infection Prevention, 1*(11), 19–23. doi:10.1177/1757177409354731

Williams, R., & Edge, D. (1996). The social shaping of technology. *Research Policy, 25*(6), 865–899. doi:10.1016/0048-7333(96)00885-2

Yin, R. (2009). *Case study research: Design and methods* (4th ed.). Los Angeles: Sage publications.

Zanaboni, P., & Wootton, R. (2012). Adoption of telemedicine: From pilot stage to routine delivery. *BMC Medical Informatics and Decision Making, 12*(1), 2–9. doi:10.1186/1472-6947-12-1 PMID:22217121

This research was previously published in the International Journal of Actor-Network Theory and Technological Innovation (IJANTTI), 8(3); edited by Ivan Tchalakov, pages 26-47, copyright year 2016 by IGI Publishing (an imprint of IGI Global).

APPENDIX

List of Abbreviations

ANT: Actor-Network Theory; VRE: Vancomycin Resistant Enterococci; PCR: Polymerase Chain Reaction; TRIP: Translating infection prevention evidence into practice; NI: Nosocomial infection; IPC: Infection prevention and control; CNIQ: Committee of nosocomial infections of Quebec; NPHIQ: National public health institute of Quebec; SN: Senior nurse; Cons: Consultant nurse; MM: Middle Manager; SM: Senior manager; CN: Chief nurse.

Supporting Information Legends

Checklist S1: Research checklist COREQ

Competing Interests

The authors declare that they have no competing interests.

Authors' Contributions

This paper is part of RA's PhD project investigating: how translating technology in professional practices optimizes the implementation of infection prevention and control interventions? RA conceived the idea for this paper, reviewed the literature and led the drafting of this paper. SLK developed the TRIP framework. MPG and GR intervened in the study design. MPG, GR and SLK provided critical revisions. All authors have read and approved the final manuscript.

Chapter 21
Why Do People Resist Patient Portal Systems?
An Application of the Dual Factor Model of IT Usage

Bahae Samhan
Illinois State University, USA

ABSTRACT

The implementation of patient portal systems (PPS) has potential benefits to both healthcare providers and their patients. However, evidence shows that PPS are being resisted by patients. Little research in IS has addressed this phenomenon. To understand PPS resistance, this study uses the dual factor model of IT usage to develop an integration of the user resistance model (URM) with the universal theory of acceptance and use of technology (UTAUT). Survey data were used to test the integrative model. A total of 265 responses were gathered from patients at a large international hospital. The data were analyzed using structural equation modeling (EQM). The results revealed resistance to change as an inhibitor to intentions to use the PPS. This study demonstrated the importance of integrating resistance to change with the technology use research, especially in healthcare settings. Moreover, this study is considered to be among the few studies in IS to incorporate patients' perspectives regarding new healthcare technology.

1. INTRODUCTION

Today's technology advancements not only offer strong infrastructure and enable the use of online applications to provide special services, but also create change in the way people perform their routine tasks. In healthcare, patients are being introduced to online patient portals. A Patient Portal is a secure online application that gives patients access to their personal health information from anywhere and at anytime. Using a secure username and password, patients can carry out a number of online activities, such as managing appointments, requesting medication refills, and communicating with healthcare providers about urgent health concerns and imaging and lab results. This is part of the new paradigm of global

DOI: 10.4018/978-1-5225-6198-9.ch021

healthcare that aims to offer anytime and anywhere access to healthcare services (Anwar et al., 2015). However, despite the numerous potential benefits of patient portal systems (PPS), there is evidence of patients' resistance toward PPS (Samhan, 2015).

A better understanding of patients' resistance can help decision makers at healthcare organizations take appropriate measures to limit PPS resistance and any subsequent effects. Further, understanding patients' resistance toward PPS will help designers build new systems that function optimally and have better adoption rates. However, there are major gaps in our understanding of how patients adapt to change from the traditional way of retrieving data and communicating with healthcare providers to using the new PPS. Prior IS research has a dearth of work on patients' resistance toward healthcare technologies. Studies have mostly focused on resistant behaviors of healthcare providers but not patients (Bhattacherjee & Hikmet, 2007; Lapointe & Rivard, 2005; Samhan & Joshi, Forthcoming). Moreover, prior IS research mainly considered technology adoption, acceptance, and use means of realizing the value of new technology (Ajzen, 1985; Davis, 1989; Taylor & Todd, 1995; DeLone & McLean, 1992, 2003). Resistant behaviors cannot be determined by the mere opposition of adoption, but need to be conceptualized distinctively (Lapointe & Rivard, 2005). Non-usage may suggest that users are still evaluating the technology prior to adoption, while resistance suggests that the technology has been considered and rejected by users (Bhattacherjee & Hikmet, 2007). The presence of resistance affects technology usage, but the lack of resistance does not necessarily increase technology usage (Hsieh et al., 2014; Lapointe & Rivard, 2005). Thus, resistance to change is a typical inhibitor to technology usage (Cenfetelli, 2004). Furthermore, prior findings (Parthasarathy & Bhattacherjee, 1998; Venkatesh & Brown, 2001) confirm that technology usage and resistance have different antecedents and are motivated differently. Therefore, there is a need to investigate intentions of use and resistance to change as two separate constructs and aim to examine the relationship between these two distinctive constructs.

This study examines the following research questions:

RQ1: Why do patients resist patient portal systems?
RQ2: How does resistance influence patients' usage decisions?

To address these questions, this study integrates technology acceptance and resistance to change literature. The remainder of the paper is organized as follows. Section 2 describes the theoretical background of this work and how URM and UTAUT are integrated within the dual factor model of IT usage. Section 3 provides theoretical discussions about the study's hypotheses and illustrates these hypotheses in a research model. Section 4 describes the investigative context of this study and section 5 shows the methodological approach used for this study. Finally, results are reported in section 6, followed by discussion about the contribution, limitations and implications of this study.

2. THEORETICAL BACKGROUND

To explain patients' resistance toward the new PPS, this study adopts the concepts of the dual factor model of IT usage (Cenfetelli, 2004) to bridge the User Resistance Model (URM) (Kim & Kankanhalli, 2009) and the Unified Theory of Acceptance and Use of Technology (UTAUT) (Venkatesh et al., 2003). The dual factor model of IT usage surmises that technology adoption can be predicted by enablers and technology resistance can be predicted by inhibitors. This theory suggests that enablers are external beliefs about the evaluated PPS that will either encourage or discourage the use of technology. On the other hand, inhibitors are defined as the factors that discourage technology usage when present, but do not necessarily encourage usage when absent. This asymmetric effect implies that inhibitors are not necessarily the opposite of enablers; rather, they are distinct constructs that may coexist.

In the context of this study, resistance to change is suggested to be an inhibitor to technology usage. The new PPS has created change in the way patients communicate with their healthcare providers and how they manage and view their personal health records. This change is suggested to contribute as an inhibitor to the adoption of a new PPS. Thus, technology use and resistance to change are examined independently within a common theoretical model. The model suggests that intention to use PPS is based on the enablers as well as the inhibitors of technology usage. In the enabling perceptions, the predictors of UTAUT are adopted: performance expectancy, effort expectancy, facilitating conditions, and social influence. In the inhibiting perceptions, the resistance to change construct from URM is adopted.

2.1. Resistance to Change

The outcome variable of the URM, resistance to change, is the suggested inhibitor of technology usage for this study. The URM suggests that decisions to resist change are affected by the Perceived Value (PV) of switching to new technology. This value can be determined by weighing the benefits of the switch against the costs of the switch. The more costs associated with the switch, the less value perceived for the change. Thus, an increase in the desire to resist the change will arise. The URM suggests that the factor self-efficacy has a direct impact on an individual's decision to resist change. In the context of this study, if patients believe they do not have sufficient knowledge and/or capabilities to use the new PPS, then they will be inclined toward resisting the change to the new system. Finally, the URM suggests social influence has a direct impact on one's decision to resist the change. In this study, sources of social influence include other patients who are using the system, family members who know about the system, or healthcare providers who talk with patients about the new system.

The URM is a comprehensive model which looks to previous literature for various antecedents for technology acceptance and resistance. It integrates concepts from the technology acceptance literature, including Attitude, Subjective Norms, and Behavioral Control from the Theory of Planned Behavior (TPB) (Ajzen, 1991). URM explains resistance to change by integrating different perspectives of user resistance, including the Status Quo Bias Theory (SQBT) (Samuelson & Zekhauser, 1988) and the Equity Implementation Model (EIM) (Joshi, 1991). The following subsections provide insights about each of these theories.

2.1.1. SQBT

SQBT helps explain patients' resistance to change due to the preference to stay with the "current state" which, in this study, is the traditional way of communicating with their healthcare providers. The SQBT suggests that resistance to change can be explained based on the evaluation of the current state of the person and the perceived future state of the person after the change. SQBT posits that resistance to change may be due to the person's preference to stay with the current state. The SQBT includes three categories: (1) Rational decision making, which includes the evaluation of costs and benefits of the change. If a person perceives the costs to be larger than benefits, then this may lead to the status quo bias. (2) Cognitive misperception, which is explained by the concept of loss aversion. Loss aversion arises when people weigh losses more than they weigh gains when evaluating the value of the change. Because of loss aversion, relatively small losses may be perceived as larger than they actually are (Kahneman & Tversky, 1979). (3) Psychological Commitment, which includes three factors: First is sunk cost, i.e., any previous commitments that may result in resistance to change. For example, if a person had put the time and effort into mastering skills that are required to complete a task, and if the future alternative way of completing the task would require a new, different set of skills, then the person would perceive the current learned skills as sunk costs. These sunk costs may negatively affect the person's decision to switch to the new status. Second, social norms refer to the perceptions of others about the change. These social norms may influence the person's status quo bias in either direction. Third, efforts to feel in control refer to people's desires to choose and control their own situations. If a person feels that s/he will lose control by changing to an unknown or unfamiliar way of working, then status quo bias would be observed.

2.1.2. EIM

EIM suggests that people do evaluate change using the concept of "net equity". EIM explains net equity as the difference between the person's perceptions of changes in his inputs (disadvantages) and changes in his outcomes (advantages) if the new proposed way of working was considered. The outcomes are measured by (increase in outcomes – decrease in outcomes) and inputs are measured by (increase in inputs – decrease in inputs). Net equity can also be determined by comparing the person's outcomes of the change with those of other people who also accepted the change. The EIM explains that when inequity (low net gain) is perceived, the person would then be resistant to the change. The greater the inequity, the greater the resistance to change would be.

These perspectives of the EIM help explain patients' resistance to change to the new PPS based on cost-benefit analysis of the change or the "net equity" associated with the change to the new system.

2.2. UTAUT

Venkatesh et al. (2003) introduced the UTAUT to explain technology usage behavior, and it has become one of the most influential models of the use and acceptance of technology. UTAUT suggests that effort expectancy and performance expectancy are two salient cognitive predictors of intention to use technology, with individuals tending to use technologies that benefit their tasks with little effort. These constructs correspond closely with concepts of ease of use and perceived usefulness from the Technology Acceptance Model (Davis, 1989).

Additionally, UTAUT suggests that social influence has an impact on technology usage intentions because users tend to care about the views of others regarding technology. This perception may steer one's intention to use the technology in either direction. UTAUT also suggests that facilitating conditions have an impact on a person's intention to use the technology.

Facilitating conditions are defined as the organizational and technical infrastructures that exist to support the use of the technology. In the context of this study, these facilitating conditions may be the technical resources, personal knowledge, and/or availability of guidance regarding the PPS.

3. HYPOTHESES

Self-Efficacy for Change refers to internal means that can enhance the achievement of control of a changed situation (Ajzen, 2002). Self-efficacy corresponds to the control concept from the SQBT. If patients believe that there will be a loss of control after the change to the new PPS, then status quo bias would be observed. This loss of control may manifest low self-efficacy. Based on URM, it is theorized that difficulties people face during change to the new system will be viewed as threats to be avoided or skills to be mastered, depending on the level of self-efficacy of the person evaluating the change (Bandura, 1995). Thus, patients with higher levels of self-efficacy would face the change with more confidence while patients with lower levels of self-efficacy would be more inclined to resist the change. Normally, patients with the knowledge and skills required to use the PPS and with previous experiences using different portals and online applications will have higher confidence levels in their ability to use the PPS without the help of others. As a result, patients will be more welcoming of the change, which leads us to the following hypothesis:

H1: Self-Efficacy will have a negative effect on Resistance to Change.

Perceived Value (PV) refers to the perceived net benefits of switching to the new PPS. PV corresponds to the net equity and net benefit concepts from EIM and SQBT, respectively. This hypothesis theorizes that people tend to like the maximization of value when making a decision (Sirdeshmukh et al., 2002). Thus, if PV of change to PPS is low, then patients are more likely to have greater resistance to the change. On the other hand, if the PV is high, then patients are likely to have lower resistance to change to the new PPS. So patients will be more involved in using the PPS if the new way of working with the system is considered of high value. Moreover, if services provided by the PPS are considered to be valuable, patient will be content with using the PPS and will be reluctant to resist the change. These discussions lead us to the following hypothesis:

H2: PV will have a negative effect on Resistance to Change.

Social Influence refers to the degree a patient perceives the value of others' beliefs regarding the use of the new PPS (Venkatesh et al., 2003). It is theorized that people have the tendency to conform to their important others' opinions because of the need for social companionship and the fear of sanction for noncompliance (Ajzen, 2002; Lewis et al., 2003). Resistant behaviors are directly affected by what other people think about change. Further, the opinion of others regarding the change may alter the original perceptions of an individual about the change (e.g., informational influence, Bunkrant & Cousineau, 1975). Positive opinions about change can reduce resistant behaviors. Similarly, positive opinions lead to greater perception of the benefits of the change. Patients' resistant behaviors are directly affected by the views of others regarding the change to the new PPS. Additionally, the desire of patients to fit in among the group of people who are using the PPS may induce a change in their belief or behavior. Kelman (1958) distinguished three types of conformity: compliance, internalization and identification. Compliance occurs when patients accept influence because they hope to achieve a favorable reaction from others. In the context of this study, adopting the PPS occurs because of the expected gain of specific approval, or to avoid specific disapproval from other PPS users.

Internalization occurs when patients accept influence because the content, ideas, and actions of the PPS are consistent with their values. For example, a patient might accept the new PPS because he or she already has favorable opinions about other processes that have been digitized, such as online banking or email systems. Identification occurs when patients accept influence because they want to establish or maintain a satisfying self-defining relationship to other PPS users.

Peers are an important source of influence in the adoption of PPS (Samhan, 2017). When patients receive positive feedback from other patients who are using the PPS, they will likely be encouraged to use the system, accepting the change. Similarly, negative opinions about the PPS may lead to discouragement of using the PPS, resisting the change. This phenomenon also applies to opinions received from other important non-patient users such as doctors and nurses. If a healthcare provider, who is highly recognized by the patient, recommended using the PPS, then there is a higher chance that the patient will actually use the PPS and accept the change. These discussions lead us to the following two hypotheses:

H3: Social Influence (positive opinions about the change) will have a negative effect on Resistance to Change.

H4: Social Influence (positive opinions about the change) will have a positive effect on Intentions to Use PPS.

The UTAUT has identified four main antecedents to use intentions: Performance Expectancy, Effort Expectancy, Facilitating Conditions, and Social Influence. Performance Expectancy refers to the degree to which patients believe that using the PPS will help to attain gains in performing desired tasks (Venkatesh et al., 2003). There are five constructs from different models that capture the concept of Performance Expectancy: perceived usefulness (Davis, 1989), extrinsic motivation (Venkatesh et al., 2003), job fit (Thompson et al., 1991), relative advantage (Moore & Benbasat, 1991), and outcome expectations (Compeau et al., 1999). Effort Expectancy refers to the degree of ease associated with the use of the PPS (Venketash et al., 2003). There are three constructs from different models that capture the concept of Effort Expectancy: perceived ease of use (Davis, 1989), complexity (Thompson et al., 1991), and ease of use (Moore & Benbasat, 1991). Facilitating Conditions construct in this study refers to the degree to which a patient believes that the technical infrastructure exists to support the use of the PPS. There are three constructs from different models that capture the concepts of Facilitating Conditions: perceived behavioral control (Ajzen, 1991; Taylor & Todd, 1995a), facilitating conditions (Thompson et al., 1991), and compatibility (Moore & Benbasat, 1991).

Social influence in our study is considered as an antecedent to use intentions. However, it was not hypothesized that Resistance to Change will affect social influence for two main reasons. First, while Cenfetelli's dual factor model of IT usage posited that inhibitors can influence IT usage indirectly through enablers that serve as mediators, this indirect influence is unidirectional and enablers do not have any corresponding effect on inhibitors (Bhattacherjee & Hikmet, 2007; Samhan, 2017). The unidirectional effect was concluded by the effect of Social Influence on Resistance to Change. Second, the Social Influence construct captures others' opinions of the PPS. By hypothesizing that Resistance to Change has an effect on Social Influence, I will be testing the effect of patient decisions to resist change based on the opinions of others about the PPS. This is a counterintuitive argument that cannot be justified within the theory in use.

Based on the dual factor model of IT usage, it is theorized that there is a unidirectional effect from inhibitors to enablers. Thus, resistance to change will have an effect on all of effort expectancy, performance expectancy, facilitating conditions, and intentions to use. These effects can also be justified in the context of our study. When patients resist the change to the PPS, they are showing little or no interest in using the PPS. These behaviors imply that patients have negative perceptions toward: (1) the PPS accomplishing their tasks (performance expectancy), (2) ease of use (effort expectancy), or (3) the existence of technical infrastructure to support their use (facilitating conditions). These negative perceptions cannot be confirmed simply because the patients are resisting the change and have not used the PPS. This theory relates to the philosophical concepts of "absence of evidence" (Alton et al., 1995) and "arguments from ignorance" (Walton, 1992; Walton, 2010).

Resistance to change also impacts the antecedents of intention to use, where patients may have actually used the system at an earlier stage and decided that they do not want to use it any further. This later decision to resist the change compounds their initial negative perceptions of effort expectancy, performance expectancy, and the facilitating conditions.

In summary, the asymmetric effects of inhibitors on enabling perceptions of PPS usage suggest that Resistance to Change will influence enablers in a negative manner, which leads us to the following hypotheses:

H5: Resistance to Change has a negative effect on Performance Expectancy.
H6: Resistance to Change has a negative effect on Effort Expectancy.
H7: Resistance to Change has a negative effect on Facilitating Conditions.

The technology acceptance literature suggests that performance expectancy is a salient cognitive determinant of usage behaviors. In the context of this study, patients will want to use the PPS if they believe it will benefit them by allowing them to complete the task they wish to accomplish (Venkatesh et al., 2003).

Patients intend to use the PPS if they believe it enables them to communicate with their healthcare providers more quickly and easily. Additionally, perceptions regarding the quality of communication are important factors. If patients believe that the PPS will provide high quality communications with healthcare providers, then they will be inclined to use the PPS. This leads us to the following hypothesis:

H8: Performance Expectancy has a positive effect on Intentions to Use PPS.

Technology acceptance literature suggests that effort expectancy is a cognitive determinant of usage behaviors. Patients will want to use the PPS with minimum effort. Prior empirical work (e.g., Tulu & Horan, 2007; Venkatesh et al., 2003; Wu et al., 2008; Foon & Fah, 2011) shows that Effort Expectancy tends to be positively related to usage intentions. In the context of this study, patients are more inclined to use the PPS if they believe the system is clear and easy to use. This relates to the design features of the PPS and conforms with the work of Davis (1989), which posited that a positive relationship exists between ease of use and intention to use within the Technology Acceptance Model (TAM). Further, patients who believe that the PPS will be easy to use and will not be time-consuming will be encouraged to use the PPS. These suggestions lead us to following hypothesis:

H9: Effort Expectancy has a positive effect on Intentions to Use PPS.

The UTAUT hypothesized that Facilitating Conditions do not affect intention to use. This hypothesis was later confirmed by the empirical findings of UTAUT, which suggested that when performance expectancy constructs and effort expectancy constructs are present together in a model, facilitating conditions become non-significant in predicting intention (Venkatesh et al., 2003). However, in this study, the three constructs from the different models that pertain to Facilitating Conditions (Compatibility, Facilitating Conditions, and Perceived Behavioral Control) are expected to all have positive influence on use intentions. In today's technological era, compatibility is a deal-breaker for technology use. For example, a patient who feels very comfortable using new, advanced technological tools may be hesitant to use the new PPS only because the application is not compatible with her phone or tablet. On the other hand, a patient less experienced in IT may be motivated to use the PPS if the application is compatible with other connectable devices. Additionally, the facilitating conditions construct posited by Thompson et al. (1991), which is embedded in the Facilitating Conditions construct of the UTAUT and used in this study, focuses on the support provided to users. PPS support includes technical support and content support. The more patients believe they have access to reliable resources of support, the more they will be encouraged to use the PPS. Moreover, unlike the situation with other populations, patients in this study have full control over every aspect of the interaction with the PPS. They have the freedom to choose the time, location, duration, and device to accomplish the task of using the PPS. They even have the freedom to decide whether to use the PPS or not. This ample freedom allows patients to react differently to the word "control" in the survey items pertaining to the construct, unlike the ways that general employees responded when asked about the control they have over a mandated technology for work at an organization that expects outcomes to be of a certain quality. These suggestions lead us to the following hypothesis:

H10: Facilitating Conditions have a positive effect on Intentions to Use PPS.

When patients resist the change to the new PPS, they will show no interest in using the PPS. Thus, we can infer a negative relation between resistance to change and intentions to use. The introduction of a new system tends to create changes in users' environments. If the impacts of these changes are perceived to be high, then people will most likely resist the source of the disruption, that is, the new system. Bhattacherjee and Hikmet (2007) argue that such reaction can be explained by the natural human proclivity to oppose change. As a result, people will have less intention of using the system. Moreover, studies that examined the relationship between user resistance and usage (Bhattacherjee & Hikmet, 2007; Hsieh et al., 2014; Norzaidi et al., 2008a, 2008b; Poon et al., 2004; Spil et al., 2004) have provided support for the negative impact resistance to change on intention to use. Also, prior work in IS has posited that technology resistance is clearly a barrier to IT usage (e.g., Bhattacherjee & Hikmet, 2007; Weeger et al., 2011; Lapointe & Rivard, 2005). So when patients consider resisting the change to the new PPS, they in fact have implicitly expressed a lesser intention to use the PPS. Moreover, Lewin (1947) described resistance to change as a cognitive effort preventing a potential behavior. The potential behavior in this study would be the intention to use the PPS. This leads us to the following hypothesis:

H11: *Resistance to Change has a negative effect on Intentions to Use PPS.*

Finally, in this study we controlled for a number of variables, namely Age, Sex, and Education Level. Figure 1 illustrates the research model of this study on the basis of the dual factor model of IT usage (Cenfetelli, 2004), and Table 1 lists the proposed hypotheses of the study.

4. INVESTIGATIVE CONTEXT

The technology investigated in this study is a newly implemented Patient Portal System (PPS), which is available to all patients of a large public hospital located in Amman, Jordan. The PPS is considered to be an informational portal as well as a transactional system. It allows patients to view results of their lab tests, imaging reports and prescriptions, and assists with their requests to fill up on their medications, to schedule and edit appointments, and to place follow-up questions to their providers.

The PPS is available as a compatible application with most of the portable devices, which allows flexibility of access from anywhere and at anytime. To gain access to the system, patients must create an online profile. The online profile includes a secure username and password. The company developing the PPA is in the process of launching the latest update of the application to include the option of fingerprint identification to access the application. This feature will require that hosting devices have the fingerprint feature.

The hospital consists of multiple health departments and 33 specialist clinics and serves over 400 outpatients every day. The total capacity of the hospital for inpatients is 550 beds, and on average 58% of these beds are constantly occupied with patients.

Figure 1. Dual Factor Model of IT usage – Integrating UTAUT with URM

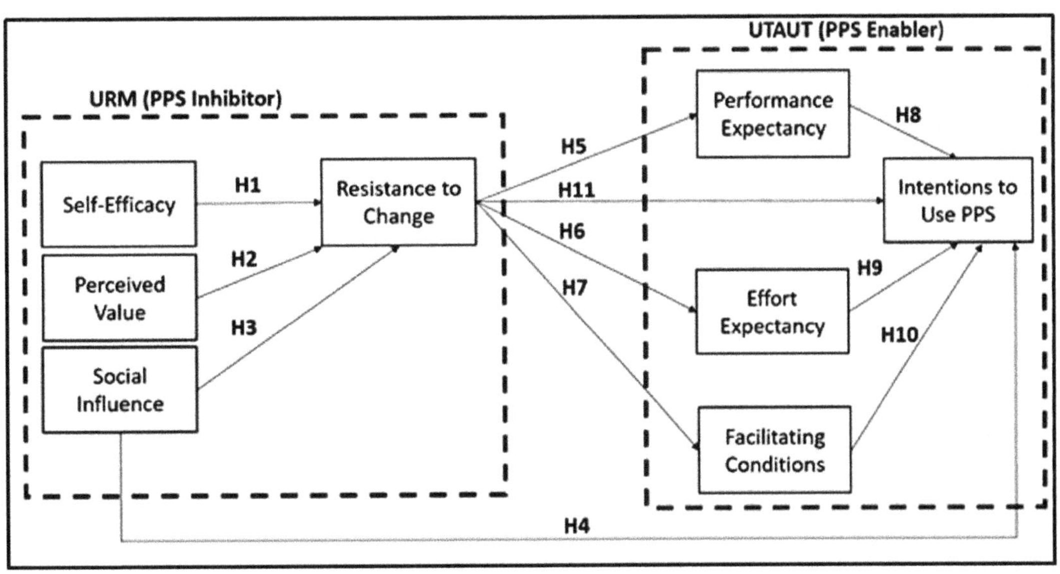

Table 1. Proposed Hypotheses

Hypothesis	Source(s)
Self-Efficacy will have a negative effect on Resistance to Change.	Kim & Kankanhalli, 2009.
PV will have a negative effect on Resistance to Change.	Joshi, 1991; Kim & Kankanhalli, 2009; Samuelson & Zeckhauser, 1988.
Social Influence (positive opinions about the change) will have a negative effect on Resistance to Change.	Kim & Kankanhalli, 2009; Samhan, 2017.
Social Influence (positive opinions about the change) will have a positive effect on Intentions to Use PPS.	Venkatesh et al., 2003; Kim & Kankanhalli, 2009; Samhan, 2017.
Resistance to Change has a negative effect on Performance Expectancy.	Cenfetelli, 2004; Bhattacherjee & Hikmet, 2007; Samhan, 2017.
Resistance to Change has a negative effect on Effort Expectancy.	Cenfetelli, 2004; Bhattacherjee & Hikmet, 2007; Samhan, 2017.
Resistance to Change has a negative effect on Facilitating Conditions.	Cenfetelli, 2004; Bhattacherjee & Hikmet, 2007; Samhan, 2017.
Performance Expectancy has a positive effect on Intentions to Use PPS.	Venkatesh et al., 2003; Davis, 1989.
Effort Expectancy has a positive effect on Intentions to Use PPS.	Venkatesh et al., 2003; Davis, 1989.
Facilitating Conditions has a positive effect on Intentions to Use PPS.	Venkatesh et al., 2003; Davis, 1989.
Resistance to Change has a negative effect on Intentions to Use PPS.	Hsieh et al., 2014; Bhattacherjee & Hikmet, 2007; Samhan, 2017.

The patient portal system is an extension of a large Electronic Health Records (EHR) system that was implemented in 2014. The vision of the hospital is to have complete paperless communication with patients in the next five years. The PPS was implemented through a governmental program which is the first national e-health initiative in Jordan. The PPS was designed and implemented by Electronic Health Solutions (EHS), which is a non-profit, innovative, technology-driven company established in early 2009. EHS is a partnership between the main healthcare stakeholders: Ministry of Health, Ministry of Information and Communications Technology, Royal Medical Health Awareness Society and Private Hospitals Association. According to EHS, 4.5 million Jordanian Dinar (JD), approximately $6.4 million, were invested in specialized resources to develop and implement the entire EHR system and the PPS.

The introduction of the PPS brought substantial changes to the process of communication between the patient and the hospital. Patients are gradually switching from using traditional ways of communicating with the hospital to the PPS. Today, 39% of all patients are registered with the PPS; however, registered patients may still use traditional ways of communicating such as calling, walking in, and requesting printed material.

Successful adoption of the PPS is very important to hospital stakeholders, mostly because there is no reliable mailing system in Jordan. Patients tend to go to the hospital in person for requests that involve printed materials, adding crowd pressure to hospital staff who process these requests. Additionally, hospital stakeholders have concerns regarding physical space at the hospital for all types of visitors, i.e., patients who need medical attention or only need some paperwork such as lab results or medical prescriptions and refills.

5. METHODOLOGY

5.1. Instrument Development

Existing validated scales were adopted for this study from the URM and UTAUT. All items were modified to fit the context of this study; however, there were important edits made to the original scales. For PV, the loss aversion items were included, which were missing from the URM's original scale. The loss aversion items were adopted from Gächter et al. (2007) and refer to the psychological fact that humans tend to value loss more than gain. Thus, this relates closely to PV items (Kim & Kankanhalli, 2009). Measurement items were anchored on five-point Likert scales (1 = strongly disagree, 5 = strongly agree) and were reviewed by IS researchers as part of a larger study. The final version of the questionnaire was developed for the main study, as shown in Exhibit 1 of the Appendix.

5.2. Instrument Validation

To validate the measurement scale, the psychometric properties of the survey were assessed through the application of Explanatory Factor Analysis (EFA) and Confirmatory Factor Analysis (CFA) using MPlus 7.1. Cronbach's α was computed using SPSS 20. After the pilot data were collected, I refined the items to keep only those with statistical significant loading larger than (0.7). Cronbach's α reliability tests for all constructs exceeded (0.8) (Fronell & Larcker, 1981). After data were collected for the main study, CFA analysis was conducted a later time. Consistent with the prior results, all items had significant loadings greater than (0.7) except for the dichotomous factor indicator Loss Aversion within Perceived Value, which had a negative loading with no statistical significance ($p = 0.23 > 0.05$). Thus, Loss Aversion items were dropped from the scale. The CFA analysis provided strong support for this measurement model, which suggested that the items under each of the constructs adequately measured the constructs. The items' loadings and the model fit statistics for the CFA are listed in Exhibit 2 of the Appendix.

5.3. Sample and Data Collection

Data were collected in two ways: (a) A paper-based survey was handed out to each patient who walked into the front desk area of the hospital's main building, where all of the major circulations take place; (b) a link to the survey was sent to all patients registered with the PPS. By doing so, I intended to avoid having response bias from data gathered only from walk-in patients, which would primarily be responses from patients who may not be PPS users. Similarly, collecting data only from online survey through the PPS would result in responses just from patients who use the PPS. By integrating both methods, I was able to provide control for response bias.

Data were collected in two phases: a pilot and a main study. For the pilot, a total of 113 responses were collected from patients of the dermatology unit at the hospital. After conducting Explanatory Factor

Analysis (EFA) and Confirmatory Factor Analysis (CFA), it was necessary to alter some of the items to address psychometric issues related to the study measures. Additionally, new items were added to better capture the concepts of some constructs. First, the SFC1 item was reworded to match the fashion of the other questions of the same construct. This resulted in an increase in its loading into the main study. Additionally, item SFC4 was added to the main study's survey to capture the levels of confidence among patients in their ability to change to the new PPS. Third, concepts of loss aversion were included within the PV construct. Lastly, item PVL4 was added to capture patients' perceptions of value after considering any limitations of the PPS. Loss aversion had a negative loading in the main study and no statistical significance ($p = 0.11 > 0.05$), so the Loss Aversion item was dropped from the scale.

After refining the survey items based on the pilot data's psychometric properties, we collected data for the main study from patients of the entire hospital. Patients who completed the first survey (i.e., pilot study) were specifically asked not to complete the survey a second time.

Main study data were collected in one day from the hospital using the printed surveys. Online responses were collected within three days. The estimated total number of surveys distributed (printed and online) was 600. The total responses were 338 (56%). After discarding all missing data, only 265 (44%) responses were useable. Descriptive statistics of the 265 respondents are reported in Table 2.

To check for issues of common method variance (CMV), the marker-variable technique (Williams et al., 2010) was used. To do so, the Educational Level (EL) control variable was selected as the marker variable because it is theoretically unrelated to the remaining variables in the study. It had a mean absolute correlation with the remaining items of interest in this study of (0.076). After calculating the CMV-adjusted correlation by partialling out the effect of the mean absolute correlation of EL with other items from the unadjusted bivariate correlations among the remaining items, the mean change in correlations due to this adjustment was (0.068). The CMV-adjusted correlations were not significantly different from the unadjusted correlations, which indicates no CMV bias in the observed data sample.

Table 2. Descriptive Statistics of Respondents

Demographic Variables		Data
Gender	Male	167 (63.01%)
	Female	98 (36.90%)
Age (Mean = 43.07, S.D. = 6.92)	<30	40 (15.09%)
	31 - 40	93 (35.09%)
	41 - 50	77 (29.05%)
	51 - 60	31 (11.70%)
	>60	24 (09.05%)
Education Level	Graduate	46 (17.35%)
	Under Graduate	60 (22.64%)
	Associate Degree	24 (09.05%)
	High school	82 (30.94%)
	No High School	53 (20.00%)
	No Education	0 (0.00%)
Total		265 (100%)

6. RESULTS

The research model was tested by applying Structural Equation Modeling (SEM) using Mplus version 7.1. We applied the maximum likelihood estimator with robust standard errors (MLR). Because the model is not saturated (i.e., not including all possible regression paths), I evaluated the model fit indicators depicted in Table 3.

Following the suggestions of Bollen (1998) regarding the evaluation of Chi-Square (X2), we calculated the Normed X2 (NC=4.23). It is suggested that NC value between (2.0 and 6.0) indicates reasonable fit. CFI is (0.912), which conforms to the rule of thumb, where values greater than roughly (.90) may indicate a reasonably good fit of the researcher's model (Hu & Bentler, 1999). The Root Mean Square Error of Approximation (RMSEA) is (0.014). According to the rule of thumb of Browne and Cudeck (1993), a RMSEA value between (0.05) and (0.08) suggests reasonable error of approximation. The standardized path coefficients and their levels of significance are depicted in Figure 2.

Table 3. Research Model Fit Indices

Indicator	Value
X2	228.42
df	54
P-Value	0.000
Calculated NC	4.23
CFI	0.912
RMSEA	0.014
90 CI	0.089 – 0.139
SRMR	0.069

Figure 2. SEM Analysis of Research Model

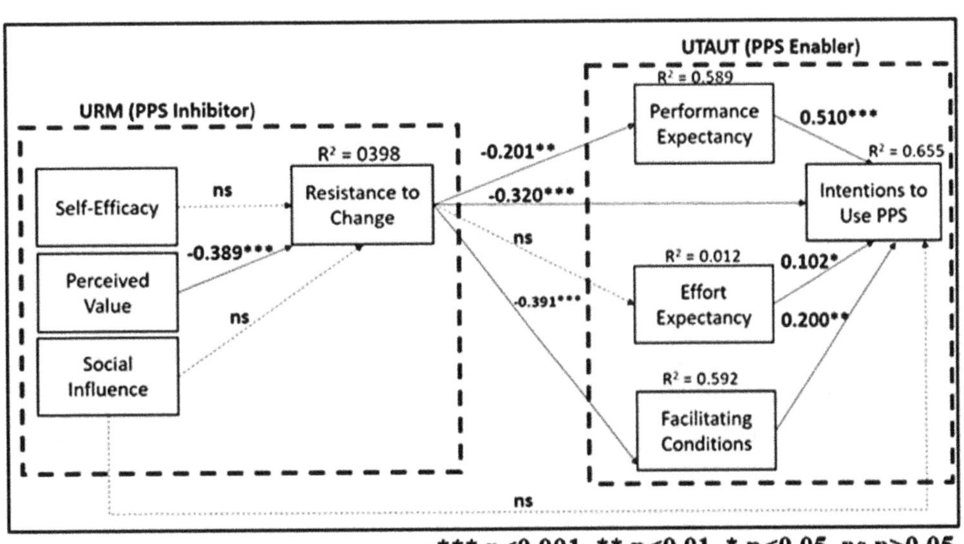

6.1. Contributions

The research model explained 65% of the variance in the dependent variable, Intention to Use PPS, while variance explained in Resistance to Change was 40%. Examining individual path coefficients, we find that majority of the initial hypotheses have been supported. Eleven of the eighteen hypothesized paths in the research model were statistically significant. The directionality of each significant path (positive or negative) was also hypothesized, providing overall support to our research model.

Resistance to Change has negatively predicated all of Performance Expectancy, Facilitating Conditions, and Intentions to Use PPS. This indicates support to the dual factor model by confirming that the inhibitor to PPS usage (Resistance to Change) has negative influence on enablers to PPS usage (Performance Expectancy, Facilitating Conditions, and Intentions to Use). However, Resistance to Change had no significant effect on Effort Expectancy. This indicates that there is no influence of patient resistant behaviors on their levels of effort perceptions (for example, ease of use levels). So patients who are resisting the change to the new PPS may still be considering the system as easy to use; however, their decision to resist the system has no influence on their perceptions. These findings confirmed that Resistance to Change does indeed have a biasing effect on patients' perceptions of PPS Performance Expectancy and Facilitating Conditions. The degree of this bias may depend on the enabler perceptions being evaluated. The results indicate that Resistance to Change biases patients' perceptions of PPS's Facilitating Conditions more than Performance Expectancy. Similarly, findings show that some enablers and/or inhibitors may have less predictive power on usage intentions. For example, Performance Expectancy had a significant effect on use intentions ($p<0.01$) while Effort Expectancy had a marginally significant effect on Intentions to Use PPS ($p < 0.10$).

These findings suggest that patients who resist the change are not likely to use the PPS. Thus, key PPS stakeholders are encouraged to mitigate resistance to change in order to increase levels of adoption. According to the findings in this study, resistance to change can be reduced by increasing levels of PV. Patients need to perceive more benefits of the PPS and less of its costs in order to maintain a high PV of the PPS and, in turn, minimize their resistance to change and increase use intention. Prior studies provide practical suggestions to increase PV and minimize change resistance, e.g., educating users on the potential benefits of the PPS and having key stakeholders take advantage of existing research on similar sociotechnical issues (Tang et al., 2006).

Another interesting result is that all of Performance Expectancy, Effort Expectancy, and Facilitating Conditions influence patient Intention to Use PPS. According to Venkatesh et al. (2003), prior research generally found that Facilitating Conditions have no significant effect on Intention to Use when Effort Expectancy constructs are accounted for in the model. One plausible reason for this finding is the context of the study. The compatibility of the PPS with patient devices and patient sense of control over the new system play major roles in influencing their use intention. These constructs are different in studies on employees who use identical devices to accomplish similar tasks and share the same level of control over the technology they are using.

The results show that Social Influence has no significant effect on Intentions to Use PPS or Resistance to Change. This again may be explained by the context of the study. Patients tend to use the PPS alone, with no communication with other patients regarding the system, mostly because patients normally do not have links of communication with one another. Also, because of privacy concerns, patients might not talk to others about the PPS to avoid discussions surrounding their own health issues. Thus, there is a lack of significant social influence being accounted for among patients.

Self-Efficacy for Change had no significant effect on Resistance to Change. This indicates that patient levels of computer literacy and familiarity with the PPS do not predict resistant behavior. This could be explained by the fact that technology advancements, such as mobile apps, are widespread and almost everyone today has adequate knowledge to use these applications. As a result, self-efficacy was not considered to be an issue swaying patient decisions to resist the change.

Lastly, age was the only control variable that had significant effect on both the outcome variable – Intentions to Use PPS – and the Resistance to Change construct. Educational level and sex had no significant effects.

Overall, the findings support the initial expectation that patient Intentions to Use PPS are predicted by both enabler (e.g., Performance Expectancy) and inhibitor (e.g., Resistance to Change) perceptions.

6.2. Limitations

This work has its limitations. First, enablers and inhibitors in this study were limited to those included in the URM and UTAUT. Future research is encouraged to include other enablers and/or inhibitors of usage that were not included in this study, e.g., perceived threat (Bhattacherjee & Hikmet, 2007) as an inhibitor or enjoyment/satisfaction (Wixom & Todd, 2005) as an enabler. Second, this study was conducted using a specific sample (patients). This may limit the generalizability of the findings. To replicate this study, researchers should conduct similar studies in different contexts with adverse sampling to make the work more generalizable. Third, since the data for this study was collected from a single source, validation concerns may arise. Future research can apply a longitudinal study or collect data from multiple sources to validate the findings. Finally, this study did not find social influence effects on use intentions nor resistance to change in this study. We encourage future research to consider different types of social influence when testing its effects, e.g., doctors' opinions about the system.

6.3. Implication for Research

This work has a number of implications for IS research. First, this is one of the first studies to include patient perceptions of Healthcare Information Technology (HIT). Prior IS research has focused mainly on HIT from the healthcare providers' perspective. Today's technology advancements are allowing patients to become major stakeholders by using HIT such as PPS, Self-monitoring devices, and telemedicine applications. Starting a stream of research that focuses on patients is a major advancement for the IS research on HIT. Second, this work provides empirical testing of the URM in a different context, which aids in the generalizability of the URM and opens the door for potential applications of the URM in a wide range of disciplines. Third, this is one of the only papers to test UTAUT on a technology that involves voluntary usage. Because of widespread technology applications, hedonic and voluntary types of usage are becoming more popular and are shaping new ways of communication that lead to interesting research questions. Although UTAUT includes Voluntariness of Use as a moderator between Social Influence and Behavioral Intentions, providing evidence that, as a whole, explains voluntary usage will encourage researchers to build on these findings and provide more advancement to this area of research. Fourth, this work provides findings on an individual level. Most prior studies have tested HIT resistance using group or organizational levels (Samhan & Joshi, 2015). Thus, this work is enriching the IS literature with findings from a novel sampling of patients. Finally, findings of this study posited that Facilitating Conditions can have positive effects on Intentions to Use while still accounting for the direct effects of Effort Expectancy and Performance Expectancy. This provides solid grounds for future research to hypothesize the effects of important Facilitating Conditions, such as compatibility, without the need to control for other important predictors.

6.4. Implications for Practice

This work has also a number of important implications for practice. First, providing a better understanding of technology resistance in general and HIT resistance, in particular, would help decision makers take adequate actions to intervene and prevent system failures on the basis of user resistance (Kim & Kankanhalli, 2009). Second, since HIT designers normally focus on system features with limited considerations of users' opinions (Bhattacherjee & Hikmet, 2007), providing a better understanding of patients' resistance toward PPS may help design systems that are acceptable to the average user and still remain functionally good. Providing insights into patient reactions to the PPS helps system designers build new systems and/or update existing systems with an increase in acceptance rates (Poon et al., 2004). Third, this study provides important stakeholders of the PPS with measurement tools that help track current and expected levels of user resistance in target healthcare organizations. Fourth, because age was the only control variable that had significant effect on both Intention to Use PPS and Resistance to Change, this sheds light on the possibility of providing seniors with other means of communication with the hospital that may be more suitable to that particular population. Finally, this work provides findings regarding technology resistance from an individual-level perspective. These findings provide better insights into people of interest such as project managers and system designers with specific focus on how end-users react individually rather than inferring results from studies conducted to measure these behaviors at organizational or group levels.

REFERENCES

Ajzen, I. (1985). From intentions to actions: A theory of planned behavior. In J. Kuhl & J. Beckmann (Eds.), Action-control: From cognition to behavior. Heidelberg: Springer. doi:10.1007/978-3-642-69746-3_2

Ajzen, I. (1991). The theory of planned behavior. *Organizational Behavior and Human Decision Processes*, *50*(2), 179–211. doi:10.1016/0749-5978(91)90020-T

Anwar, M., Joshi, J., & Tan, J. (2015). Anytime, anywhere access to secure, privacy-aware healthcare services: Issues, approaches and challenges. *Health Policy and Technology*, *4*(4), 299–311. doi:10.1016/j.hlpt.2015.08.007

Archer, C. (2015). Technology's impact on organizational changes. *Smallbusiness.chron.com*. Retrieved from http://smallbusiness.chron.com/technologys-impact-organizational-changes-20377.html

Bandura, A. (1986). *Social foundations of thought and action: A social cognitive theory*. Englewood Cliffs, NJ: Prentice Hall.

Bhattacherjee, A., & Hikmet, N. (2007). Physicians resistance toward healthcare information technology: A theoretical model and empirical test. *European Journal of Information Systems*, *16*(6), 725–737. doi:10.1057/palgrave.ejis.3000717

Bollen, K. A. (1998). Structural equation models. In P. Armitage & T. Colton (Eds.), *Encyclopedia of Biostatistics* (pp. 4363–4372). Sussex, UK: Wiley.

Browne, M., & Cudeck, R. (1993). Alternative ways of assessing model fit. In K. A. Bollen & J. S. Long (Eds.), *Testing Structural Equation Modeling* (pp. 136–162). Newbury Park, CA: Sage.

Burnham, T., Frels, J., & Mahajan, V. (2003). Consumer switching costs: A typology, antecedents, and consequences. *Journal of the Academy of Marketing Science*, *31*(2), 109–203. doi:10.1177/0092070302250897

Campbell, S., Shield, T., Rogers, A., & Gask, L. (2004). How do stakeholder groups vary in a Delphi technique about primary mental health care and what factors influence their ratings? *Quality & Safety in Health Care*, *13*(6), 428–434. doi:10.1136/qshc.2003.007815 PMID:15576704

Cenfetelli, R. (2004). Inhibitors and enablers and dual factor concepts in technology usage. *Journal of the Association for Information Systems*, *5*(11), 472–492.

Compeau, D., Higgins, C. A., & Huff, S. (1999). Social Cognitive Theory and individual reactions to computing technology: A longitudinal study. *Management Information Systems Quarterly*, *23*(2), 145–158. doi:10.2307/249749

Corp, I. B. M. (2013). *IBM SPSS Statistics for Windows, Version 22.0*. Armonk, NY: IBM Corp.

Davis, F. (1989). Perceived usefulness, perceived ease of use, and user acceptance of Information Technology. *Management Information Systems Quarterly*, *13*(3), 319–340. doi:10.2307/249008

Electronic Health Solutions (EHS). (n. d.). Retrieved from http://www.ehs.com.jo/en

Foon, Y., & Fah, B. C. Y. (2011). Internet banking adoption in Kuala Lumpur: An application of UTAUT model. *International Journal of Business and Management*, *6*(4), 161.

Gächter, S., Johnson, E. J., & Herrmann, A. (2007). *Individual-level loss aversion in riskless and risky choices. Inst. Study Labor Discuss. Pap. 2961*. Bonn, Germany: Inst. Study Labor.

Hartwick, J., & Barki, H. (1994). Explaining the role of user participation in Information System use. *Management Science*, *40*(4), 40–465. doi:10.1287/mnsc.40.4.440

Hsieh, P.-J., Lai, H.-M., & Ye, Y.-S. (2014). Patients' acceptance and resistance toward the health cloud: An integration of technology acceptance and status quo bias perspectives. *Proceedings of PACIS*, Chengdu, China.

Hu, L., & Bentler, P. M. (1999). Cutoff criteria for fit indexes in covariance structure analysis: Conventional criteria versus new alternatives. *Structural Equation Modeling*, *6*(1), 1–55. doi:10.1080/10705519909540118

Joshi, K. (1991). A model of user perspective on change: The case of Information Systems Technology implementation. *Management Information Systems Quarterly*, *15*(2), 229–242. doi:10.2307/249384

Kahneman, D., & Miller, D. T. (1986). Norm theory: Comparing reality to its alternatives. *Psychological Review*, *93*(2), 136–153. doi:10.1037/0033-295X.93.2.136

Kahneman, D., & Tversky, A. (1984). Choices, values, and frames. *The American Psychologist, 39*(4), 341–350. doi:10.1037/0003-066X.39.4.341

Kelman, H. (1958). Compliance, identification, and internalization: Three processes of attitude change. *The Journal of Conflict Resolution, 2*(1), 51–60. doi:10.1177/002200275800200106

Kim, H., & Kankanhalli, A. (2009). Investigating user resistance to information systems implementation: A status quo bias perspective. *Management Information Systems Quarterly, 33*(3), 567–582.

Lapointe, L., & Rivard, S. (2005). A multiple model of resistance to Information Technology implementation. *Management Information Systems Quarterly, 29*(3), 461–491.

Lewin, K. (1947). Frontiers in group dynamics: Concept, method, and reality in social sciences, social equilibria, and social change. *Human Relations, 1*(1), 5–41. doi:10.1177/001872674700100103

Lewis, W., Agarwal, R., & Sambamurthy, V. (2003). Sources of influence on beliefs about information technology use: An empirical study of knowledge workers. *Management Information Systems Quarterly, 27*(4), 657–678.

Moore, G., & Benbasat, I. (1991). Development of an instrument to measure the perceptions of adopting an Information Technology innovation. *Information Systems Research, 2*(3), 192–222. doi:10.1287/isre.2.3.192

MPLUS (Version 6.11). [Computer Software]. Los Angeles, CA: Muthén & Muthén.

Parthasarathy, M., & Bhattacherjee, A. (1998). Understanding postadoption behavior in the context of online services. *Information Systems Research, 9*(4), 362–379. doi:10.1287/isre.9.4.362

Poon, E., Blumenthal, D., Jaggi, T., Honour, M., Bates, D. W., & Kaushal, R. (2004). Overcoming barriers to adopting and implementing computerized physician order entry systems in U.S. hospitals. *Health Affairs, 23*(4), 184–190. doi:10.1377/hlthaff.23.4.184 PMID:15318579

Samhan, B. (2017). Patients' resistance towards Health Information Technology: A perspective of the dual factor model of IT usage. *Proceedings of the 50th Hawaii International Conference on System Sciences*, Big Island, HI. 10.24251/HICSS.2017.412

Samhan, B., & Joshi, K. D. (2015). Resistance of Healthcare Information Technologies: Literature review, analysis, and gaps. System Sciences (HICSS), Kawai, HI. *Proceedings of the 48th Hawaii International Conference*. IEEE.

Samhan, B., and Joshi, K.D. (in Press). Understanding electronic health records resistance: A revealed causal mapping approach. *Int. J. Electronic Healthcare, 9*(2/3).

Samuelson, W., & Zeckhauser, R. (1988). Status quo bias in decision making. *Journal of Risk and Uncertainty, 1*(1), 7–59. doi:10.1007/BF00055564

Sherif, M. (1936). *The psychology of social norms*. New York: Harper & Row.

Sirdeshmukh, D., Singh, J., & Sabol, B. (2002). Consumer trust, value and loyalty in relational exchange. *Journal of Marketing, 66*(1), 15–37. doi:10.1509/jmkg.66.1.15.18449

Spil, T., Schuring, R. W., & Michel-Verkerke, M. B. (2004). Electronic prescription system: Do the professionals use it? *International Journal of Healthcare Technology and Management*, *6*(1), 32–55. doi:10.1504/IJHTM.2004.004824

Tang, P. C., Ash, J. S., Bates, D. W., Overhage, J. M., & Sands, D. Z. (2006). Personal health records: Definitions, benefits, and strategies for overcoming barriers to adoption. *Journal of the American Medical Informatics Association*, *13*(2), 121–126. doi:10.1197/jamia.M2025 PMID:16357345

Taylor, S., & Todd, P. A. (1995a). Assessing IT usage: The role of prior experience. *Management Information Systems Quarterly*, *19*(2), 561–570. doi:10.2307/249633

Taylor, S., & Todd, P. A. (1995b). Understanding Information Technology Usage: A Test of Competing Models. *Information Systems Research*, *6*(4), 144-176.

Thompson, R., Higgins, C.A., & Howell, J.M. (1991). Personal Computing: Toward a Conceptual Model of Utilization. *MIS Quarterly*, *15*(1), 124-143.

Tulu, B., Burkhard, R., & Horan, T. A. (2007). Continuing use of medical information systems by medical professionals: Empirical evaluation of a work system model. *Communications of the AIS*, *18*(31), 641–656.

Venkatesh, V., & Brown, S. A. (2001). A longitudinal investigation of personal computers in homes: Adoption determinants and emerging challenges. *Management Information Systems Quarterly*, *25*(1), 71–102. doi:10.2307/3250959

Venkatesh, V., Morris, M. G., Davis, G. B., & Davis, F. D. (2003). User acceptance of Information Technology: Toward a unifying view. *Management Information Systems Quarterly*, *27*(3), 425–478.

Weeger, A., Gewald, H., & Vriesman, L. J. (2011). Do risk perceptions influence physician's resistance to use Electronic Medical Records? An exploratory research in German hospitals. *Proceedings of the American Conference on Information Systems* (AMCIS). Detroit, MI: Association for Information Systems (AIS).

Williams, L. J., Hartman, N., & Cavazotte, F. (2010). Method variance and marker variables: A review and comprehensive CFA marker technique. *Organizational Research Methods*, *13*(3), 477–514. doi:10.1177/1094428110366036

Wixom, B. H., & Todd, P. A. (2005). A theoretical integration of user satisfaction and technology acceptance. *Information Systems Research*, *16*(1), 85–102. doi:10.1287/isre.1050.0042

Wu, Y., Tao, Y., & Yang, P. C. (2008). The use of unified theory of acceptance and use of technology to confer the behavioral model of 3G mobile telecommunication users. *Journal of Statistics and Management Systems*, *11*(5), 919–949. doi:10.1080/09720510.2008.10701351

This research was previously published in the International Journal of Healthcare Information Systems and Informatics (IJHISI), 12(4); edited by Joseph Tan, pages 68-86, copyright year 2017 by IGI Publishing (an imprint of IGI Global).

APPENDIX

Table 4. Exhibit 1: Measurement Items

Construct	Item	Wording
Perceived Value	PVL1	Considering any effort and time I have to spend, accessing my data on the patient portal is worthwhile.
	PVL2	Considering any losses I incur, the patient portal is of good value.
	PVL3	Considering any hassles I have to experience, accessing my data on the patient portal is beneficial to me.
	PVL4	Considering the limitations of the patient portal, I find it advantageous.
Social Influence	SOI1	People who influence my behavior think that I should use the portal.
	SOI2	People who are important to me think that I should use the portal.
	SOI3	I use the system because of the proportion of people I know who use the system.
	SOI4	People who use the patient portal have a sense of prestige.
	SOI5	People who use the patient portal have a high profile.
	SOI6	Using the patient portal is a status symbol.
Self-Efficacy for Change	SFC1	I am able to use the patient portal easily based on my own knowledge, skills and abilities.
	SFC2	I am able to use the patient portal without the help of others.
	SFC3	I am able to use the patient portal reasonably well on my own.
	SFC4	I have confidence in my ability to use the patient portal without any difficulties.
Resistance to Change	RES1	I would rather not comply with the change to the new way of working with the patient portal.
	RES2	I oppose the change to the new way of working with the patient portal.
	RES3	I do not agree with the change to the new way of working with the patient portal.
Performance Expectancy	PE1	Using the patient portal enables me to accomplish desired tasks quickly.
	PE2	Using the patient portal makes it easier to complete desired tasks.
	PE3	I find the patient portal useful in completing desired tasks.
	PE4	Using the patient portal significantly increases the quality of the task completed.
Effort Expectancy	EE1	Learning to operate the patient portal is easy for me.
	EE2	My interaction with the patient portal is clear and understandable.
	EE3	Using the patient portal does not takes too much time.
	EE4	Working with the patient portal is not so complicated; it is not difficult to understand what is going on.
Facilitating Conditions	FC1	I have control over using the patient portal.
	FC2	I have the resources necessary to use the patient portal.
	FC3	I have the knowledge necessary to use the patient portal.
	FC4	The patient portal (website and mobile app) is compatible with other systems and devices I use.
	FC5	Guidance is available to me on how to use the patient portal.
	FC6	Using the patient portal fits with my style of accomplishing tasks.
Intention to Use	IU1	I intend to use the patient portal now and continue to use it in the future.
	IU2	I predict I would use the patient portal now and continue to use it in the future.
	IU3	I plan to use the patient portal now and continue to use it in the future.

Table 5. Exhibit 2: Instrument Validation

Items Loadings and CFA Model Fit Statistics						
Construct	**Item**	**Std. Loading**	**Mean**	**STD**	**Cronbach's α**	**CFA Model Fit Statistics**
PVL	PVL1	0.953	3.249	1.808	0.912	X2: 19.673 df: 2 P < 0.001 CFI: 0.912 RMSEA: 0.127 CI: 0.122 - 0.322 SRMR: 0.044
	PVL2	0.969				
	PVL3	0.956				
	PVL4	0.976				
SFC	SFC1	0.761	3.537	0.987	0.913	X2: 9.905 df: 2 P < 0.05 CFI: 0.963 RMSEA: 0.104 CI: 0.011 - 0.101 SRMR: 0.004
	SFC2	0.888				
	SFC3	0.977				
	SFC4	0.740				
SOI	SOI1	0.925	3.632	0.899	0.965	X2: 17.413 df: 2 P < 0.01 CFI: 0.981 RMSEA: 0.135 CI: 0.072 - 0.131 SRMR: 0.029
	SOI2	0.927				
	SOI3	0.902				
	SOI4	0.846				
	SOI5	0.913				
	SOI6	0.985				
	OGS2	0.956				
	OGS3	0.958				
	OGS4	0.833				
RES	RES1	0.979	3.544 2.998	1.011 1.111	0.943 0.926	X2: 9.942 df: 2 P < 0.05 CFI: 1.00 RMSEA: 0.000 CI: 0.000 - 0.002 SRMR: 0.000
	RES2	0.976				
	RES3	0.964				
PEF	PE1	0.945	2.989	1.115	0.972	X2: 7.683 df: 2 P < 0.05 CFI: 0.965 RMSEA: 0.089 CI: 0.000 - 0.133 SRMR: 0.018
	PE2	0.946	3.11	1.113	0.977	
	PE3	0.943	3.97	0.998	0.999	
	PE4	0.946	2.99	0.897	0.983	
EFE	EE1	0.998	3.31	0.989	0.976	X2: 13.969 df: 2 P < 0.001 CFI: 0.987 RMSEA: 0.188 CI: 0.103 – 0.206 SRMR: 0.023
	EE2	0.997	3.44	0.911	0.956	
	EE3	0.979	3.21	0.999	0.967	
	EE4	0.989	3.30	0.988	0.934	

continued on following page

Table 5. Continued

Items Loadings and CFA Model Fit Statistics						
Construct	**Item**	**Std. Loading**	**Mean**	**STD**	**Cronbach's α**	**CFA Model Fit Statistics**
FAC	FC1	0.991	3.24	1.089	0.887	X2: 9.800 df: 2 P < 0.05 CFI: 0.936 RMSEA: 0.014 CI: 0.011 - 0.121 SRMR: 0.008
	FC2	0.929	2.99	1.003	0.981	
	FC3	0.927	2.87	1.002	0.899	
	FC4	0.923	3.01	1.321	0.901	
	FC5	0.913	3.13	1.091	0.921	
	FC6	0.965	3.01	0.998	0.879	
Intention to Use	IU1	0.901	2.99	1.115	0.913	X2: 10.420 df: 2 P < 0.05 CFI: 1.00 RMSEA: 0.000 CI: 0.000 - 0.00 SRMR: 0.001
	IU2	0.899	3.01	1.112	0.942	
	IU3	0.976	3.01	1.133	0.954	

Chapter 22
User Behavioral Intention Toward Using Mobile Healthcare System

Pantea Keikhosrokiani
School of Computer Sciences, Universiti Sains Malaysia, Malaysia

Norlia Mustaffa
School of Computer Sciences, Universiti Sains Malaysia, Malaysia

Nasriah Zakaria
Medical Education Department, College of Medicine, King Saud University, Saudi Arabia & School of Computer Sciences, Universiti Sains Malaysia

Ahmad Suhaimi Baharudin
School of Computer Sciences, Universiti Sains Malaysia, Malaysia

ABSTRACT

This chapter introduces Mobile Healthcare Systems (MHS) and employs some theories to explore the behavioral intention of Smartphone users in Penang, Malaysia to use MHS. A survey was conducted in the form of questionnaire to Smartphone users in Penang, Malaysia for the duration of three weeks starting in September 2013. A total number of 123 valid surveys out of 150 were returned, which is equivalent to a response rate of 82%. The authors use Partial Least Squares (PLS) for analyzing the proposed measurement model. The factors that are tested are self-efficacy, anxiety, effort expectancy, performance expectancy, attitude, and behavioral intention to use. The results indicate which factors have a significant effect on Smartphone users' behavioral intention and which factors are not significant. The results assist in assessing whether MHS is highly demanded by users or not, and will assist in development of the system in the future.

DOI: 10.4018/978-1-5225-6198-9.ch022

INTRODUCTION

Patient monitoring and intelligent health care system can prevent chronic diseases on time. With the advancement of wireless technology and mobile-based systems, the process of monitoring patients becomes ubiquitous. Mobile Health System (MHS) provides the facility of constant monitoring of patients in case of emergency. Medical professionals are concerned about risk and uncertainty of new technologies in healthcare. Thus, they usually tend to adapt new technologies first until those technologies become mature. Introduction of information technology (IT) in healthcare has a deep effect on the performance of the healthcare systems. Many researchers have studied IT adoption and diffusion recently. Adoption is defined as the decision to accept, or invest in a new technology (Masrom & Hussein, 2008). Accessing to the value of the IT to the projects or organization is one of the main objectives for those IT researchers. IT resources can improve overall effectiveness of the organizations. IT adoption and usage by individual users is another key objective considered by researchers.

The main goal of this article is to conduct a survey in order to evaluate user behavioral intention toward using MHS. In consideration of user's behavior toward using MHS, we will conduct a survey among Smartphone users in Penang, Malaysia. The main construct of this study will be user behavioral intention to use the MHS. Behavioral intention to use is a measure of the strength of individual intention to perform a specified behavior which is use of a new technology or system (Davis, Bagozzi, & Warshaw, 1989). For this study, we proposed to use a research model based on three models i.e. Unified Theory of Acceptance and Use of Technology (UTAUT), Theory of Planned Behavior (TPB) and Theory of Acceptance Model (TAM). We will use Partial Least Squares (PLS) for analyzing the proposed measurement model. Firstly, construct validity that comprises convergent and discriminant validities will be used for this purpose. In addition, the structural equation modeling technique will be performed to evaluate the causal model. The factors that will be tested are self-efficacy, anxiety, effort expectancy, performance expectancy, attitude, and behavioral intention to use. The results will indicate which factors have significant effect on Smartphone user's behavioral intention and which factors are not significant. The results will assist in assessing whether MHS is highly demanded by users or not, as well as assist us in development of the system in the future.

LITERATURE REVIEW

This section discusses briefly the literature review approach and background of the study. In the background section, theoretical bases along with development of appropriate hypothesis will be mentioned. Based on the previous researches, research framework for this study is designed which includes formulating conceptual model, research hypotheses, and methodology.

Several literature reviews cover the usage of mobile technologies in healthcare. Moreover, many researchers have studied acceptance of mobile technology in healthcare or user behavioral intention to use of new technologies in healthcare. However, those studies have been conducted with different aims, and it is not easy to compare their results. In order to find important factors of user acceptance or user behavioral intention to use of mobile technology in healthcare, we have done a systematic literature search in several databases such as MEDLINE, PubMed, ScienceDirect, SpringerLINK, IEEE Xplore and so on. We searched using the terms "mobile healthcare system", "acceptance of mobile technology in healthcare", "behavioral intention to use of mobile device in healthcare." In the next step, we have combined the search results and shortlisted the most recent and relevant papers with the aim to evaluate user behavioral intention to use of mobile device in healthcare.

MOBILE HEALTHCARE SYSTEM (MHS)

With growth of mobile device penetration as well as worldwide enhancement of wireless networks, the wireless infrastructure can support many healthcare systems. This can provide ubiquitous healthcare, which means healthcare services available at anytime at anywhere. The wireless technologies offer various services such as reliable access along with transmission of medical data, tracking patient's current location, support of patient mobility, providing quick services in emergency and so on (Varshney, 2007). Location management can add value to healthcare by offering a location-based healthcare services using location tracking technologies such as global positioning systems (GPS), cellular networks, wireless local area networks (LANs), and radio frequency identification (RFID) (Tu, Zhou, & Piramuthu, 2009; Varshney, 2003). Various remote devices are utilized in wireless healthcare systems such as wireless sensors for blood oxygen saturation monitoring (Asada, Shaltis, Reisner, Sokwoo, & Hutchinson, 2003; Sokwoo, Boo-Ho, Kuowei, & Asada, 1998), infrared and radio-based locator badges (Stanford, 2002), using wearable devices for long-term health monitoring (Jovanov, O'Donnel, Morgan, Priddy, & Hormigo, 2002), remote heart monitoring systems (Ernest, et al., 2011; Ross, 2004), healthcare monitoring for patients with hypertension and arrhythmia (Keikhosrokiani, Mustaffa, Zakaria, & Sarwar, 2012), providing emergency services based on patient's current location (El-Masri & Saddik, 2012; Keikhosrokiani, et al., 2012), wireless telemetry systems for EEG epilepsy monitoring (Liszka, et al., 2004), real-time monitoring of patients in the home environment (Korhonen, Parkka, & Van Gils, 2003), and lastly, using GlucoBeep device for checking and transmitting the patient's diabetic information (Jansà, et al., 2006). In addition, there are other studies have been conducted in order to improve emergency healthcare systems (Bergrath, et al., 2011; Li, Cheng, Lu, & Lin, 2011).

The use of wireless technologies in healthcare can provide useful treatment for patients in hospitals, thus the outcome of the healthcare will increase effectively (Gravelle & Siciliani, 2008). Moreover, using mobile device and wireless technologies in healthcare can reduce the long-term cost and increase productivity in healthcare systems. In order to support patient's healthcare need, wireless patient monitoring systems must be designed for the homes, nursing homes and hospitals (Kevin & Yuan-Ting, 2003; Pollard, Rohman, & Fry, 2001; Varady, Benyo, & Benyo, 2002). Utilizing mobile device and wireless technologies in healthcare can improve the quality of healthcare. Hence, exploring the use of mobile devices is a major issue for medical personnel, nursing staff, and patients in hospitals.

THEORIES ON TECHNOLOGY ACCEPTANCE MODELS (TPB, TAM, AND UTAUT)

Theory of Planned Behavior (TPB) was introduced in 1985 by Ice Ajzen and it was also known as social cognition model (SCM) (Ajzen, 1985). TPB and Technology Acceptance Model (TAM) are originally based on Theory of Reasoned Action (TRA) that is a model for predicting and explaining human behavior in different domains (Fishbein & Ajzen, 1975). TPB adds perceived behavioral control to the TRA model where perceived behavioral control is determined by availability of skills, resources, and opportunities to achieve outcomes. It is closely linked to self-efficacy belief concept (Ajzen, 1991). The concept of self-efficacy is concerned with people's belief in their ability to produce effects. TPB illustrates attitude (AT), subjective norm (SN), and perceived behavioral control (PBC) as direct determinants of intentions that will influence behavior. The focus of this study is on the attitude (AT) as well as behavioral intention (BI) in TPB model. AT refers to "the degree to which the person has a favorable or unfavorable evaluation of the behavior in question" (Ajzen, 1989). Whereas BI refers to "the subjective probability of one's engagement in any behavior." The stronger the behavior, the more likely is the execution of the behavior (Fishbein & Ajzen, 1975)

TAM is an information system theory that depicts how user accepts and uses a computer-based technology. It was developed by Davis Jr, in order to explain computer-based behavior (Davis Jr, 1986). As mentioned before, TAM is the extension of TRA (Fishbein & Ajzen, 1975). The main purpose of TAM is to provide a basis for tracing the impact of external variables on internal beliefs, attitudes, and intentions. In TAM model attitude towards behavior in using the system is a central feature. Two most effective factors on attitude are perceived usefulness (PU) as well as perceived ease of use (PEOU). PU refers to "the degree to which a person believes that using a particular system would enhance his or her job performance." PEOU refers to "the degree to which a person believes that using particular system would be free from effort." Behavioral intention to use (BI) the system is directly determined by the person's attitude toward using the system as depicted in TAM model. BI is defined as "a measure of the strength of one's intention to perform a specific behavior, that is, use an information system" (Davis Jr, 1986).

Unified theory of acceptance and use of technology (UTAUT) was developed by Venkatesh, Morris and Davis, (2003) in order to merge previous TAM related studies. UTAUT is combination of TAM, TRA, TPB, TAM-TPB, motivational model, innovation diffusion theory, and social cognitive theory. It is developed for use in information and communication technology research. It seeks to clarify user intention to use an information system (IS) and consequent usage behavior. The theory consists of four main constructs including performance expectancy, effort expectancy, social influence, and facilitating condition. These constructs are direct determinants of usage behavior (Venkatesh, Morris, Davis, & Davis, 2003).

The focus of this study is on performance expectancy (PE) as well as effort expectancy (EE) in UTAUT model. PE refers to "the degree to which an individual believes that using the system will help him/her to attain gains in job performance"; whereas, EE is defined as "the degree of ease associated with the use of the systems."

In line with this, a study has been conducted in order to understand student perceptions in using course management software. The studied course management software is Blackboard which is a web-based tool that becomes popular and important application in higher education (Marchewka, Liu, & Kostiwa, 2007). This study utilized UTAUT model as well as adding self-efficacy and anxiety constructs to the model. Self-efficacy (SE) and anxiety (Ax) are another two more important constructs that can be added to UTAUT (A Iahad, Rahim, Zairah, & Oye, 2012). SE is defined as "perceptions of user's ability to use new technology in the accomplishment of the task." Ax can be defined as "the degree to which a user's fear, concern and hesitation of damaging the device, being embarrassed to use the new system" (A Iahad, et al., 2012).

CONCEPTUAL MODEL AND RESEARCH HYPOTHESIS

Definitions and Research Model

Based on the foregoing discussion, this study integrates TPB, TAM, and UTAUT with two additional variables including self-efficacy and anxiety affecting behavioral intention to use MHS among Smartphone users. Table 1 summarized the definition of each construct and the conceptual research model is illustrated in Figure 1.

The conceptual model depicts four independent variables (self-efficacy, effort expectancy, performance expectancy, and anxiety) which affect dependent variable that is behavioral intention to use MHS. There is a mediating effect of the attitude towards using MHS between three independent variables (self-efficacy, effort expectancy, and performance expectancy) and dependent variable (behavioral intention to use MHS).

Table 1. Definition of the constructs

Construct	Definition
Self-Efficacy (SE)	The Smartphone user's perceptions of his or her ability to use MHS in the accomplishment of healthcare task.
Effort Expectancy (EE)	The degree to which a Smartphone user believes that the use of MHS would be free of physical and mental effort.
Performance Expectancy (PE)	The degree to which a Smartphone user believes that the use of MHS would enhance his or her performance.
Attitude towards Using MHS (AT)	A Smartphone user's positive or negative feeing about using MHS.
Anxiety (Ax)	The degree to which a Smartphone user's fear, concern and hesitation of damaging the device, being embarrassed to use MHS.
Behavioral Intention to Use (BI)	The degree to which a Smartphone user's motivation intend to use the MHS.

Figure 1. Conceptual model for behavioral intention to use MHS

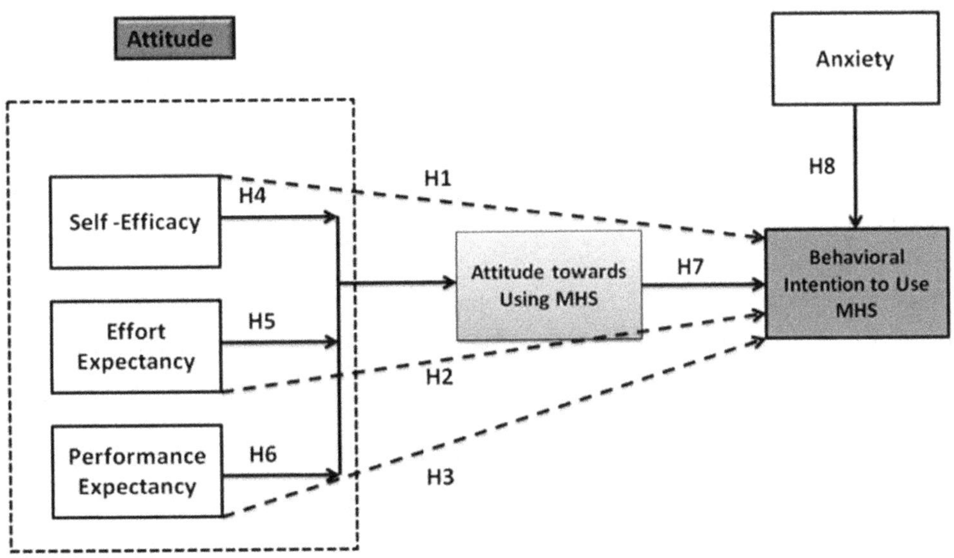

The following section discusses the development of relevant hypothesis for this study. Majority of information system model do not specifically regard differences between stationary computing and mobile devices in technology acceptance models. Thus, in order to draw the mobile context and study on technology acceptance models, Biljon, et al., (2007) designed a Mobile Phone Technology Acceptance Model (MOPTAM) (Biljon & Kotz, 2007). The need for considering control-based beliefs to understand personal behavior related to the adoption of mobile healthcare was indicated in many studies (Hung & Jen, 2012), (Wan Ismail, Chan, Buhari, & Muzaini, 2012), (I.-L. Wu, Li, & Fu, 2011), (Holden & Karsh, 2010), (Ozok, et al., 2009), and (J.-H. Wu, Wang, & Lin, 2007). User's belief in using mobile technologies can be related to individual categories such as perceived self-efficacy in using mobile technology, or organizational category such as technology and management support as well as training.

Based on the above discussion as well as literature review that have been conducted, there are three direct variables i.e. self-efficacy (SE), effort expectancy (EE), and performance expectancy (PE) for determining behavioral intention to use of MHS (Marchewka, et al., 2007). For the purpose of this study, there are three linkages between these three variables and behavioral intention to use the MHS. Therefore, three hypotheses are proposed as follow:

H1: Self-efficacy has a positive effect on Smartphone user's behavioral intention to use the MHS.
H2: Effort expectancy has a positive effect on Smartphone user's behavioral intention to use the MHS.
H3: Performance expectancy has a positive effect on Smartphone user's behavioral intention to use the MHS.

Moreover, based on TAM and TPB, as mentioned previously, attitude has a direct impact on behavioral intention to use MHS. Based on the literature review, self-efficacy, effort expectancy, and performance expectancy can be considered as direct variables for determining attitude (Yu, Ha, Choi, & Rho, 2005). For this study, we established that there are three linkages between these three variables and attitude towards using MHS. Therefore, three hypotheses are proposed as follows:

H4: Self-efficacy has a positive effect on Smartphone user's attitude towards using MHS.
H5: Effort expectancy has a positive effect on Smartphone user's attitude towards using MHS.
H6: Performance expectancy has a positive effect on Smartphone user's attitude towards using MHS.

Based on the theories such as TRA, TPB, and TAM, it is ascertained that attitude has a positive effect on the behavioral intention to use of the new technology. Thus, the hypothesis is set as follows:

H7: Smartphone user's attitude towards using MHS has a positive effect on the behavioral intention to use of MHS.

Mobile healthcare anxiety here refers to "the main reason for prevention of computer technology adoption is defined as trepidation, fear, concern and hesitation of damaging the device, being embarrassed" (Heinssen, Glass, & Knight, 1987). Jen and Chao (2008) defined mobile healthcare anxiety as "a high anxious response towards interaction with the mobile patient safety information system" which has a negative effect on the quality of care that physician provides for the patient. Therefore, it has a negative effect on user' s attitude towards using that healthcare system (Jen & Chao, 2008). Based on the previous studies, mobile patient safety information systems play a role in improving patient services and reducing patient risk, as well as showing that use of these communication systems may unintentionally be a factor in physician anxiety (Jen & Chao, 2008). Marchewka, et al. (2007) used anxiety as one of the effective factor on behavioral intention to use in UTAUT model for his research (Marchewka, et al., 2007). We can conclude that less anxiety means user has a positive intention towards using MHS and finally would like to use it. Thus, the following hypothesis is set:

H8: The level of mobile healthcare anxiety has a negative effect on the behavioral intention to use MHS.

RESEARCH METHODOLOGY AND DESIGN

Instrumentation

A survey is used as an instrument to collect data for this study. It was conducted for the duration of three weeks starting in September 2013. The survey was conducted by distributing the questionnaire to the respondents who are Smartphone users in Penang, Malaysia. The questionnaire was designed in the form of four categories of questions addressing demographic, Smartphone users' experience using Smartphone, chronic disease experience, and assessment part. The type of questions in the first three parts is selective while in the assessment part the response is collected using a five point Likert-type scale from one, being "strongly disagree" to five, being "strongly agree".

DATA ANALYSIS AND RESULTS

Demographics

This section presents descriptive analyses of demographic data providing a profile of respondents including frequency values. All the tables in this section present a profile of the surveyed respondents with regard of age, gender, and education, Smartphone users' experience using Smartphone and their experience about chronic disease. In this study, we are using basic descriptive analysis with frequency table to provide a demographic profile of respondents. The second part of data analysis is to understand Smartphone users and their opinion about usefulness of Smartphone for healthcare. Third part is analyzing about experience of chronic disease; whether the respondents experience it themselves or someone in their family face the chronic disease. All questions required to be answered, so there is no missed question. Respondents were Smartphone users from Penang, Malaysia, with the total number of 150. A total number of 123 valid surveys out of 150 were returned which is equivalent to a response rate of 82%.

Table 2 illustrates demographic data of the respondents. As Malaysia is a multicultural country, the respondents comprise of different ethnicity namely Malay, Chinese, and Indian with the percentage of 34.1%, 39.0% and 26.8%, respectively. Based on Table 2, majority of the respondents were female with 56.1% while the rest is 43.9% male participants in the survey. Most of the respondents were between 26 to 33 years old and no participant whose age is less than 18 years old. For the academic qualification part, the respondents mostly graduated at least at undergraduate level (60.2%). The respondents' income mainly is in between 1,000 to 3,000 (Ringgit Malaysia) and 3,001 to 5,000 (Ringgit Malaysia) with the percentage of 38.2 and 37.4 respectively.

Table 2. Demographic data

Item	Option	Frequency	Percent (%)
Race	Malay	42	34.1
	Chinese	48	39.0
	Indian	33	26.8
Age	Under 18	-	-
	18-25	34	27.6%
	26-33	64	52.0%
	34-41	19	15.4%
	42 and above	69	4.9%
Gender	Male	54	43.9%
	Female		56.1%
Education	High School	2	1.6%
	Diploma	-	-
	Undergraduate	74	60.2%
	Postgraduate	39	31.7%
	Professional Degree	8	6.5%
Income	Less than RM 1,000	16	13.0%
	Between RM 1,000 to RM 3,000	47	38.2%
	Between RM 3,001 to RM 5,000	46	37.4%
	Between RM 5,001 to RM 10,000	14	11.4%

Measurement Model

Partial Least Squares (PLS) is a powerful structural equation modeling technique that uses a nonparametric and component-based approach for estimation (W. Chin, Marcolin, & Newsted, 1996; W. W. Chin, 1998a, 1998b). PLS reduce demands on sample size and residual distributions. The sample size needed for PLS is 10 times of the number of indicators for the most complex construct. When there are three variables and each includes 2-4 measuring items, the sample size of 123 respondents might not be enough to use Analysis of Moment Structures (AMOS; another type of analytical tool) for a goodness of mode-fit. PLS is a good analytical tool for fitting the requirement of a small sample size. Thus, PLS is suitable in this study for analyzing the measurement model. Firstly, construct validity test that comprises of convergent and discriminant validities is used and this is followed by hypotheses testing.

- **Convergent Validity:** Convergent validity of a construct is assessed by few criteria including: (1) All of the item loading in a construct must be more than 0.5 (Janz & Prasarnphanich, 2003; Saadé, 2007) that suggest item reliability while the Cronbach's alpha values must exceed 0.6 for the construct reliability (Nunnally, 2010). The reliability of individual item was assessed by examining the loading of items on their respective constructs that are illustrated in Table 3. The loading must be greater than 0.5 to indicate significant variance was shared between each item and the construct. The only loading that is below 0.5 is Ax2 (0.491) that is recommended to be removed

Table 3. Assessment of convergent validity

Construct	Items	Item Loading	Composite Reliability	AVE	Cronbach'α
AT	AT1	0.933	0.936	0.880	0.863
	AT2	0.943			
Ax	Ax1	0.976	0.807	0.599	0.802
	Ax2	0.491			
	Ax3	0.776			
BI	BI1	0.912	0.922	0.797	0.873
	BI2	0.919			
	BI3	0.846			
SE	SE1	0.902	0.875	0.778	0.717
	SE2	0.862			
EE	EE1	0.856	0.715	0.562	0.232
	EE2	0.626			
PE	PE1	0.903	0.914	0.730	0.717
	PE2	0.958			
	PE3	0.688			
	PE4	0.844			

from the model. As shown in Table 3, reliability was evaluated using the Cronbach's alpha. Most of the constructs appear to have a good degree of reliability since each computed statistic is above 0.60. The only questionable construct is Effort Expectancy (EE) as the respective test statistic falls below 0.60. (2) Composite reliability (CR) for a construct should exceed 0.6 and can be interpreted like Cronbach's alpha coefficient (Bagozzi & Yi, 1988). Based on Table 3, CR for all of the constructs is more than 0.6. (3) Next important factor is average variance extracted (AVE) for a construct that should be greater than 0.5 (Bagozzi & Yi, 1988). After removing Ax2 from the model, calculation was done again and AVE for anxiety changed from 0.599 to 0.776. AVE for all of the constructs is more than 0.5 except the AVE of effort expectancy (EE) and anxiety (Ax) that are 0.562 and 0.599 respectively. These AVEs are not strong enough but still acceptable (depicted in Table 3).

- **Discriminant Validity:** "Discriminant validity is a measure of the extent to which the construct is not a reflection of some other variable." Table 4 depicts discriminant validity of the construct using squared root of the AVE for each construct. This value should be more than the correlations between it and all other constructs. The results illustrate a high degree of discriminant validity as shown in Table 4.

Statistical Analysis and Hypothesis Testing

In order to examine the causal relationships among the constructs, a structural model was built. For evaluating the structural model, two steps are needed here. The first step is estimation of standardized path coefficients and their statistical significance for testing the hypotheses. Unfortunately, a significance

Table 4. Assessment of discriminant validity

	AT	Ax	BI	EE	PE	SE
AT	***0.938***					
Ax	0.031	***0.881***				
BI	0.768	-0.154	***0.893***			
EE	0.444	0.133	0.588	***0.750***		
PE	0.392	-0.041	0.476	0.665	***0.854***	
SE	-0.064	-0.524	0.046	0.073	0.014	***0.882***

Note: The bold-italic numbers in the diagonal row are square roots of AVE

test or confidence interval estimation is not provided by PLS. In this step, we used Bootstrapping analysis to estimate path coefficients, statistical significance, and relevant parameters, such as means, standard errors, item loadings, and item weights. The second step is calculation of the coefficient of determination (R2) for endogenous variables in order to assess the predictive power of the structural model. The R-square values (0.222) for attitude towards using MHS and 0.717 for behavioral intention to use MHS suggest a good model fit as shown in Figure 2.

Table 5 illustrates the results of hypotheses testing. Although, the significance of the proposed relationship between self-efficacy and performance efficacy with Smartphone user's attitude toward using MHS is not supported, effort expectancy ($\beta = 0.340$, $p < 0.05$) poses significant influence on attitude. This supports H5 and confirms the importance of effort expectancy in the attitude towards using MHS.

As shown in Table 5, the results support the prediction that attitude ($\beta=0.616$, $p<0.01$) and anxiety ($\beta=-0.265$, $p<0.05$) have statistically significant relationships with behavioral intention to use of MHS. Thus, hypothesis H7 and H8 are supported. However, effort expectancy ($\beta = 0.369$, $p < 0.01$) is reported

*Figure 2. Results of structural model. Value on path: standardized coefficients (β), R2: Coefficient of determination, **p<0.01, *p<0.05.*

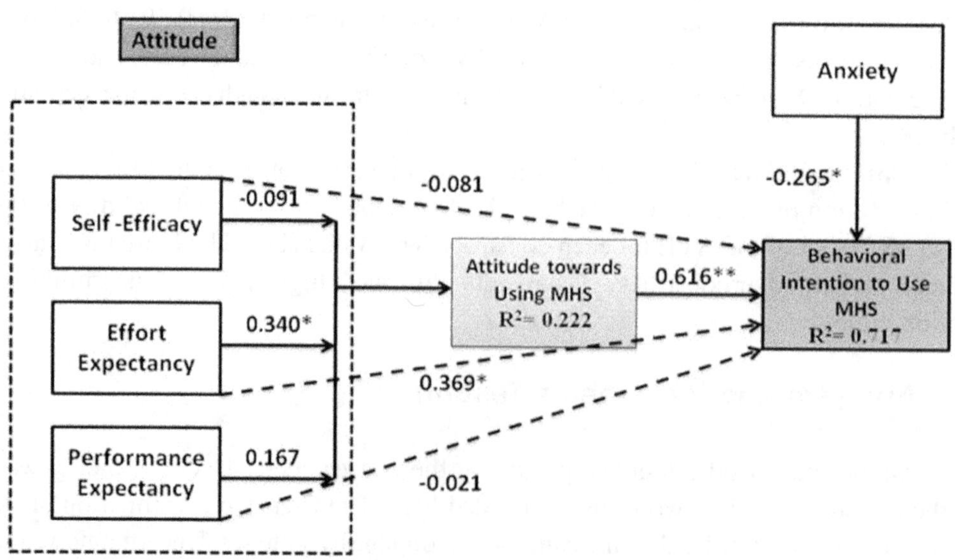

Table 5. Results of hypotheses testing

Path	Beta	Standard Error (SE)	T-Value	P-Value
AT → BI	0.616	0.063	9.722	p<0.01
Ax → BI	-0.265	0.126	2.100	p<0.05
EE → AT	0.340	0.132	2.571	p<0.05
EE → BI	0.369	0.093	3.982	p<0.01
PE → AT	0.167	0.136	1.226	Not significant
PE → BI	-0.021	0.096	0.215	Not significant
SE → AT	-0.091	0.099	0.915	Not significant
SE → BI	-0.081	0.078	1.028	Not significant
SE→AT→BI	-0.056	0.060	-0.936	Not Significant
EE→AT→BI	0.209	0.089	2.349	P<0.01
PE→AT→BI	0.103	0.086	1.194	Not Significant

to be important antecedents of behavioral intention to use and H2 is supported. In contrast, the constructs of self-efficacy (β=-0.081) as well as performance expectancy (β=-0.021) indicates no significant influence on behavioral intention to use MHS.

Moreover, we evaluate the mediating effect of the attitude between three independent variables (self-efficacy, effort expectancy, and performance expectancy) and dependent variable (behavioral intention to use). Based on our evaluation, attitude mediate the relationship between effort expectancy and behavioral intention to use MHS where β=0.209 and p<0.01.

CONCLUDING REMARKS

This study proposed a revised model based on TPB, TAM, and UTAUT models in order to investigate what determined Smartphone's user behavioral intention toward using MHS. The results indicated that effort expectancy, anxiety, and attitude are important determinants to Smartphone user's behavioral intention to use MHS. This current model indicates high explanatory power for behavioral intention to use MHS with R^2=0.717 while similar study by Yi et al. (2006) indicates lower level of predictive power (Yi, Jackson, Park, & Probst, 2006) with R^2=0.57 and another similar study by Chau and Hu (2002) yields R^2=0.43 (Chau & Hu, 2002). This high explanatory power is due to effort expectancy, attitude, and anxiety that are significant in their influence; however, some factors such as self-efficacy and performance expectancy were not supported. The test results showed that Smartphone user's intention to use MHS can be predicted by effort expectancy, attitude and anxiety (Marchewka, et al., 2007). Moreover, we found that attitude can mediate the effect of effort expectancy on behavioral intention to use MHS.

Although, users would like to use MHS, the intention to use this kind of systems was not very high as users' demand was not analyzed carefully (Reuss, Menozzi, Büchi, Koller, & Krueger, 2004). Prior instruction and practice is needed to improve the users' confidence toward using MHS. Thus, they will be more likely to use this kind of systems. In addition, MHS designers must be careful more about user requirement analysis in order to meet their expectation.

Even though, this study provides insight into the factors affecting on intention to use MHS, there are some limitations in the findings. The introduction and implementation of the new MHS is time consuming and it cannot be possible to employ a new MHS and measure the effects immediately. Nevertheless, our study provides useful information for understanding the determinants of implementing new MHS. However, this research finding is only based on our study; therefore, the readers must generalize the findings and inferences from other studies related to MHS as well.

Studies related to the evaluation and the use of new technologies in healthcare can assist healthcare providers to design better systems while considering users' expectations. Many other important factors can be considered in the use of MHS and the factors can be studied further. Promoting the use of MHS among healthcare professionals as well as Smartphone users will result in the improvement of general healthcare and allocation of healthcare resources.

ACKNOWLEDGMENT

The authors would like to thank the anonymous reviewers and the editor for their insightful comments and suggestions.

REFERENCES

Ajzen, I. (1985). From Intentions to Actions: A Theory of Planned Behavior. In J. Kuhl & J. Beckmann (Eds.), *Action Control* (pp. 11–39). Springer. doi:10.1007/978-3-642-69746-3_2

Ajzen, I. (1989). Attitude, structure, influence and role expectations on shopping center patronage intentions. *Journal of the Academy of Marketing Science*, *24*(3), 208–218.

Ajzen, I. (1991). The theory of planned behavior. *Organizational Behavior and Human Decision Processes*, *50*(2), 179–211. doi:10.1016/0749-5978(91)90020-T

Asada, H. H., Shaltis, P., Reisner, A., Sokwoo, R., & Hutchinson, R. C. (2003). Mobile monitoring with wearable photoplethysmographic biosensors. *Engineering in Medicine and Biology Magazine, IEEE*, *22*(3), 28–40. doi:10.1109/MEMB.2003.1213624 PMID:12845817

Bagozzi, R., & Yi, Y. (1988). On the evaluation of structural equation models. *Journal of the Academy of Marketing Science*, *16*(1), 74–94. doi:10.1007/BF02723327

Bergrath, S., Ro¨rtgen, D., Rossaint, R., Beckers, S. K., Fischermann, H., Brokmann, J. C., & ... (2011). Technical and organisational feasibility of a multifunctional telemedicine system in an emergency medical service - an observational study. *Journal of Telemedicine and Telecare*, *17*(7), 371–377. doi:10.1258/jtt.2011.110203 PMID:21933897

Biljon, J. v., & Kotz, P. (2007). *Modelling the factors that influence mobile phone adoption*. Paper presented at the 2007 annual research conference of the South African institute of computer scientists and information technologists on IT research in developing countries. New York, NY.

Chau, P. Y., & Hu, P. J.-H. (2002). Investigating healthcare professionals' decisions to accept telemedicine technology: an empirical test of competing theories. *Information & Management*, *39*(4), 297–311. doi:10.1016/S0378-7206(01)00098-2

Chin, W., Marcolin, B., & Newsted, P. (1996). *A partial least squares latent variable modeling approach for measuring interaction effects: results from a Monte Carlo simulation study and voice mail emotion/adoption study*. Academic Press.

Chin, W. W. (1998a). Commentary: Issues and opinion on structural equation modeling. *Management Information Systems Quarterly*, vii–xvi.

Chin, W. W. (1998b). *The partial least squares approach for structural equation modeling*. Academic Press.

Davis, F. D. Jr. (1986). *A technology acceptance model for empirically testing new end-user information systems: Theory and results*. Massachusetts Institute of Technology.

Davis, F. D., Bagozzi, R. P., & Warshaw, P. R. (1989). User acceptance of computer technology: a comparison of two theoretical models. *Management Science*, *35*(8), 982–1003. doi:10.1287/mnsc.35.8.982

El-Masri, S., & Saddik, B. (2012). An Emergency System to Improve Ambulance Dispatching, Ambulance Diversion and Clinical Handover Communicationâ-A Proposed Model. *Journal of Medical Systems*, *36*(6), 3917–3923. doi:10.100710916-012-9863-x PMID:22673893

Ernest, K., Lamei, C., Mohamed, S., Shakshuk, M., Badreldin, I., & ElBabli, I. (2011, 21-24 Feb. 2011). *A ZigBee-based telecardiology system for remote healthcare service delivery*. Paper presented at the Biomedical Engineering (MECBME). New York, NY.

Fishbein, M., & Ajzen, I. (1975). *Belief, Attitude, Intention, and Behavior: An Introduction to Theory and Research*. Boston: Addison-Wesley.

Gravelle, H., & Siciliani, L. (2008). Optimal quality, waits and charges in health insurance. *Journal of Health Economics*, *27*(3), 663–674. doi:10.1016/j.jhealeco.2007.08.004 PMID:18191254

Heinssen, R. K., Glass, C. R., & Knight, L. A. (1987). Assessing computer anxiety: Development and validation of the Computer Anxiety Rating Scale. *Computers in Human Behavior, 3*(1), 49-59. doi: 10.1016/0747-5632(87)90010-0

Holden, R. J., & Karsh, B.-T. (2010). The technology acceptance model: its past and its future in health care. *Journal of Biomedical Informatics*, *43*(1), 159–172. doi:10.1016/j.jbi.2009.07.002 PMID:19615467

Hung, M.-C., & Jen, W.-Y. (2012). The Adoption of Mobile Health Management Services: An Empirical Study. *Journal of Medical Systems*, *36*(3), 1381–1388. doi:10.100710916-010-9600-2 PMID:20878452

Iahad, A, N., Rahim, A., Zairah, N., & Oye, N. (2012). Acceptance and usage of ICT by university academicians using UTAUT model: A case study of University of Port Harcourt, Nigeria. *Journal of Emerging Trends in Computing and Information Sciences*, *3*(1), 81–89.

Jansà, M., Vidal, M., Viaplana, J., Levy, I., Conget, I., Gomis, R., et al. (2006). Telecare in a structured therapeutic education programme addressed to patients with type 1 diabetes and poor metabolic control. *Diabetes Research and Clinical Practice, 74*(1), 26-32. doi: 10.1016/j.diabres.2006.03.005

Janz, B. D., & Prasarnphanich, P. (2003). Understanding the antecedents of effectiveness knowledge management: The importance of a knowledge-center culture. *Decision Sciences, 34*(2), 351–384. doi:10.1111/1540-5915.02328

Jen, W.-Y., & Chao, C.-C. (2008). Measuring mobile patient safety information system success: An empirical study. *International Journal of Medical Informatics, 77*(10), 689-697. doi: 10.1016/j.ijmedinf.2008.03.003

Jovanov, E., O'Donnel, A., Morgan, A., Priddy, B., & Hormigo, R. (2002). *Prolonged telemetric monitoring of heart rate variability using wireless intelligent sensors and a mobile gateway.* Paper presented at the Engineering in Medicine and Biology, 2002. 24th Annual Conference and the Annual Fall Meeting of the Biomedical Engineering Society EMBS/BMES Conference. New York, NY.

Keikhosrokiani, P., Mustaffa, N., Zakaria, N., & Sarwar, M. I. (2012). A Proposal to Design a Location-Based Mobile Cardiac Emergency System (LMCES). *Studies in Health Technology and Informatics, 182*, 83–92. PMID:23138083

Kevin, H., & Yuan-Ting, Z. (2003). Implementation of a WAP-based telemedicine system for patient monitoring. *IEEE Transactions on* Information Technology in Biomedicine, *7*(2), 101–107.

Korhonen, I., Parkka, J., & Van Gils, M. (2003). Health monitoring in the home of the future. *IEEE Engineering in Medicine and Biology Magazine, 22*(3), 66–73. doi:10.1109/MEMB.2003.1213628 PMID:12845821

Li, S.-H., Cheng, K.-A., Lu, W.-H., & Lin, T.-C. (2011). Developing an Active Emergency Medical Service System Based on WiMAX Technology. *Journal of Medical Systems, 36*(5), 3177–3193. doi:10.100710916-011-9809-8 PMID:22109672

Liszka, K. J., Mackin, M. A., Lichter, M. J., York, D. W., Dilip, P., & Rosenbaum, D. S. (2004). Keeping a beat on the heart. *IEEE Pervasive Computing, 3*(4), 42–49. doi:10.1109/MPRV.2004.10

Marchewka, J. T., Liu, C., & Kostiwa, K. (2007). An application of the UTAUT model for understanding student perceptions using course management software. *Communications of the IIMA, 7*(2), 93–104.

Masrom, M., & Hussein, R. (2008). *User Acceptance of Information Technology: Understanding Theories and Models: Venton Publishing (M)*. Sdn. Bhd.

Nunnally, J. C. (2010). *Psychometric theory*. India: McGraw-Hill Education.

Ozok, A. A., Gurses, A., Wu, H., Nelson, M., Moen, D., & Wei, J. (2009). Usability and User Acceptance for Personal Health Records: A Perspective from Healthcare Citizens. In A. A. Ozok & P. Zaphiris (Eds.), *Online Communities and Social Computing* (Vol. 5621, pp. 690–699). Springer. doi:10.1007/978-3-642-02774-1_74

Pollard, J. K., Rohman, S., & Fry, M. E. (2001). *A Web-based mobile medical monitoring system.* Paper presented at the Intelligent Data Acquisition and Advanced Computing Systems: Technology and Applications. New York, NY.

Reuss, E., Menozzi, M., Büchi, M., Koller, J., & Krueger, H. (2004). Information access at the point of care: what can we learn for designing a mobile CPR system? *International Journal of Medical Informatics, 73*(4), 363–369. doi:10.1016/j.ijmedinf.2004.02.003 PMID:15135755

Ross, P. E. (2004). Managing care through the air [remote health monitoring]. *IEEE Spectrum, 41*(12), 26-31.

Saadé, R. G. (2007). Dimensions of perceived usefulness: Toward enhanced assessment. *Decision Sciences Journal of Innovative Education, 5*(2), 289–310. doi:10.1111/j.1540-4609.2007.00142.x

Sokwoo, R., Boo-Ho, Y., Kuowei, C., & Asada, H. H. (1998). *The ring sensor: A new ambulatory wearable sensor for twenty-four hour patient monitoring.* Paper presented at the Engineering in Medicine and Biology Society. New York, NY.

Stanford, V. (2002). Using pervasive computing to deliver elder care. *Pervasive Computing, IEEE, 1*(1), 10–13. doi:10.1109/MPRV.2002.993139

Tu, Y.-J., Zhou, W., & Piramuthu, S. (2009). Identifying RFID-embedded objects in pervasive healthcare applications. *Decision Support Systems, 46*(2), 586–593. doi:10.1016/j.dss.2008.10.001

Varady, P., Benyo, Z., & Benyo, B. (2002). An open architecture patient monitoring system using standard technologies. *IEEE Transactions on* Information Technology in Biomedicine, *6*(1), 95–98.

Varshney, U. (2003). Location management for mobile commerce applications in wireless Internet environment. *ACM Transactions on Internet Technology, 3*(3), 236–255. doi:10.1145/857166.857169

Varshney, U. (2007). Pervasive Healthcare and Wireless Health Monitoring. *Mobile Networks and Applications, 12*(2-3), 113–127. doi:10.100711036-007-0017-1

Venkatesh, V., Morris, M. G., Davis, G. B., & Davis, F. D. (2003). User Acceptance of Information Technology: Toward a Unified View. *Management Information Systems Quarterly, 27*(3).

Wan Ismail, W. K., Chan, P. H. K., Buhari, N., & Muzaini, A. (2012). *Acceptance of Smartphone in Enhancing Patient-Caregivers Relationship* (Vol. 7). Academic Press.

Wu, I.-L., Li, J.-Y., & Fu, C.-Y. (2011). The adoption of mobile healthcare by hospital's professionals: An integrative perspective. *Decision Support Systems, 51*(3), 587–596. doi:10.1016/j.dss.2011.03.003

Wu, J.-H., Wang, S.-C., & Lin, L.-M. (2007). Mobile computing acceptance factors in the healthcare industry: A structural equation model. *International Journal of Medical Informatics, 76*(1), 66–77. doi:10.1016/j.ijmedinf.2006.06.006 PMID:16901749

Yi, M. Y., Jackson, J. D., Park, J. S., & Probst, J. C. (2006). Understanding information technology acceptance by individual professionals: Toward an integrative view. *Information & Management, 43*(3), 350–363. doi:10.1016/j.im.2005.08.006

Yu, J., Ha, I., Choi, M., & Rho, J. (2005). Extending the TAM for a t-commerce. *Information & Management, 42*(7), 965–976. doi:10.1016/j.im.2004.11.001

KEY TERMS AND DEFINITIONS

Anxiety: The degree to which the person has a favorable or unfavorable evaluation of the behavior in question.

Attitude: The degree to which the person has a favorable or unfavorable evaluation of the behavior in question.

Effort Expectancy: The degree of ease associated with the use of the systems.

Mobile Healthcare System: The system that is able to Transfer biomedical signals (blood pressure, cardiac performance, insulin level etc.) over a long distance using wired or wireless communication technologies.

Performance Expectancy: The degree to which an individual believes that using the system will help him/her to attain gains in job performance.

Self-Efficacy: The concept of self-efficacy is concerned with people's belief in their ability to produce effects.

Theory of Acceptance Model: Is a model for predicting and explaining human behavior for accepting new technologies in different domains.

Theory of Planned Behavior: Is a model for predicting and explaining human behavior in different domains.

Unified Theory of Acceptance and Use of Technology: This theory is the combination of few theories and it seeks to clarify user intention to use an information system (IS) and consequent usage behavior.

User Behavioral Intention to Use: Is a measure of the strength of individual intention to perform a specified behavior, which is use of a new technology or system.

This research was previously published in Healthcare Informatics and Analytics edited by Madjid Tavana, Amir Hossein Ghapanchi, and Amir Talaei-Khoei, pages 128-143, copyright year 2015 by Medical Information Science Reference (an imprint of IGI Global).

Chapter 23
A Survey of Using Microsoft Kinect in Healthcare

Roanna Lun
Cleveland State University, USA

Wenbing Zhao
Department of Electrical and Computer Engineering, Cleveland State University, USA

INTRODUCTION

Launched in 2010, the Microsoft Kinect device is one of the most popular game controllers in recent years. Kinect allows users to naturally interact with a computer or game console by human body movement or voice command. At a very affordable price ranging from $99 to $200, Kinect is built with a color camera, infrared depth sensor, and a multi-array microphone, as shown in Figure 1. In late 2011, Microsoft released the Software Development Kit (SDK) for Kinect, which enables users to develop sophisticated computer-based human body tracking applications on both C# and C++ programming platforms (Jana 2012). Through the SDK, Kinect provides skeleton tracking of the positions of 20 articulated joints of the human body (Shotton et al., 2013), shown in Figure 2.

Figure 1. The Kinect sensor

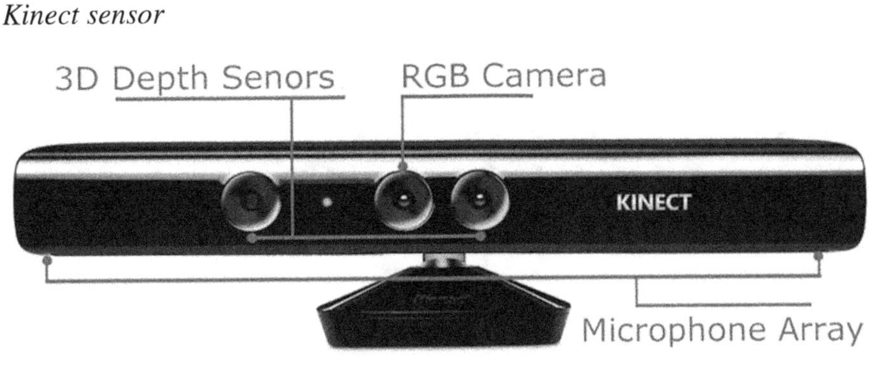

DOI: 10.4018/978-1-5225-6198-9.ch023

The low-cost, and the availability of SDK for the Kinect sensor has attracted many researchers to investigate its applications beyond the video gaming industry, particular in the healthcare realm (for example, Brian et al., 2012). As the aging population rapidly grows in the United States, demands of healthcare services, especially physical therapy and rehabilitation services, have grown enormously in recent years. To meet the increasing demands and reduce the cost of services, providers are often looking for computers and other equipment that can assist them in providing services to patients in an affordable, convenient, and user-friendly environment. As a low-cost, portable, accurate, nonintrusive, and easily set up motion detecting sensor, Kinect enables researchers to develop computer-based vision control without using traditional input devices, e.g. mouse, keyboard, or joystick. This revolutionary technology makes it possible for Kinect to meet the challenge of providing high quality evaluations and interventions at an affordable price for healthcare services, as seen from the works surveyed in this article.

BACKGROUND

Computer-vision based human body motion capturing and analysis technologies have become increasingly popular in healthcare, especially in the applications that require tracking human activities. However, such motion analysis systems are generally used in specific laboratories rather than in more convenient locations, e.g. physician's office, or even the patient's home, due to their size, high cost, and difficulty of setup. Most importantly the use of wearable sensors is intrusive to patients and it is very difficult for patient with hand impairment to set up wearable sensors.

With the launch of the Kinect sensor, researchers have paid great attention to applying the Kinect technology to healthcare related services, such as physical therapy exercises, sports rehabilitation games, motor skill training, monitoring patient activities, medical operating room assistance, and other clinical areas, which will be surveyed in the next section.

Figure 2. Kinect color, skeleton, and depth images

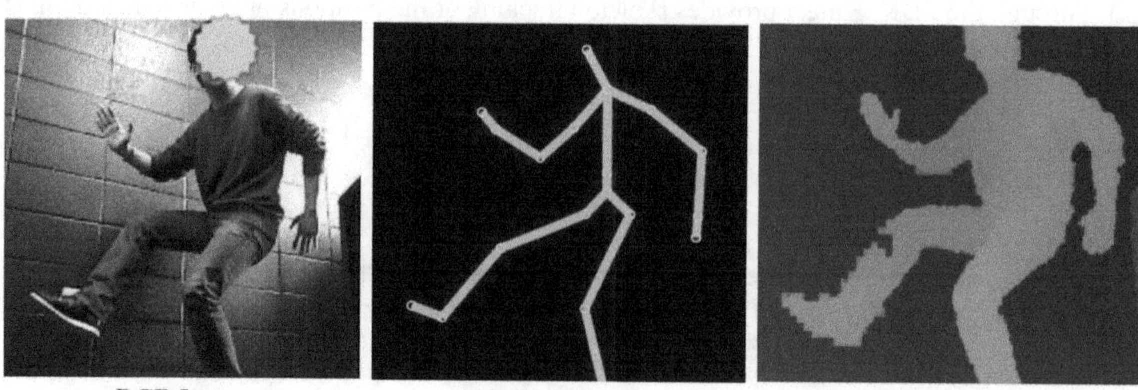

RGB Image — Skeleton Image — Depth Image

APPLICATIONS REVIEW

In this section, we review the applications of the Kinect technology in a number of healthcare domains. A summary of the applications is provided in Table 1. The detailed description of the applications is given in individual subsections.

Physical Therapy and Rehabilitation Applications

Kinect has been widely used in physical therapy exercises, sports rehabilitation, and post-surgery training. Research has proved that Kinect can be a viable physical rehabilitation tool. Compared to traditional motion tracking and capturing programs that use reflective markers attached to users' bodies, the Kinect technology enables the development of solutions with nonintrusive natural user interfaces and with real-time visual feedback to users.

Using Kinect's interactive interface with audio and video can get patients more motivated and engaged in physical activities, particular for regular and repetitive movements in physical rehabilitation exercises. Such activities have been stated as lacking interest for patients in traditional training programs (Lloyd-Jones et al., 2010). Chang, Chen, and Huang (2011) developed an interactive Kinect-based system to assist therapists in rehabilitating students in public school settings. The system is able to detect the student's joint position, and record movement data to allow therapists to review students' rehabilitation progress quickly. It also includes an interactive visual interface to enhance students' motivation, interest, and perseverance to engage in physical rehabilitation. The result has demonstrated that Kinect's natural interactive visual computer interface can significantly attract more users. In the study of real-time ges-

Table 1. A summary of Kinect applications in healthcare

Application Domain	Specific Applications	References
Physical Therapy and Rehabilitation	Kinect-based physical rehabilitation systems	Chang et al., 2011; Rahman et al., 2013; Saini et al., 2012
	Virtual reality based games for balance training and upper body rehabilitation	Lange et al., 2011; Lange et al., 2012; Gotsis et al., 2012
	Kinect based game for Alzheimer patients	Cervantes et al., 2012
	A hand rehabilitation system	Metcalf et al., 2013
	Providing feedbacks on the quality of exercises	Velloso et al., 2013; Su, 2013
Clinical Environment	Kinect-based systems for high sterile operation rooms	Johnson et al., 2011; Gallo et al., 2011; Bigdelou et al., 2012
Fall Detection	Fall motion detection using Kinect depth images	Mastorakis and Makris, 2012
	Fall detection based on randomized decision forest algorithm	Bian et al., 2012, Girshick et al., 2011, Shotton et al., 2013
	A comparison study on using Kinect and mark-based systems for fall detection	Obdrzalek et al., 2012
	Fall detection and abnormal event detection on stairways	Parra-Dominguez et al., 2012
	Overcoming occlusions for fall detection	Rougier et al., 2011
	Use of two Kinects for fall detection	Zhang et al., 2012

ture recognition, Bigdelou, Benz, Schwarz, and Navab (2012) concluded that most users are fond of the dynamic interaction with system. Applications using Kinect technology are overwhelmingly regarded as friendly visual interaction programs.

A research group at the University of Southern California Institute for Creative Technologies (ICT) has successfully applied the Kinect device to the study of virtual reality simulation technology for clinical purposes. Through extensive evaluation, assessment, and analysis, the researchers at the ICT have shown that Kinect technology can make a major contribution to the quality of traditional intervention with computer-based training games specializing in mental health therapy, motor skills rehabilitation, cognitive assessment, and clinical skills exercises. Lange et al. (2011) developed a virtual reality simulation game used for balance training of adults with neurological injury. Further they developed a video game, "JewelMine," that consists of a set of static balance training exercises in rehabilitation settings (Lange et al., 2012). Their colleagues Suma, Lange, Rizzo, Krum, and Bolas (2011) developed the Flexible Action and Articulated Skeleton Toolkit (FAAST), which is middleware that facilitates integration of full-body control with virtual reality applications and video games using Kinect.

Focusing on cognitive rehabilitation for Alzheimer patients, Cervantes, Vela, and Rodríguez (2012) developed a video game that replaces the traditional techniques used by therapists with the Kinect device. The interactive body-motion controlled game increases patients' motivation to participate in exercise. An interactive rehabilitation system for disabled children was developed by Rahman, Qamar, Ahmed, Rahman, and Basalamah (2013). This application allows activities to be synchronized with on-line virtual counterparts. Patients can follow the exercise at home in the absence of the therapist because visual guidance for correctly performing the exercise is provided. Gotsis et al. (2012) demonstrated a mixed reality game for upper body exercise and rehabilitation. The exercise protocol was adopted from an evidence-based shoulder rehabilitation program for individuals with spinal cord injury.

Metcalf et al. (2013) studied an application for real-time monitoring and assessing hand movement based on Kinect. Hand rehabilitation is often needed for stroke patients' recovery. Medical professionals require detailed and accurate measurement of the hands in order to assess progress and functional recovery. The system is capable of measuring finger joint kinematics in real time for home-based clinical rehabilitation programs. This system is designed in two steps: 1) identify anatomical landmarks and classification between grip types; 2) calculate hand joint angles from a kinematic model. The initial phase detects two different modes: spread hand mode and pincer grip mode. Then the landmarks go through validation procedures first by using ground truth estimation between reviewers and the associated algorithms to assess the accuracy of landmark definition, and second by joint angles compared to a laboratory-based gold standard motion capture system and validated kinematic measurement technique. And on the other side, patients require real-time feedback on their hand posture if they participate in a virtual environment rehabilitation gaming platform. Although this novel technique can robustly track and accurately measure the movements of the hand, it only provides a limited number of gestures that are only suitable for a specific rehabilitation application.

The real-time feedback process is critical in ensuring the quality and effectiveness of physical rehabilitation exercises. Patients have to perform movements in a specific manner to achieve the objectives of the exercises. Any incorrect movement, or movement that does not reach the required coordination, could be potentially harmful rather than beneficial in a training program. On the other hand, medical experts need to assess exercise outcomes to make sure the desired movements have been accurately completed. Velloso, Bulling, and Gellersen (2013) presented the *Motion*MA system that can provide real-time feedback. This system creates a three-step-communication loop: 1) experts (medical professionals) demonstrate movements, and the system extracts a model for such movements; 2) novice users (patients) repeat the movements from the extracted model and are able to see real-time action during the exercises; and 3) *Motion*MA automatically generates feedback for evaluating the given execution to help patients assess and improve their performance in real time.

In order to reinforce the quality of home-based real-time rehabilitation exercise, Su (2013) proposed an approach using the Dynamic Time Warping (DTW) algorithm and Fuzzy Logic for ensuring that exercise is performed correctly. First it allows a patient to perform a prescribed exercise in the presence of a professional. Then the patient's exercise is recorded as a base for evaluation of exercise at home. The outcomes of the evaluation can be used as a reference for the patient to validate his/her exercise and to prevent adverse events. A summary report of the outcomes may also be uploaded to a cloud setting for physicians to monitor progress and adjust the prescription. The DTW algorithm is used to compare two sequences that have different time lengths in order to determine the similarity between the standard and the patient's exercise movements. For the rehabilitation exercise, the trajectory and speed mainly are based on experience and subjective evaluation. Therefore traditional logic theory is not practical for evaluating the quality of rehabilitation exercise by trajectory and speed. In this application, Fuzzy Logic is employed to build a fuzzy inference of a physician's subjective evaluation.

Saini et al. (2012) presented an approach to providing stroke patients with instruction and guidelines to perform rehabilitation exercises. A "watchdog" developed in this system provides the restriction that patients cannot exceed the limited angle, and ensures that a patient who is exercising at home will not exceed the prescribed level or perform for a longer period than necessary. A centralized system stores the entire patient's data that will be used to analyze the patient's movements and activities. The biofeedback system allows the user to give feedback to the therapist, and the therapist will evaluate the patient's movement and give further instruction or direction. The biofeedback is displayed for the patient as well as the therapist. The main benefit for biofeedback is that patients can directly communicate with the therapist for further instruction for improving their movements. This strategy is beneficial for stroke patents because patients will do exercises correctly.

Clinical Environment Applications

A key technical challenge in using Kinect in clinical environment, e.g. in a medical operating room, is how to recognize the gestures interactively, accurately, robustly, and effectively. Touch-free interaction with a computer makes the Kinect device suitable for clinical environments in highly sterile conditions (Johnson, O'Hara, Sellen, Cousins, & Criminisi, 2011). Their study showed that the touch-free motion recognition Kinect device provides an appealing opportunity to control medical images or image-guided devices without using the hands. Some researchers investigated gesture recognition systems based on the Kinect technology to address needs in medical surgical rooms.

Gallo, Placitelli, and Ciampi (2011) developed an open source interface using static postures and dynamic gesture recognition techniques to explore CT, MRI, or PET images in operating room. Utilizing hand and arm gestures, this application is able to execute basic tasks such as image selection, zooming, translating, rotating, and pointing; and some complex tasks, e.g. manual selection for controlling medical digital images effectively. The interaction follows the event – state – action paradigm and it is developed specifically for fast operation at video frame rates. The limitation of this interface is that the number of postures and gestures used for recognition is fixed. The extension of the gesture control interface to enable interaction with immersive medical imaging environments must be addressed in future development.

Bigdelou, Benz, Schwarz, and Navab (2012) applied Kinect 3D gesture-based interface technology to an intra-operative medical image viewer in a surgical environment. They used skeleton input to recognize the gesture type and the relative poses within that gesture. The recognized gesture and poses can be used for an interface for controlling a medical image viewer. The training phase is used to define and customize gestures from skeleton images, and during the interaction phase, the actual gesture and the relative pose within the gesture is estimated by discriminative kernel regression mapping. The quantitative experiments show this method can simultaneously recognize several gestures to a high degree of accuracy. Again, the number of gestures recognized is limited to 16 due to computational complexity for human body movement.

Fall Detection Applications

The Kinect device has been used in a real-time fall monitoring and detection system for a wide range of falls, including backward, forward, and sideways; on stairways; in occlusion places; in hospital wards; and others.

In a study of detecting human body falling motion, Mastorakis and Makris (2012) adopted the use of Kinect depth images instead of skeleton data to robustly and efficiently track body motion. In their approach, a fall is detected by analyzing the 3D bounding box's width, height, and depth, i.e., the width, height, and depth of the human posture. The algorithm is provided by OpenNI. This approach outperforms the articulated model, which requires significantly more computational power. Since a human fall is a fast activity, a high frame rate in real-time systems is required to avoid missing detections. Furthermore, this algorithm has proved stable and can even detect movement when half of the subject's body is occluded by surrounding objects.

The Randomized Decision Forest (RDF) algorithm has been widely used in detecting human body falls due to its high accuracy in recognition and reliable skeleton images provided by the Kinect device. Several groups of researchers (Bian, Chau, & Magnenat-Thalmann, 2012; Girshick, Shotton, Kohli, Criminisi, & Fitzgibbon, 2011; Shotton et al., 2013) have successfully used RDF in fall detection. The techniques used recognize skeleton shape deformation caused by the human body falling. However, due to changes in the orientation of the body during movement, the accuracy of recognition is reduced. Bian, Chau, and Magnenat-Thalmann (2012) presented an algorithm to improve the accuracy of skeleton extraction by rotating the human body trunk based on the pose in the previous frame. Considering that the human fall is a fast activity, RDF computational performance is critical. Other researchers demonstrated an alternative approach to address this performance issue, which will be discussed later in this article.

Obdrzalek et al. (2012) presented a method that the accuracy of body tracking to a great extent depends on the type of poses being observed, the distance from the sensor and possible occlusions. They conducted a data acquisition process from captured human poses using the marker-based motion capture system developed by Microsoft Kinect. The result shows that Kinect works with much denser but less precise 3D data comparing to the marker-based system. They also observed that the markerless skeleton tracking of Kinect depends solely on dense depth information and thus frequently fails due to occlusions. For physical exercise of the elderly population, the Kinect skeleton tracking struggles with occluding body parts or objects in the scene, e.g. a chair. The occlusion is indeed an open issue for the use of Kinect technology in healthcare applications.

In the research work of Parra-Dominguez, Taati, and Mihailidis (2012), a method particularly to detect falls and other abnormal events on stairs was proposed. This work can automatically estimate the walking speed and extract a set of features that encode human motion during stairway descent.

Ni, Nguyen, and Moulin (2012) developed a computer vision based system to prevent potential falls in hospital ward environments. This system is able to automatically detect the event of a patient getting out of a bed. The nursing staff is alerted immediately to provide assistance once the getting up event is detected. The detection algorithm combines multiple features from multiple modalities via an MKL framework to achieve high accuracy and efficiency.

Rougier, Auvinet, Rousseau, Mignotte, and Meunier (2011) introduced an approach to address occlusion for detecting human body falls. The method is based on human centroid height relative to the ground and body velocity. With the help of computing 3D personal velocity just before the occlusion, this method can accurately detect falls by measuring human centroid height, because the vast majority of falls end on the ground or near the ground.

Zhang, Liu, Metsis, and Athitsos (2012) proposed a system that employs two Kinect sensors to detect human falls. The statistical method is based on information about how the person moves during the last few frames. The major contribution of this research work is that it collects all training data from a specific viewpoint, and then collects all the test data from a different viewpoint. They believe the new protocol is a useful approach for measuring the robustness of the system by spatial displacement of camera images. Stone and Skubic (2012) developed a system for capturing variation of stride-to-stride gait in home environments for elderly adults. By measuring the changes in gait, falls can be predicted. If the motion of joints of a specific body subject is detected in an unusual time sequence, the prevention message is generated as a caution.

FUTURE RESEARCH DIRECTIONS

The algorithms used for processing images obtained from the Kinect device play a very important role for human body motion analysis in healthcare applications. In the study of automatically tracking and detecting abnormal events, Parra-Dominguez, Taati, and Mihailidis (2012) evaluated different binary classification algorithms, e.g. Random Decision Forest (RDF), Logistic Regression, and Naive Bayes. The results show that RDF outperforms the other algorithms during the recognition process, and it is more suitable for real-time application. As a result, more algorithms need to be evaluated in the future to achieve the ultimate goal of performance requirements. Velloso, Bulling, and Gellersen (2013) suggested exploring more complex curve analysis techniques.

Although some existing applications already provide feedback for physical rehabilitation, there is still a huge range of possibilities of how this feedback may be improved (Velloso, Bulling, & Gellersen, 2013). The potential improvements could be multi-modal interaction, closer loop between the expert and the user, automated assessment, and so forth.

Enhancing the entertaining and amusing elements of the system is possible for future improvement (Chang et al., 2011). It would make rehabilitation more enjoyable if more peers (or users) are engaged in rehabilitation activities simultaneously. Game-based exercises could increase patients' motivation for therapy (Saini et al., 2012). Continuously exploring 3D interaction techniques in virtual environments (Cervantes et al., 2012) will be the focus in the next generation for patients involved in physical exercise programs.

Home-based remote healthcare service might become more pervasive for out-patient physical therapy and rehabilitation. With the development of advanced 3D sensors and cloud gaming software (O'Connor, Davy, & Jennings, 2013), it is very possible that more home-based rehabilitation training exercises will be available online (Rahman, Qamar, Ahmed, Rahman, & Basalamah, 2013) or in a cloud server, which would be more convenient for professional therapists wishing to instruct, monitor, and assess patients' activities.

The limited viewing range of the Kinect sensor has made it is difficult to detect and track full body movements. In a study of automatically tracking and detecting abnormal events on stairs in elderly living centers, Parra-Dominguez, Taati, and Mihailidis (2012) discovered that tracking hip joint points from the skeleton frame in Kinect is more stable than tracking feet and ankles, due to the fact that feet and ankles are sometimes partially occluded from Kinect view. On the other hand, applications in healthcare services require full body motion analysis. Therefore, full-body detection is a critical challenge for the future development of Kinect applications in healthcare.

One of the main drawbacks of using Kinect in healthcare applications is that Kinect uses a very non-anthropometric kinematic model with variable limb lengths (Obdrzalek et al., 2012). In general postures, the variability of the pose estimation is about 10 cm. Future algorithms should address occlusions and self-occlusions, unconventional body postures, and use of wheelchairs or walkers.

CONCLUSION

We believe that the research and development on using Kinect technology in healthcare will gain more momentum. The demand of Kinect-based applications is high, due to Kinect's low cost and portability, and its accurate and robust motion detection capability. In this article, we have surveyed the current applications of using the Kinect technology in healthcare. Furthermore, we outlined a number of open research issues that could overcome the limitations of the current Kinect technology.

REFERENCES

Bian, Z. P., Chau, L. P., & Magnenat-Thalmann, N. (2012). Fall detection based on skeleton extraction. In *Proceedings of the 11th ACM SIGGRAPH International Conference on Virtual-Reality Continuum and its Applications in Industry* (pp. 91-94). ACM.

Bigdelou, A., Benz, T., Schwarz, L., & Navab, N. (2012). Simultaneous categorical and spatio-temporal 3d gestures using Kinect. In *3D User Interfaces (3DUI), 2012 IEEE Symposium* (pp. 53-60). IEEE.

Bó, A. P. L., Hayashibe, M., & Poignet, P. (2011, August). Joint angle estimation in rehabilitation with inertial sensors and its integration with Kinect. In *Engineering in Medicine and Biology Society*, EMBC, 2011 Annual International Conference of the IEEE (pp. 3479-3483). IEEE.

Cervantes, J. C., Vela, F. L. G., & Rodríguez, P. P. (2012). Natural interaction techniques using Kinect. In *Proceedings of the 13th International Conference on Interacción Persona-Ordenador* (p. 14). ACM.

Chang, Y. J., Chen, S. F., & Huang, J. D. (2011). A Kinect-based system for physical rehabilitation: A pilot study for young adults with motor disabilities. *Research in Developmental Disabilities*, *32*(6), 2566–2570. doi:10.1016/j.ridd.2011.07.002 PMID:21784612

Gallo, L., Placitelli, A. P., & Ciampi, M. (2011). Controller-free exploration of medical image data: Experiencing the Kinect. In *Computer-Based Medical Systems (CBMS)*, 2011 24th International Symposium (pp. 1-6). IEEE.

Girshick, R., Shotton, J., Kohli, P., Criminisi, A., & Fitzgibbon, A. (2011, November). Efficient regression of general-activity human poses from depth images. In *Computer Vision (ICCV)*, 2011 IEEE International Conference (pp. 415-422). IEEE.

Gotsis, M., Lympouridis, V., Turpin, D., Tasse, A., Poulos, I. C., Tucker, D., & Jordan-Marsh, M. (2012). Mixed reality game prototypes for upper body exercise and rehabilitation. In Virtual Reality Workshops (VR), 2012 IEEE (pp. 181-182). IEEE.

Jana, A. (2012). *Kinect for Windows SDK Programming Guide*. Packt Publishing Ltd.

Johnson, R., O'Hara, K., Sellen, A., Cousins, C., & Criminisi, A. (2011). Exploring the potential for touchless interaction in image-guided interventional radiology. In *Proceedings of the SIGCHI Conference on Human Factors in Computing Systems* (pp. 3323-3332). ACM.

Lange, B., Chang, C., Suma, B., & Newman, A. Rizzo, & Bolas, M. (2011). Development and evaluation of low cost game-based balance rehabilitation tool using the Microsoft Kinect sensor. In *Proceeding of the Annual International Conference of IEEE Engineering in Medicine and Biology Society*, EMBC, vol. 2011, (pp. 1831–4).

Lange, B., Koenig, S., McConnell, E., Chang, C., Juang, R., Suma, E., & ... (2012). *Interactive game-based rehabilitation using the Microsoft Kinect. Virtual Reality Short Papers and Posters (VRW)*. IEEE.

Lloyd-Jones, D., Adams, R., Brown, T., Carnethon, M., Dai, S., De Simone, G., & ... (2010). Heart disease and stroke statistics—2010 update: A report from the American Heart Association. *Circulation*, *121*, 46–215. doi:10.1161/CIRCULATIONAHA.109.192667

Mastorakis, G., & Makris, D. (2012). Fall detection system using Kinect's infrared sensor. *Journal of Real-Time Image Processing*, 1–12.

Metcalf, C. D., Robinson, R., Malpass, A. J., Bogle, T. P., Dell, T. A., Harris, C., & Demain, S. H. (2013). *Markerless Motion Capture and Measurement of Hand Kinematics: Validation and Application to Home-Based Upper Limb Rehabilitation. Biomedical Engineering, IEEE Transactions, pp. (2184-2192)*. IEEE.

Ni, B., Nguyen, C. D., & Moulin, P. (2012). RGBD-camera based get-up event detection for hospital fall prevention. In *Acoustics, Speech and Signal Processing (ICASSP)*, 2012 IEEE International Conference (pp. 1405-1408). IEEE.

O'Connor, C., Davy, A., & Jennings, B. (2013). Controlling the transfer of Kinect data to a cloud-hosted games platform. In *Proceeding of the 23rd ACM Workshop on Network and Operating Systems Support for Digital Audio and Video* (pp. 55-60). ACM.

Obdrzalek, S., Kurillo, G., Ofli, F., Bajcsy, R., Seto, E., Jimison, H., & Pavel, M. (2012). Accuracy and robustness of Kinect pose estimation in the context of coaching of elderly population. In Engineering in *Medicine and Biology Society (EMBC)*, 2012 Annual International Conference of the IEEE (pp. 1188-1193). IEEE.

Parra-Dominguez, G. S., Taati, B., & Mihailidis, A. (2012). 3d human motion analysis to detect abnormal events on stairs. In *3D Imaging, Modeling, Processing, Visualization and Transmission (3DIMPVT)*, 2012 Second International Conference (pp. 97-103). IEEE.

Rahman, A., Qamar, A. M., Ahmed, M. A., Rahman, M. A., & Basalamah, S. (2013). Multimedia interactive therapy environment for children having physical disabilities. In *Proceedings of the 3rd ACM conference on International conference on multimedia retrieval*, pp. 313-314.

Rougier, C., Auvinet, E., Rousseau, J., Mignotte, M., & Meunier, J. (2011). Fall detection from depth map video sequences. In Toward Useful Services for Elderly and People with Disabilities (pp. 121-128). Springer Berlin Heidelberg.

Saini, S., Rambli, D. R. A., Sulaiman, S., Zakaria, M. N., Shukri, M., & Rohkmah, S. (2012, June). A low-cost game framework for a home-based stroke rehabilitation system. In *Computer & Information Science (ICCIS)*, 2012 International Conference (Vol. 1, pp. 55-60). IEEE.

Shotton, J., Sharp, T., Kipman, A., Fitzgibbon, A., Finocchio, M., Blake, A., & Moore, R. (2013). Real-time human pose recognition in parts from single depth images. *Communications of the ACM*, *56*(1), 116–124. doi:10.1145/2398356.2398381

Stone, E. E., & Skubic, M. (2012). Passive, in-home gait measurement using an inexpensive depth camera: Initial results. In *Pervasive Computing Technologies for Healthcare (PervasiveHealth)*, 2012 6th International Conference (pp. 183-186). IEEE.

Su, C. J. (2013). Personal rehabilitation exercise assistant with Kinect and Dynamic Time Warping. *International Journal of Information and Education Technology*, *3*(4), 448–454. doi:10.7763/IJIET.2013.V3.316

Suma, E. A., Lange, B., Rizzo, A., Krum, D. M., & Bolas, M. (2011). Faast: The flexible action and articulated skeleton toolkit. In *Virtual Reality Conference (VR), 2011 IEEE* (pp. 247-248). IEEE.

Velloso, E., Bulling, A., & Gellersen, H. (2013). MotionMA: Motion Modelling and Analysis by Demonstration. In *Proceedings of the SIGCHI Conference on Human Factors in Computing Systems (CHI '13)* (pp. 1309-1318).

Zhang, Z., Liu, W., Metsis, V., & Athitsos, V. (2012, November). A viewpoint-independent statistical method for fall detection. In *Pattern Recognition (ICPR)*, 2012 21st International Conference (pp. 3626-3630). IEEE.

ADDITIONAL READING

Bannach, D., Lukowicz, P., & Amft, O. (2008). Rapid prototyping of activity recognition applications. *Pervasive Computing, IEEE, 7*(2), 22–31. doi:10.1109/MPRV.2008.36

Chen, Y., Baran, M., Sundaram, H., & Rikakis, T. (2011). A low cost, adaptive mixed reality system for home-based stroke rehabilitation. Engineering in Medicine and Biology Society, EMBC, 2011 Annual International Conference of the IEEE, pp.1827-1830.

Moeslund, T. B., & Bajers, F. (1999). *Computer Vision-Based Human Motion Capture — A Survey*. Waltham, MA: Academic Press.

Myers, C., Rabiner, L., & Rosenberg, A. (1980). Performance tradeoffs in dynamic time warping algorithms for isolated word recognition. Acoustics, Speech and Signal Processing. *IEEE Transactions on, 28*(6), 623–635.

Reinhard, K., & Tee, G. (2008). Understanding Human Motion: A Historic Review. *Human Motion, Springer, 2008*, 1–22.

Sakoe, H., & Chiba, S. (1978). Dynamic programming algorithm optimization for spoken word recognition. *Acoustics, Speech and Signal Processing. IEEE Transactions on, 26*(1), 43–49.

Webb, J., & Ashley, J. (2012). *Beginning Kinect programming with the Microsoft Kinect SDK*. Austin, TX: Apress. doi:10.1007/978-1-4302-4105-8

"Xbox 360 + Kinect," Microsoft Corporation, 2012. [Online]. Available: www.xbox.com/en-us/kinect

Yamato, J., Ohya, J., & Ishii, K. (1992, June). Recognizing human action in time-sequential images using hidden Markov model. In *Computer Vision and Pattern Recognition*, 1992. Proceedings CVPR'92, 1992 IEEE Computer Society Conference (pp. 379-385). IEEE.

KEY TERMS AND DEFINITIONS

Dynamic Time Warping (DTW): An algorithm for measuring similarity between two temporal sequences that may vary in time or speed. DTW calculates an optimal match between two given sequences (e.g. time series) with certain restrictions. The sequences are "warped" non-linearly in the time dimension to determine a measure of their similarity independent of certain non-linear variations in the time dimension. This sequence alignment method is often used in time series classification.

Fuzzy Logic: A form of many-valued logic. Fuzzy logic deals with reasoning that is approximate rather than fixed and exact. Compared to traditional true or false values, fuzzy logic variables may have a truth value that ranges in degree from 0 to 1. Fuzzy logic has been extended to handle the concept of partial truth, where the truth value may range between completely true and completely false.

Kalman Filter: An algorithm that uses a series of measurements observed over time, containing random variations and other inaccuracies, and produces estimates of unknown variables that tend to be more precise than those based on a single measurement alone.

Logistic Regression (LR): A type of probabilistic classification model used for predicting the outcome of categorical dependent variables based on one or more predictor variables (features).

Naive Bayes: A simple probabilistic classifier based on applying Bayes' theorem with strong (naive) independence assumptions.

OpenNI: An open source SDK used for the development of 3D sensing middleware libraries and applications for depth sensors such as Kinect.

Randomized Decision Forest (RDF): An ensemble learning method for classification and regression that can operate by constructing a multitude of decision trees at training time and outputting the class that is the mode of the classes output by individual trees.

This research was previously published in the Encyclopedia of Information Science and Technology, Third Edition edited by Mehdi Khosrow-Pour, pages 3279-3287, copyright year 2015 by Information Science Reference (an imprint of IGI Global).

Index

A

B

C

D

E

F

G

H

I

K

L

M

N

O

P

Q

R

S

T

U

V

W

Stay Current on the Latest Emerging Research Developments

Become an IGI Global Reviewer for Authored Book Projects

The overall success of an authored book project is dependent on quality and timely reviews.

In this competitive age of scholarly publishing, constructive and timely feedback significantly decreases the turnaround time of manuscripts from submission to acceptance, allowing the publication and discovery of progressive research at a much more expeditious rate. Several IGI Global authored book projects are currently seeking highly qualified experts in the field to fill vacancies on their respective editorial review boards:

Applications may be sent to:
development@igi-global.com

Applicants must have a doctorate (or an equivalent degree) as well as publishing and reviewing experience. Reviewers are asked to write reviews in a timely, collegial, and constructive manner. All reviewers will begin their role on an ad-hoc basis for a period of one year, and upon successful completion of this term can be considered for full editorial review board status, with the potential for a subsequent promotion to Associate Editor.

If you have a colleague that may be interested in this opportunity, we encourage you to share this information with them.

www.ingramcontent.com/pod-product-compliance
Lightning Source LLC
La Vergne TN
LVHW060355300926
849162LV00006B/1271
* 9 7 8 1 5 2 2 5 6 1 9 8 9 *